43rd International Deep Drawing Research Group (IDDRG) Conference 2024

IOP Conference Series: Materials Science and Engineering
Volume 1307

Melbourne, Australia
12-15 March 2024

ISBN: 978-1-7138-9814-6
ISSN: 1757-8981

Printed from e-media with permission by:

Curran Associates, Inc.
57 Morehouse Lane
Red Hook, NY 12571

Some format issues inherent in the e-media version may also appear in this print version.

Institute of Physics
Dirac House, Temple Back
Bristol BS1 6BE UK

Phone: 44 1 17 929 7481
Fax: 44 1 17 920 0979

techtracking@iop.org

Additional copies of this publication are available from:

Curran Associates, Inc.
57 Morehouse Lane
Red Hook, NY 12571 USA
Phone: 845-758-0400
Fax: 845-758-2633
Email: curran@proceedings.com
Web: www.proceedings.com

TABLE OF CONTENTS

IOP Conf. Series: Materials Science and Engineering **1307** (2024) 011001 doi:10.1088/1757-899X/1307/1/011001

43rd International Deep Drawing Research Group (IDDRG) Conference

Sustainable Sheet Forming - Circular Economy
Melbourne, Australia

Australia, 12-15 March 2024

1. Introduction

The 43rd International Deep Drawing Research Group (IDDRG) conference was held in Melbourne, Australia, on 12–15 March. The conference was co-organised by Deakin University and the Korean Society for Technology of Plasticity. The conference chairs were: Prof . Bernard

IOP Conf. Series: Materials Science and Engineering **1307** (2024) 011001 doi:10.1088/1757-899X/1307/1/011001

Rolfe (Deakin University), A/Prof. Matthias Weiss (Deakin University), and Prof Jeong Whan Yoon (KAIST/Deakin University).

During IDDRG 2024, all engineering aspects of sheet metal forming were discussed, such as behaviour of workpiece materials, formability, tools, tribology, process robustness, simulation methods and experiments. The general theme of IDDRG 2024 was *Sustainable Sheet Forming – Circular Economy*.

International Deep Drawing Research Group

The International Deep Drawing Research Group (IDDRG) was started in the late 1950s as an organization of national groups devoted to the study of sheet metal forming including forming processes, materials, formability issues, tooling, tribology and many other interesting aspects of sheet metal forming research and industrial practices.

At the IDDRG, the leading sheet metal specialists and researchers from all over the world gather for fruitful discussions on challenging technical topics in forming technology. The conference is designed to provide an open and inspiring environment for knowledge development, discussions and interactions.

Conference topics

The theme of the IDDRG 2024 conference was Sustainable Sheet Forming – Circular Economy. In addition to the main theme of the conference, topical sessions with the focus on the challenges and current trends in the field of sheet metal forming, such as: fossil-free steel, sustainable/advanced/smart materials, multi-material concepts/solutions, forming/formability, lightweight structures, digital transformation, digitalization, smart plants, smart press shops, cold and hot forming and trimming (incl. single and multi-step forming and trimming), tribology, simulations (incl. material models, topology optimization, tribology, springback compensation, and geometry assurance), testing, testbeds, springback, shape accuracy, geometry assurance, surface quality, tools and dies, and additive manufacturing.

There were six additional mini-symposiums organised by internationally renowned experts held at IDDRG 2024:

- 1. Material, Process Monitoring and Quality Control in Roll Forming Processes, organized by Matthias Weiss (Deakin University, Australia);
- 2. Forming, Joining and Failure of Fuel Cell and Battery Materials, organized by Michael Pereira (Deakin University, Australia), and Jeong Yoon (KAIST/Deakin University, Korea/Australia);
- 3. Machine Learning Driven Forming, organized by Hyunsung Choi (Korea Institute of Materials Science, Korea);
- 4. Flexible Metal Forming, organized by Namsu Park (Korea Institute of Industrial Technology, Korea);
- 5. Innovative Metal Forming Technology, organized by Professor Richard Yang (Western Sydney University, Australia);
- 6. Advanced Materials Characterization, organised by Madhav Baral (University of Kentucky, USA).

It was the first time to consider fuel cell and battery cell forming, however, given the global conversion to electric vehicles, it is assumed this forming topic may continue at future IDDRG conferences.

Keynotes

Nine keynote speeches were given by academic and industrial experts:

- Professor Yannis P. Korkolis, Technical University of Dortmund, *Validation and identification of constitutive models for sheet metal forming simulations*
- Professor Takayuki Hama, Kyoto University, *Cup drawing of hexagonal close-packed sheet metals: experiment and crystal plasticity finite-element simulation*
- Professor Ton van den Boogaard, University of Twente, *Digital Twins, Data Analysis, Artificial Intelligence – What's in it for us?*
- Professor Heng Li, Northwestern Polytechnical University, Xi'an, China, *Towards defect-free tube forming: mechanisms, modelling and control*
- Professor Ming Wang Fu, The Hong Kong Polytechnic University, *Deformation behaviours and mechanisms of extra-thin sheet metals and metallic films affected by size effects*
- Albert Sedlmaier, Managing Director data M Sheet Metal Solutions GmbH, *Unravelling the Art of Roll Forming with advanced CAE techniques*
- Dr Bart Carleer, Corporate Technical Director AutoForm, *How stamping simulation contributes to sustainability*
- Professor Dr.-Ing Wolfram Volk, Technical University of Munich, *Advanced material characterization and validation strategies for sheet metal forming applications*
- Professor Jeong Whan Yoon, Korea Advanced Institute of Science and Technology and Deakin University, *Advances in Materials Characterization for Constitutive and Fracture Modelling*

Venue

The venue for the conference was the Melbourne City campus of Deakin University: Deakin Downtown.

Chairs

- Prof Bernard Rolfe, Deakin University
- A/Prof Matthias Weiss, Deakin University
- Prof Jeong Whan Yoon, KAIST / Deakin University

Organising committee

- Prof Bernard Rolfe, co-chair, Deakin University
- A/Prof Matthias Weiss, co-chair, Deakin University
- Prof Jeong Whan Yoon, co-chair, KAIST/Deakin University
- Dr Jingsi Jiao, general secretary, Deakin University
- Dr Peng Neo Zhang, coordinator of reviews and proceedings, Deakin University
- A/Prof Michael Pereira, Deakin University
- Dr Mariana Paulino, Deakin University
- A/Prof Yanan Wang, Deakin University
- Jodi Morgan, Deakin Events Manager
- Georgina Fatouros, Deakin Events
- Natalie Onions, Deakin Events
- Preeta Panicker, Deakin Events

Scientific Committee

Nader Asnafi
(Örebro University, Sweden)
Kester Clarke
(Colorado School of Mines, USA)
Matthew Doolan
(UNSW Sydney, Australia)
Patrick Duroux
(ArcelorMittal, France)
Gašpar Gantar
(The Faculty of Environmental Protection, Slovenia)
Eiji Izuka
(JFE Steel Corporation, Japan)
Zhengyi Jiang
(University of Wollongong, Australia)
Celalettin Karadogan
(DYNAmore GmbH, Germany)
Toshihiko Kuwabara
(Tokyo University of Agriculture and Technology, Japan)
Myoung-Gyu Lee
(Seoul National University, South Korea)
Nico Langerak
(Tata Steel Europe, The Netherlands)
Mathias Liewald
(University of Stuttgart, Germany)
Zsolt Lukács
(University of Miskolc, Hungary)
Marion Merklein
(FAU Erlangen-Nürnberg, Germany)
Dirk Mohr
(ETH Zurich, Switzerland)
K. Narasimhan
(IIT Bombay, India)
Mariana Paulino
(Deakin University, Australia)
Michael Pereira
(Deakin University, Australia)
Pasi Peura
(Tampere University, Finland)
Bernard Rolfe
(Deakin University, Australia)
Eneko Saenz de Argandoña
(Mondragon University, Spain)
Abel Santos
(University of Porto, Portugal)
Zhang Shi-Hong
(Institute of Metal Research, China)
Sandrine Thuillier
(University of South Brittany, France)
Wolfram Volk
(Technical University of Munich, Germany)
Leopold Wagner
(voestalpine Stahl GmbH, Austria)
Yanan Wang
(Deakin University, Australia)
Matthias Weiss
(Deakin University, Australia)
Mike Worswick
(University of Waterloo, Canada)
Prasad Yarlagadda
(Queensland University of Technology, Australia)
Jeong Yoon
(Deakin University/KAIST Australia/South Korea)

Editors

Prof Bernard Rolfe, Deakin University
A/Prof Matthias Weiss, Deakin University
Prof Jeong Yoon, KAIST / Deakin University
Dr Peng Neo Zhang, Deakin University

Number of presentations, published scientific papers/first authors, and participants

In total 145 participants from 17 countries attended IDDRG 2024. From these, approximately 30% of participants have industrial affiliations, 6% are from Government labs, and 64% are from academic institutes. The chart below shows the number of participants by country.

In overall, after pre-selection of the abstracts, 59 full paper submissions were made to IDDRG 2024 by 240 authors from 12 countries. Of these submissions, 52 manuscripts were selected based on single-blind reviews by the external reviewers and the scientific committee. The chart below shows the number of authors by country.

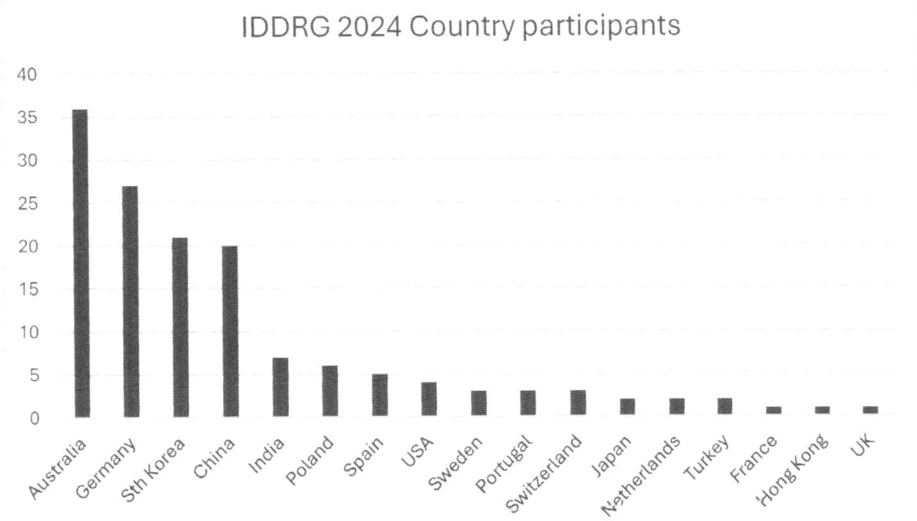

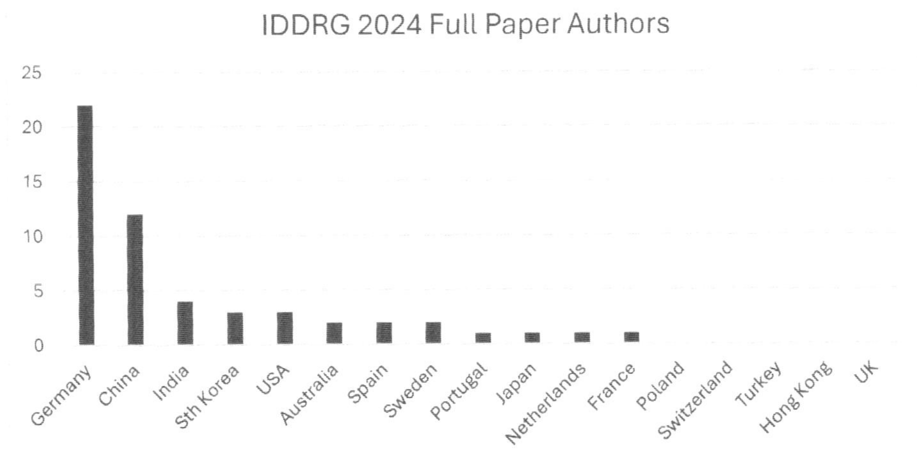

Technical/Industrial tours

Social tours were arranged for over 40 conference participants on 15th March to tour the Great Ocean Road. Following this over 50 conference participants did a tour of Deakin University's facilities associated with manufacturing at Deakin's Waurn Ponds campus.

Acknowledgements

The conference organizing committee would like to express their gratitude to all who supported IDDRG 2024, including the participants, reviewers, keynote speakers, scientific committee members, local organizing committee members and the conference sponsors and exhibitors.

IOP Conf. Series: Materials Science and Engineering **1307** (2024) 011001 doi:10.1088/1757-899X/1307/1/011001

Sponsors and exhibitors

Platinum sponsors:

AUTOFORM
Forming Reality

Silver sponsors:

posco

Bronze sponsors:

SIMUSERV

Post Script:

Many participants asked about the origin of the name "Deakin University", is it related to an indigenous word or meaning? The answer is that Alfred Deakin was an Australian politician who helped promote the federation/creation of Australia in 1901. He was Prime Minister of Australia three times (1903, 1905, 1909). Deakin was born in the colony of Victoria, which became the state of Victoria in federated Australia. Deakin University is founded in Geelong, Victoria, Australia.

IOP Conf. Series: Materials Science and Engineering **1307** (2024) 011002 doi:10.1088/1757-899X/1307/1/011002

Peer review statement

All papers published in this volume of IOP Conference Series: Materials Science and Engineering have been peer-reviewed through processes administered by the proceedings Editors. Reviews were conducted by expert referees to the professional and scientific standards expected of a proceedings journal published by IOP Publishing.

- Type of peer review: Single Anonymous
- Conference submission management system: The CVENT submission management system and Morressier
- Number of full paper submissions received: 59
- Number of full paper submissions sent for review: 59
- Number of submissions accepted: 52
- Acceptance Rate (Submissions Accepted / Submissions Received × 100): 88.13%
- Average number of reviews per paper: 2
- Total number of reviewers involved: 37

Contact persons for queries:
Name: Bernard Rolfe
Email: Bernard.Rolfe@deakin.edu.au
Affiliation: Deakin University, Australia

Name: Matthias Weiss
Email: Matthias.Weiss@deakin.edu.au
Affiliation: Deakin University, Australia

Name: Peng (Neo) Zhang
Email: Peng.Neo.Zhang@deakin.edu.au
Affiliation: Deakin University, Australia

Name: Jeong Whan Yoon
Email: j.yoon@kaist.ac.kr
Affiliation: KAIST, South Korea / Deakin University, Australia

IOP Conf. Series: Materials Science and Engineering **1307** (2024) 012001 doi:10.1088/1757-899X/1307/1/012001

Tribological investigations of water-based lubricants for application in the deep drawing process

B-A Behrens, S Hübner, J Wehmeyer, P. Müller, S Yarcu*
Leibniz Universität Hannover, Institute of Forming Technology and Forming Machines, An der Universität 2, 30823, Garbsen, Germany.
* Corresponding contact: +49 511 762 3835, s.yarcu@ifum.uni-hannover.de

Abstract. This paper describes the tribological investigation of water-based lubricants in the context of a deep drawing process. Therefore, different methods were conducted in order to determinate the suitability of these lubricants for a deep drawing process. Three lubricant manufacturers each provided their lubricants as part of this work. The sheet materials investigated are an aluminum material (AA6014) and two steel materials (DC04, DP800), each with a thickness of 1.5 mm. All sheet materials were examined in the as-delivered condition as part of this work. The hypothesis for this research proposal is that it is possible to achieve comparable tribological properties in sheet metal forming using water-based lubricants as when using mineral oil-based lubricants. The aim is to optimize a tribological system for water-based lubricants when used as additional lubrication in order to replace mineral oil-based lubricants and, as a result, to shorten the representative process chain by one process step (cleaning). In the course of this work, topographical measurements of the sheet materials were carried out in order to investigate the lubricant holding capacity of the sheet materials. Furthermore, strip drawing tests were performed to determine the friction coefficients of the different lubricant-sheet combinations. The final step was to conduct deep-drawing tests to determine the limits of use of the water-based lubricants by continuously increasing the holding-down force.

Keywords: Tribology, water-based lubricants, sustainable production

1 Introduction

In times of scarce raw materials, the concept of alternative lubricants occupies a special role. In this context, water-based lubricants offer many advantages over mineral oil-based lubricants in terms of disposal, handling and savings. However, water alone is not suitable for lubrication in industrial environments due to its physical and biological limitations, so additives must be added. For this reason, lubricants based on mineral oils and additives are mainly used in today's practice. Various manufacturers already have water-based lubricants in their catalog. These lubricants are described as particularly sustainable because they are easy to dispose of compared to mineral oil-based lubricants.

However, it is not specifically researched how the combination of the individual additives affect the forming behavior of the sheet materials. The current literature deals with the applicability of water-based lubricants only on the basis of friction values and wear. There are still some open questions and problems to be clarified regarding the application context. This includes the behavior of water-based lubricants when used as additional lubrication on an oiled surface and to what extent this influences the tribosystem. Some lubricant manufacturers offer water-based lubricants in their range. In order for these to be considered for SMEs and their manufacturing process, it is necessary to research their applicability and process limits. Furthermore, it is not known how the lubricants are best applied. In addition, the behavior of the lubricants for application on real components is not known. Moreover, it is still open whether high-strength steel materials are also suitable for these lubricants in addition to common deep-drawing grades. In the scientific-technical context, there is therefore still a lot of research need with regard to the applicability.

Water-based lubricants offer great potential in terms of reducing cleaning steps, which have to be carried out with great effort for lubricants based on mineral oil. In this context, the influence of these lubricants on the wear of the tools is an aspect with high weighting. Especially for the small and medium-sized suppliers and component manufacturers from the non-automotive sectors such as white goods, ventilation system technology and wheel and roller technology, the use of water-based lubricants can be of particular interest.

2 State of the art

Water is a lubricant that is already used in many areas of nature. Due to its physical and biological limitations, it is not suitable for lubrication in industrial environments on its own and must therefore be additivated. As a result, lubricants based on mineral oil are mainly used in practice [1].

A study by PtU at the University of Darmstadt also found that a minor water influence, which occurs in practice due to weather conditions, can reduce the friction value by up to 60% in a common tribosystem [2].

In the USA, water-based lubricants are used in certain industrial areas. In 1972, a patent was filed by the Atlantic Richfield Company for a lubricant based on water-soluble amine salts of styrene-maleic anhydride copolymer resins and half esters with a water content of 10-50%. The lubricant is used in metalworking, but this is not further specified. Another patent from 1981 describes a lubricant with a water content of 65-75% and a solid lubricant content of 18-25%, consisting of a mixture of molybdenum disulfide and graphite. This lubricant is intended for use in rolling bearings. It should also be noted that the water-based lubricants used in the USA are adulterated with chloroparaffins, which are harmful to humans and the environment [3,4]. These have been banned within the EU since 2002 [5].

Various manufacturers already offer water-based lubricants in their product range. These lubricants are described as particularly environmentally friendly, as their disposal is uncomplicated. Moreover, it is possible to reduce lubricant consumption through suitable dosing and application techniques (spraying or rolling) [6,7].

In the last decade, investigations have been carried out on the tribological properties of water-based lubricants. To use water as a lubricant, additives must be added to protect against corrosion, increase viscosity, and reduce wear. Lubricants are usually supplemented with polymers, esters, and certain EP additives. The combination of various other additives with water has already been investigated in various research projects [8,9,10,11].

TOMALA et al. examined the influence of the respective concentration of additives from ethanolamines, glycols, and ethylamines on the tribological properties of water. In the studies, it was found that certain concentrations of additives significantly improve the tribological properties of water. The investigation of the tribological properties was carried out using the pin-on-disk test with a surface pressure of 290 MPa. The impressions on the disk were then optically measured, and the wear rate was estimated using software [8]. In further research projects, the use of new additives, the so-called "nanosheets," was thoroughly examined, and their positive effect on the friction value (pin-on-disk test) and wear was established, but without reference to forming technology [9,10,11].

Regarding the suitability of water-based lubricants for sheet metal forming, little research has been conducted according to the current state of technology. BAY et al. investigated various environmentally friendly and mineral oil-based lubricants for sheet metal forming using bead drawing tests, bending under tension, and strip drawing tests. The lubricants examined included three mineral oil-based lubricants, two of which were adulterated with chloroparaffins, two fatty acid-based lubricants, and one water-based lubricant with a water-soluble wax. Under laboratory conditions, it was shown that the water-based lubricant could achieve comparable tribological properties compared to mineral oil-based lubricants. However, the water-based lubricant was not used as additional lubrication as envisaged in this research project. No investigations were conducted on the limits of its applicability in sheet metal forming. Wear was not considered at all [Bay08]. An overview of the developments of environmentally friendly lubricants in sheet metal forming can be found in [12].

SÜSSE et al. investigated oil-free water-based lubricants (with a wax dispersion) for their formability. They also examined the influence of lubricant additives on deep drawability. They found that water-based lubricants are suitable for deep drawing through systematic additives. However, in the present application, water-based lubricants are applied as additional lubrication to already oiled surfaces. As a result, completely different tribological conditions prevail. Furthermore, this study also lacks statements on wetting ability [13].

Other lubricants that contain water are emulsions and dry lubricants. The following will elaborate on these to clarify the difference to water-based lubricants.

2.1 Emulsions

An emulsion is a lubricant consisting of a water-oil mixture. This mixture's special property is its high cooling performance due to its high water content. In sheet metal forming, the ratio of an emulsion between oil and water is typically 1-10 or 1-20. Due to their low oil content, emulsions must be applied extensively (3 g/m2) to prevent adhesion between the tool and the workpiece during deep drawing. Compared to mineral oil-based lubricants, emulsions exhibit inferior tribological properties. Furthermore, the disposal of emulsions is costly, and they are prone to germ formation when stored, leading to frequent maintenance. Due to the complicated disposal process, emulsions are no longer used in the automotive industry, as the costs for saving cleaning steps do not balance out with those of disposal [14].

2.2 Dry Lubricants

Dry lubricants (Drylube) are characterized by their temperature and pressure resistance. They are grip-stable on surfaces, thus providing a coating on thin sheets. This coating protects the surface of the sheets during transport from the rolling mill to the press shop. Especially for aluminum sheet materials, the surface condition plays a significant role [15]. Dry lubricants differ in the following characteristics:

- Water-containing polymer dispersions (WP) / water-based systems (SW) are applied to sheets and then dried using UV radiation or heat, resulting in a dry polymer layer that can be removed in a water-alkaline manner.
- Water-free systems are used for components like axle parts or packaging materials that do not require post-treatment. These lubricants are electrostatically applied at a suitable temperature.

Dry lubricants must meet various requirements across the entire process chain, from application in the rolling mill to forming in the press shop. The application is typically done by spraying or with the Coil-Coating process. A uniform distribution is necessary during application to avoid lubricant-free areas and wetting disturbances. The friction behavior during forming should not be impaired compared to mineral oils. There should be no cold welding between the workpiece and the tool, and the lubricant should not accumulate on the tool side, negatively affecting the sheet's surface condition [15].

The IFU (Stuttgart) is pursuing an entirely different approach to optimizing the tribosystem in sheet metal forming. To avoid both adverse environmental aspects and complicated cleaning steps with mineral oil-based lubricants, volatile media such as CO_2 and N_2 are used as lubricants [16].

IOP Conf. Series: Materials Science and Engineering **1307** (2024) 012001 doi:10.1088/1757-899X/1307/1/012001

3 Experimental setup

3.1 Materials

DC04, DP800, and AA6014 are important sheet materials in the automotive industry due to their respective form and strength properties. DC04 is highly formable, enabling complex shaping. DP800 provides high tensile strength and energy absorption in collisions. AA6014 is a lightweight and corrosion-resistant aluminum alloy with a good surface quality (Table 1). In this paper, a thickness of 1.5 mm for all materials was examined.

Table 1: Properties of the examined materials [17]

Properties	AA6014	DC04	DP800
Fracture strain A_{80} [%]	25	45,2	19,3
Yield strength $R_{p0,2}$ [MPa]	90	156	500
Tensile strength R_m [MPa]	195	284	796
Strain hardening coefficient n	0,29	0,18	0,12

3.2 Surface analysis

Surface analysis measurements of roughness parameters are conducted with a Keyence digital microscope (Figure 1) and analyzed through the VR-3000 G2 Series software, providing the ability to assess volume, area, and line parameters. The surface topography is of significant interest as it characterizes both the material content and the void volume of a surface (Figure 2). This can be used to determine the ratio of roughness to the lubricant's holding capacity of the material.

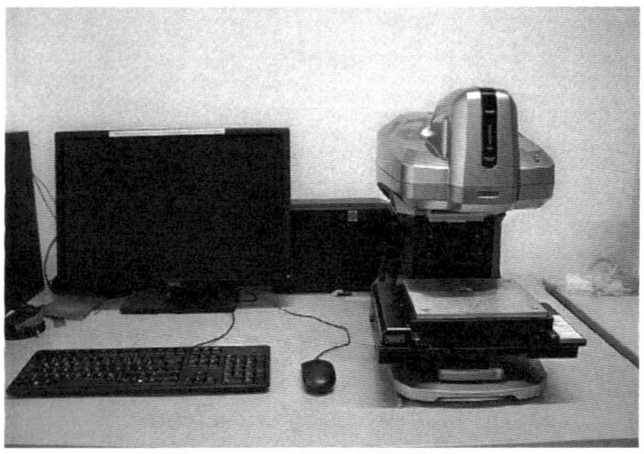

Figure 1: Keyence VR-3200 for measuring surface properties

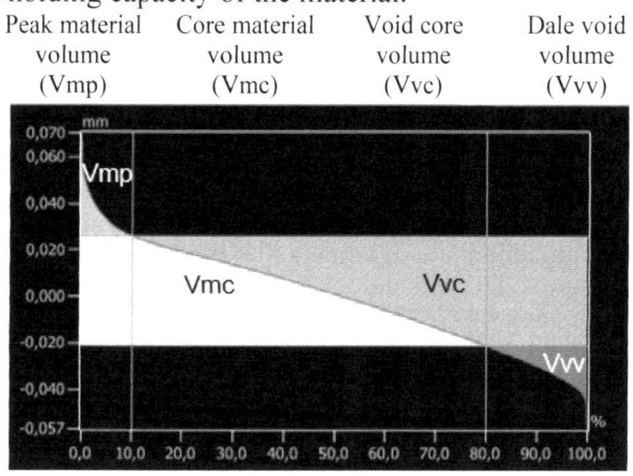

Figure 2: Example of a material ratio curve for the material DC04

3.3 Strip drawing test

The friction coefficient measurements are conducted utilizing the strip tensile test facility at IFUM (Figure 3), with subsequent evaluation via LabVIEW software. During the measurement process, holding force and tensile force are measured as a function of distance, facilitating the calculation of the average value from the resultant values. Values are inserted into respective formulas. The mean value is calculated using only the values within a specific travel range. This range excludes the start and end of the measurement process, which may experience the most significant vibrations.

Figure 3: Strip drawing machine of the IFUM

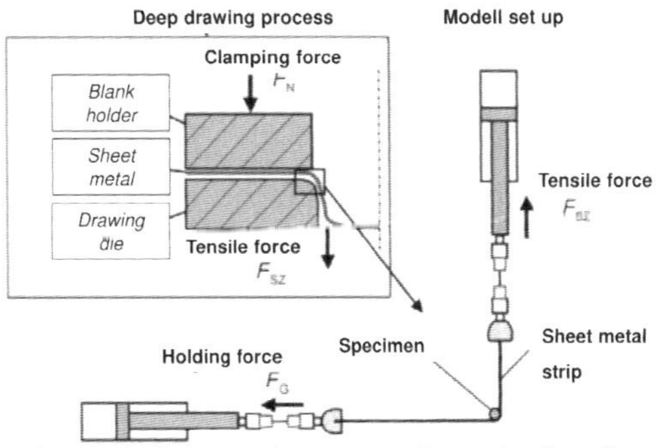

Figure 4: Schematic set up of a strip drawing test

3.4 Deep drawing test

Deep drawing tests were conducted to assess the effect of forming lubricants on the tribological system during forming processes. Technical abbreviations were explained at first use. The tests included forming processes with increased surface pressure in the radius region and friction in the flat surface using strip drawing tests. A hat-shaped profile, measuring 50 mm in depth and 80 mm in width, was manufactured in a hydraulic forming press from the company Dunkes using a rectangular tool with a bead insert to examine the friction-forming interaction. As a result, beads clamp the sheet metal strip at the flange, increasing resistance to flow out of the flange area. The variable height of the bead can balance geometric retention force and friction. In these tests, the tool utilized the following metrics: drawing bead radius of 6 mm, bead height of 3 mm, drawing ring radius of 8 mm, and a drawing clearance of 1.8 mm (Figure 5). The objective was to analyze the correlation between friction and forming and to determine the maximum holding force that the lubricants being studied can sustain before failing. The hold-down force increased gradually during the investigation until the press reached its maximum capacity of 500 kN.

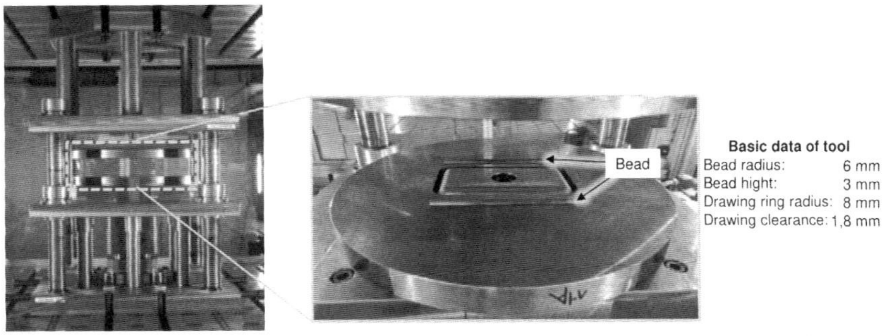

Figure 5: Rectangular tool used (left), detail of rectangular tool with bead inserts (right)

4 Results

4.1 Surface analysis

The surface characteristics of sheet materials DP800, DC04, and AA6014 were analyzed. The interest focused on the topography since it provided insight into the materials' content and void volume. The findings were utilized to establish the correlation between roughness and lubricant absorption capacity. Table *2* displays the surface characteristics of AA6014, DC04, and DP800 sheet materials. The average roughness Sa and Ra depict the surface structure, showing that AA6014 has the lowest roughness. Vmp indicates the volume of the highest structures on the surface, and AA6014 has the highest Vmp. Vmc

IOP Conf. Series: Materials Science and Engineering **1307** (2024) 012001 doi:10.1088/1757-899X/1307/1/012001

represents the deepest structure volume on the surface, and DP800 has the highest Vmc. Technical abbreviations are explained on first use. The Vvc, or material-free core volume, measures the lubricant holding capacity of a material's cavities or pores, with DP800 boasting the highest Vvc and thus the optimal capacity. Meanwhile, the Vvv, or scoring volume, gauges the material's surface space unavailable for lubricants due to depressions, scratches, or scoring, with DC04 recording the highest Vvv and therefore the worst lubricant holding capacity.

Table 2: Results of the surface analysis

Surface properties	**AA6014**	**DC04**	**DP800**
Mittlere Rauheit R_a [µm]	0,5	0,6	1
Peak material volume (Vmp) [ml/m^2]	1,3	0,9	0,5
Core material volume (Vmc) [ml/m^2]	26,1	21,8	34,8
Void core volume (Vvc) **[ml/m^2]**	**33,2**	**25,2**	**36,2**
Dale void volume (Vvv) [ml/m^2]	2,9	1,9	2,8

4.2 Strip drawing test

This Table 3 presents an overview of strip pull tests with deflection, wherein the tested lubricants were applied to the materials. Each lubricant was tested three times (3x). Lubricants are categorized under LBI, Raziol, Fuchs, and Bekuform, according to their respective manufacturers. The delivery condition refers to the state of the sheet after it leaves the rolling mill and is treated with Prelube. Furthermore, the lubricants are segmented into charge 1 to 3, which allows manufacturers to make necessary adjustments after each charge. Water-based lubricants are distinguished by a blue label, while oil-based lubricants are marked with brown. The Bekuform 169 functions as a high-performance forming lubricant and serves as a benchmark for water-based lubricants.

Table 3: Test matrix of the examined materials

Lubricant producer	**AA6014**	**DP800**	**DC04**	**Lubricant Charge 1**	**Lubricant Charge 2**	**Lubricant Charge 3**	**Base**
LBI	3x	3x	3x	LB3464	LB3464-2	LB3464-3	Water-based
	3x	3x	3x	LB3666	-	-	
	3x	3x	3x	LB3467	-	-	
Raziol	3x	3x	3x	-	Raziol-VM	Alpha 7	
Fuchs	3x	3x	3x	-	Fuchs 4	Fuchs 6	
	3x	3x	3x	-	Fuchs 5	-	
Bekuform	3x	3x	3x	Bekuform 169	-	-	Oil-based
As-delivered condition	3x	3x	3x	Prelube	-	-	Reference

In the figure shown, a comparative analysis is presented regarding the friction coefficients of three materials, The materials were tested with different lubricants, and the bars depict the friction coefficient for each material-lubricant pairing, indicating the standard deviation through error bars, providing insight into the data spread. The examined lubricants are the as-delivered condition, Bekuform 169, a series of Fuchs lubricants (2 through 6), LB series (3464, 3466, 3467, and variants 3464-2, 3464-3), and

Raziol products (VPAW50, VM, Alpha 7). The friction coefficients display different trends depending on the combination of lubricant and material, with water-based lubricants sometimes proving more effective than their oil-based counterparts in reducing friction. AA6014 often has lower friction coefficients in most lubrication conditions, while DP800 exhibits higher coefficients with particular lubricants, particularly in its as-delivered state and with Bekuform 169. The presented graph emphasizes the crucial role of lubricant selection in optimizing frictional performance in metal forming processes.

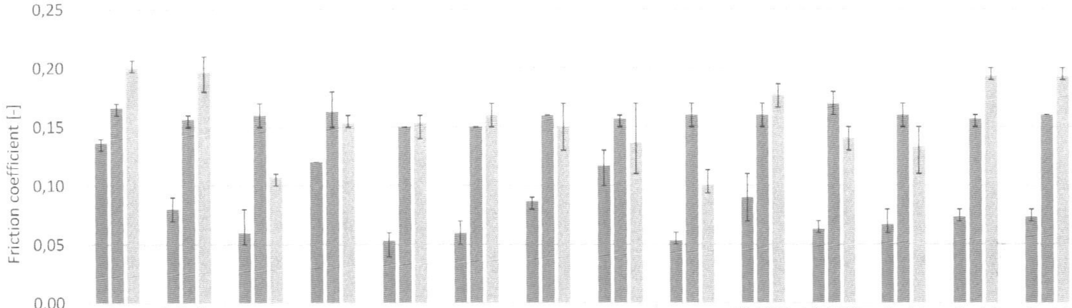

	As-delivered	Bekuform 169	Fuchs 2	Fuchs 4	Fuchs 5	Fuchs 6	LB 3464	LB 3466	LB 3467	LB 3464-2	LB-3464-3	Raziol VPAW50	Raziol VM	Raziol Alpha 7
AA6014	0,14	0,08	0,06	0,12	0,05	0,06	0,09	0,12	0,05	0,09	0,06	0,07	0,07	0,07
DC04	0,17	0,16	0,16	0,16	0,15	0,15	0,16	0,16	0,16	0,16	0,17	0,16	0,16	0,16

Figure 6: Results of the strip drawing test with deflection

4.3 Deep drawing test

For AA6014 alloy, it was established that the baseline hold-down force in an unlubricated state stood at 70 kN. The application of Bekuform 169 notably escalated this requirement to 290 kN, highlighting a substantial enhancement in lubrication performance (Figure7). The study further revealed disparities across lubricant charge, with some exhibiting lower efficacy than others. Specifically, while certain Fuchs lubricant charge paralleled the performance of Bekuform, LBI lubricants generally fell short, albeit with some charge achieving comparable results. Raziol lubricants displayed consistent performance across their charge, although varied when benchmarked against other lubricants.
The analysis extended to DC04 material, where the essential hold-down force without lubrication was identified as 110 kN. Here, all Fuchs Lubricant charge demonstrated up to a 20% reduction in performance relative to Bekuform. LBI lubricants varied, with a notable dip in performance for specific charge, whereas the third charge aligned closely with Bekuform's efficacy. Raziol's second charge mirrored Bekuform's lubricity, with other charge achieving similar outcomes to the Prelube condition.
Turning attention to the high-strength steel DP800, the study found that Prelube necessitated a hold-down force of 290 kN, which surged to 500 kN with Bekuform 169, indicating a significant increase in lubrication performance. The examination of Fuchs and LBI lubricants unveiled a spectrum of effectiveness, with certain charge nearing or falling below Bekuform's performance. Remarkably, the third charge of LBI lubricants matched Bekuform 169's hold-down force requirement, while Raziol lubricants exhibited a gradation in performance across their charge.
This investigation underscores the criticality of tailored lubricant selection to mitigate friction and wear, thereby enhancing stamping performance and preserving material integrity. It asserts that a strategic alignment of lubrication techniques with specific material characteristics and processing conditions is essential to elevate process efficiency and avert material failure. The findings advocate for a discerning approach to lubricant application, paving the way for advancements in metal forming technologies.

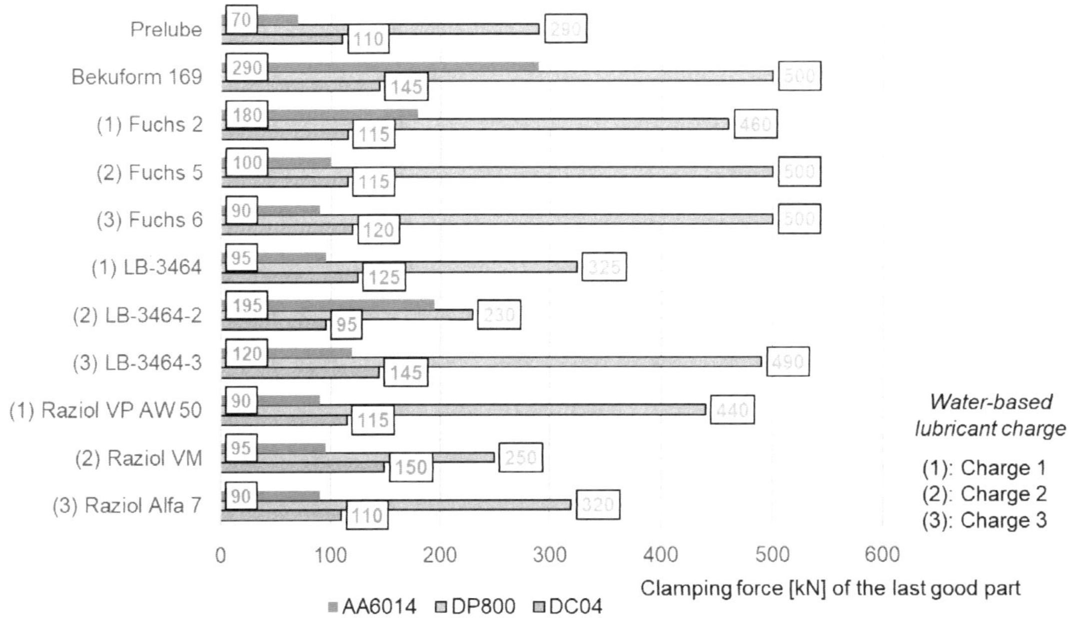

Figure 7: Results of the strip drawing test with deflection

5 Summary and conclusion

This paper presents a comprehensive study on the utilization of water-based lubricants in the deep drawing process, with a focus on tribological analysis. The paper provides a detailed account of the test results and insights gained from the study. The study assessed the suitability of different lubricants provided by three manufacturers using a variety of methods. The study materials were aluminum (AA6014) and two types of steel (DC04 and DP800), each having a thickness of 1.5 mm. The study aimed to determine if water-based lubricants could attain tribological qualities in sheet metal forming similar to mineral oil-based lubricants. The lubricant holding capacity of the materials was evaluated through topographical measurements, while friction coefficients were assessed through strip drawing tests. Deep drawing tests were also conducted to determine the maximum application capacities of water-based lubricants by gradually increasing the hold-down force.

The study confirmed that water-based lubricants can be optimized to match the tribological performance of mineral oil-based lubricants in sheet metal forming. This has the potential benefits of reducing the process chain by eliminating the cleaning step usually required for oil-based lubricants. The study discovered that there was variability in the friction coefficients based on the types of lubricant-material combinations tested, with instances where water-based lubricants outperformed oil-based ones. This has set the groundwork for future research aimed at developing these lubricants further for use in a wider range of industrial applications.

Acknowledgements

The authors thank the Industrial collaborative research (IGF) and the European Research Association for Sheet Metal Working (EFB) for the financial support of the research project "Application of water-based lubricants for use in sheet metal forming"(IGF-Code: 21287N).

References

[1] J. Dietrich: „Praxis der Umformtechnik - Umform- und Zerteilverfahren, Werkzeuge, Maschinen"; Springer Vieweg Auflage 8; Seite 20-22; Wiesbaden; 2018

[2] Groche, P.; Müller, C.; Reuter, D.: „Wasser senkt Reibwert"; PtU, Technische Universität Darmstadt; Umformtechnik; Meisenbach Verlag GmbH; Bamberg, S. 35-36, (4), ISSN 0300-3167; 2012

[3] Gower, H.D.; Gower, B.G.; "Water based Lubricants"; United States Patent Office; Patent No. 3,645,897; Atlantic Richfield Company; New York, 1972

[4] Laepple, W. "Water based lubricant"; United States Patent Office; Patent No. 4,303,537; Dow Corning GmbH Munich; 1981

[5] Directive 2002/45/EC of the European Parliament and of the Council of June 25, 2002on20. amending Council Directive 76/769/EEC as regards restrictions on the marketing and use of certain dangerous substances and preparations (short-chain chlorinated kerosenes)

[6] LbI oilfree GmbH; Produktdatenblatt: LUB form; LUBform von LBI oil free (lbi-oilfree.de) 01.11.2023

[7] Raziol Zibulla und Sohn GmbH; https://www.raziol.com/de/raziol-umformschmierstoff/raziol-fluid#anwendung-des-raziol-fluids; 01.08.2023

[8] A. Tomala, A. Karpinska, W.S.M. Werner, A. Olver, H. Störi: "Tribological properties of additives for water-based lubricants"; Wear Vol. 269 pp. 804-810; Elsevier; 2010

[9] Yongxia Wang, Yanyan Du, Jinan Deng, Zhaopeng Wang: "Friction reduction of water based lubricant with highly dispersed functional MoS2 nanosheets"; Colloids and Surfaces A: Physicochemical and Engineering Aspects Vol. 562 pp. 321-328; Elsevier; 2019

[10] Hongmei Xie, Suihu Dang, Bin Jiang, Lin Xiang, Shan Zhou, Haoran Sheng, Tianhao Yang, Fusheng Pan: "Tribological performances of SiO2/graphene combinations as water-based lubricant additives for magnesium alloy rolling"; Applied Surface Science Vol. 475 pp. 847-856; Elsevier; 2019

[11] Hiroshi Kinoshita, Yuta Nishina, Aidil Azli Alias, Masahiro Fujii: "Tribological properties of monolayer graphene oxide sheets as water-based lubricant additives"; Carbon Vol. 66 pp. 720-723; Elsevier; 2014

[12] Bay, N.; Azushima, A.; Groche, P.; Ishibashi, I.; Merklein, M.; et. al.: „Environmentally Benign Tribo-systems for Metal Forming"; CIRP Annals – Manufacturing Technology, Vol. 59 (2019), 760-780

[13] Süße, D.; Reichert, J.; Kötter, H.; Eickemeyer, J.; Schürer, A.-C.: „Ölfreie Schmierstoffe in der Blechumformung – Entwicklung und Prüfung"; Tagungsband T- 021 EFB Kolloquium Blechbearbeitung; 2001

[14] Meiler, M.: „Großserientauglichkeit trockenschmierstoffbeschichteter Aluminiumbleche im Presswerk"; Universität Erlangen-Nürnberg; Meisenbach Verlag Bamberg; 2005

[15] Wager, S.; Kleinert, H.; Zimmermann, R.: „Trockenschmierstoffe für die Blechumformung"; wt Werkstattstechnik online; Jahrgang 92; S. 488-495; 2002

[16] Reichardt, G.; Liewald, M.: "Investigation on friction behaviour of deep drawing radii using volatile media as lubricant substitutes"; International Conference on Sheet Metal SHEMET 2019, Procedia Manufacturing Vol. 29 (2019), 193-200

IOP Conf. Series: Materials Science and Engineering **1307** (2024) 012002 doi:10.1088/1757-899X/1307/1/012002

Investigation of the influence of cobalt content and grain size of WC-Co cemented carbides on wear when shear cutting electric sheet metal

Agnes Schrepfer[1], Florian Steinlehner[1] and Wolfram Volk[1]

[1]Chair of Metal Forming and Casting, Technical University of Munich, Germany

*Agnes.Schrepfer@tum.de

Abstract. In the production of electrical machines, shear cutting of electrical steel laminations before further processing into rotors and stators represents an essential manufacturing process. Investigations have shown that using abrasively worn active elements while shear cutting increases the hysteresis losses in the electrical steel sheet due to introduced deformations and resulting residual stresses. Therefore, knowledge of suitable tool materials is of great importance for producing high-quality electric motors. Due to its high silicon content, electrical steel has a strong abrasive effect on the tools, which is why Tungsten carbide cobalt (WC-Co) cemented carbides are commonly used. In this work, continuous stroke tests are carried out on an electric sheet with punches made of different WC-Co cemented carbides, differing in cobalt content and grain size. The punches are examined with increasing wear regarding cutting-edge geometry, surface condition and microstructure. The resulting cutting parts are inspected for cutting surface characteristics. In addition, the influence of geometry and tribology of the punches regarding residual stresses in the material is investigated by FEM-Simulation. The investigations will provide information on the wear development and causes when shear cutting electrical sheets and the effects on electric losses.

1. Introduction

The production of rotors and stators plays a vital role in the manufacturing process of an electric motor. It is essential to keep the energy losses introduced through each production process as low as possible and achieve the electric motor's highest possible magnetisability. In an industrial environment, rotors and stators are mainly assembled from stacked electrical steel laminates. To produce large quantities, the electrical laminates are shaped using shear cutting. However, this process is associated with the deterioration of the magnetic properties of the shear-cut materials. [1]

This is due to the introduction of residual stresses into the sheet metal by the shear-cutting process in the area of the cut surface. During the cutting process, the material is deformed until the maximum shear stress in the sheet reaches the material-dependent shear fracture limit, the deformation capacity is thus exhausted and cracking occurs. The plastic deformation also causes the material near the cut surface to harden. [2] Due to the strong deformation of the sheet material, strains occur in the material, leading to stresses in the zone affected by shear cutting. Some of these stresses remain even after the cutting process and the sheet metal has been relieved from external forces. These stresses are called residual stresses. In magnetized sheets, these residual stresses hinder the alignment of the magnetic domains in the direction of the external magnetic field. The higher these residual stresses are in the electric sheet metal, the more the alignment of the domains is prevented, leading to reduced efficiency. [3]

In addition, the depth of this affected zone also affects the total losses. Several process parameters influence the level and characteristics of the residual stresses in the zone affected by shear cutting. It was already shown that the greater the cutting clearance, the distance between the two cutting elements punch and die, the greater the residual stresses that can be measured. [4] showed that a larger cutting clearance leads to greater losses when guillotining electrical steel. Weiss confirmed this for shear cutting of electrical steel and showed that the same applies to using worn cutting edges compared to sharp-edged ones [5].

Due to its comparatively high silicon content, electrical steel generally has a highly abrasive effect on the active elements. For this reason, WC-Co cemented carbides are commonly used as materials for active elements for shear cutting electrical steel sheets for their excellent wear resistance compared to conventional tool steels. Most conventional cemented carbide grades comprise a high volume proportion of the hexagonal WC phase embedded in a soft and viscous Co-binding phase. [6] The high wear resistance thus results from the combination of the high hardness due to the WC matrix in combination with sufficient toughness imparted by the Co-binder [7]. [8] and [9] showed that the WC-Co cemented carbide's hardness directly influences its wear volume. Using different test methods, they showed that a lower hardness simultaneously leads to a higher wear volume on the tested samples. According to [10], the hardness of the material Co-low_GS-fine ecreases with increasing cobalt content. As a logical conclusion, [11] recognized that the wear volume increases with increasing cobalt content.

Comparing the fracture toughness of WC-Co cemented carbides, [12] showed that as the volume fraction of the carbide phase increases, fracture toughness decreases. Additionally, the grain size of the carbides also influences the mechanical properties of the materials. Grades with fine carbides exhibit higher hardness and lower fracture toughness than those with a coarser microstructure. [10]

This paper aims to evaluate the suitability of different carbides for shearing electrical steel sheets based on their composition through experimental investigations. For this purpose, continuous stroke tests are carried out with carbide punches made of materials with different cobalt contents and grain sizes. An overall number of one million strokes are carried out, and the wear profile and the surface condition on the cutting edge of the punch are recorded. In addition, the effects on forming residual stresses in the sheet metal using different cemented carbides are evaluated by observation of cut surfaces and FE Analysis.

2. Materials

2.1 WC-Co cemented carbides for active elements

Four different WC-Co cemented carbides were analyzed for their suitability for use as active elements in the processing of electrical steel sheets. From a chemical point of view, these differ mainly in terms of their cobalt content and the tungsten carbide grain size. This affects mechanical properties such as hardness, transverse rupture strength and fracture toughness. Table 1 shows the chemical composition and the mechanical properties of the investigated materials which were provided by the manufacturers. It can be seen, that the material's fracture toughness increases with the proportion of cobalt binder. This generally reduces the tendency of the active element to chipping from the cutting edge. In contrast, a high cobalt content reduces the material's hardness, negatively affecting resistance to abrasive wear. Micrographs of the different carbide grades are shown in figure 1. The differences in the size of the carbides and the amount of binder are particularly evident here.

Table 1. Chemical composition and mechanical properties of the four analyzed WC-Co carbides.

Grade	Binder content	Average grain size	Hardness HV10	Transverse Rupture Strength	Fracture Toughness
Co-high_GS-fine	12,0 %	fine	1403	3200 N/mm^2	12,5 $MPa*m^{1/2}$
Co-mid_GS-finest	10,0 %	finest	1678	>4200 N/mm^2	
Co-low_GS-finest	6,0 %	finest	1927	>3800 N/mm^2	
Co-low_GS-fine	6,0 %	fine	1823	3200 N/mm^2	9,9 $MPa*m^{1/2}$

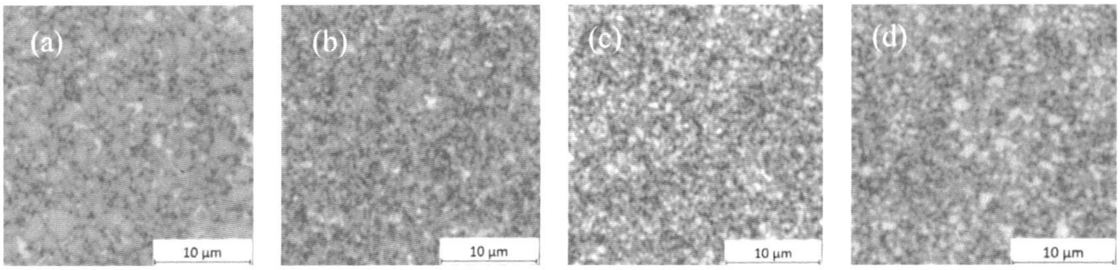

Figure 1. Preparation of material Co-low_GS-fine (a), Co-low_GS-finest (b), Co-mid_GS-finest (c) and Co-high_GS-fine (d).

2.2 Electric sheet material

The NO30-16 electrical steel sheet with a thickness of 0.3 mm was used for the shear-cutting tests. It was coated on one side with a PE75 W baking varnish. The sheet was inserted into the cutting tool so that the coated side was oriented to the die side to reduce die wear. The contact of the punch with the uncoated side of the sheet, however, increases the abrasive effect of the sheet on the punch. The chemical composition of the sheet material is shown in table 2.

Table 2. Chemical composition of electric steel sheet.

Alloying element in mass percent (wt%)							
Fe	**C**	**Si**	**Mn**	**P**	**S**	**Cr**	**Al**
94,90	0,02	1,70	0,19	0,02	0,01	0,04	1,3

3. Experimental Setup

3.1 Shear cutting test setup

Continuous stroke tests were carried out to investigate the wear behavior of the different carbides. The tool was operated on a Bruderer BSTA 1600 (Bruderer AG, Frasnacht, Swiss) high-speed press. Punches with cutting-edge diameters of 1 mm were used to be able to see wear as quickly as possible. A small cutting clearance of 15 µm, which corresponds to a relative cutting clearance of 5 % concerning the sheet thickness of 0.3 mm, was also selected to further increase wear. The tool (figure 2 a) was operated with a stroke rate of 150 strokes per minute, which is within the range of stroke rates used in the production environment for producing rotors and stators. The punches were removed at regular intervals and the condition of the cutting edges was checked and measured using a laser confocal microscope Keyence VK-X150. Four measuring points were distributed around the circumference of the cutting edge. Images of the stamp surfaces for visual inspection were taken at equal intervals using a digital microscope Keyence VHX 200. The tests were carried out until a total number of one million strokes was reached. Two to four punches of the same material were used for the wear tests.

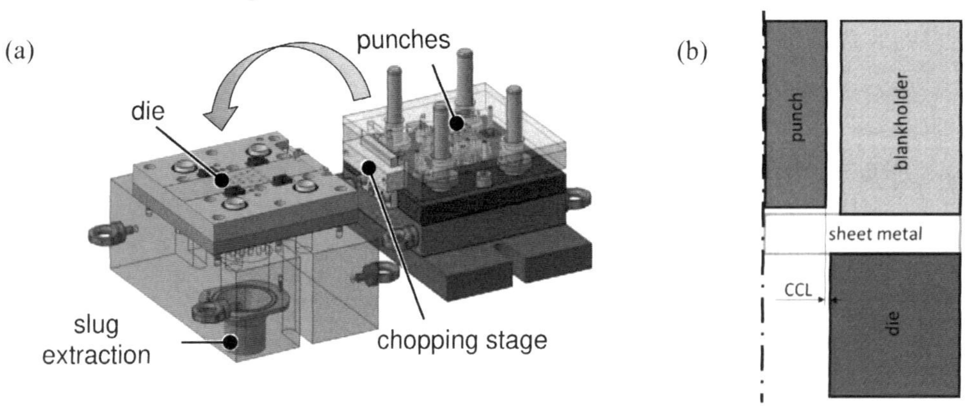

Figure 2. (a) Shear cutting tool; (b) draft of the cutting active elements.

3.2 Mechanical FEA setup

A 2D simulation model of the shear-cutting process was set up to categorize the influence of the determined wear profiles on the characteristics of the residual stresses in the sheet metal. The wear profiles on the punches recorded after zero and one million strokes were therefore modeled within the simulation model. A springback simulation following the shear-cutting simulation shows the different residual stress distributions in the sheet metal after removing all external forces. To minimize the computing time, the symmetry of the structure was used and only half the model was built with symmetry boundary conditions, similarly to figure 2 b. Element sizes of 0.005 mm to 0.001 mm were selected in the shear zone of the sheet metal; outside this area, a coarser meshing of 0.03 mm to 0.08 mm was set. The entire sheet consisted of plain strain elements. A ductile damage model was chosen as damage criterion. The purpose of analysing the process using FEA is to qualitatively assess the residual stresses remaining in the sheet metal at different wear states of the punch. The stress-strain curves and the damage criteria were therefore calibrated using the measured force curves. A comparison of the resulting cut surfaces is not always expedient when cutting electrical steel, as the fracture behaviour varies depending on the arrangement of the grains in the sheet. The influence of grain position and size was not considered in the FEA analysis. All active elements in contact with the sheet - punch, die, and blank holder - were modeled as rigid bodies.

4. Results and Discussion

4.1 Shear cutting tests

During the continuous stroke tests, the different carbides showed different wear patterns. Both punches made out of Co-low_GS-fine failed relatively early on at a number of 250.000 and 300.000 strokes, see figure 3. Both punches broke out over a large area. The first failure due to breakage from the cutting edge of a punch made of Co-low_GS-finest was recorded at 450.000 strokes. Further breakouts followed in two other Co-low_GS-finest punches after 700.000 strokes. One of four stamps from material Co-low_GS-finest lasted one million strokes without registered damage. In comparison, all punches out of Co-mid_GS-finest and Co-high_GS-fine could be used for up to one million strokes without damage.

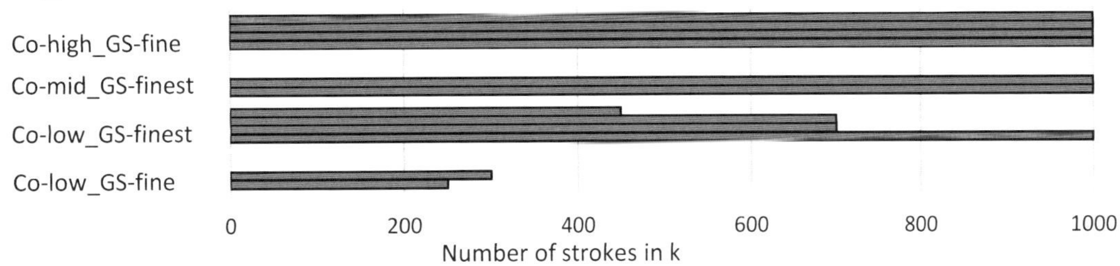

Figure 3. Number of strokes until failure or reaching 1 million for stamps made of carbides Co-high_GS-fine, Co-mid_GS-finest, Co-low_GS-finest and Co-low_GS-fine.

Figure 4 a shows the wear progression from zero to one million strokes of a punch made from Co-high_GS-fine. It can be seen that the wear increases linearly with the number of strokes. All profiles show an almost uniform material removal over large areas on the lateral surface of the punch with a slight increase in removal towards the cutting edge. In comparison, only minimal face wear can be recognized even after one million strokes. The effect of this type of wear on the deformation and residual stresses in the sheet metal is therefore similar to an increase in the cutting clearance.

A significant increase in material removal in the range of 500 µm, which corresponds to the immersion depth, and 400 µm from the cutting edge is due to the formation of a rollover on the sheet and the resulting reduced friction between the sheet and punch. The increased wear on the lateral surface can be attributed to the small cutting clearance, as the smaller the cutting clearance, the greater the radial springback of the punching grid. This increases the friction work to be performed during the exchange and return stroke of the cutting punch, favoring abrasive wear on the lateral surface of the punch. [13]

In addition, the work hardening of the sheet material on the cut surface leads to an increased abrasive effect during the punch return stroke on the punches' lateral surface.

All the cemented carbide punches analyzed showed the same effect of significantly greater lateral surface wear than face wear, regardless of the number of strokes tested. However, material Co-low_GS-finest with its higher hardness exhibits a wear progress that differs significantly from Co-high_GS-fine, for all punches investigated. Figure 4 b shows, that while the cutting-edge profile changes noticeably from zero to 250.000 strokes, further progressions are close together and wear progress is no longer recognizable. Continuous abrasive wear is therefore less than with the Co-high_GS-fine. It can be concluded that after initial removal of the top layer, the high hardness of the tungsten carbide matrix prevents further progression of abrasive wear.

Figure 4 c shows the absolute wear volume after one million strokes of the three materials lasting until reaching one million strokes. It shows that the Co-low_GS-finest carbide material, which has the highest maximum hardness of the carbides compared, shows the lowest abrasive wear. However, three out of four punches failed prematurely and this material is therefore only marginally suitable for use in high-volume production. This early failure can be attributed to the low cobalt content in the material, just as with the premature failure of material Co-low_GS-fine. A comparison of the carbides Co-low_GS-fine and Co-high_GS-fine shows a significant difference in fracture toughness with the same grain size but a significantly different binder content.

The carbide Co-high_GS-fine showed the opposite performance. All four punches out of Co-high_GS-fine analyzed could be used for up to one million strokes, which can be attributed to its high fracture toughness due to the high binder content. However, comparing the wear profiles revealed the greatest abrasive wear on the lateral surface. This high wear in turn harms the magnetic properties of the cut sheets. A similar picture emerges when analyzing the wear profiles for material Co-mid_GS-finest with medium hardness values and a medium proportion of cobalt binder compared to the other materials studied. The analyzed punches made of material Co-mid_GS-finest wore more than the harder material Co-low_GS-finest but less than the softer material Co-high_GS-fine. Considering the amount of wear and the service life of the individual punches, material Co-mid_GS-finest therefore shows the best performance in terms of failure.

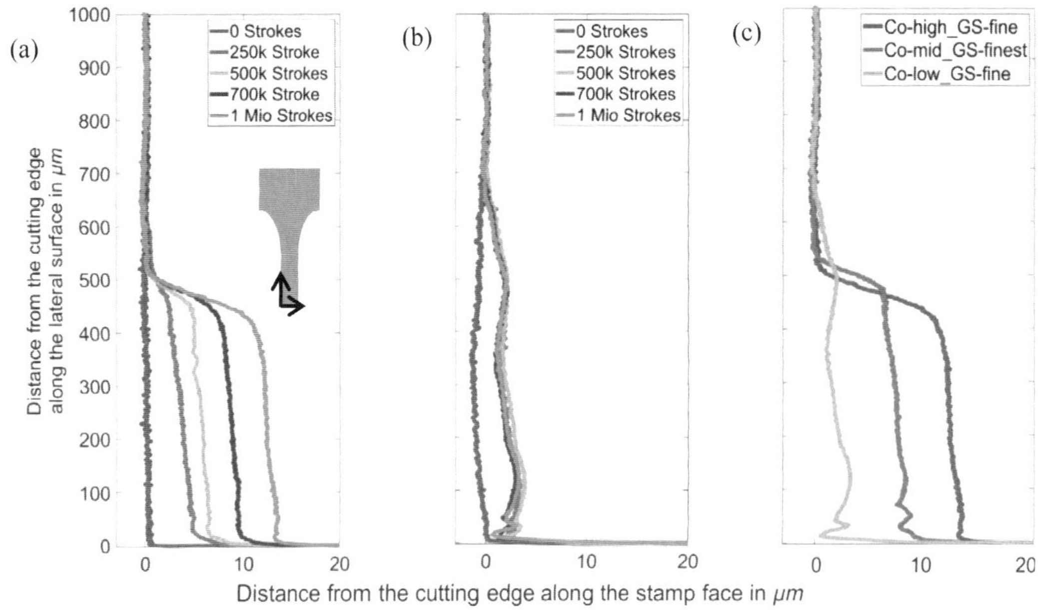

Figure 4. Wear progression from zero to one million strokes of a punch made of Co-high_GS-fine (a) and Co-low_GS-finest (b); Comparison of the wear profiles of punches made of Co-high_GS-fine, Co-mid_GS-finest and Co-low_GS-finest after one million strokes (c).

A closer look at the surfaces of the punches after one million strokes is possible in figure 5 with a comparative representation of the punch cutting edges at different magnifications. The recorded wear geometries from figure 4 can also be seen here. In particular, the difference in the transitions in the 400 μm to 500 μm distance range, in which the lateral surface wear increases significantly, can be seen when comparing figures 5 a, b and c. While the Co-high_GS-fine punch in figure 5 a shows a mark around the circumference of the punch, Co-mid GS-finest (figure 5 b) shows less and Co-low_GS-finest (figure 5 c) shows almost no transition. In the subsequent area closer to the cutting edge, grooves can be seen in the lateral direction on the punches, which are again most pronounced on the punch made of Co-high_GS-fine in figure 5 a.

The wear directly on the cutting edge can be classified more precisely using figures 5 d, e and f. The punch made of Co-high_GS-fine is damaged on the surface in a larger area than those made of Co-mid_GS-finest and Co-low_GS-finest, with the smallest area. The size and appearance of the chippings and releases of the material on the cutting edge differ significantly (figures 5 g, h and i). While many small defects occur with Co-high_GS-fine, these become significantly larger but at the same time fewer from Co-mid_GS-finest to Co-low_GS-finest. This can be attributed to the lower hardness of Co-high_GS-fine and the associated continuous abrasive wear progress due to the leaching of the cobalt binder and the subsequent triggering of carbides. With Co-mid_GS-finest, but especially with Co-low_GS-finest, larger flakes primarily break out of the lateral surface, which can be linked to the lower fracture toughness due to the lower cobalt content. Almost no material removal can be seen near the cutting edge of the Co-low_GS-finest punch, which can be explained by the maximum surface pressure while shear cutting, which is located slightly above the cutting edge. An influence of the size of the WCs is not recognizable in the comparison of the surfaces of the cemented carbides.

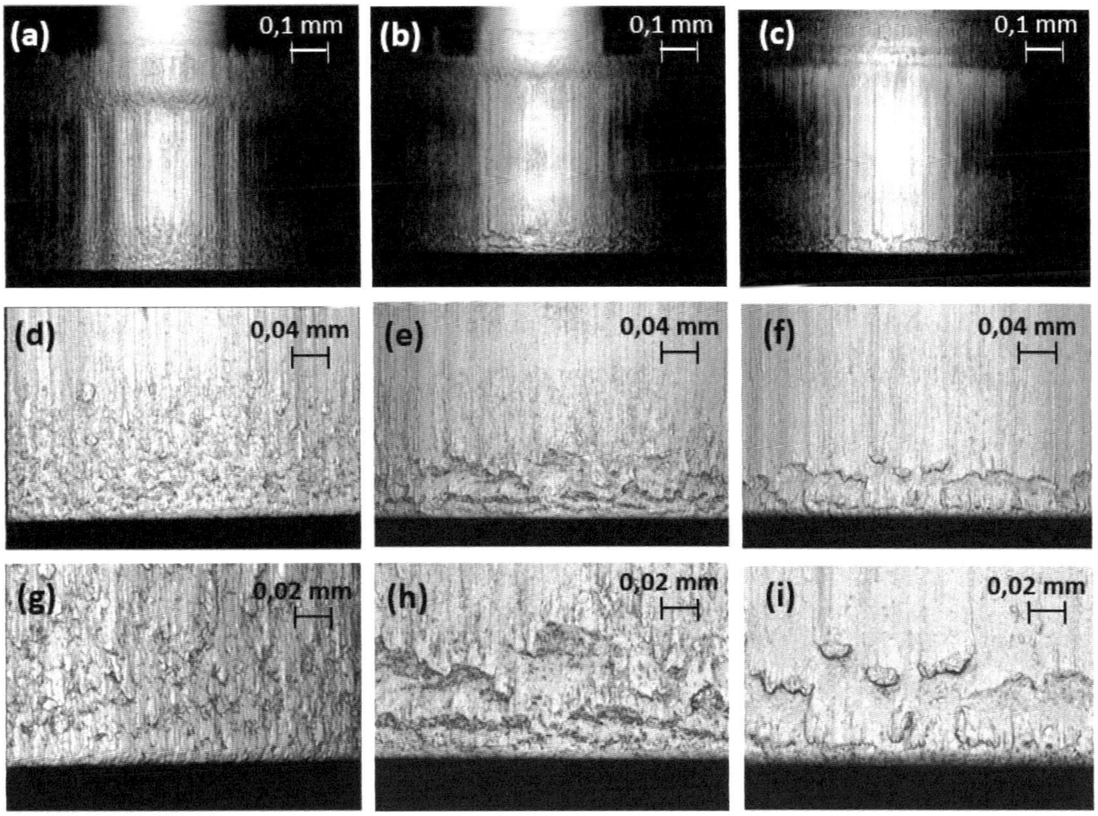

Figure 5. Lateral surfaces of punches out of Co-high_GS-fine (a, d, g), Co-mid_GS-finest (b, e, h) and Co-low_GS-finest (c, f, i) after 1 Mio. strokes at different magnifications.

These different resulting cutting-edge profiles due to different wear heights and types directly influence the resulting cut surface on the cut electric sheet. Figure 6 shows the influence of the condition of a material Co-high_GS-fine punch on the geometric characteristics of the shear influence zone. The cut surface b has a significantly greater rollover height and a larger fracture angle. What can also be recognized is a stronger material Co-low_GS-fine eformation in the area of the cutting edge, which is clearly shown by the dislocation lines. It is expected that the wear level of material Co-high_GS-fine will result in a deterioration of the electromagnetic properties of the cut sheet after one million strokes.

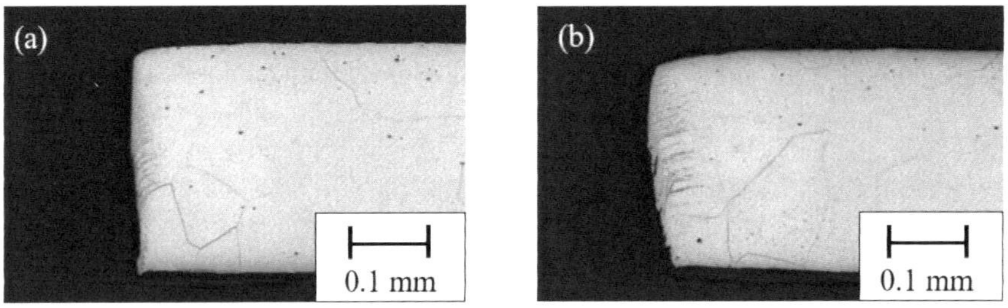

Figure 6. Micrograph of the electric sheet cut with sharp (a) and after 1 Mio. strokes worn (b) punch out of material Co-high_GS-fine.

4.2 Mechanical FEA results

A qualitative assessment of the remaining residual stresses is possible on the basis of figure 7. Figure 7 a graphically shows the result of the simulation of a shear-cutting process with a sharp-edged punch by mapping the zone affected by the shear-cutting of the electric steel sheet. In comparison, figures b, c and d show the zones affected by shear cutting on the sheet after cutting with the worn punches out of Co-low_GS-finest, Co-mid_GS-finest and Co-high_GS-fine. An increase in the wear volume (from a to d) is accompanied by a change in the cutting surface. In particular, the rollover width, the edge rollover height and the fracture angle increase, as already shown in figure 6.

For analyzing the residual stresses concerning the electromagnetic properties consideration of the von Mises stresses is not suitable, as it is impossible to distinguish between tensile and compressive stresses. The mixed stress value $\sigma_{MAX,in\text{-}plane,abs}$ is used for a simpler comparison of the different wear states. Here, the largest local in-plane absolute principal stress is used for the comparison. The algebraic sign is not taken into account for the selection. However, it is taken into account when displaying the resulting values.

When looking at the resulting tensions it is recognizable that a wear volume in the range of several micrometers already influences the level of compressive and tensile stresses. The tensile stresses increase on the upper side of the sheet as well as in the area near the fracture surface with increasing cutting-edge wear on the punch. In the center of the sheet, the compressive stresses in particular increase in the direction of the rollover on the upper side of the sheet. The stress conditions at the bottom of the sheet remain virtually unchanged. In addition, the size of the zone affected by shear cutting increases with increasing tool wear. The electromagnetic properties of sheets cut with similarly worn active elements should have already deteriorated noticeably.

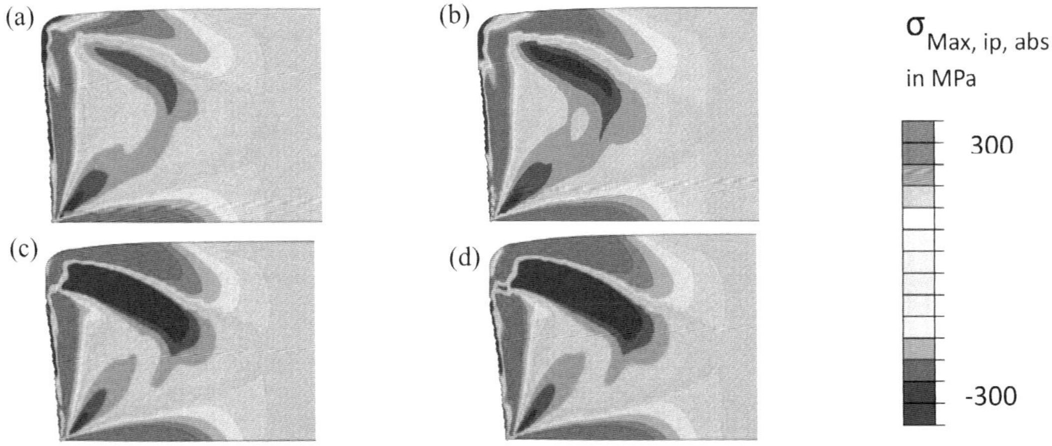

Figure 7. Residual stress distribution for wear profile of (a) sharp tool, (b) worn punch out of Co-low_GS-finest, (c) worn punch out of Co-mid_GS-fine, (d) worn punch out of Co-high_GS-fine.

5. Conclusion

The performed tests showed that the chemical composition of the WC-Co cemented carbides significantly influences the wear resistance and suitability for use in the production of large quantities of electrical steel laminations. Stamps made from the carbide material Co-mid_GS-finest with a cobalt binder content of 10 % and medium hardness values showed the best performance as a combination of durability and wear resistance. Carbides with higher hardness failed early on due to lower fracture toughness. In contrast, punches made of comparatively softer carbides could be used for a long time due to their high fracture toughness. However, they showed greater abrasive wear over their service life. In future, users will have to compromise between high hardness, low wear and thus the production of low-loss rotors and stators and the most economical production of high quantities by using softer cemented carbides that guarantee the longevity of the tools.

References

[1] Emura M, Landgraf F, Ross W and Barreta J 2003 *J. Magn. Magn. Mater.* **254-255** 358-360
[2] Hoffmann H, Neugebauer R and Spur G 2012 *Handbuch der Fertigungstechnik* vol 2 (Munich: Carl Hanser Verlag) p 681
[3] LoBue M, Sasso C, Basso V, Fiorillo F and Bertotti G 2000 *J. Magn. Magn. Mater.* **215-216** 124-126
[4] Baudouin P, Wulf M, Kestens L and Houbaert Y 2003 *J. Magn. Magn. Mater.* **256** 32-40
[5] Weiss HA, Leuning N, Steentjes S, Hameyer K, Andorfer T, Jenner S and Volk W 2017 *J. Magn. Magn. Mater.* **421** 250-259
[6] Santhanam AT, Tierney P and Hunt JL 1990 Cemented Carbides, in: *Metals Handbook* vol 2 (Ohio: ASMI Materials Park) p 950
[7] González R, Echeberria J, Sanchez JM and Castro F 1995 *J. Mater Sci* **30** 3435-39
[8] Gee MG, Roebuck B and Gant AJ 2005 *Proc. 16 th Int. Plansee Seminar High Performance PM Materials (Reutte)* vol 2 pp 1235-49
[9] Vilhena LM, Fernandes CM, Soares E, Sacramento J, Senos AMR and Ramalho A 2016 *Wear* **346-347** 99-107
[10] Milman YV, Luyckx and Northorp IT 1999 *Int. J.Refract Hard Met* **17** 39-44
[11] Renqi M, Shaodong J, Haiyan C and Chenyang S 2017 *IOP Conf. Ser.: Mater. Sci. Eng.* **207**
[12] Sheikh S, M'Saoubi R, Flasar P, Schwind M, Persson T, Yang J and Llanes L 2015 *Int. J.Refract Hard Met* **49** 153-160
[13] Buchmann K. 1962 *Beitrag zur Verschleißbeurteilung beim Schneiden von Stahlfeinblechen* (Wiesbaden: Springer Fachmedien GmbH) pp 78-82

IOP Conf. Series: Materials Science and Engineering **1307** (2024) 012003 doi:10.1088/1757-899X/1307/1/012003

Lubricant savings in sheet metal forming through thermally oxidized wear protection layers

P. Müller[1]*, **U. Holländer**[2], **I. Lendiel**[2], **H.-J. Maier**[2], **S. Hübner**[1], **J. Wehmeyer**[1], **T. Fünfkirchler**[1] and **B.-A. Behrens**[1]

[1] Institute *of Forming Technology and Machines, Leibniz University Hannover, Garbsen, Germany*
[2] Institute *of Materials Science, Leibniz University Hannover, 30823, Garbsen, Germany*

*mueller@*ifum.uni-hannover.de*

Abstract. Increasing demands on quality, variety of shapes and complexity of deep-drawn components with simultaneous price stability make the development of innovative solutions in the field of deep-drawing indispensable. For ecological and economic reasons, there is a great interest in reducing the amount of lubricant required in forming processes to a minimum. In this work, tool coatings for forming tools are to be investigated and further developed in order to use their friction-reducing properties on the tribo-system and thus enable a reduction in the use of lubricants. For this purpose, in this research strip drawing tests are carried out to investigate the friction-influencing properties of different coatings and their potential for lubricant saving. Therefore, cylindrical forming heads are provided with different coating systems. Strips of aluminum sheet are drawn around the forming heads at an angle of 90°. The amount of lubricant applied is varied. Friction values are recorded and analyzed for each system. Wear mechanisms are identified, examined and evaluated macroscopically and microscopically. For this purpose, energy dispersive X-ray spectroscope (EDX) and scanning electron microscope (SEM) examinations are carried out. A statement on the potential of the investigated coatings for lubricant saving is derived from the results.

1. Introduction

The tribological conditions in sheet metal forming influence process parameters such as temperature, process forces, tool wear and the robustness of the production process [1]. The quality of the manufactured component is primarily determined by the type and quantity of lubricant used, the topography of the sheet metal and tool surface and by process-dependent factors such as the blankholder force [2], [3]. Due to the increased demands on sheet materials, modern deep-drawing tools must be able to exhibit high strength, toughness and wear resistance. Multi-stage deep-drawing tools for the step-by-step forming of sheet metal are mainly used when the complexity of the component or a high overall drawing ratio no longer allow forming in a single stage [4]. An optimization of the tribological boundary conditions for multi-stage tools with regard to the reduction of friction lowers the number of necessary drawing steps and extends the process limits, so that the increased complexity is kept as low as possible compared to deep drawing in one step [5], [6]. One possibility for optimization often provides for additional lubrication of the sheet metal beyond the initial factory lubrication in order to improve the tribological system [7]. This additional lubrication is particularly relevant in the subsequent stages, since the oil film on the sheet is already greatly reduced as a result of the first stage, which sometimes leads to the occurrence of solid or mixed friction in the process [8]. The disadvantage of additional lubrication, however, is increased lubricant consumption, which should be kept as low as possible for ecological and economic reasons [9]. Optimizations with regard to the surface microstructure show another possibility for improving the tribological system. They lead to a reduction of friction and wear values [10]. A friction reduction through the application of suitable tool coatings shows advantages in terms of adhesion reduction on the one hand, on the other hand the application of surface structures to a forming tool also shows an increase in tool life as well as improved lubrication conditions [11]. Within the framework of the DFG priority program "SPP 1676 - Dry Forming in Forming Technology", dry sheet metal forming by means of selective oxidation of tool surfaces was fundamentally investigated. For this purpose, a process was developed for the thermal generation of an oxide layer in the submicrometric range under inert gas. Wear investigations were able to show that iron oxide systems have a friction-reducing effect and are also wear-resistant [12]. Further investigations also suggested that the combination of an oxide layer and minimal oiling of the workpiece can significantly improve the tribology in the deep drawing process [9]. The sequential deposition of amorphous silicon dioxide also stabilizes the coating system and brings good dry lubrication properties with high hardness into the tribological system. The layer thicknesses of the oxide systems achieved so far amounted to approx. 300 nm and must be adapted to conventionally applied tool coatings (~ 2 µm) in the future. Since the deposition of the oxide layer is slower the thicker it becomes, the production of a comparable layer thickness is a challenge that requires the development of alternative methods with faster growth rates and higher activation energies. In the context of this research, various coating systems are applied to cylindrical forming heads in order to investigate its effects on reducing the lubricant requirement in a forming process. Strip drawing tests with deflection on aluminum strips are carried out. The amount of lubricant is varied. The influence of the coatings on the tribological system, the friction values and the tool wear is analyzed and evaluated. Wear mechanisms are detected and evaluated by means of EDX and SEM measurements.

2. Materials and Methods

As testing material strips of Aluminum (EN AW-5657) in a sheet thickness of 0.8 mm are used. This is an alloy with small amounts of magnesium to increase the strength. EN AW-5657 is suitable for applications that require shiny surfaces. The alloy is weldable and hot and cold formable. Typical applications for this material are cladding for household appliances, car parts, power and window shutters as well as parts for the cosmetics industry. The mechanical properties of EN AW-5657 are shown in Table 1 [13]:

Table 1. Mechanical properties of EN AW-5657.

Young's modulus [N/mm²]	Tensile strength [N/mm²]	Yield strength [N/mm²]
68,000	140 170	80

The conventional tool steel 1.2379 is used for the cylindrical forming heads. This is a ledeburitic high-performance cutting steel, which is moderately machinable, wear-resistant, hardenable, low-distortion and nitridable. The mechanical properties of 1.2379 are shown in Table 2 [14]:

Table 2. Mechanical properties of 1.2379

Young's modulus [N/mm²]	Density [g/cm²]	Thermal conductivity [W/mK]
210,000	7.7	20

Two different heat treatment methods were used to modify the surfaces of the test specimens. Firstly, an oxidizing heat treatment under strongly reduced oxygen partial pressure was carried out in a continuous inert gas furnace at 500°C, producing thin surface layers of Fe_2O_3 (Oven 1 without SiH_4). Using nitrogen as inert gas with approximately 0,1 ppm oxygen a further reduction of the O_2-level can be achieved by adding small amounts of SiH_4 (monosilane) (Oven 2 with SiH_4) [12].

On the other hand, an ultra-short time laser treatment (LB1) of the test specimens in air was tested, which was carried out with a Nd:YAG laser under variation of the laser parameters like pulse repetition frequency, power and trave speed of beam, (see Fig. 1). This type of surface treatment resulted in a characteristic geometric microstructure of sparse depressions due to microscopic melting processes. With this microstructure, an increased lubricant retention capacity and thus more effective lubrication is aimed for.

Generally in all cases a periodical structure was created on the surface in accordance to the applied distances of 100 µm between discrete lasered lines and the laser spot diameter of about 50 µm, cf. light microscopic images in figure 1.

However, scanning electron microscopic analysis of the lasered sample, Fig. 2, reveal different topographic structures on the microscopic scale strongly depending on the laser process parameters. A pulse repetition rate of 7 kHz yield remelted surfaces with a smoothly undulating surface, while a treatment with 4 kHz applying the same beam travel speed (75 mm/s) result already in a fissured microstructure. The structural depth appears to be very shallow in all cases, amounting to only a few micrometers.

From a tribological point of view, a slightly undulating surface structure should be expected to exhibit lower sliding friction in the tool application.

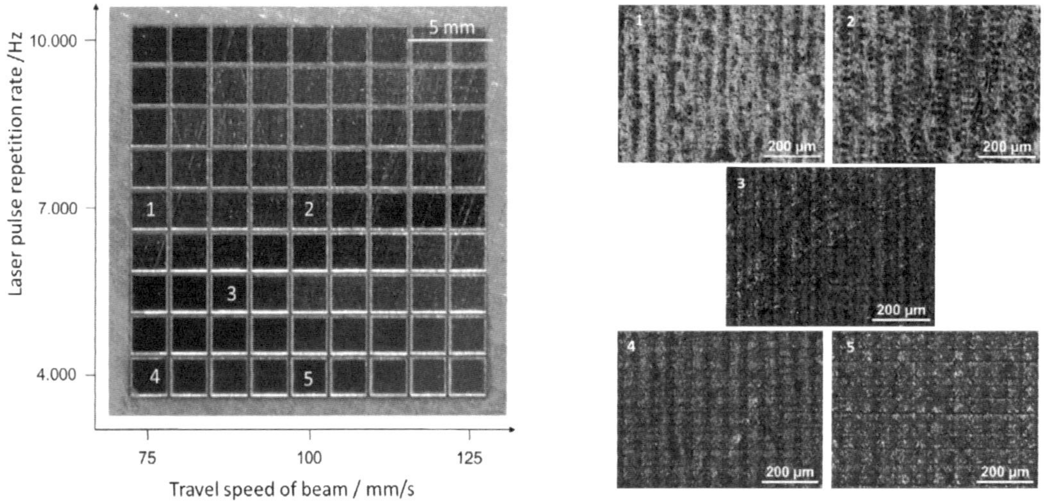

Figure 1. Thermal oxidation and surface structuring of 1.2379 by LASER treatment (Nd:YAG, 1064 nm)

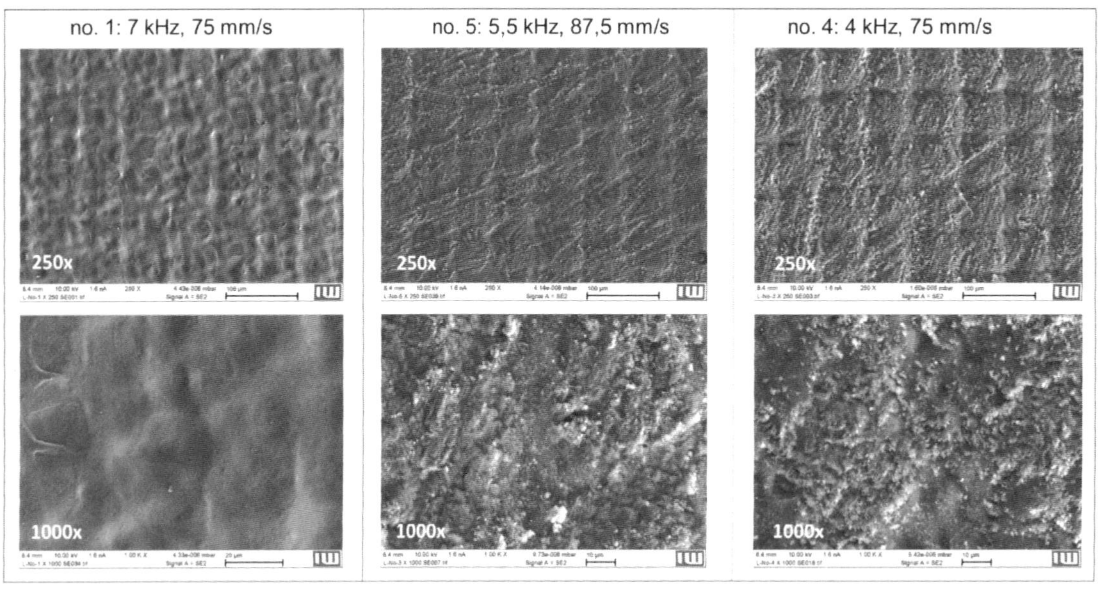

Figure 2. SEM images using a secondary electron detector from lasered surfaces with different laser parameters

3. Experimental Setup

3.1. Strip drawing test

In the strip drawing test, strips of sheet metal are drawn around cylindrical forming heads at an angle of 90°. The forming heads are initially lubricated with a defined amount of lubricant using a spraying system manufactured by the company Raziol GmbH. Renoform DYO 5006 from the company FUCHS LUBRICANTS GmbH with a viscosity of 126 mm²/s at 40 °C is used as lubricant. A film of defined lubricant mass is sprayed onto the sheet metal strips via a nozzle (see Fig. 3). The amount of lubricant is sprayed onto the sheet over a preset pulse duration. The quantity applied is checked beforehand using a precision scale to ensure that the lubricant application is reproducible.

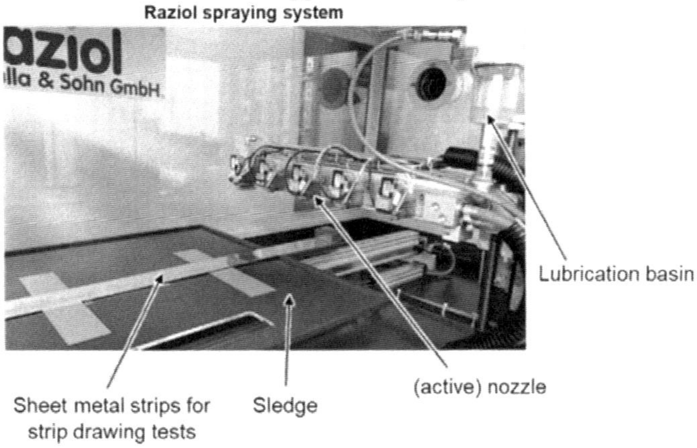

Figure 3. Oil spraying system.

The strip drawing test stand replicates the load conditions experienced at the edge of a deep drawing die during drawing. Therefore, a sheet metal strip is drawn around a cylindrical test specimen. The counterholder force F_C is maintained at constant level, while the tensile force F_D is adjusted based on factors like the sheet material and the friction between the test specimen and the sheet strip. The test parameters are set constant to a drawing speed of v_{Draw} = 10 mm/s and a constant counterholder force of F_C = 2 kN. The test specimen is tempered to a constant temperature of 40 °C via a heating cartridge with integrated thermocouple. The friction values can be calculated from the force ratios of counterholder to drawing force by a formulaic relationship according to [8]. The experimental setup is shown in Fig. 4:

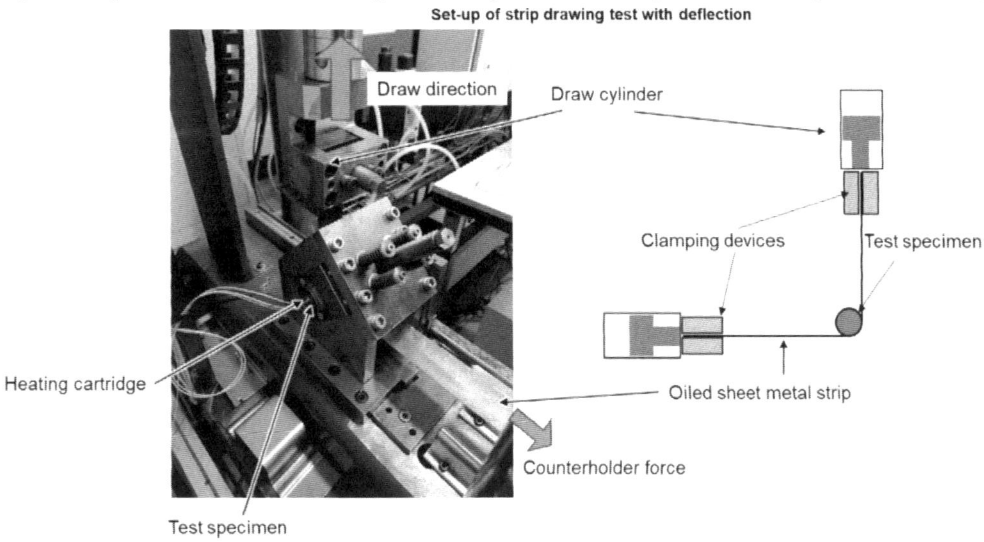

Figure 4. Strip drawing test stand.

Three lubrication conditions are investigated. The strips are tested with oil amounts of 12 g/m^2, 1 g/m^2 and without lubrication. The strips have a thickness of 0.8 mm, a width of 46 mm and a length of 1000 mm. The forming heads have a diameter of 16 mm and are hardened to a value of 57 HRC.

3.2. Test specimen

Fig. 5 visualizes the optical changes of the surface modified specimens. Both the laser treatment (LB1, corresponds to surface 2 in Fig. 1) and the heat treatment under nitrogen without silane (Oven 1 without SiH_4), lead to a brown discoloration of the surface resulting from the formation of a thin alpha iron oxide coating (thickness 0,5-1 µm). The reduction of the oxygen partial pressure by silane addition leads to a significantly lower oxidation of the surface, which do not yet have closed layers and microscopically consist of discrete αFe_2O_3 crystallites. Optically, these layers appear transparent.

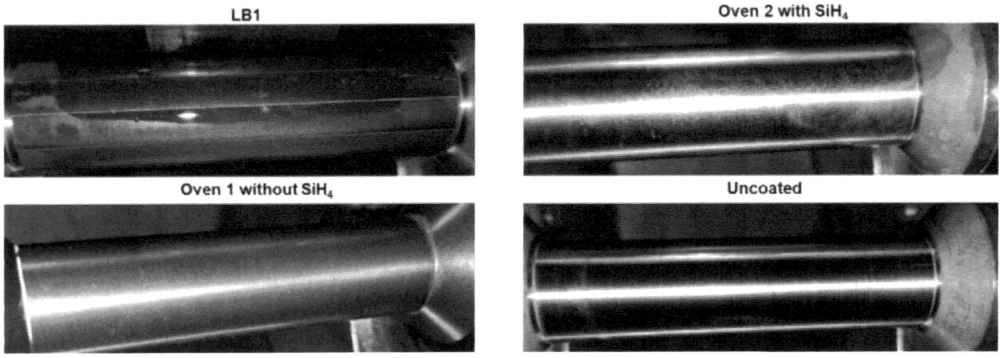

Figure 5. Experimental surfaces.

4. Results

4.1. Influencing the friction value

Fig. 6 shows the evaluation of the friction values with an oil amount of 12 g/m^2 for all forming heads. The bars show the mean friction values over three tests with the corresponding standard deviation.

The dot diagrams show the measured friction values for each test carried out in the order of the test sequence. Here it can be seen that the friction values increase after each stroke for all examined surfaces. This indicates that the surfaces are subjected to wear, which worsens the contact conditions and increases the friction. The forming heads "Oven 2 with SiH_4" show the lowest averaged friction value of $\mu = 0.17$ and therefore lays with about 19 % below the friction values of the uncoated reference with an average friction value of $\mu = 0.21$. The highest averaged friction value of $\mu = 0.25$ is achieved with the forming heads "Oven 1 without SiH_4". It should be noted that the friction values of the layer "Oven 1 without SiH_4" clearly deteriorate over the testing procedure from $\mu = 0.2$ in stroke 1 to $\mu = 0.31$ in stroke 3. This suggests that the wear process has already been initiated to a greater extent. Since the forming heads "LB1" show no significant difference to the uncoated reference, these will not be considered in the following.

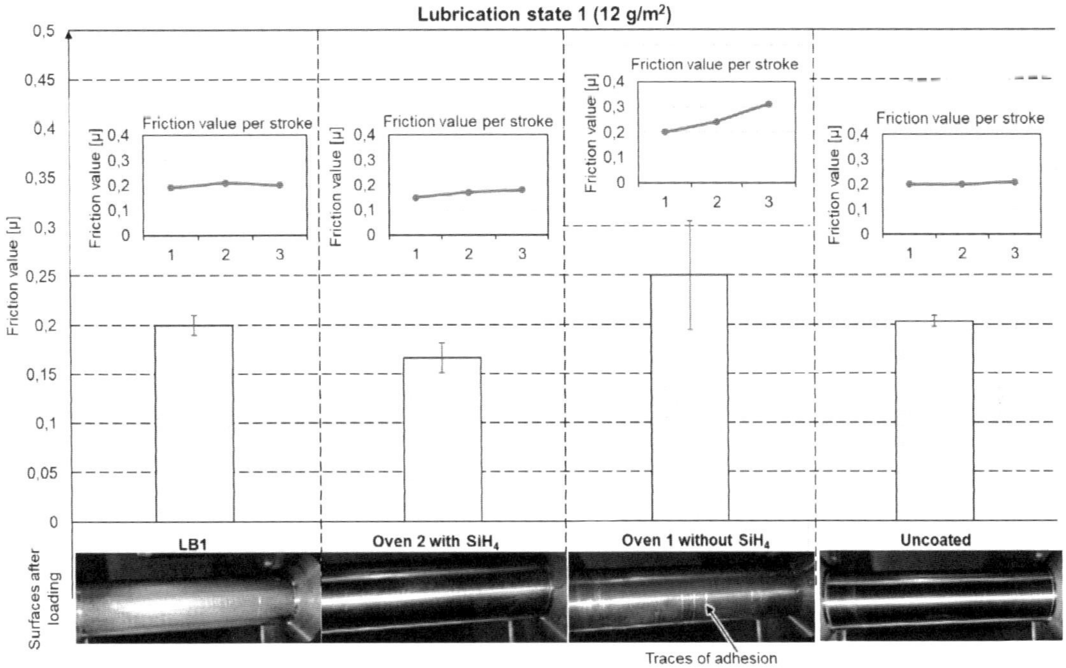

Figure 6. Friction values for 12 g/m² DYO5006

Fig. 7 shows the evaluation of the friction values for tests with an oil amount of 1 g/m^2. Therefore, the influence of a reduced lubricant quantity is analyzed. In this case, also, the surface "Oven 2 with SiH_4" with an averaged friction coefficient of $\mu = 0.195$ turns out to be the most promising one and is close to the averaged friction coefficient of the uncoated reference surface with $\mu = 0.205$. The increase of the friction coefficients per stroke is also not significant for "Oven 2 with SiH_4" and reaches from $\mu = 0.19$ to $\mu = 0.21$. For the uncoated reference the friction coefficient reaches from $\mu = 0.195$ at the first stroke to $\mu = 0.22$ at the third stroke. The surface "Oven 1 without SiH_4", on the other hand, shows a significantly worse performance with an average friction coefficient of $\mu = 0.28$ and, with a significant increase in the friction values from $\mu = 0.22$ in stroke1 to $\mu = 0.35$ in stroke 3, also strongly advancing wear.

Furthermore, experiments are carried out without lubricant to investigate the dry friction properties of the coating systems. For this purpose, "Oven 2 with SiH_4", the most promising coating so far, is compared with the uncoated surface of the reference forming head. The results show, that the sheet metal strips crack in the second stroke for both surfaces.

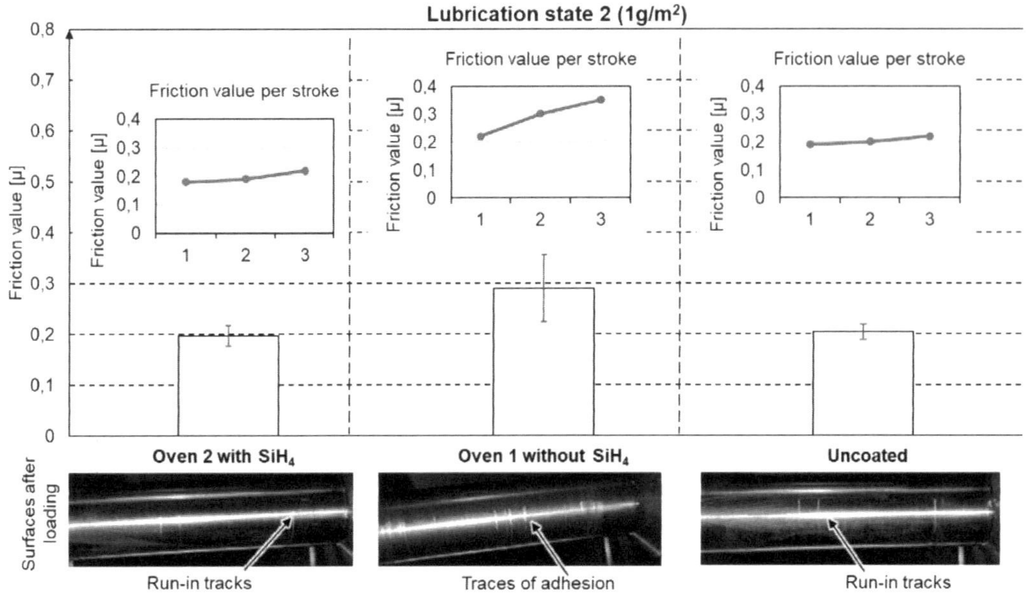

Figure 7. Friction values for 1 g/m² DYO5006

The failure pattern can be seen in Fig. 8. This is caused by a massive increase in wear during the dry tests. Due to the decreasing of surface quality, the frictional resistance between the forming head and the aluminum sheets increases until the drawing force becomes so great that the sheet tears.

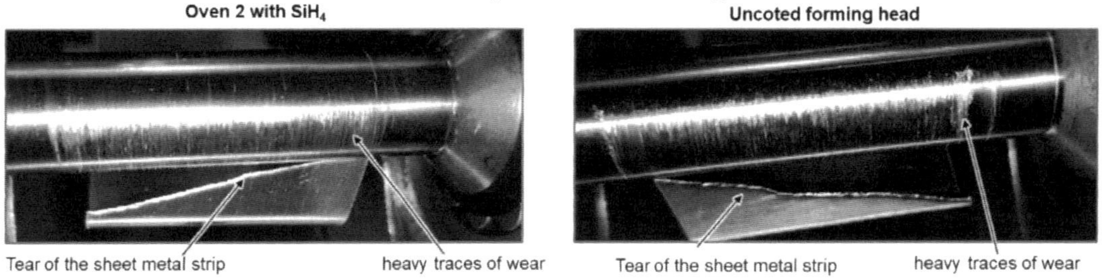

Figure 8. Test specimen after forming without lubrication.

4.2. Wear mechanism

As part of the work, SEM (with acceleration voltage 20kV) and EDX studies were also carried out. These were used to investigate the dominant wear mechanisms for the various coating systems. The surface of forming head with coating LB1, "Oven 2 with SiH_4", "Oven 1 without SiH_4" and Uncoated forming head was studied. Using the BSE (back scattered electrons) and corresponding images, the difference in composition between the components of the sample was visualized in different shades of grey (Fig. 9). Since a substance with larger atoms (for example, iron Z=26) scatters electrons much more strongly compared to a substance consisting of light atoms (for example, aluminium Z=13), they create a stronger signal and are lighter in the SEM image (Fig. 9). The SEM results show, that for all coatings macroscopic stripes appear in the tensile direction of the forming zone. These stripes were examined by EDX to determine their elemental composition and the presence of aluminium, iron, chromium and other elements. The various deposits of the elements are visualised in Fig. 9 in the coloured illustrations, which all show the same enlarged area presented in der SEM picture on the left. The EDX measurements show that these stripes consist primarily of aluminium, which indicates that sheet material is deposited on the forming heads. The dominant wear mechanism is therefore an adhesive material transfer. This leads to a deterioration of the surface roughness and thus to a progressive increase in the frictional force. Qualitative differences in the material transfer from the aluminium band to the surface of the tool steel cannot initially be derived from the SEM examinations.

In the case of the surface-modified tool samples, a measurable change in the high adhesion tendency of the aluminium without lubricant cannot be demonstrated yet on a microscopic scale.

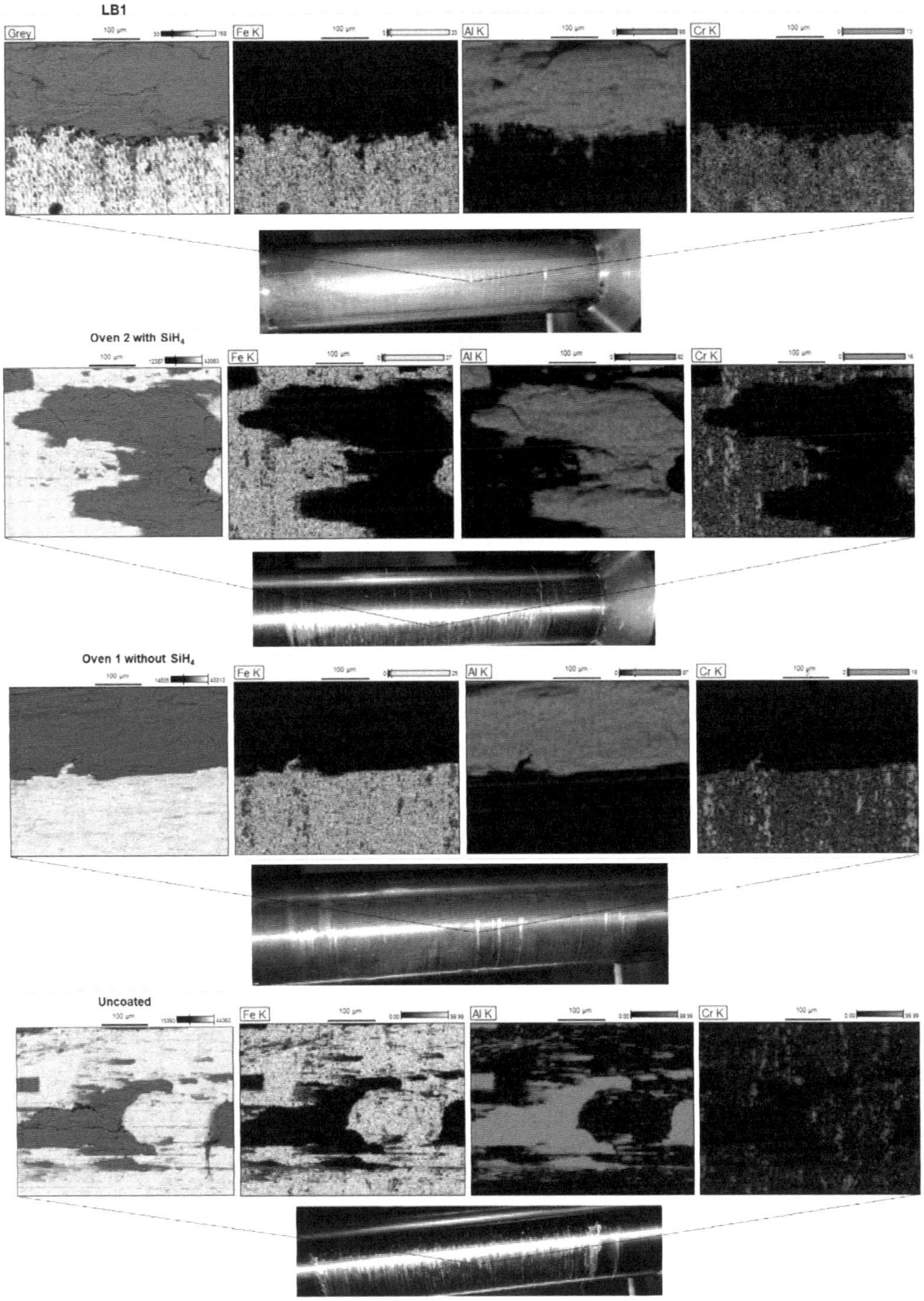

Figure 9. Surface measurements of worn samples.

5. Conclusion and Outlook

Within the scope of this research, strip drawing tests were carried out with differently coated specimens in order to investigate the influence of tool coatings on the tribological conditions of a deep drawing process. It could be shown that the coating "Oven 2 with SiH_4" causes a reduction of the friction factor compared to an uncoated tool surface as reference under saturated tool lubrication conditions. For experiments with reduced lubricant quantity, a slight improvement of the tribological conditions could be achieved with this layer compared to a conventional tool surface. However, a complete elimination of the need for lubricants could not be achieved. Adhesion is the primary wear mechanism identified for all surfaces. In future research, other layer systems such as further modified surfaces via laser treatment will be investigated, in order to focus on lubricant retaining structures for tool steel surfaces. The most promising variants will be applied on a deep-drawing die in a forming tool to investigate and evaluate their performance under real process conditions.

Acknowledgments

The IGF project "Lubricant savings in sheet metal forming through thermally oxidized wear protection layers" of the research association EFB e.V. is funded under the funding number AiF 21683N via the German Federation of Industrial Research Associations (AiF) as part of the programme for the promotion of joint industrial research (IGF) by the Federal Ministry for Economic Affairs and Energy on the basis of a resolution of the German Bundestag.

6. References

[1] Behrens, B.-A.; Meijer, A.; Stangier, D.; Hübner, S.; Biermann, D.; Tillmann, W.; Rosenbusch, S.; Müller, P.: Static and oscillation superimposed ring compression tests with structured and coated tools for Sheet-Bulk Metal Forming. JMP, Volume 55, pp. 78-86, 2020

[2] Behrens, B.-A., Maier, H.J., Hübner, S., Bonk, C., Almohallami,A., Lummer, C.,Schein, P., Scheland, H., Micke-Camuz, M.: Wear Behavior of MoS_2 Lubricant Layers during Sheet Metal Forming, Journal of Manufacturing Processes, 55, pp. 78–86, 2020

[3] Wagner, S.: Tribology in Drawing Car Body Parts. In: International Body Engineering Conference and Exposition, 1999

[4] Ravi Kumar, D., Manoha, M.: Determination of Process Parameters in Multi-Stage Hydro-Mechanical Deep Drawing by FE Simulation. In: Journal of Physics: Conf. Series 896. 36th IDDRG Conference – Materials Modelling and Testing for Sheet Metal Forming, 2017

[5] Kunze, T., Mousavi, A., Stucky, Th., Böttcher, F., Roch, T., Brosius, A., Lasagni, A.: Tribological Optimization of Dry Forming Tools. In: Dry Metal Forming Open Access Journal Vol.5. pp. 46-49, 2016

[6] Üstünyagiz, E., Altan, T.: Design of Progressive Die Sequence by Considering the Effect of Friction, Temperature and Contact Pressure. In: Key Engineering Materials. pp. 232-239, 2018

[7] Altan, T., Erman Tekkaya, A.: Sheet Metal Forming – Fundamentals, Chapter 7 Friction and Lubrication. Ohio: ASM International, 2012

[8] Lachmayer, R.; Behrens, B.-A.; Ehlers, T.; Müller, P.; Althaus, P.; Oel, M.; Farahmand, E.; Gembarski, P.C.; Wester, H.; Hübner, S. Process-Integrated Lubrication in Sheet Metal Forming. J. Manuf. Mater. Process., *6*, 121, 2022

[9] Behrens, B.-A., Yilkiran, D., Schöler, S., Özkaya, F., Hübner, S., Möhwald, K.: Wear investigation of selective α-Fe_2O_3 oxide layers generated on surfaces for dry sheet metal forming, Procedia Manufacturing, 15, pp. 923–930, 2018

[10] Groche, P., Callies, T.: Tribology in Sheet Metal Forming with Regard to Challenges in Lightweight Construction. In: Advanced Materials Research., pp. 93-100, 2005

[11] Groche, P., Köhler, M.: Development and application of functional surfaces. In: Engineering Manufacture, 2005

[12] Yilkiran, D.; Wulff, D.; Almohallami, A.; Özkaya, F.; Bouguecha, A.; Hübner, S.; Möhwald, K.; Maier, H. J.; Behrens, B.-A.: Wear behaviour of thermally oxidised tool surfaces as low-friction separation layers for dry sheet metal forming, Wear 376-377, 1789-1803, 2017

[13] MakeItFrom: Material Properties Database; AA 5000 Series (Aluminum-Magnesium Wrought Alloy), 2009

[14] Stauberstahl: 1.2379 / X155CrVMo12-1 Datenblatt, Stauber GmbH Metalltechnologie, Meckesheim, 2023

IOP Conf. Series: Materials Science and Engineering **1307** (2024) 012004 doi:10.1088/1757-899X/1307/1/012004

Investigation of Temperature Impact on Friction Conditions in Running Production of Automotive Body Components

A Barlo[1*], M Sigvant[1,2], J Pilthammar[1,3]
[1]Dept. of Mechanical Engineering, Blekinge Institute of Technology, Karlskrona, Sweden
[2]AutoForm Engineering Sweden AB, Olofström, Sweden
[3]Volvo Cars Dept. 81013 Manufacturing Technology Development, Olofström, Sweden

E-mail: `Alexander.Barlo@bth.se`

Abstract. During the running production of automotive body components drifts in the process window is seen causing problems with non-conforming parts. Up until now, these drifts have been counter-acted based on the knowledge and experience of the press line operators. This experience-based process control will however become more troublesome in the future as recycled material grades will undoubtedly present larger in-coil variations in material parameters and effect also the friction conditions from component to component.

The following study will present two cases from production of the Volvo XC60. For the two cases, the initial simulations made for the components showed a safe part, but during running production failure occurred suspected to be due to temperature effects in the tribology system. The study will furthermore present updated simulations considering developing thermal effects to replicate the failures, as well as present both standard and thermal simulations of the adjustments made in production.

1. Introduction

When manufacturing automotive sheet metal components, one must be aware of the tribological conditions within the process. Several authors, including Sigvent et al [1] and Hol et al [2], have outlined the importance of dynamic friction models (dependent on pressure, velocity and straining) and the coating on the material. An addition to the discussion on tribology recently raised by Waanders et al [3] and Kott et al [4] is the impact of temperature increases on friction behaviour during running production.

The following paper will present some practical examples showing what happens when the tribology causes problems, how to resolve these issue by changing oils and sheet coatings, and how the phenomena can be captured through Finite Element (FE) simulation. Two practical examples will be presented: a wheelhouse outer and front door inner from the second generation of the Volvo XC60. The geometries are presented in Figure 1 and 2 respectively. Both of these components are manufactured from a VDA239-100 CR4 mild steel, originally with a hotdip galvanized (unexposed) coating (GI 50/50-U). The lubrication RP4107S was originally chosen to be applied to the sheet in both cases, but during running production both components displayed fractures as in the highlighted regions of Figures 1 and 2.

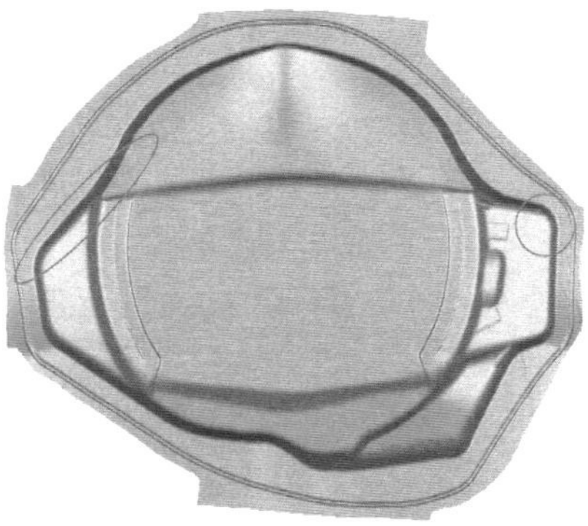

Figure 1. Volvo XC60 wheel house outer final geometry. The red circles illustrates the main area where failure has been observed in reality.

Figure 2. Volvo XC60 front door inner geometry after initial forming operation. The red circle illustrates where failure has been observed in reality.

To combat the fractures during running production outlined in Figures 1 and 2 changes to the lubricant was made for the wheelhouse outer using a PLS100T pre-lube and changes to both the coating and lubricant were made for the door inner using a 3802 pre-lub and a zinc-magnesium (unexposed) coating. The different types of oils and coatings used for the two components are outlined in Table 1.

Part	Material	Coating	Pre-lube	Process
Wheelhouse Outer	CR4	GI 50/50-U	RP4107S PLS100T	Double action
Front Door Inner	CR4	GI 50/50-U ZM 50/50-U	RP4107S 3802	Single action

Table 1. Parts, coatings, and lubrications included in the study. Besides the presented data, each component will be analysed at two different process stages - after one stroke and 1000 strokes (original tribology system) or when stationary conditions have been reached (new tribology system).

The cases presented in Table 1 will be simulated to show that the fractures observed in production can be reproduced, and the changes to coatings and lubricant will present a safe panel in production as well. To illustrate the effects of temperature, simulations will be created using the 'standard' approach i.e. being modelled in a single-stroke configuration, as well as a simulation that considers the temperature increase in the tools over 1000 strokes for the original GI 4107S tribology system and stationary conditions for the updated tribology systems.

2. Simulation Setup

To capture the effects of the increased temperatures that arises during running production, the AutoForm R11 solver *Cold Forming with Tempereture Effects* was used. The following section will describe the setup of the simulation base model (standard cold forming simulation) and the 'running production' mode of the simulation (thermal effects).

2.1. Base Models

Since both components are manufactured from the same VDA239-100 CR4 mild steel, a single material card is used for both cases. For the modelling of the material behaviour, the BBC2005 constitutive model with a non-integer exponent was used, as this model and approach represents the material behaviour more accurately [5] than the standard values based on the crystallographic material structure proposed by Banabic [6]. The hardening behaviour of the material was assumed to be isotropic, and the hardening curve was constructed using tensile test data and a power law extrapolation. The nine material inputs to the constitutive model can be found in Table 2, and the hardening curve and final yield surface can be found in Figure 3.

σ_0 [MPa]	σ_{45} [MPa]	σ_{90} [MPa]	σ_b [MPa]	R_0	R_{45}	R_{90}	R_b	M
156.6	160.0	156.0	187.0	1.805	1.336	1.876	0.982	4.5

Table 2. Material data for the BBC2005 constitutive model.

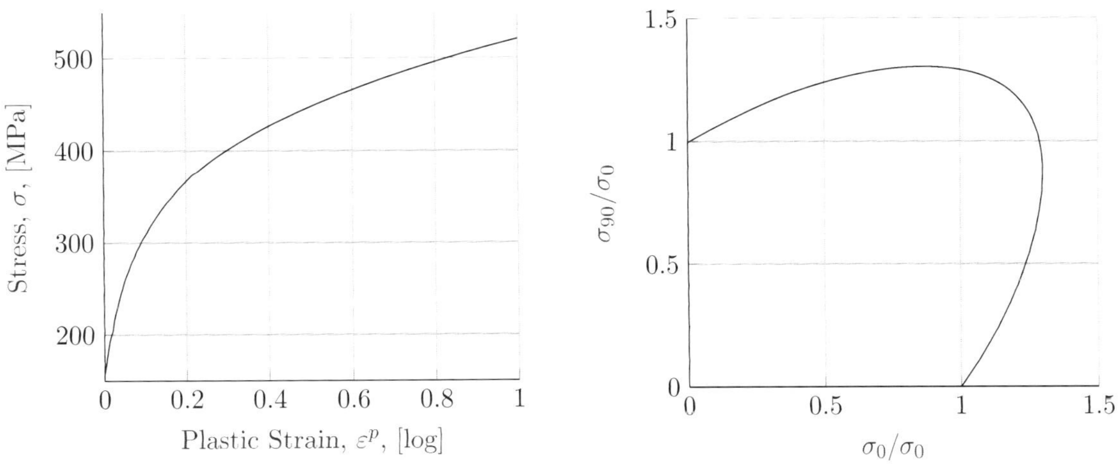

Figure 3. Hardening curve and yield surface of the VDA239-100 CR4 mild steel.

For the modelling of the tribology systems, TriboForm is used to generate dynamic friction models based on pressure, velocity, strain, and temperature, with all models presented in the paper based on the TriboForm database. To get an initial understanding of the different pre-lubes and coatings presented in Table 1, Figures 4 and 5 presents a comparison of the tribology systems used for the wheelhouse outer and door inner respectively.

2.2. Thermal Model

The influence of the temperature increase is expected to be on the coefficient of friction throughout the production series. Figure 6 presents the friction coefficient as a function of temperature for the three tribology systems discussed in this paper. The figure presents the original system for the two components (GI RP4107S) where a significant temperature dependency can be seen. Going to the two new tribology systems (GI PLS100T and ZM 3802) overall lower coefficients of friction are seen as well as a more stable coefficient with developing temperatures.

IOP Conf. Series: Materials Science and Engineering **1307** (2024) 012004 doi:10.1088/1757-899X/1307/1/012004

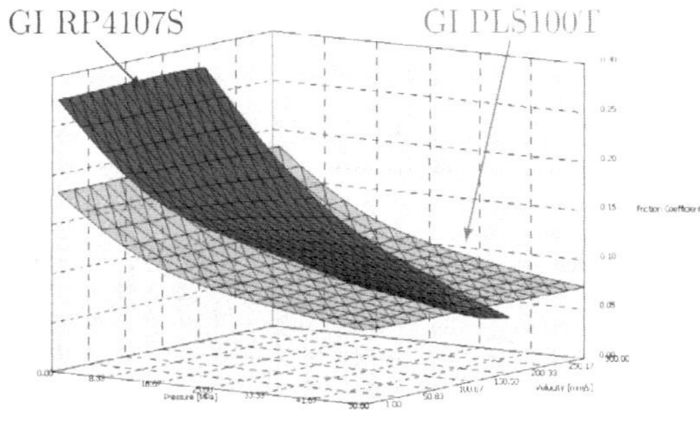

Figure 4. Comparison of the friction coefficient as a function of pressure and velocity for the RP4107S and PS100T pre-lube with a GI coating.

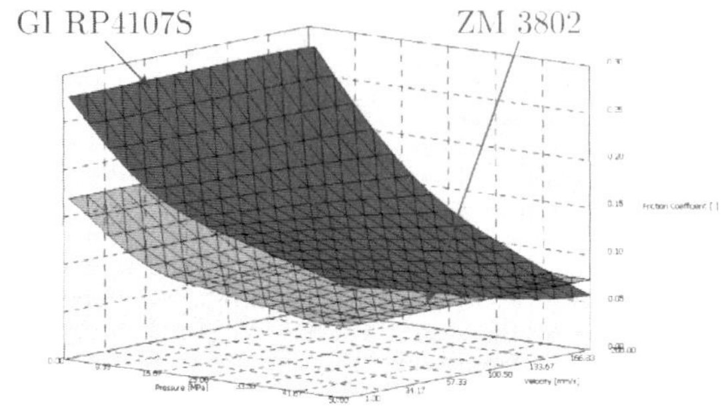

Figure 5. Comparison of the friction coefficient as a function of pressure and velocity for the RP4107S (GI coating) and 3802 (ZM coating) pre-lube.

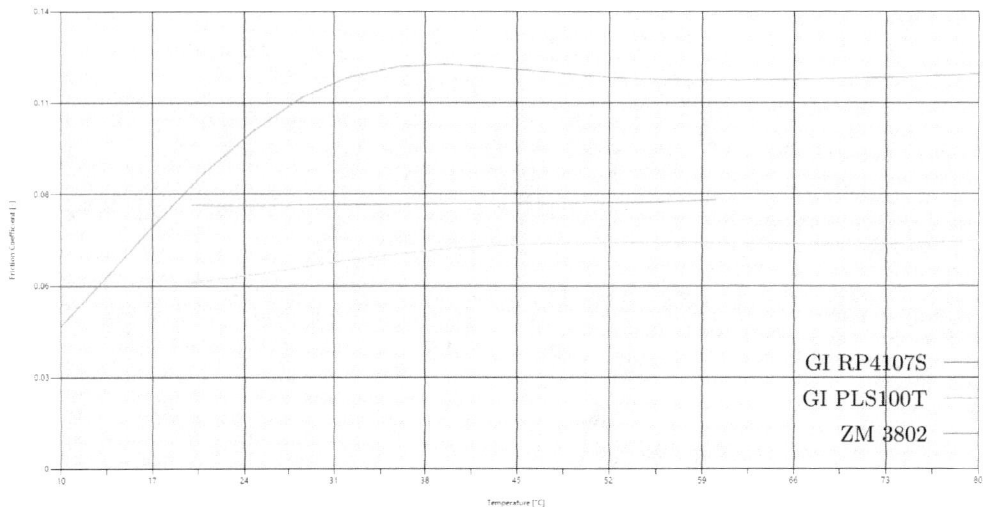

Figure 6. Friction coefficient as a function of temperature for the three combinations of pre-lube and coatings at a contact pressure of 25 MPa, a strain of 0.2, and a velocity of 300 mm/s.

For the setup of the simulation of multiple components, the Heat Transfer Coefficient (HTC) from the sheet to the tool and the tool to the surroundings needs to be specified. For the heat transfer from sheet to tool, a HTC of 7.0 mW/(mm^{2}K) is used, but is conditioned as a pressure dependent HTC. The pressure dependent HTC model scales the HTC based on the contact pressure between the sheet and tool. For this study, both the scaling and HTC has been doubled compared to the AutoForm standard as it is assumed that the coatings used for cold forming are not as efficient insulators as the AluSi-based coatings used for conventional hot forming [7]. The HTC scaling factor as a function of contact pressure can be found in Figure 7. For the heat transfer to ambient, the ambient temperature is set to 20 °C and a constant HTC is set to 0.02 mW/(mm^{2}K) (AutoForm standard).

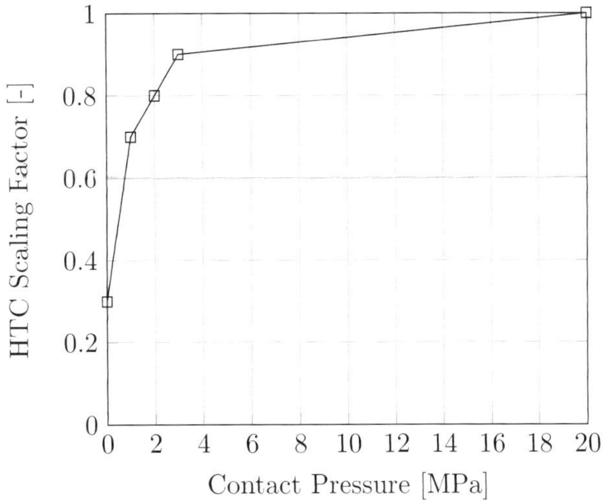

Figure 7. Scaling factor of the Heat Transfer Coefficient between sheet and tool as a function of contact pressure.

3. Volvo XC60 Front Door Inner

To replicate the fracture presented in Figure 2, the thermal simulation was set to run 1000 strokes to observe the situation in the middle of a batch. Both the single-stroke and thermal simulation was set up to run the production-rate of 14 strokes/min in a continuous stride.

The component in its original designed state (GI RP4107S tribology system) was initially simulated to replicate the issues. Figures 8 and 9 presents the advanced formability fringe plot for the single-stroke and thermal simulation respectively. Since considering complex geometries, the advanced formability functionality in AutoForm is used, as it accounts for non-linear strain paths and is based on the generalized forming limit concept (GFLC) proposed by Volk & Suh [8]. Figure 8 presents the single-stroke simulation, where a failure zone is indicated. The indicated zone does however not align with the fractures observed in reality, and is furthermore located on the addendum why this component from a manufacturing feasibility point-of-view can be accepted if a fully green simulation cannot be obtained without substantial changes to either process or component geometry. Figure 9 presents the same plot after 1000 strokes where the simulation has been repeated with elevated tool temperatures. This simulation shows a failure similar to the one observed in reality. In the area of the fracture, a temperature increase ΔT between $1-3°$C is observed compared to the single-stroke. The temperature increase is not very large in the area of interest, but it is enough to push the component over the edge, displaying how sensitive complex geometries can be to even minor changes in friction conditions. Another interesting aspect to observe is the development of the tool temperature. Figure 10 presented the temperature of the binder after one stroke (left) and 1000 strokes (right). The figure presents a substantial increase in temperature which also could be a contributing factor to the failure seen in the component.

To bring the component back into the process window during the entire produciton series, the GI RP4107S tribology system was replaced with the ZM 3802 system. Figures 11 and 12 presents the advanced formability fringe plots for the updated design at single-stroke and at stationary conditions respectively. The single-stroke simulation presents a completely safe component, and the thermal simulation show a fracture on the addendum. These results align well with reality where a stable process is present.

IOP Conf. Series: Materials Science and Engineering **1307** (2024) 012004 doi:10.1088/1757-899X/1307/1/012004

Figure 8. Front door inner with GI RP4107S tribology system after single stroke.

Figure 9. Front door inner with GI RP4107S tribology system after 1000 strokes.

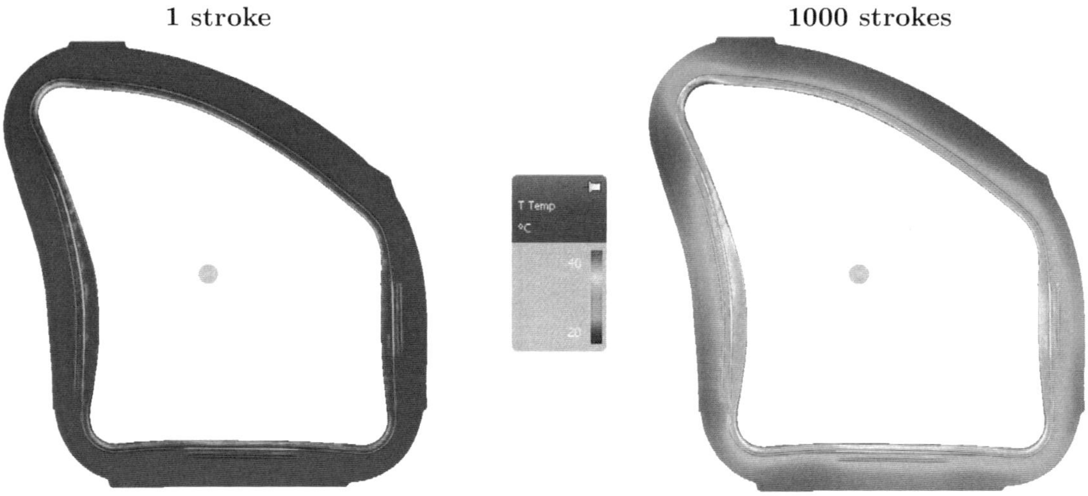

Figure 10. Binder temperature after 1 stroke (left) and 1000 strokes (right) with the GI RP4107S tribology system.

Figure 11. Front door inner with ZM 3802 tribology system after single stroke.

Figure 12. Front door inner with ZM 3802 tribology system at stationary conditions.

4. Volvo XC60 Wheelhouse Outer

To replicate the the fracture presented in Figure 1, the thermal simulation was again set to run 1000 strokes to observe the situation in the middle of a batch. Both the single-stroke and thermal simulation was set up to run the production-rate setup of 10 strokes/min in a continuous stride. The process is modelled as a double-action process in AutoForm, but only one velocity profile is allowed to be defined. Therefore, the velocity profile of the inner ram was used where an assumption of a pure sinusoidal behaviour corresponding to 10 strokes/min was made. The component in its original designed state (GI RP4107S tribology system) was initially simulated to replicate the issues. Figures 13 and 14 presents the advanced formability fringe plot for the single-stroke and thermal simulation respectively. Figure 13 presents the results from the single-stroke simulation where a safe part is seen. Figure 14 presents the same plot after 1000 strokes where the simulation has been repeated with elevated tool temperatures (slight increase in temperature of 1°C between one and 1000 strokes is seen). The simulation shows a failure in the component, however not in the same location as observed in reality. One possible explanation for this deviation could be that either the sheet, tool, or both are subject(s) to thermal expansion. With thermal expansion of either tool or sheet (or both) the double-action nature of the press would most likely result in an over-closing of the of the tooling causing damage to the sheet, thereby provoking the fracture seen in reality. Other potential factors that could be the cause of the incorrectly predicted failure location could be that the tools are assumed to be rigid and not elastic, differences between CAD tool surfaces and production tools, or effects of shimming.
To bring the component back into the process window during the entire produciton series, the GI RP4107S tribology system was replaced with the GI PLS100T system. Figures 15 and 16 presents the advanced formability fringe plots for the updated design at single-stroke and at stationary conditions respectively. Both components show a safe component, which align well with reality where a stable process is present.

5. Conclusion

The presented study showed the impact of including thermal influence on the tribology system for two automotive components. For the first component, a front door inner from the Volvo XC60 stamped in a single-action press, the study showed how the use of the AutoForm R11 solver *Cold Forming with Temperature Effects* was able to accurately predict failure after 1000 strokes, where the standard single-stroke simulation was not for the initial tribology system (GI RP4107S). To test the practical solution in simulation, a second set of simulations were run for the new tribology system (ZM 3802) where both simulations marked safe components, as seen in reality. A similar study was made for the second component, a wheelhouse outer from the Volvo XC60 stamped in a double-action press, however the tested solver was not able to accurately predict the failure location observed in reality. This deviation from reality is expected to be cause by thermal expansion in the physical system, which is something that needs to be investigated further. Also for this component, another set of simulations was run for a new tribology system (GI PLS100T) where both simulations marked safe components, as seen in reality. Overall, the presented study on the inclusion of thermal effects on the tribology systems shows promising results (especially for single-action operations) that could predict potential problems in running production, however further research is needed on the deviations seen for the double-action operation.

Acknowledgments

Open Access funding was provided by Blekinge Institute of Technology. This study was also funded by VINNOVA in the Eureka SMART program (grant number 2021-03144).

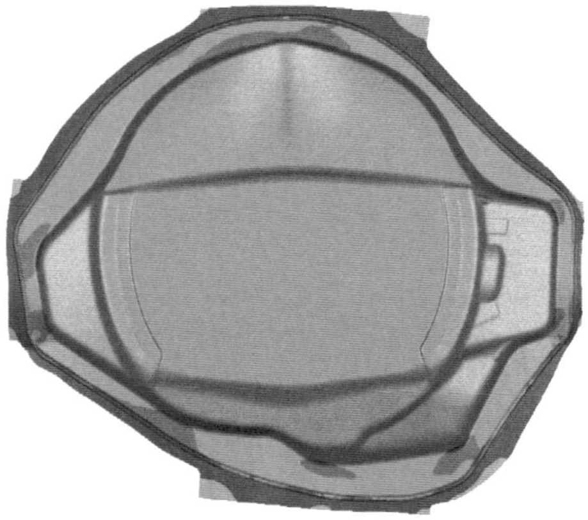

Figure 13. Wheelhouse outer component with GI RP4107S tribology system after single stroke.

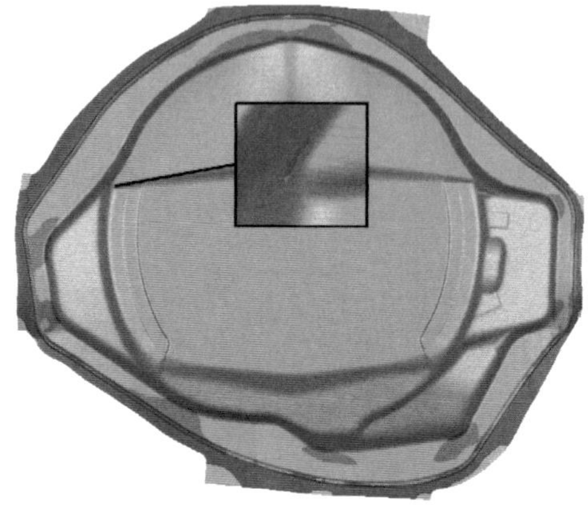

Figure 14. Wheelhouse outer component with GI RP4107S tribology system after 1000 strokes.

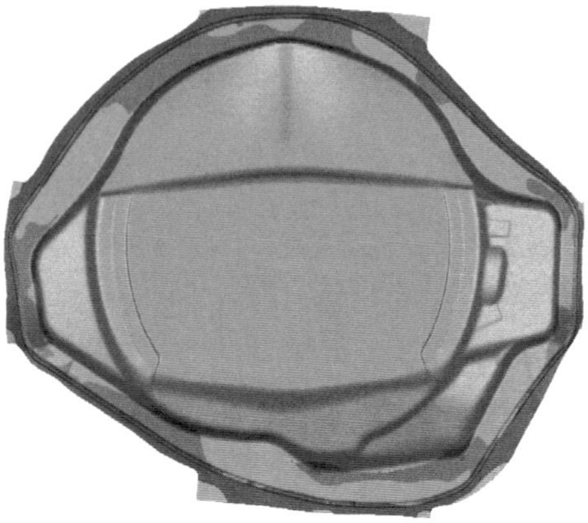

Figure 15. Wheelhouse outer component with GI PLS100T tribology system after single stroke.

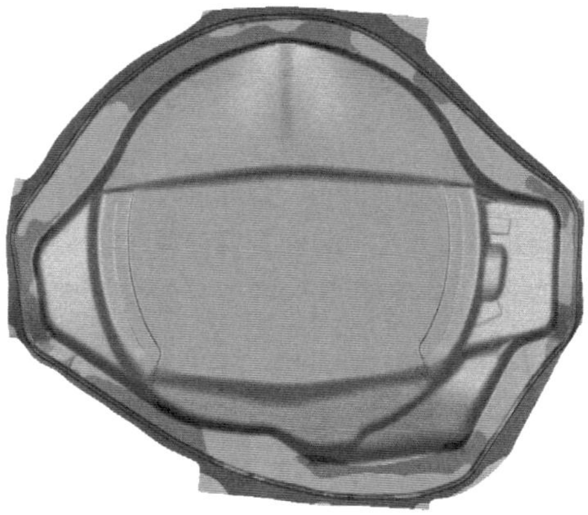

Figure 16. Wheelhouse outer component with GI PLS100T tribology system at stationary conditions.

References

[1] Sigvant, M *et al.* (2018) Friction in Sheet Metal Forming Simulations: Modelling of New Sheet Metal Coatings and Lubricants. *IOP Conf. Ser.: Mater. Sci. Eng.* **418** pp. 012093.
[2] Hol, J *et al.* (2012) Advanced friciton modeling for sheet metal forming *Wear* **286-287** pp. 66-78.
[3] Waanders, D *et al.* (2020) Temperature dependent friction modelling: The influence of temperature on product quality. *Procedia Manufacturing* **47** pp. 535-540.
[4] Kott, M *et al.* (2020) Controllability of Temperature Induced Friction Effects during Deep Drawing of Car Body Parts with High Drawing Depths in Series Produciton. *Procedia Manufacturing* **47** pp. 553-560.
[5] Pilthammar, J., D. Banabic and M. Sigvant 2021 BBC05 with non-integer exponent and ambiguities in nakajima yield surface calibration *Int. J. Mater. Form.* **14**, pp. 577-592.
[6] Banabic, D (2010) Sheet Metal Forming Processes - Constitutive Modelling and Numerical Simulation. 1st edn. ISBN: 978-3-540-88112-4. Springer London.
[7] Dessain, C *et al* (2011) Experimental Estimation of Heat Transfer Coefficients in the Hot Stamping Process of Usibor® 1500P *Proceedings of the 3rd International Conference on Hot Sheet Metal Forming of High-Performance Steel, Kassel, Germany, June 13-17, 2011* pp. 293-301.
[8] Volk, W & Suh, J (2013) Prediction of formability for non-linear deformation history using generalized forming limit concept (GFLC) *AIP Conference Proceedings* **1567**, pp. 556-561.

Characterization of microstructural evolution pattern toward nano-scale in commercial-purity titanium by incremental sheet forming

W Q Li[1], Z D Chang[2] and J Chen[1*]

[1] School of Materials Science and Engineering, Shanghai Jiao Tong University, Shanghai 200030, China.

[2] School of Mechanical Engineering, Shanghai Jiao Tong University, Shanghai 200030, China.

* *jun chen@sjtu.edu.cn*

Abstract. In the present investigation, a commercial-purity titanium sheet was formed by a flexible sheet metal forming method, incremental sheet forming (ISF), to investigate its microstructure evolution. The deformation microstructures across the wall of formed parts were systematically examined by optical microscopy (OM), electron backscatter diffraction (EBSD), transmission electron microscopy (TEM) and high resolution TEM (HRTEM). Microstructural evolution from millimeter- to nano-scale was explored, with special attention paid to the refinement below 100 nm. A general pattern of structural evolution begins with the formation of uni- and multi-directional twins, which is accompanied by the initiation and development of ultrafine lamellar structure and final evolution of nano-lamellar structure and nano-equiaxed structure. A twinning-dominated process that was supplemented with dislocation slip-dominated one governed the microstructural evolution toward nano-regime. The strain-induced evolution of nano-lamellar and nano-equiaxed structures were discussed, of which the underlying refinement mechanism was analyzed. This technique was promising to offer an optional route for realizing the strengthening of metal parts via synthesizing nanostructures, showing potential scientific and technological importance in fabricating the high-strength metal formed productions for further meeting industrial applications.

1. Introduction

The application of titanium and its alloys in biomedical field has a well-established area owing to their excellent biocompatibility, corrosion resistance and low Young's modulus [1-3]. Commercial-purity titanium (CP Ti) is chemically insert and biologically more compatible than Ti6Al4V, while coarse-grained Ti has low abrasion and wear resistance due to its low strength. Therefore, it is of great significance to enhance the strength of CP Ti via refining the microstructures. As is known to all, heat treatment is not a well option for strengthening Ti, since the strengthening effect of both the second phase and precipitates are relatively weak. However, it is has been demonstrated that plastic deformation-induced grain refinement can greatly improve the strength of pure Ti. A large number of plastic deformation methods have been developed to refine the grain in the bulk Ti, such as cold rolling [4], accumulative roll bonding [5], equal channel angular pressing [6] and high pressure torsion [2], which can reduce the grain size down to the ultrafine-regime by 100-200 nm. Recently, a series of surface plastic deformation techniques like rotationally accelerated shot peening [7], ultrasonic surface rolling process [8] and surface mechanical attrition treatment [9, 10], have been proposed to fabricate the nano-sized structures with the size range below 100 nm. However, the detailed process of structural refinement below 100 nm in a wide range of length scale needs to be clarified.

Recently, we have adopted a technique that is able to create gradient plastic deformation in the processing area during shaping the sheet into formed parts at room temperature, i.e. incremental sheet forming (ISF).The surface of processing area was fixed by the surrounding backing plates together with two lower-auxillary sheets and sheared by a WC forming tool with one hemispherical tool tip along the particular toolpath, induced gradient deformation enhanced the grain refinement and the understanding of the structural evolution in a wide range of length scale. The present work is aimed at exploring the successive process of grain subdivision toward the nano-scale and the refinement in the regime below 100 nm was emphasized.

2. Experimental

A commercial-purity titanium (CP Ti, Grade I) sheet was used in this work with the size of 180 mm × 180 mm × 1.0 mm and the following chemical composition (wt%): 0.2% Fe, 0.08% C, 0.03% N, 0.015% H, 0.18% O and balance Ti. Its microstructure was consisted of equiaxed α-grain with a mean size of 17 μm. The ISF and toolpath have been described in **Figure 1**. Ti sheet together with two lower-auxillary sheets (mild steel DC 05, 180 mm x 180 mm x 0.8 mm in size) were fixed by the surrounding backing plates, and tool tip moved at a constant feed velocity of v and feed depth of d along the toolpath that consisted of multilayer processing paths (layer-1, layer-2, layer-3……layer-N) with the similar path lines (spacing size of a) and contour lines in each layer. Until finishing the layer-160 processing, a thin-walled formed part was fabricated with the shape of square cavity. In the present investigation, the ISF process was conducted at room temperature with processing parameters v=2000 mm/min, d=0.2 mm and a=0.1 mm.

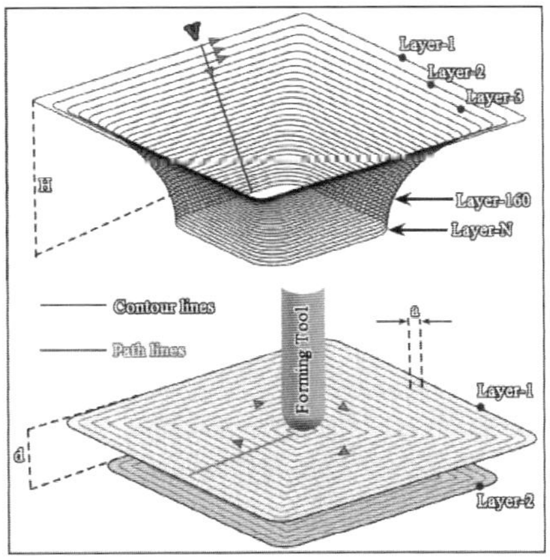

Figure 1. The toolpath set-up.

Microstructural characterization was conducted on the cross section by employing optical microscopy (OM), electron backscatter diffraction (EBSD), transmission electron microscopy (TEM) and high resolution TEM (HRTEM). OM observation was performed on an Axio Imager M 2m light microscope. EBSD examination was operated on a Mira SEM equipped with an EDAX instruments symmetry EBSD system. TEM and HRTEM characterizations were operated in the FEI Talors F2000X at a voltage of 200 KV. The samples for OM and EBSD observations were prepared by the standard metallographic preparation procedures and the following vibratory polishing with a frequency of 70 Hz for 3 hours to obtain a mirror surface, etched by etching liquid composed of 2% HF, 10% HNO_3 and 88% H_2O. The cross-sectional specimens for TEM and HRTEM characterizations were prepared by the focused ion beam (FIB) milling technique on the Gaia double-beam SEM.

3. Results

3.1 OM observation of deformation layer

Typical deformed microstructures is observed within the thickness of 300 μm by OM image in **Figure 2**, where the depth-dependent microstructure could be divided into three regions including twinning region, twin-dislocation transition region and nanostructured region. Twinning region consists of slightly deformed coarse grains coexisted with a large amount of deformation twins, which locates at the depth range from 200 to 300 μm. Twin-dislocation transition region is characterized by severely elongated grains with the low-density of DTs at a depth range of 60-200 μm. Nanostructured region is the topmost layer with a depth below 60 μm and typical of extended deformation bands parallel with the shear direction, in which DTs are rarely observed and forming extremely fine structures could need to be examined by TEM.

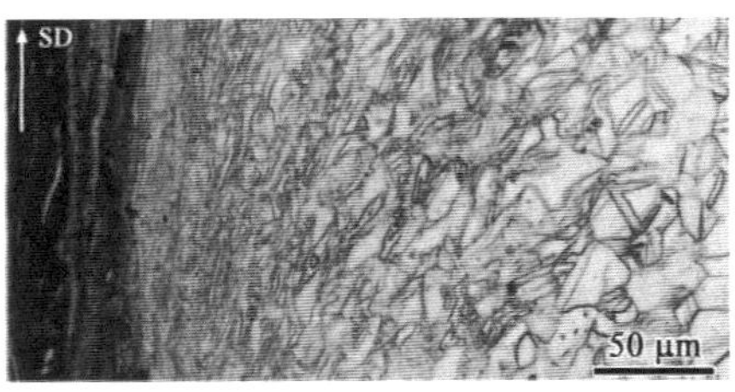

Figure 2. OM observation of deformed structures.

3.2 Electron microscopy characterization of deformation layer.

Twinning region at the depth of 220 μm is characterized by which the grains are slightly elongated along the shear direction and generate typical deformation microstructures inside the grain interiors in **Figure 3.(a)**, featured by three deformation characteristics as follows: i) orientation changes reflected by color variations in G_1 and G_2 grains, depending upon the movement and accumulation of dislocations[11]; ii) low-density unidirectional deformation twins expanded the entire G_3 grain due to activating monosystem twinning; and iii) multi - directional deformation twins in a high density interserted with each other, ascribed to the activation of multisystem twining. Obviously, here plastic deformation is governed by twinning with different behaviors, and the following accompanied with slip. The twinning- and slipping-induced microstructures could be detailed characterized by TEM as shown in **Figure 3(b) and (c)**. The microstructures are composed of parallel or interlaced bands with the thickness varying from tens of nanometer to a few microns. According to the inserted selected area electron diffraction pattern taken from areas containing two intersected bands in Figure 3(b), mirror spots can be observed with respect to the (10-12) plane, indicative of a {10-12} twin system. These intersected twins subdivide the area into the block bordered by twin boundaries and a number of dislocation piles up in the intersection position. In addition, the intersected bands could be observed in the other forms without the twin relationship as seen in Figure 3(c). Their boundaries are dense dislocation walls (DDWs) and not as straight as those of the twins and dislocation pile-ups are formed near the boundaries, which demonstrate that the formation of bands can be dominated by dislocation slip.

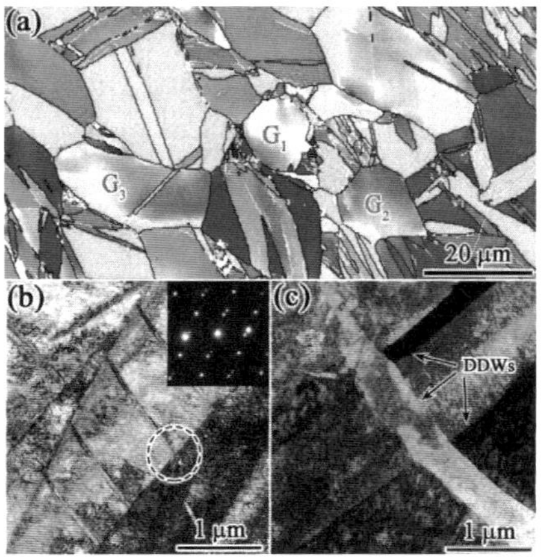

Figure 3. EBSD IPF image (a) and TEM observations (b, c) at 220 μm deep.

Twin-dislocation region contains low-density deformation twin segments and affluent dislocation substructures, as characterized by TEM at the depth of 140 μm in **Figure 4**. Typical dislocation structures are parallel DDWs with deviation from SD by 20°. At this depth, DDWs become very extended and straight, also named as lamellar boundary, having a heterogeneous boundary spacing marked by DDWs-I and DDWs-II, in which the boundary spacing of DDWs-I is twice than that of DDWs-II. It is worth noting that all DDWs have the interval below 1000 nm, namely ultrafine lamellar (UFL) structures. Surprisingly, the twin segment is casually observed in between the adjacent DDWs with the coarse spacing of 600-800 nm, as the electron beam is parallel with [10-26] reflected by the inserted SAED pattern.

IOP Conf. Series: Materials Science and Engineering **1307** (2024) 012005 doi:10.1088/1757-899X/1307/1/012005

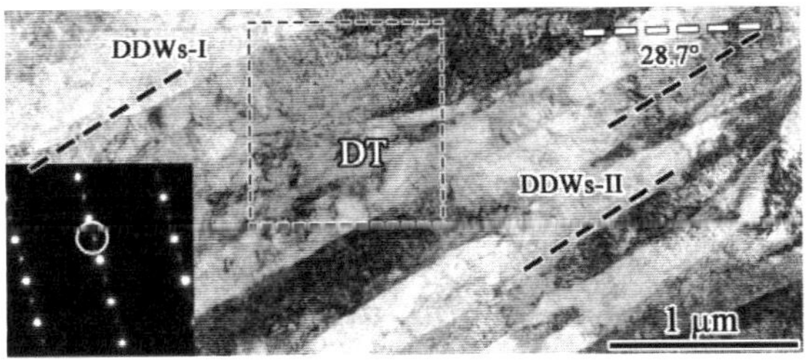

Figure 4. TEM bright-field image and the inserted SAED pattern at 140 μm deep. Yellow dashed line and black dashed line indicates SD and DDWs, respectively.

Nanostructured region with the structural size below 100 nm at the depth below 60 μm is featured by nanolamellar structures (NL) and nano-equiaxed (NE) structures. In the depth of 50 μm in **Figure 5(a)**, NL is formed with the interconnecting dislocation boundaries (IDBs, marked by black arrows) and the long and straight lamellar boundaries inside the lamellae, which show a similarity in morphology as those ultrafine lamellar (UFL) structures in pure nickel and IF steel deformed by severe plastic deformation[12, 13]. However, the difference between them is that NL has low angle boundaries as reflected by the inserted SAED pattern consisted of discontinuous diffraction circles. In addition, NL has the lamellar boundaries roughly parallel with SD having a deviation angle of 14.8°, and the boundary spacing is spanned from 50 nm to 200 nm, having a mean size of 96 nm in terms of the histogram of boundary spacing in **Figure 5(b)**. Such structural size is close to that deformed via traditional plastic deformation [2, 4-6].

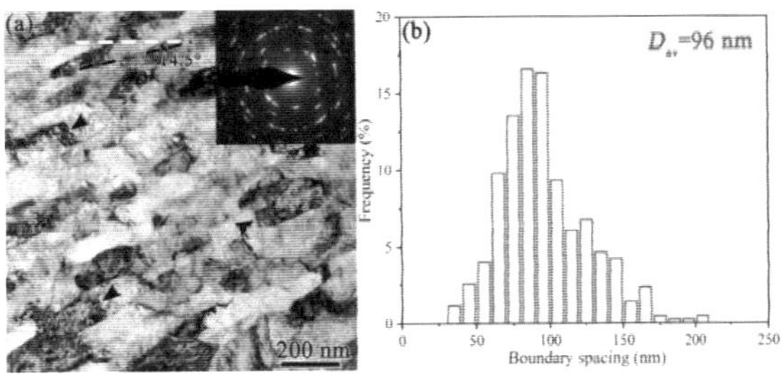

Figure 5. TEM observation (a) and histogram of boundary spacing (b) at 50 μm deep.

Further reaching the processed surface of 5 μm deep as shown in **Figure 6(a)**, the microstructures are composed of NL and NE structures. Here NL becomes extremely thinning. The lamellar boundaries are long and parallel, showing strong deformation texture according to the discontinuous diffraction circles in the inserted SEAD pattern. This indicates that most lamellar boundaries are low angle boundaries, which resembles that reported in the SMGT Ni[14, 15] and IF steel[16]. Here NL has the lamellar boundary spacing below 50 nm in **Figure 6(b)**, with an average size of 18 nm that is one order of magnitude smaller than the structural size induced by traditional severe plastic deformation [2, 4-6]. Furthermore, as shown in **Figure 6(c)**, NE structures have a random orientation and many high angle boundary reflected by continuous diffraction circles in the inserted SAED pattern. Further observation by HRTEM in **Figure 6(d)**, the diameter of NE structure is roughly 5-8 nm that is the attainable minimum structural size in deformed Ti.

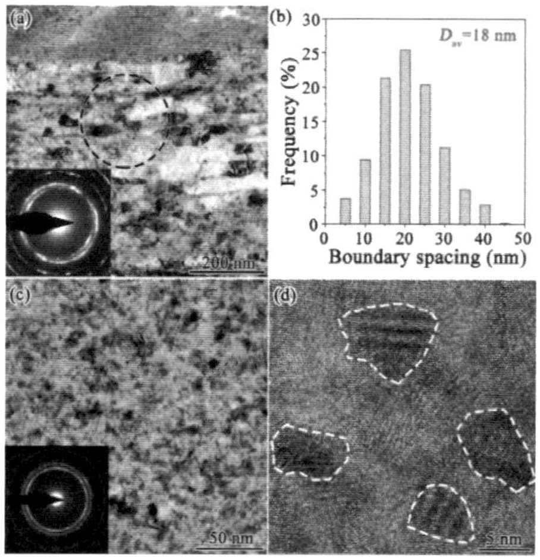

Figure 6. TEM observation (a), histogram of boundary spacing (b), local-magnified TEM image (c) and HRTEM image (d) from the red dotted rectangle in (a) at the depth of 5 μm.

4. Discussion

The examined microstructural features at different depths with various levels of strain in the ISF Ti sample can be summarized in **Table 1**. As the strain and strain rate were increased, the general pattern of microstructural evolution toward nano-regime of CP Ti during ISF involved:

i) formation of uni- and multi-directional twins;

ii) development of UFL structure;

iii) evolution of NL structure and NE structure;

Table 1. Summary of TEM observations at different depth on the ISF Ti sample.

Depth (μm)	Microstructures	Structural features
220	Parallel and intersected bands	DTs and DDWs
140	Layered structures	Lamellae and twin segments
50	NL	Low-angle misorientation and mean boundary spacing of 96 nm
5	Mixed NL and NE	NL: preferential orientation with mean boundary spacing of 18 nm; NE: random orientation with structural size of several of nanometers

There are five independent slip systems required for polycrystalline materials to undergo homogeneous plastic deformation, and thus the operations of twin systems are necessary in order to maintain the deformation compatibility because of merely four independent slip systems in Ti [17]. The main twinning planes in Ti are {10-12}, {11-21}, {11-2-2} and {10-11} determined by deformation conditions [18]. In the present work, twinning-dominated deformation is observed at the low level of plastic strain, namely twinning region, in which the observed twinning planes is the {10-12} type that is most frequent during room-temperature deformation. As the strain is increased, twinning can rotate the parent grain lattice to an orientation that favors other twinning behavior and dislocation slip, leading to the further deformation. The typical structural features are twin-twin intersection and formation of DDWs. The different twinning systems are activated along the multiple directions in parent grains, leading to twin intersections. Twins with different orientations interact with each other to significantly refine the parent grains, meanwhile slip is active due to a remarkable increase of low angle grain boundary (LAGB) fraction. The formation of LAGBs for instance DDWs is owing to the dislocation tangle, which promote the structural subdivision.

As the strain further increases to a certain level, the dislocation activity will gradually predominant plastic deformation, while the presence of a large number of twins and their intersection will hinder their movement, which results in a high dislocation density occurring at twin boundaries that is observed at the transition region between slip and twinning. At this stage, deformation twins ceases due to the very small grain size that is demonstrated by the previous investigation on the critical resolved shear stress of twinning increased with decreasing grain size in HCP materials [19, 20]. An increasing dislocation density is observed, which gives rise to the formation of lamellae including DDWs and IDBs. The area was bordered by parallel DDWs with the spacing between 100-1000 nm and vertical IDBs, namely forming UFL structure, and such microstructural feature is very similar to those BCC and FCC metals deformed at large strain [13, 14].

Formation of the nanostructures in the topmost surface layer of 60 μm (nanostructured region) is the mixture of NL structure and NE structure. The unique microstructures depends on the deformation conditions that facilitate two basic processes including: i) enhanced generation of dislocations and boundaries; and ii) stabilization of boundaries and dislocations generated. ISF process was anticipated to induce a shear deformation with large strain, high strain rate and high strain gradient, which is akin to other surface plastic deformation like SMGT [14-16]. Among these deformation parameters high strain rate and strain gradient differ greatly from those traditional severe plastic deformation.

The high strain rate facilitates dislocation generation and suppresses dislocation annihilation so that the pump-in dislocation density can be elevated, which enhance grain refinement in the early stage of deformation before the low-angle boundaries changes into high-angle ones. The presence of a high strain gradient needs storage of geometrically necessary dislocations (GNDs) to maintain the crystal continuity. As high strain gradient accumulated a high density of GNDs may influence the accumulation, annihilation and trapping of statistically stored dislocations according to the classical Kocks model [21], so as to enhance total dislocation density and suppress the process of dynamic recovery. Under the conditions, as the high density of dislocation was poured, the microstructures could be significantly refined down to the nano-sized, which is demonstrated by the publication that the fabrication of nanostructure was subjected to the surface plastic deformation [14-16].

5. Conclusion

1) The nanostructures were fabricated on the surface of formed parts during ISF. Typical nanostructure was composed of NL structure with the mean size of 18 nm and NE structure with a diameter ranged from 5 nm to 8 nm;

2) A general pattern of structural subdivision covering 4-5 orders of magnitude of characteristic length scale was build, involving formation of uni- and multi-directional twins, development of UFL structure and following evolution of NL as well as NE structures.

Acknowledgement

The authors are grateful for the financial support from National Nature Science Foundation of China through grant 52175347.

References

[1] Sabirov I, Perez-Prado M T, Molina-Aldareguia J M, Semenova I P, Salimgareeva G K and Valiev R Z 2011, Scr. Mater. **64** 69.
[2] Nie M Y, Wang C T, Qu M H, Gao N, Wharton J A, and Langdon T G 2014 J. Mater. Sci. **49** 2824.
[3] Yapici G, Karamen I and Maier H J 2006 Mater. Sci. Eng. A **434** 294.
[4] Zherebtsov S V, Dyakonov G S, Salem A, Salishchev G A and Semiatin S L 2011 Mater. Sci. Eng. A **528** 3474.
[5] Terada D, Inoue S and Tsuji N 2007 J. Mater. Sci. **42** 1673.
[6] Zhao X C, Yong X R, Liu X Y, Wang X Y and Langdon T G 2010 Mater. Sci. Eng. A **572** 6335.
[7] Huang A W, Yong P L, Liang N and Li Y S 2019 Mater. Charact. **149** 52.
[8] Li X, Sun B H, Guan B, Jia Y F, Gong C Y, Zhang X C and Tu S T 2021 Int. J. Fatigue **146** 6142.
[9] Yang D K, Cizek P, Fabijanic D, Wang J T and Hodgson P D 2013 Acta Mater. **61** 2840.
[10] Zhu K Y, Vassel A, Brisset F, Lu K and Lu J 2004 Acta Mater. **52** 4101.
[11] N. Hansen 2001 Metall. Mater. Trans. A **32** 2917.
[12] Zhang H W, Huang X and Hansen N 2008 Acta Mater. **56** 5451.
[13] Renk O and Pippan R 2019 Mater. Trans. **60** 1270.
[14] Liu X C, Zhang H W and Lu K 2013 Science, **342** 337.
[15] Liu X C, Zhang H W and Lu K 2015 Acta Mater. **96** 24.
[16] Liu X C, Zhang H W and Lu K 2015 Scr. Mater. **95** 54.
[17] Chichili D K, Hemker K J and Ramesh K T 1999 Acta Mater. **47** 3705.
[18] Munroe N, Tan X and Gu H 1997 Scr. Mater. **36** 1383.
[19] Yu Q, Shan Z W, Li J, Huang X, Xiao L, Sun J and Ma E 2010 Nature **463** 335.
[20] Zhu Y T, Liao X Z and Wu X L 2012 Prog. Mater. Sci. **57** 1.
[21] Kocks U F and Mecking H 2003 Prog. Mater. Sci. **48** 171.

Using tensile and compressive stress superposition during incremental forming to manipulate martensitic transformation in SS304

E M Mamros[1]*, F Maaβ[2], A E Tekkaya[2], B L Kinsey[3], and J Ha[3]

[1] Department of Mechanical Engineering, Bucknell University, Lewisburg, PA, USA

[2] Institute of Forming Technology and Lightweight Components, TU Dortmund University, Dortmund, Germany

[3] Department of Mechanical Engineering, University of New Hampshire, Durham, NH, USA

elizabeth.mamros@bucknell.edu

Abstract. Single point incremental forming (SPIF) is a flexible manufacturing process that has applications in industries ranging from biomedical to automotive. In addition to rapid prototyping, which requires easy adaptations in geometry or material for design changes, control of the final part properties is desired. One strategy that can be implemented is stress superposition, which is the application of additional stresses during an existing manufacturing process. Tensile and compressive stresses applied during SPIF showed significant effects on the resulting microstructure in stainless steel 304 truncated square pyramids. Specifically, the amount of martensitic transformation was increased through stress superposed incremental forming. Finite element analyses with advanced material modeling supported that the stress triaxiality had a larger effect than the Lode angle parameter on the phase transformation that occurred during deformation. By controlling the amount of tensile and compressive stresses superposed during incremental forming, the microstructure of the final component can be manipulated based on the intended application and desired final part properties.

1. Introduction

For creating the highly customizable components required in, e.g., the biomedical and automotive, industries, flexible manufacturing processes are desired. Incremental forming is one example of a rapid prototyping process that can achieve this objective without the use of custom tooling. The conventional single point incremental forming (SPIF) process simply requires a milling machine, hemispherical tool, and support frame for the blank to implement in an industrial environment. During SPIF, a user-defined toolpath guides the tool in a layer-by-layer manner to impose local deformation until the desired geometry is achieved. Incremental forming is a combination of bending, tension, and shear deformation mechanisms [1].

Stress superposition is a used in manufacturing to increase material formability, decrease the required forming loads, and enable customization of components [2]. Alternatively, the stress superposition strategy can be adopted to manufacture functionally graded parts using stress state sensitive microstructure changes, e.g., martensite transformation [3], in metals. These heterogeneous parts are advantageous for applications requiring highly customized parts. Stress-superposed incremental forming (SSIF) variants, such as tensile (TSSIF) and compressive (CSSIF), have been proposed to control the material and final part properties. For instance, a polyurethane (PU) die was introduced when forming aluminum alloy 5083 to superpose compressive stresses into the SPIF process, CSSIF, which affected the residual stress development [4]. For TSSIF, investigations have used a custom tensile frame to examine the effect on residual stresses [5]. Note that the phase transformation behavior of the material can be directly influenced by manipulating the stress state or other parameters, including the temperature in the forming zone, strain rate, equivalent plastic strain level, and material orientation [6], during manufacturing processes.

In this paper, SPIF and three SSIF variants, TSSIF, CSSIF, and tensile compressive stress-superposed incremental forming (TCSSIF), were investigated experimentally and numerically for a truncated square pyramid. The primary objective was to determine the effect of the different stress states during deformation on the properties, e.g., the phase transformation, of the final product. An isotropic, combined strain hardening law for constituent phases was identified to model SS304 and coupled with the Beese and Mohr 2011 martensitic transformation kinetics model [7]. This material model was implemented into Abaqus 2019 [8] for each process using a two-step approach [3]. The numerical results were validated by experimental results for all four IF processes. A Feritscope was used to measure the α'-martensite volume fraction on the formed parts. Collectively, these results support that stress superposition can be used in incremental forming to customize product properties and achieve functionally graded parts for specific applications, e.g., biomedical trauma fixation hardware.

2. Material Characterization

SS304 sheets with an initial thickness of 0.8 mm were characterized by uniaxial tension, bulge, in-plane torsion, and disk compression experiments. The resulting α'-martensite volume fractions were measured using an FMP30 Feritscope (Fischer Technology Inc.) and validated by electron backscatter diffraction. Further details regarding the material characterization experiments and Feritscope validation can be found in [3]. The Young's Modulus is approximately 170 GPa calculated from the stress-strain curve of uniaxial tension in the rolling direction (RD), and the Poisson's ratio is 0.33 based on the textbook value.

3. Constitutive Modeling

The SS304 material behavior, including phase transformation, was assumed to be isotropic for simplicity in modeling. An overview of the constitutive modeling is described below and summarized in Table 1. Additional details, including experimental data, can be found in [3].

Swift strain hardening models were used to describe the behavior of the constituent phases, i.e., γ-austenite and α'-martensite, individually. Then, these true stress-strain curves were combined in a weighted equivalent stress equation, which is a function of the phase volume fraction, i.e., the α'-martensite volume fraction denoted by $f_{\alpha'}$.

The martensitic transformation kinetics was described using the isotropic Beese and Mohr 2011 model [7]. Based on experiments, Beese and Mohr expanded the definition of the stress state to include the Lode angle parameter, $\bar{\theta}$, in addition to stress triaxiality, η. This model assumes isothermal conditions, and the maximum achievable volume fraction, c_{max}, is dictated by the temperature value. The transformation kinetics order from greatest to least with respect to equivalent plastic strain for SS304 is: shear, equibiaxial tension, uniaxial tension, and uniaxial compression [3]. Note that this order is dependent on the specific SS material and component form investigated [2].

PU 90° Shore A was chosen for the die material. To characterize this material, cyclic compression experiments were conducted according to ISO 7743 [9]. In each experiment, a cylindrical specimen with an initial height of 30 mm and initial diameter of 20.5 mm was subjected to three compressive cycles. Three repetitions were conducted to ensure consistency of the results. For modeling the PU die, a hyperelastic material model was used, which derives the stress-strain relation from the deformation energy density. According to the Mooney-Rivlin model [10], the dependence of the strain energy density (MJ/m^3) on the deformation is expressed by a polynomial equation with the strain invariants of the

Cauchy-Green strain tensor, I_1 and I_2, [11]. However, for $N = 1$, the Mooney-Rivlin model reduces to the two-parameter form with C_{10} and C_{01}. The parameters required for the simulation were calculated using Abaqus from the experimental data. Their values were $C_{10} = 1.714$ MPa and $C_{01} = -0.186$ MPa.

Table 1. SS304 isotropic strain hardening and martensitic transformation kinetics modeling summary.

Strain hardening model			
$\bar{\sigma} = f_{\alpha'} \cdot \sigma_{\alpha'} + (1 - f_{\alpha'}) \cdot \sigma_Y$			
Swift model			
$\sigma = K(\varepsilon_0 + \bar{\varepsilon})^b$			
	K (MPa)	ε_0	b
$\sigma_{\alpha'}$	1547	0.002	0.11
σ_Y	1313	0.050	0.36

Martensitic transformation model				
$\dot{c} = (c_{max} - c)nD(D\bar{\varepsilon})^{n-1}\dot{\bar{\varepsilon}}$				
Stress state dependency				
$D = D(\eta, \bar{\theta}) = (D_0 + a_\theta \bar{\theta} + a_\eta \eta)_+$				
c_{max}	n	D_0	a_θ	a_η
0.49	2.13	2.05	0.10	0.46

4. Incremental Forming Experiments

The conventional incremental forming process, SPIF, and three variants with stress superposition, TSSIF, CSSIF, and TCSSIF were investigated. For consistency purposes, the same toolpath and general experimental setup were used for all four forming processes. The specimens were laser cut from SS304 sheets of 0.8 mm thickness into the shape shown in Figure 1. The bottom surface of each specimen, i.e., non-toolside, was electrochemically etched with a dot pattern to allow for strain and geometry measurements via a photogrammetry camera after forming. An 85 mm base truncated square pyramid with a 45° wall angle was the target geometry and is shown in Figure 1.

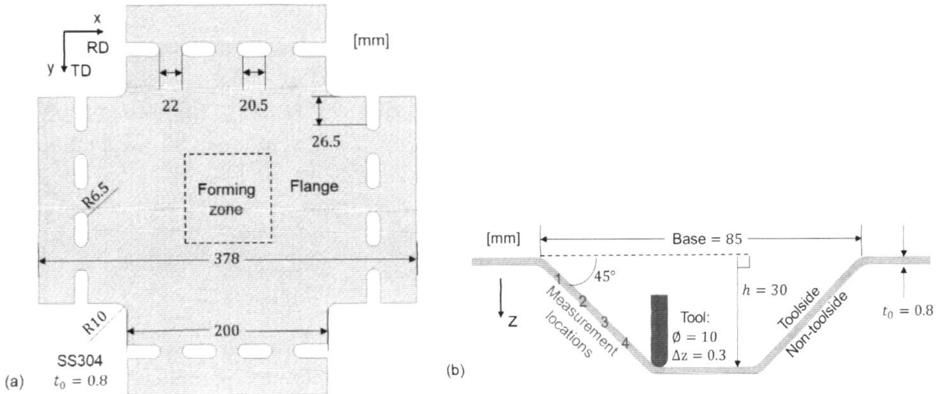

Figure 1. Incremental forming (a) blank, (b) target, and forming tool geometries.

The incremental forming experiments were conducted using a 5-axis DMU50 milling machine (DMG Mori). The machine's computer numerical controller was programmed to follow a user-defined, bidirectional toolpath, and the start point for each layer was located at the corner farthest from the two cylinders of the frame as indicated in Figure 2. For all process variants, a fixed (spindle rotation = 0 rpm), hemispherical tool with a 10 mm diameter tip was used with a step down of 0.3 mm and constant feed rate of 1500 mm/min, which resulted in a total forming time of ~811 s for the 30 mm depth. The base of the square truncated pyramid was formed first, i.e., an outside-in strategy was employed. A thin layer of deep-drawing oil (Castrol) was applied to the surface of the sheet prior to each experiment. A custom hydraulic frame was installed to act as a blankholder (Figure 2a), and during SPIF, a minimal pressure (< 5 bar) was applied to the hydraulic cylinders to fix the sheet specimens in-plane at the initial gripped position. The sheet blank was aligned such that the RD coincided with the x-axis of the setup.

The hydraulic frame shown in Figure 2a was used to superpose equibiaxial tensile stress to the SPIF process for two SSIF variants: TSSIF and TCSSIF. For both cases, the frame was set to its maximum pressure, i.e., ~200 bar for each cylinder, to apply in-plane equibiaxial tension to the sheet metal blank. Digital Image Correlation (DIC) measured that the von Mises strain (equivalent strain by von Mises yield function) achieved across the forming area is ~0.01.

To superpose compressive stress to SPIF, a PU die (190 mm x 190 mm x 50 mm) was used with the experimental setup described previously to form two SSIF variants: CSSIF and TCSSIF. A riser is placed in the center of the forming section in the tensile frame and holds the die securely (Figure 2b). The negative die cavity was milled into the surface of the die to the target geometry dimensions. Thus, a compressive phenomenon, so-called squeeze factor in DSIF, proportional to the sheet thickness, was applied at the tool contact point with the sheet during forming. A thin layer of lubricant was applied to the surface of the die cavity between experiments. Note that the die and die riser were removed from the experimental setup for SPIF and TSSIF experiments.

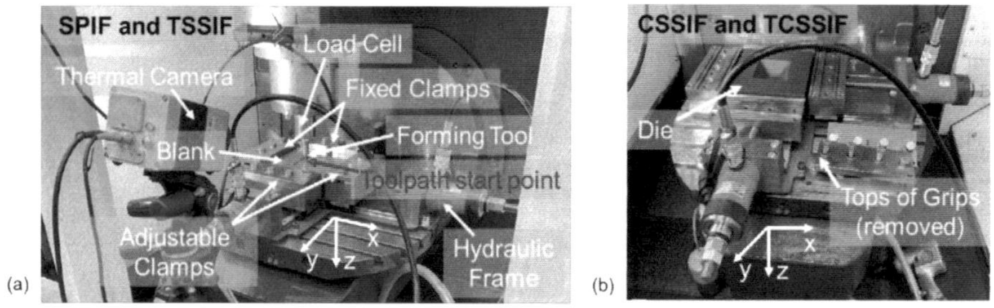

Figure 2. Experimental setup: (a) SPIF and TSSIF; and with die added for (b) CSSIF and TCSSIF.

5. Numerical modeling

An overview of the numerical model created for the SPIF process is described below and detailed further in [5]. The SPIF finite element (FE) model is used and modified as necessary for the SSIF process variants. For TSSIF and TCSSIF, an additional simulation step was required to model the tensile stress superposition. For CSSIF and TCSSIF, another component was required in the model assembly to represent the PU die. For all processes, a two-step approach was used: 1) a full incremental forming process simulation using Abaqus/Explicit 2019 with only isotropic elastic-plastic models, i.e., Hooke's law combined with von Mises yield function and Swift strain hardening, not considering phase transformation, to determine the strain in a given element at a specified location; 2) one element simulations using Abaqus/Implicit 2019 with isotropic elastic-plastic models with martensite transformation kinetics, which were implemented into a user material subroutine (UMAT), to predict the transformation at select locations (Section 5.1). Note that this was not a thermal-mechanical analysis.

The SPIF process was modeled using linear brick elements with reduced integration, C3D8R, for the sheet with three elements through the sheet thickness direction. Note that a simplified sheet geometry was used in the model that excluded part of the gripped region. A subsection of the forming area that is larger than the target geometry contained a refined mesh for increased accuracy. A transition zone is partitioned around this subsection, and the remaining flange area was coarsely meshed to increase computational efficiency(Figure 3). Mass scaling of 10^3 was used. Mesh and mass scaling convergence studies were completed. The tool was modeled as a rigid body. The toolpath used for experiments was converted from G-code to amplitudes and step times based on the length of each toolpath layer. This converted toolpath was implemented as displacement boundary conditions without time scaling. Due to the presence of lubrication on one or both sides of the sheet, depending on the chosen IF process, friction between the tool and the sheet on the toolside and between the sheet and the die on the non-toolside were ignored, i.e., the friction coefficient, $\mu = 0$.

To superpose tensile stress in the numerical model, a step for pre-stress was added to the TSSIF and TCSSIF simulations. During this step, the tool and the clamped, i.e., -x and -y, sides of the blank were fixed. Displacement boundary conditions were applied to the adjustable, i.e., +x and +y, sides of the blank. The displacement value, i.e., 1.975 mm, was chosen to create approximately 1% von Mises strain in the forming area to replicate the DIC experimental results of the loaded blank in the tensile frame.

To superpose the compressive stress in the numerical model, the PU die was included in the assembly as a deformable body. PU 90° Shore A was modeled based on the results in Section 3. The die contained 2988 elements (type C3D8R) with seven elements through the thickness and is shown in Figure 3. The interaction between the die and the sheet was modeled as kinematic contact. An additional boundary condition was applied to impose the displacement restrictions on the riser used in experiments.

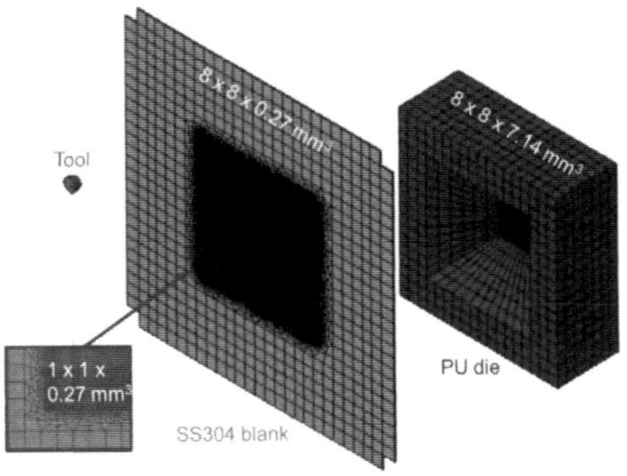

Figure 3. FE model assembly for CSSIF and TCSSIF: exploded isometric view.

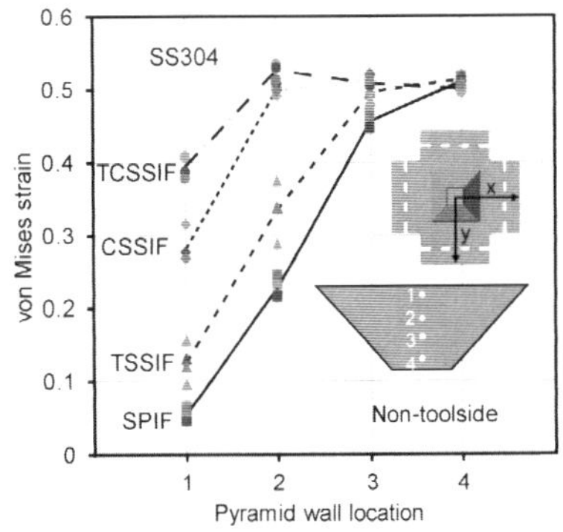

Figure 4. von Mises strain at locations 1-4 on the non-toolside (Exp.). Averages are shown as lines.

5.1. One element simulation for martensitic phase transformation

One element, implicit simulations were used to predict the phase transformation along the pyramid walls at the same locations as the Feritscope measurements. The nodal displacements from 12 elements for each incremental forming variant, i.e., four elements from the toolside, midplane, and non-toolside, from the full model simulations were exported and prescribed as the boundary conditions for the one element simulations. The isotropic strain hardening and martensitic transformation kinetics models were combined in the UMAT to describe the deformation induced phase transformation in the sheet specimens during stress-superposed incremental forming. This two-step method is both computationally efficient and provides information not readily available experimentally, e.g., the volume fractions at the midplane along the thickness direction, which cannot be measured directly using the Feritscope.

6. Results and discussion

The von Mises final strain states from experiments at locations 1-4 on the non-toolside are shown in Figure 4. The largest difference in strain between the processes was observed at location 1. TCSSIF had the highest strain, followed by CSSIF, then TSSIF, and, lastly, SPIF with the lowest strain. At location 2, the same order applies although the difference between the processes decreased. At location 3, the strains for the IF processes start to converge, and at location 4, they are nearly equal. This trend is also in agreement with the assumption that the effectiveness of the stress superposition varies throughout the forming process. Based on the trends in Figure 6, the stress superposition had the greatest effect when the strain increased significantly between locations. Explicitly, the compressive stress superposition appears to be the most effective from locations 1 to 2 (see the largest von Mises strain increase for the CSSIF case), while the tensile stress superposition effect was sustained from locations 1 to 3. As a result, the contributions of each type of stress superposition caused the von Mises strain results for the TCSSIF case. Further analysis into the hierarchy of the forming mechanisms for each process at each location is warranted to improve the understanding of these observations.

From the chosen martensitic transformation kinetics model, the stress state can be described by two quantities: the stress triaxiality and Lode angle parameter. Figure 5 shows the evolution of the stress state throughout the forming process for each IF process with respect to the equivalent plastic strain.

These results were extracted from the full model simulations and filtered based on the equivalent plastic strain increment, which represents the effective plastic deformation for the martensite transformation. Note that a material point (or element in FE simulation) deforms a very small strain increment, and the associated stress state change does not significantly impact the martensite transformation, when the tool is located far away from the element of interest, e.g., on other pyramid walls. Thus, only steps which contained an equivalent plastic strain increase $\Delta\bar{\varepsilon} > 0.001$ are plotted.

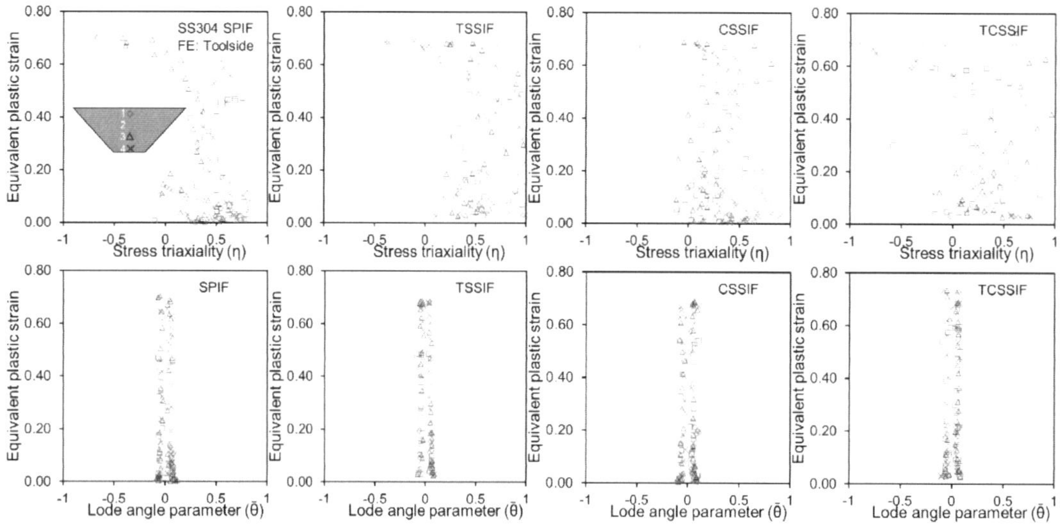

Figure 5. Stress state evolutions from full FE models of IF processes on toolside at locations 1-4.

For all four processes, the Lode angle parameter showed less variation than the stress triaxiality during forming. At low levels of equivalent plastic strain, i.e., < 0.1, the spread of Lode angle parameter values was larger than at higher levels of equivalent plastic strain, where it fluctuated between -0.2 and 0.2. A Lode angle parameter value close to 0 indicates generalized shear, which is expected due to the presence and contact of the tool on this side of the sheet. However, the stress triaxiality showed stronger evolution and different trends for each of the forming processes with respect to equivalent plastic strain. Near location 1, the stress triaxiality was positive for all four processes with values indicative of generalized shear ($\eta = 0$), uniaxial tension ($\eta = 0.33$), plane strain tension ($\eta = 0.58$), and equibiaxial tension ($\eta = 0.67$). For SPIF, near location 3, the stress triaxiality trended toward negative values, which indicate compression ($\eta = -0.33$) and biaxial compression ($\eta = -0.67$). For TSSIF, the stress triaxiality remained positive for the majority of the process but started to trend toward zero, i.e., shear, at locations 3 and 4. For CSSIF, the stress triaxiality initially trends toward zero values for all four locations, but near locations 2 and 3, it curves toward positive values. For TCSSIF, all four locations started with positive stress triaxialities, and then locations 2 and 4 trended towards negative values.

These differing trends can be attributed to the process variations but also provide information related to the effectiveness of the stress superposition throughout the process. For example, in TSSIF, the stress triaxiality remains positive throughout the majority of the process due to the tensile stress superposition rather than trending towards negative values later like SPIF. This result of stress triaxiality closer to uniaxial tension rather than uniaxial compression resulting in increased phase transformation for TSSIF compared to SPIF is consistent with the martensitic transformation kinetics trend with respect to stress state. For the CSSIF and TCSSIF, the effectiveness of the die seemingly increased at locations 3 and 4 as a result of greater contact with the tool at larger forming depths. The variation in the stress state, particularly stress triaxiality, for each process also had significant effects on the phase transformation.

To analyze the martensitic transformation, first, the ferrite number was measured on the toolside at four locations along each pyramid wall (as shown in Figure 1b) by an FMP30 Feritscope, and the α'-

martensite volume fraction, $f_{\alpha'}$, was calculated using the preprogrammed conversion factor. The same methodology was repeated for the non-toolside, and the results are shown in Figure 6. The measurements on the non-toolside were consistently lower than the toolside by ~0.1 α'-martensite but show the same trends. This observation is consistent with the deformation mechanisms of SPIF, which include increased deformation on the toolside.

Comparing Figures 4 and 6, the four processes followed the same trend with respect to one another for all four locations on the non-toolside. TCSSIF had the highest von Mises strain and phase transformation, followed by CSSIF, then TSSIF, and lastly SPIF. Between locations 1 and 2, the von Mises strain and α'-martensite volume fraction increased for all processes except for TCSSIF, which experienced a slight decrease in phase transformation. This is likely the result of the decreasing effectiveness of the tension stress superposition as the forming process progressed, which is reflected in the stress triaxiality evolution towards compression. From location 3 to location 4, the von Mises strain and α'-martensite volume fractions converged for all processes. Increased von Mises strain leads to increased phase transformation; contrastingly, increased temperature inhibits phase transformation. Further investigations are warranted to determine the weighted contributions of these factors.

To explain the variation in the Feritscope measurements for the four pyramid walls, two possible explanations are the material anisotropy and the frame in the experimental setup. Since two sides of the frame are fixed and the loading is only applied to the remaining two sides, the loading is not perfectly equibiaxial tension with variations in the biaxial stress state applied on the four sides. Although not considered in this work, anisotropy with respect to martensitic transformation is possible [7].

The α'-martensite volume fractions predicted by the one element simulations and measured from experiments are shown in Figure 6. The error bars on the experimental results represent the maximum error, which includes the variation between the four walls of the truncated square pyramids, the repeated Feritscope measurements at each location, and the multiple samples for each process. In addition to the toolside and non-toolside predictions, the FE model also provided data related to the midplane of the pyramid walls, which cannot be measured directly using the Feritscope.

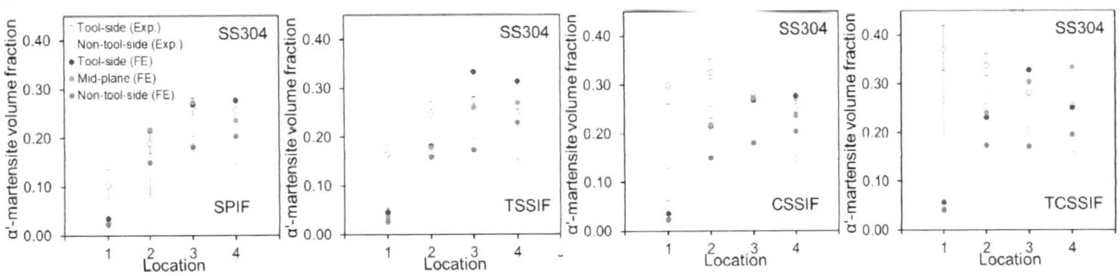

Figure 6 Comparison of α'-martensite volume fraction measurements from Feritscope (Exp.) with predictions from one element simulations (FE).

In general, the one element simulations provided fair predictions of the α'-martensite volume fractions compared to the experimental values without requiring significant additional computational expense. The implemented martensitic transformation kinetics model is able to capture increasing phase transformation trends with respect to the pyramid wall location well, e.g., SPIF locations 1-3. A limitation of the model is its inability to account for inhibited phase transformation at elevated temperatures, e.g., at location 4 for SPIF, TSSIF, and CSSIF. This two-step method could be optimized to be used with different martensitic transformation kinetics models to improve the predictions.

7. Conclusions and future work

The stress superposition of in-plane tensile and normal to the sheet compressive stresses during SPIF to manufacture a truncated square pyramid using SS304 was investigated for four IF variants: SPIF, TSSIF,

CSSIF, and TCSSIF. Final strain state, stress state evolution, and phase transformation results were analyzed and compared between process variants and experimental and numerical analyses.

A two-step FE method was utilized to further investigate the phase transformation mechanisms affecting the IF processes. SPIF, TSSIF, CSSIF, and TCSSIF revealed different trends with respect to the stress state evolution during the forming process. Particularly, the stress triaxiality showed greater variation than the Lode angle parameter, which remained close to generalized shear for locations 1 to 4. The different stress state evolutions between the processes affected the phase transformation at each location. However, the α'-martensite volume fraction was ultimately impacted by several effects, including process temperature, equivalent plastic strain level, and stress state, which needed to be considered collectively. A Feritscope was used to measure the α'-martensite volume fraction on the formed truncated square pyramids. At locations 1 and 2, TCSSIF had the greatest transformation, followed by CSSIF, then TSSIF, and SPIF had the least amount of transformation. At locations 3 and 4, the four processes converged to similar volume fractions of α'-martensite. Future work is planned to establish a better understanding of the relationship between the martensitic transformation and residual stress development in SS304 during SSIF. However, the current results support that functionally graded, formed parts can be created from these manufacturing processes for implementation into several industries. For example, biomedical trauma fixation hardware requires patient-specific geometries and heterogeneous material properties, e.g., increased strength near the fixture locations of an implant.

References

[1] Martins P A F, Bay N, Skjoedt M and Silva M B 2008 Theory of single point incremental forming *CIRP Annals* **57** 247–52

[2] Tekkaya A E, Groche P, Kinsey B and Wang Z G 2023 Stress superposition in metal forming *CIRP Annals* **72** 621-644

[3] Mamros E M, Maaß F, Hahn M, Tekkaya A E, Ha J and Kinsey B L 2023 Martensitic transformation of SS304 truncated square pyramid manufactured by single point incremental forming *(submitted)*

[4] Maaß F, Hahn M and Tekkaya A E 2021 Adjusting residual stresses by flexible stress superposition in incremental sheet metal forming *Arch Appl Mech* **91** 3489–99

[5] Maaß F, Hahn M and Tekkaya A E 2022 Setting Residual Stresses in Tensile Stress-superposed Incremental Sheet Forming *Proceedings of the 22nd International Conference on Material Forming* ESAFORM 2022

[6] Feng Z, Mamros E M, Ha J, Kinsey B L and Knezevic M 2021 Modeling of plasticity-induced martensitic transformation to achieve hierarchical, heterogeneous, and tailored microstructures in stainless steels *CIRP Journal of Manufacturing Science and Technology* **33** 389–97

[7] Beese A M and Mohr D 2011 Effect of stress triaxiality and Lode angle on the kinetics of strain-induced austenite-to-martensite transformation *Acta Materialia* **59** 2589–600

[8] Hibbitt D, Karlsson B and Sorensen P 1978 Abaqus (Dassault Systèmes Simulia Corporation)

[9] ISO 7743:2017 2017 *Rubber, vulcanized or thermoplastic — Determination of compression stress-strain properties* (ISO)

[10] Mooney M 1940 A Theory of Large Elastic Deformation *Journal of Applied Physics* **11** 582–92

IOP Conf. Series: Materials Science and Engineering **1307** (2024) 012006 doi:10.1088/1757-899X/1307/1/012006

[11] Stommel M, Stojek M and Korte W 2018 Mechanisches Werkstoffverhalten und -modellierung *FEM zur Berechnung von Kunststoff- und Elastomerbauteilen* (Carl Hanser Verlag GmbH & Co. KG) pp 5–144

Superplastic deformation behavior of low Al–added medium Mn steels with and without Ti and Mo elements

Pufan Yuan[1], Hongtao Zhang[1], Qi Chao[2], Chenyu Sun[1], Bowen Jia[1], Shuai Tang[3], Haile Yan[1], Peter Hodgson[4], Minghui Cai[1,3,*]

[1]School of Materials Science & Engineering, Northeastern University, Shenyan, 110819, China
[2]Center for Key Light-weight Materials, Nanjing Tech University, Nanjing 211816, China
[3]State Key Laboratory of Rolling and Automotion, Northeastern University, Shenyang 110819, China
[4]Institute for Frontier Materials, Deakin University, Geelong 3216, Australia

Corresponding author: Tel.: +86 83691570; E-mail: caimh@smm.neu.edu.cn

Abstract. Fe-Mn-Al-C medium Mn steels were found to reveal extraordinary superplasticity and have significant potential for forming the complex structural parts due to high strength, excellent ductility and material cost. In present study, the effect of tensile deformation temperature on the superplastic deformation behavior of a cold-rolled low Al-added medium Mn steel was studied. A maximum tensile elongation of approximately 1170% was obtained at 745 °C and 10^{-2} s^{-1}, which is to our best knowledge the highest low-temperature-high-strain-rate superplasticity in medium Mn steels. Also, the role mechanism of microalloying elements such as Ti and Mo was revealed to further enhance the strength level of superplastic medium Mn steels. In view of the representative microstructural features, such as equiaxed grains, random texture, and sluggish grain growth, grain boundary sliding is thought to be the dominative mechanism during the high-strain-rate superplastic deformation.

1. Introduction

Superplasticity refers to the remarkable ability of certain materials to undergo substantial tensile elongation, often exceeding 400%, prior to failure in a generally isotropic manner [1]. Superplastic forming (SPF) of sheet metal has been used to produce complex patterns and integrated structures that are light-weight and stronger than the assembled components. In conventional metals, grain boundary sliding (GBS) typically governs superplasticity, with accommodation processes like dislocation slip and grain boundary diffusion playing key roles [2, 3]. Notably, medium Mn steels have recently exhibited extraordinary superplasticity, reaching up to ~1314%, accompanied by more intricate superplastic flow behavior [4]. The outstanding superplastic properties of medium Mn steels can be attributed to the high thermal stability of the ultrafine duplex microstructure comprising ferrite (α) and austenite (γ) at elevated temperatures. This stability is believed to result from the sluggish uphill partitioning of Mn and Al atoms between the two phases [5,6]. Achieving this unique superplastic behavior involves intercritical deformation, leading to the presence of a soft ferrite matrix in the microstructure and a significant reduction in the yield strength of final formed parts. The addition of Al is crucial for superplastic deformation in medium Mn steels. Steels lacking Al exhibit lower elongation (~270% at 645°C), mainly attributed to the low tensile temperature [4]. Interestingly, reducing Al content from 2.3 wt% to ~1.0 wt% has been shown to enhance low-temperature superplasticity in medium Mn steels [7]. Additionally, microalloying elements like Nb, Mo, Ti, and V are known to significantly improve the yield strength of medium Mn steels through mechanisms such as precipitation hardening and grain refinement strengthening [8–11].

Despite substantial progress in understanding the impact of microalloying elements on the strength and ductility of medium Mn steels, their specific role in the superplastic flow behavior of these steels remains less clear. In this context, we conducted a study on two cold-rolled medium Mn steels, one with Ti-Mo addition and the other without, to elucidate the influence of Ti-Mo on precipitation, dynamic recrystallization (DRX), and superplastic deformation behavior.

2. Experimental procedure

Two 50 kg ingots with compositions of Fe–5.6Mn–1.1Al–0.2C (wt%, termed 5.6Mn) and Fe–5.6Mn–1.1Al–0.2C–0.1Ti–0.22Mo (wt%, termed 5.6MnTiMo) were prepared by induction melting. According to the results obtained using Thermo–Calc TCFE 9 database (Fig.1), both steels had an intercritical annealing temperature range from 450 to 474 °C (A_1) to 795–797 °C (A_3). The ingots were forged into a slab with a cross sectional area of 40 mm× 100 mm. After homogenization at 1200 °C for 2 h, the ingots were hot rolled to 5.0 mm thickness, and then subjected to cold rolling to obtain a sheet of ~2.0 mm thickness.

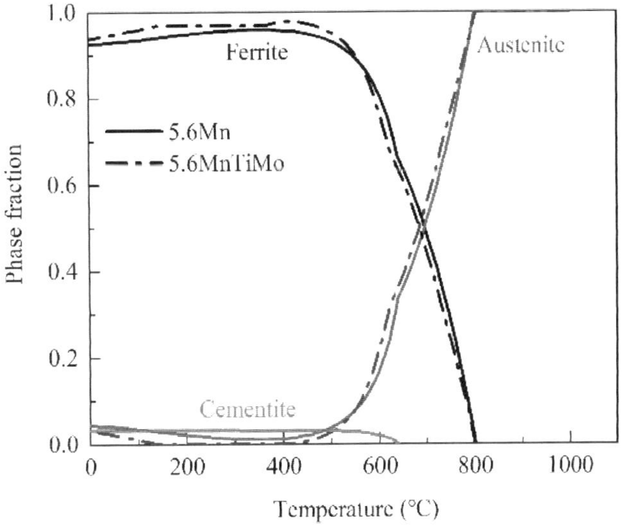

Fig.1 The equilibrium phase diagram of both medium Mn steels

Superplastic tensile tests were carried out on a universal testing machine (MTS, CMT5303) equipped with a constant strain rate of 10^{-2} s^{-1}. The deformation temperatures (685 °C and 745 °C) were chosen in the $\alpha + \gamma$ dual–phase temperature range. Microstructural observations were performed using a field–emission scanning electron microscope (FE–SEM, Supra, SSX–550) equipped with an electron backscattered diffractometer (EBSD). The accelerating voltage and step size for SEM–EBSD operation were 15 kV and 370 nm, respectively. Specimens for EBSD were electro–polished using a 15% perchloric acid alcohol solution at 20 V for 20 s to remove the damaged layer. Data acquisition and post–processing were performed using HKL Channel 5 software.

3. Results and discussion

Fig. 2a and b illustrate the engineering stress–engineering strain and true stress–strain curves for both steels at an initial strain rate of 10^{-2} s^{-1} at 685 °C and 745 °C during superplastic deformation. The peak stress of both steels decreased with increasing tensile temperature. At both tensile temperatures, the 5.6MnTiMo steel exhibited the higher peak stress compared to the 5.6Mn steel. However, the superplastic elongation of the 5.6MnTiMo steel was lower than that of the 5.6Mn steel. These results suggest that the addition of Ti–Mo enhanced strength with limited loss of elongation. The phase maps of both steels before and after superplastic deformation at 685 °C are shown in Fig. 2c and d. When the specimen was reheated to 685 °C, it possessed a mixed microstructure consisting of retained γ, α and/or α'. Both steels exhibited a similar microstructure consisting of ultrafine equiaxed grains and coarsened, elongated grains prior to tensile deformation. However, after tensile failure, finer, elongated α grains with more low–angle grain boundaries (indicated by red lines) were observed in the 5.6MnTiMo steel, indicating that the addition of Ti-Mo inhibited the occurrence of DRX and further grain growth. To further understand the microstructural features during superplastic flow, EBSD geometrically necessary dislocation (GND) maps at various strains were employed, as depicted in Fig. 3. Under high–temperature tensile deformation, the annealed duplex microstructure underwent both dynamic transformation and dynamic recrystallization (DRX) [12–14]. It was observed that equiaxed recrystallized grains coarsened with increasing tensile strain for both steels. This phenomenon indicates that the grains tended to grow during superplastic deformation, leading to local hardening and impeding the initiation of local necking. Notably, the 5.6MnTiMo steel exhibited finer α and γ grains compared to the 5.6Mn steel, indicating that the addition of Ti–Mo significantly restrained the grain growth rate for both α and γ phases. This observation suggests that Ti–Mo addition plays a crucial role in controlling microstructural evolution during superplastic deformation, contributing to the refined grain structure observed in the 5.6MnTiMo steel.

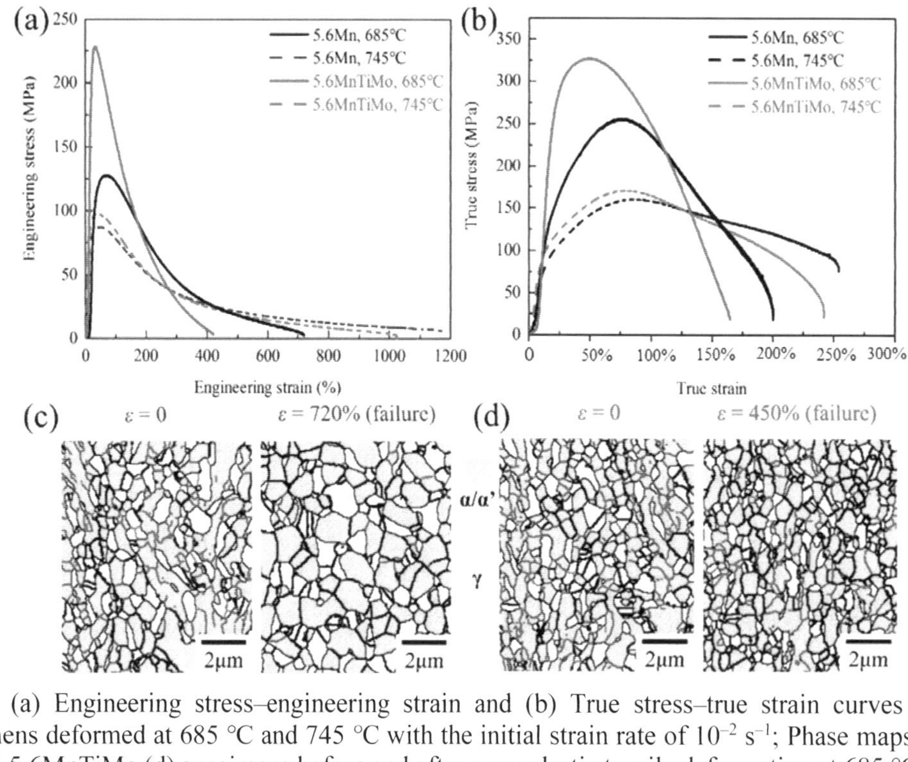

Fig. 2 (a) Engineering stress–engineering strain and (b) True stress–true strain curves of tensile specimens deformed at 685 °C and 745 °C with the initial strain rate of 10^{-2} s^{-1}; Phase maps of 5.6Mn (c) and 5.6MnTiMo (d) specimens before and after superplastic tensile deformation at 685 °C.

The flow behavior of superplastic medium Mn steels is intricately linked to both the grain size and its morphology, as well as the microstructural stability driven by the dominant grain boundary sliding (GBS) mechanism [15]. The addition of Ti–Mo has substantial effects on microstructural evolution by influencing precipitation, phase transformation, and grain refinement, ultimately impacting the GBS behavior and its associated accommodation mechanisms. For a more in-depth analysis of the precipitation behavior, TEM micrographs of nano–precipitates for the 5.6MnTiMo steel post-tensile failure at 685 °C and 745 °C are presented in Fig.4. Notably, nano-sized spherical particles were observed near grain boundaries and on dislocations. It is also seen that the number of nano–precipitates decreased with increasing temperature. The (Ti, Mo)C particles play a role in slowing down grain growth, attributed to the Zener pinning effect. This phenomenon significantly enhances the thermal stability of the duplex microstructure [16]. It has been reported that the local segregation of Mn and Mo atoms at ferrite/austenite interfaces contributes to the stabilization of fine–grained microstructure by a reduction of the interphase mobility [9].Consequently, the 5.6MnTiMo steel exhibits a much finer duplex microstructure. When deformed at 685 °C, fine and dense (Ti, Mo)C particles impeded the DRX process by pinning the dislocations and hindering the migration of subgrain boundaries to form the HAGBs, leading to the much lower elongation for 5.6MnTiMo steel. It is confirmed that the adverse influence of microalloying elements on superplasticity in the present study was attributed to the pinning effects on the movement of grain boundaries and dislocations. The nano-sized particles are believed to inhibit both GBS and DRX.

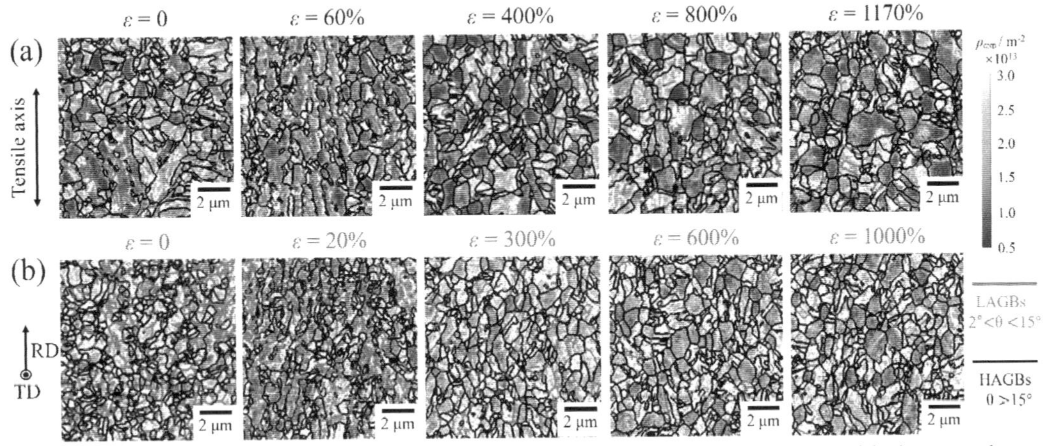

Fig. 3 EBSD geometrically necessary dislocation (GND) maps overlapped with low–angle grain boundaries (LAGBs) and high angle grain boundaries (HAGBs) of tensile specimens deformed up to various engineering strains at 745 °C: (a) 5.6Mn steel; (b) 5.6MnTiMo steel. RD and TD represent rolling direction and transverse direction, respectively.

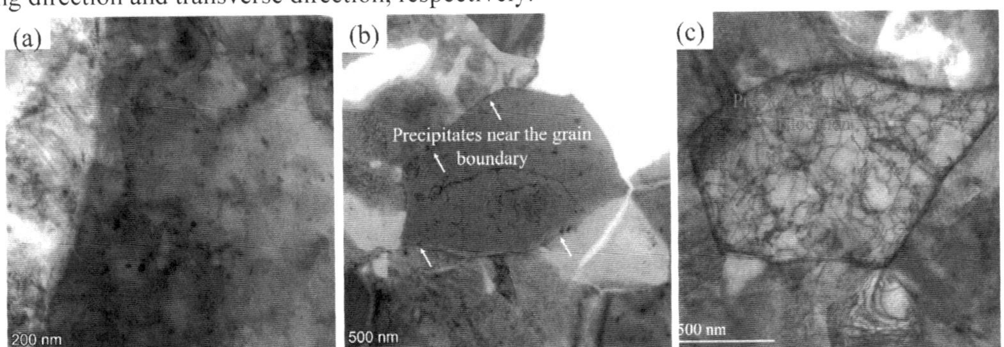

Fig. 4 TEM images of the precipitation region in 5.6MnTiMo steel after tensile failure at 685 °C and 745 °C: (a) precipitates at 685 °C; (b) precipitates near the grain boundary at 745 °C; (b) precipitates near the dislocations at 745 °C.

The high–temperature deformation mechanism was determined by examining the strain rate sensitivity exponent (m) and activation energy of deformation (Q) at a given tensile temperature. The strain–rate jump tests were conducted at 685 °C and 745 °C with five different strain rates of 10–1, 10^{-2}, 5×10^{-3}, 10^{-3} and 10^{-4} s^{-1}. The m values, calculated by $m = \partial \ln\sigma / \partial \ln\dot{\varepsilon}$ [17], are listed in Table 1. The values of m were larger than 0.3 except for the case of the strain rate of 10^{-1} s^{-1} for both steels. This suggests that the flow behavior was mainly controlled by GBS during tensile deformation at 685 °C and 745 °C for both steels. Furthermore, the activation energy Q for superplastic deformation was determined by plotting the logarithmic true stress vs. the inverse absolute temperature (Fig.5). According to the flow curves at various temperatures with a strain rate of 10^{-2} s^{-1}, the Q values were estimated to be 122~153 and 193.8~284.7 kJ/mol for the 5.6Mn and 5.6MnTiMo steels, respectively. In contrast, the 5.6MnTiMo steel exhibited higher values of Q implying that the addition of Ti and Mo increases the resistance to GBS during superplastic deformation.

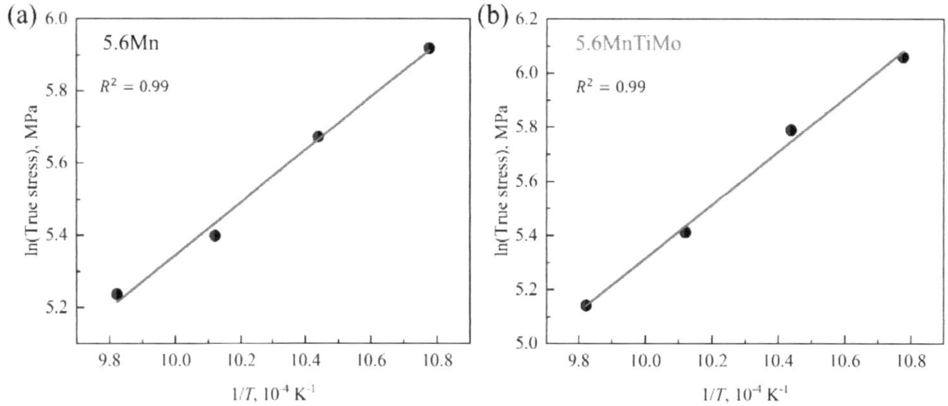

Fig. 5 lnσ-1/T curve of (a) 5.6Mn and (b) 5.6MnTiMo steels.

Table 1 Strain rate sensitivity exponent m determined by using three–stage strain rate jump test for 5.6Mn and 5.6MnTiMo steels

	Initial strain rate, s^{-1}	10^{-4}~10^{-3}	10^{-3}~5×10^{-3}	5×10^{-3}~10^{-2}	10^{-2}~10^{-1}
5.6Mn	*m* (685 °C)	0.52	0.50	0.48	0.23
	m (745 °C)	0.49	0.58	0.45	0.23
5.6MnTiMo	*m* (685 °C)	0.31	0.33	0.36	0.20
	m (745 °C)	0.35	0.39	0.43	0.22

4. Conclusions

The influence of microalloying elements on the superplastic behavior of medium Mn steels has been thoroughly investigated in the current study. The addition of Ti–Mo resulted in an improvement in the strength level of the final formed parts, accompanied by a limited reduction in superplastic elongation. A notable superplastic elongation of approximately 1000% was achieved at 745 ℃ with a strain rate of 10^{-2} s^{-1} in the newly designed Ti - Mo alloyed steel, which is comparable to the ~1170% observed in the base 5.6Mn steel. GBS has been identified as the predominant mechanism during superplastic deformation in the medium Mn steel under consideration. The presence of nano-sized precipitates was found to inhibit both DRX and GBS.

Acknowledgements

The authors gratefully acknowledge the financial support of the Natural Science Foundation of China (Nos. 51975111/52274379), Fundamental Research Funds for the Central Universities (No. 2023GFZD14) and Key R &D and Promotion Special Project of Henan Province (No. 212102210444). Special thanks are due to the instrumental/data analysis from Analytical and Testing Center, Northeastern University.

References

[1] A.H. Chokshi, A.K. Mukherjee, T.G. Langdon, Superplasticity in advanced materials, Materials Science and Engineering: R: Reports. 1993, 10: 237–274.

[2] T.G. Langdon, Seventy-five years of superplasticity: historic developments and new opportunities, Journal of Materials Science. 2009, 44: 5998–6010.
[3] M. Kawasaki, T.G. Langdon, Principles of superplasticity in ultrafine-grained materials, Journal of Materials Science. 2007, 42: 1782–1796.
[4] J. Han, S.H. Kang, S.J. Lee, M. Kawasaki, H.J. Lee, D. Ponge, D. Raabe, Y.K. Lee, Superplasticity in a lean Fe-Mn-Al steel, Nature Communications. 2017, 8: 751.
[5] Y. Maehara, T.G. Langdon, Superplasticity of steels and ferrous alloys, Materials Science and Engineering: A. 1990, 128: 1–13.
[6] W. Cao, C. Huang, C. Wang, H. Dong, Y. Weng, Dynamic reverse phase transformation induced high-strain-rate superplasticity in low carbon low alloy steels with commercial potential, Scientific Reports. 2017, 7: 9199.
[7] H.T. Zhang, M.H. Cai, W.J. Zhu, S.H. Sun, H.L. Yan, S.J. Yao, Y.K. Luan,S. Tang, P. D. Hodgson, Low–temperature superplastic deformation of cold–rolled Fe–5.6Mn–1.1Al–0.2C steel, Metall. Mater. Trans. A. 53 (2022) 3869–3880.
[8] R.D.K. Misra, J. Hu, I.V.S. Yashwanth, V.S.A. Challa, L.-X. Du, G.-S. Sun, H. Xie, Phase reverted transformation-induced nanograined microalloyed steel: Low temperature superplasticity and fracture, Materials Science and Engineering: A. 2016, 668: 105–111.
[9] Y.K. Lee, J. Han, Current opinion in medium manganese steel, Materials Science and Technology. 2015, 31: 843–856.
[10] E. Alabort, P. Kontis, D. Barba, K. Dragnevski, R.C. Reed, On the mechanisms of superplasticity in Ti–6Al–4V, Acta Materialia. 2016, 105: 449–463.
[11] H. Masuda, E. Sato, Diffusional and dislocation accommodation mechanisms in superplastic materials, Acta Materialia. 2020, 197: 235–252.
[12] T.G. Langdon, Grain boundary sliding revisited: Developments in sliding over four decades, Journal of Materials Science. 2006, 41: 597–609.
[13] F.R. Cao, J.Z. Cui, J.L. Wen, and L. Fang, Mechanical Behaviour and Microstructure Evolution of Superplastic Mg-8.4 wt pct Li Alloy and Effect of Grain Size and Phase Ratio on Its Elongation, Journal of Materials Science & Technology. 2000, 16: 55–58.
[14] S.S. Sohn, D.G. Kim, Y.H. Jo, A.K. da Silva, W. Lu, A.J. Breen, B. Gault, D. Ponge, High-rate superplasticity in an equiatomic medium-entropy VCoNi alloy enabled through dynamic recrystallization of a duplex microstructure of ordered phases, Acta Materialia. 2020, 194: 106–117.
[15] W. Wang, M. Yang, D. Yan, P. Jiang, F. Yuan, X. Wu, Deformation mechanisms for superplastic behaviors in a dual-phase high specific strength steel with ultrafine grains, Materials Science and Engineering: A. 2017, 702: 133–141.
[16] Y. Han, J. Shi, L. Xu, W.Q. Cao, H. Dong, TiC precipitation induced effect on microstructure and mechanical properties in low carbon medium manganese steel, Mater. Sci. Eng. A 530 (2011) 643–651.
[17] M. Kawasaki, T.G. Langdon, Principles of superplasticity in ultrafine–grained materials, J. Mater. Sci. 42 (2007) 1782–1796.

Research on Damage Evolution in Ultra-thin Sheet Material under Deep Drawing and Ironing Process

Nan Gu[1], Jiuchang Zhang[1], Wen Zhang[1,2], Xincun Zhuang[1,*], Zhen Zhao[1,3]

1 Institute of Forming Technology and Equipment, Shanghai Jiao Tong University, Shanghai, China
2 State Key Laboratory of Mechanical System and Vibration, Shanghai Jiao Tong University, Shanghai, China
3 National Engineering Research Center of Die & Mold CAD, Shanghai Jiao Tong University, Shanghai, China

* E-mail: georgezxc@sjtu.edu.cn

Abstract. In this paper, the effect of damage induced in deep drawing and ironing processes on the subsequent service performance of deep drawn cups was investigated. To obtain the cups with different amount of deformation damage, three kinds of drawing tests including one-stage deep drawing, two-stage deep drawing, and two-stage deep drawing and ironing were conducted using a 0.3 mm interstitial-free steel sheet. With help of the DF2015 ductile fracture model, the damage evolution of the drawn cups was calculated. Moreover, ring specimens cut from the side walls of the drawn cups were tested in a specially designed tension platform. The load F_{max} and displacement x corresponding to the maximum load point, and the difference in displacement dx from the maximum load point to the fracture initiation point of ring tensile specimens were used as indicators to evaluate the effect of accumulated damage on the subsequent service performance of the parts.

1. Introduction

Deep drawing process is a sheet metal forming process widely used in automotive, aerospace and other fields. Based on the global requirement of energy saving and carbon emission reduction, the demand for automotive parts light-weighting technology has become higher. The process of deep drawing followed by ironing can reduce the thickness of parts, presenting an effective way of light-weighting. Material experiences large deformation in deep drawing and ironing process, which results in non-negligible deformation damage to the material[1]. The knowledge of damage evolution and its effect on subsequent service performance of drawn cups is of great important.

Many researchers have investigated the deep drawing process of sheet material by combining theoretical analysis and numerical simulation. Zhang et al. [2] carried out a two-stage deep drawing test to study the effect of different hardening models and the strain path change on the prediction accuracy of deep drawing process. Gorji and Mohr [3] performed experiments on the deep drawing of triangular parts and evaluated tension-shear fracture in the forming process using simulations combining Yld-2000 plasticity with Hosford–Coulomb fracture model. Kim and Hong [4] developed a stress triaxiality failure diagram for Al3003 sheet and used the generalized incremental stress state dependent damage model to predict shear fracture in multi-stage forming of the rectangular cup. Basak and Panda [5] predicted the failure limits of AA5052 sheets by various uncoupled fracture criteria and performed experiments containing two-stage deep drawing to evaluate the forming limits under nonlinear strain paths. Based on the Lemaitre model, Mueller and Herrig [6] investigated the effect of different loading paths on the damage evolution and distribution in the material during classical deep drawing, multistep deep drawing and reverse deep drawing processes.

During the sheet metal forming process damage such as micro-holes and micro-cracks can occur in the interior of parts under a combined effect of material properties, external loads and microstructure. Similar to work-hardening and residual stresses, forming induced damage also affects the mechanical properties of the product [7]. Metal forming is focused on creating products with known mechanical properties [8], so it is necessary to explore the effect of damage during the forming process on product properties. Besserer et al. [9] investigated the damage evolution and fatigue behavior of orbital forming parts and concluded that damage has a detrimental effect on fatigue life. Tekkaya et al. [10] excluded the interference of strain hardening and residual stresses and demonstrated experimentally for the first time the separate effect of damage on the fatigue strength of extrusions and bends. Fayolle et al. [11] accurately predicted the damage modes of self-piercing riveted joints by using the damage obtained through the continuum damage mechanics model as an input to the mechanical strength simulation of the parts.

The aim of this study is to investigate the damage evolution during deep drawing and ironing processes of ultra-thin sheets, as well as the effect of damage induced in the forming processes on the mechanical properties of the cylindrical cups. A 0.3 mm interstitial-free (IF) steel was selected as the studied material. Firstly, different deep drawing forming tests containing one-stage deep drawing (OSDD), two-stage deep drawing (TSDD) and two-stage deep drawing and ironing (TSDDI) were carried out. Finite element simulations were conducted to obtain the damage evolution of the material under the three drawing processes by utilizing the fracture locus determined in previous work [12]. Subsequently, tension tests on ring specimens cut from the deep-drawn cylindrical cups were designed and carried out to evaluate the effect of damage on the subsequent service properties of the material.

2. Experiments

2.1. Deep drawing and ironing test

The material studied in this paper is 0.3mm IF steel, and the mechanical properties of the material in the rolling direction measured by uniaxial tensile experiments are shown in Table 1.

Table 1. Basic mechanical properties of the material

	Young's modulus E (GPa)	Yield stress σ_y (MPa)	Tensile strength σ_{max} (MPa)	Elongation δ (%)
0.3mm IF	202.17	260.09	328.63	37.34

Three kinds of deep drawing tests containing OSDD, TSDD, and TSDDI were carried out to involve different amounts of deformation and damage in the drawn cups. The schematic diagrams and die structures of the three deep drawing processes are shown in Fig. 1. The final diameter of the cups obtained from OSDD and TSDD was the same. Meanwhile, the cups with different thickness reduction rates (TRRs) were achieved in TSDDI using various punches with different diameters and a fixed die geometry. The material of punches and dies for the deep drawing tests was Cr12MoV, and the Rockwell hardness of the material after quenching treatment was about HRC 55~58. The deep-drawing dies were assembled on the SANS CMT5305 testing machine. Before starting the tests, sandpaper and alcohol were used to remove rust and oil and other debris from the surface of the specimen, and zinc stearate lubricant was applied to the tool surfaces and specimens. The tests were carried out on blanks with diameters of 44 mm, 46 mm, and 48 mm at a loading speed of 10 mm/min and more than three parallel tests were conducted for each size of blank. In OSDD and the first stage of TSDD and TSDDI, three nitrogen gas springs were used to provide a blank holder force of approximately 11kN; no blank holder was required in the second stage of TSDD and TSDDI.

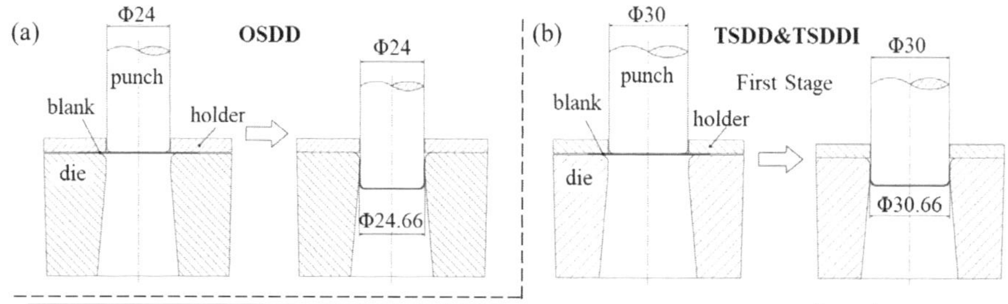

(c)

Unit: mm	First process		Second process			
	OSDD	TSDD& TSDDI	TSDD	TSDDI (with three TRRs) 13.3%	23.3%	33.3%
Punch diameter	24	30	24	27.88	27.94	28
Mold clearance	0.33		0.33	0.26	0.23	0.2
Punch fillet radius	3		2			
Die fillet radius	3.5		1.5			
Die working length	5		3 (with die inclination=45°)			

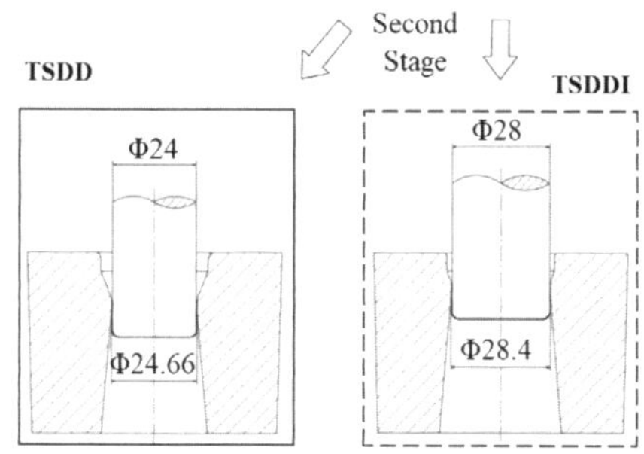

Figure 1. Die structures of deep drawing processes: (a) OSDD, (b) TSDD and TSDDI; (c) summary of tool feature dimensions.

To analyse the deformation and damage evolution during deep drawing process, finite element simulation was performed based on ABAQUS/Explicit software. Considering the geometric symmetry of the models, two-dimensional axisymmetric deep drawing forming models were constructed. The blank was meshed as the axisymmetric linear reduction element CAX4R with a length of 0.06 mm and the tools were set as analytical rigid bodies. The damage was defined using the DF2015 model, in which the equivalent plastic strain, stress triaxiality and maximum shear stress are assumed to control the void nucleation, growth and coalescence. The expression of DF2015 model is presented in Eqs. (1) and (2).

$$\left(\frac{2}{\sqrt{L^2+3}}\right)^{C_1}\left(\left\langle\frac{f(\eta,L,C)}{f(1/3,-1,C)}\right\rangle\right)^{C_2}\bar{\varepsilon}_{\mathrm{f}}=C_3 \qquad \langle x\rangle=\begin{cases}x & \text{when } x\geq 0\\ 0 & \text{when } x<0\end{cases} \tag{1}$$

$$f(\eta,L,C)=\eta+C_4\frac{(3-L)}{3\sqrt{L^2+3}}+C=0 \tag{2}$$

where η, L, and $\bar{\varepsilon}_{\mathrm{f}}$ are stress triaxiality, Lode parameter, and fracture strain, respectively; C_i (i =1-4) and C are material constants needed to be identified. The material constants were calibrated using the surface fitting method between DF2015 model and the fracture limit obtained from Nakajima test. The detailed calibration procedure can be referred to the previous work[12]. The values are C_1=2.4516, C_2=0.5641, C_3=0.7419, C_4=-0.6764 and C=2/3. For non-proportional strain paths, the accumulated damage (D) can be obtained from the integral form of DF2015 model as seen in Eq. (3). The fracture occurs when D = 1.

$$D=\frac{1}{C_3}\int_0^{\bar{\varepsilon}_{\mathrm{f}}}\left(\frac{2}{\sqrt{L^2+3}}\right)^{C_1}\left(\left\langle\frac{f(\eta,L,C)}{f(1/3,-1,C)}\right\rangle\right)^{C_2}\mathrm{d}\bar{\varepsilon}_{\mathrm{p}} \tag{3}$$

2.2. Ring specimen tension test

To evaluate the damage of the cups produced in different process routines, tension testing of ring specimen was designed and carried out. As shown in Fig. 2(a), the ring-shaped specimen was cut from the middle section of the side wall in the deep drawn cup since the deformation in this region is more uniform. Fig. 2(b) demonstrates the ring tension test platform, where the D-shaped blocks were placed inside the specimen and clamped with the upper and lower holdings to ensure that the scalar segment of the ring specimen was free from contact. The loading speed of ring specimen tension test was set as 0.36 mm/min. Through comparing the force and displacement curves, the effect of damage on the mechanical properties of the cups was analysed.

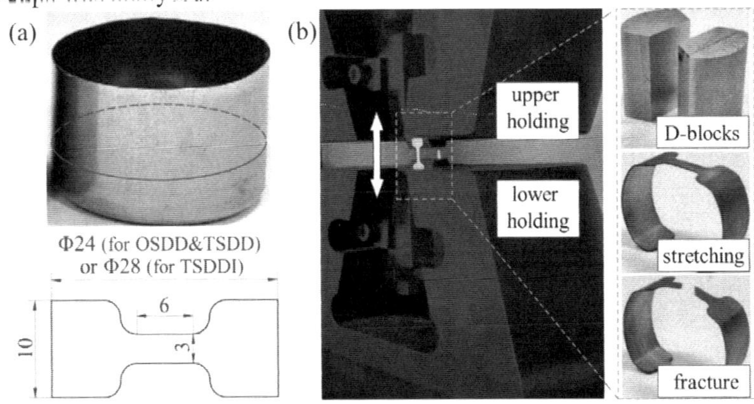

Figure 2. Design of ring specimen tension test: (a) specimen geometry; (b) test device.

3. Results and Discussion

3.1. Formability in deep drawing and ironing process

3.1.1. OSDD and TSDD

In the OSDD and TSDD tests, the blanks with diameters of 44 mm, 46 mm and 48 mm were utilized, and the diameters of the final cups were the same. Fig. 3(a, b) and Fig. 3(c-e) show the experimental and numerical load-displacement curves under OSDD and TSDD, respectively. Fig. 3(f) demonstrates the evolution of the maximum damage value against equivalent strain in different tests. It should be noted that for the 48 mm diameter blank, fracture occurred during OSDD because the drawing ratio is out of the material's limit, while under TSDD the blank was successfully formed through redistributing the deformation. By comparing the load under OSDD and TSDD, it can be found that the maximum load of the sheet under two-stage deep-drawing forming was significantly reduced, which is conducive to reducing the tensile stress transmitted by the side wall of the cup and improving the safety margin of sheet metal forming. Meanwhile, the deformation of the cup side wall during TSDD slightly increased compared with OSDD, while the material damage was reduced.

Fig. 3(g, h) present the damage and equivalent plastic strain distribution of a 44 mm diameter blank under OSDD and TSDD, respectively. The maximum damage value of the specimen in OSDD was located around the round corner of the punch, and the dangerous cross-section in TSDD was shifted from the round corner to the side wall. In TSDD, the material in the cup side wall experienced work-hardening, which enhanced the loading capacity and reduced the risk of fracture. Comparing with OSDD, the deformation in TSDD was decomposed into two steps, and thereby the damage accumulation speed was alleviated. It can also be seen that as the blank diameter decreased, the maximum drawing force decreased, leading to the reduction of the damage values.

3.1.2. TSDDI

In TSDDI tests, the 48 mm diameter blanks were employed, and three TRRs of 13.3%, 23.3% and 33.3% were involved in the ironing stage. Fig. 4(a-c) show the load-displacement curves in the ironing stage with different TRRs. Fig. 4(d) demonstrates the maximum damage value evolution against equivalent strain under different TRRs and Fig. 4(e) presents the damage and equivalent plastic strain distributions of a 33.3% TRR blank. According to the equivalent plastic strain distribution and damage evolution of the material, it can be seen that the maximum damage value of the TSDDI cup still located at the rounded corner of the punch in the first stage deep drawing.

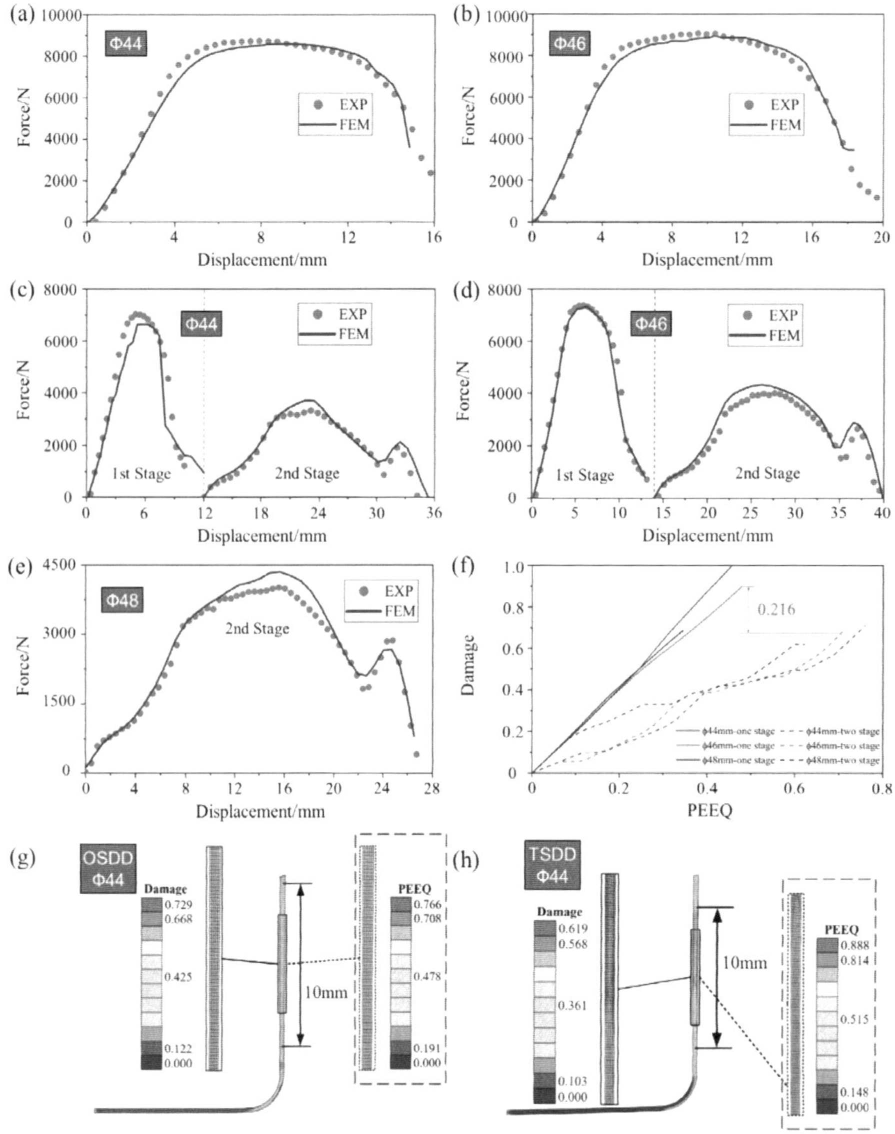

Figure 3. Load-displacement curves of OSDD tests: (a) blank with a diameter of 44 mm and (b) blank with a diameter of 46 mm; load-displacement curves of TSDD tests: (c) blank with a diameter of 44 mm, (d) blank with a diameter of 46 mm, (e) blank with a diameter of 48 mm; (f) damage evolution in different tests; damage and equivalent plastic strain distributions of 44mm diameter sheets: (g) after OSDD test and (h) after TSDD test.

Because in the ironing forming process the die clearance is smaller than the blank thickness, the thickness direction of the blank is subjected to compressive stress. Hydrostatic compressive stress can reduce the additional tensile stress caused by the inhomogeneous deformation, thus making the deformation more homogeneous. As the TRR increased, the maximum load as well as the plastic deformation increased, leading to the increase in the damage value. The tensile stress transmitted to the cup side wall was easier to exceed the loading capacity under larger TRRs.

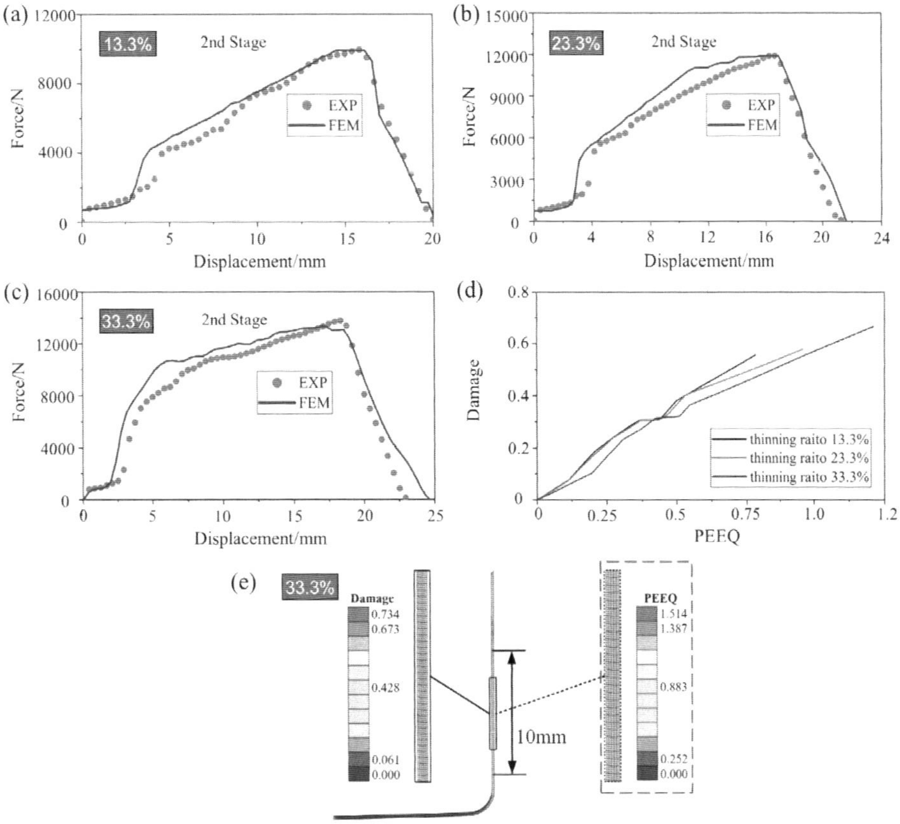

Figure 4. Load-displacement curves of TSDDI tests: (a) blank with 13.3% TRR, (b) blank with 23.3% TRR and (c) blank with 33.3% TRR; (d) damage evolution of blanks with different TRRs; (e) damage and equivalent plastic strain distributions of blank with 33.3% TRR.

3.2. Results of damage evaluation tests

Fig. 5(a, b) respectively show the load-displacement curves of the ring specimens obtained from the cups under OSDD, TSDD, and TSDDI tests. It can be seen that the load displacement curves contain an initial slow rise stage, a rapid rise stage and a drop stage. The initial slow rise stage shows a brief near-linear growth, corresponding to the straightened process of the deformation region on the ring specimen. The rapid rise stage is characterized by a rapid growth of the load until reaching the maximum load point, indicating that the tensile specimen enters a stable deformation stage; the drop stage represents that the tensile specimen starts diffused necking and localized necking, and the load begins to decrease until the specimen fractures. In this study, the load F_{max} and displacement x corresponding to the maximum load point of the ring tension test, as well as the displacement difference dx between the maximum load point and the fracture onset point are considered as indicators for measuring the loading capacity and deformation capacity of the plates, in which F_{max} and x are considered to be tensile strength related and dx is considered to be ductility related. The average values of damage and deformation at the middle position on the cup side wall and the measures of subsequent loading capacity F_{max}, x and dx of the cups corresponding to the different tests are listed in Table 2.

For OSDD and TSDD, the drawn cups had the same diameter, and the damage and deformation strain could be considered as the main influence factors for the difference in the mechanical properties of the ring tension specimens. For the same forming process, i.e., either OSDD or TSDD tests, a blank with a smaller initial diameter has a lower damage value and lower deformation, resulting in a higher F_{max}, x and dx, and a better tensile strength, ductility and loading capacity. In comparing the different forming methods, it is found that before reaching the peak force, due to the comparable deformation degree of ring specimens of the OSDD and TSDD cups, the load-displacement curves have similar evolution trends. However, for blanks of the same diameter after the two kinds of deep drawing forming processes, the three measurement indicators are higher in TSDD. This result is attributed to the fact that the proper distribution of the drawing coefficients in TSDD improves the forming capacity of the material, leading to a lower damage value after deep drawing and an increase in the subsequent deformation capacity of the part. Meanwhile, for both OSDD and TSDD, it is observed that the lower the damage value, the higher the measurement index F_{max} and x, i.e., the better its subsequent tensile strength.

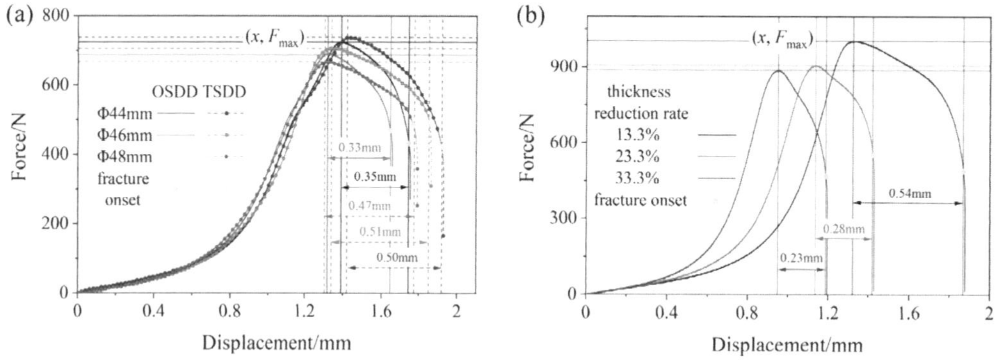

Figure 5. Load-displacement curves of ring-shaped specimens cut from cups in (a) OSDD and TSDD tests, and (b) TSDDI tests.

Table 2. Summary of damage values, deformation at the middle position on the side wall and F_{max}, x and dx under ring tension test of each deep drawn cup

Forming process		Specimen size/mm	Thickness /mm	Damage	Equivalent strain	Stress /MPa	Under ring tension test		
							F_{max}/N	x/mm	dx/mm
OSDD		44	0.292	0.492	0.459	497.31	724	1.39	0.35
		46	0.286	0.541	0.521	515.25	688	1.32	0.32
TSDD		44	0.298	0.457	0.489	506.29	738	1.42	0.50
		46	0.299	0.518	0.562	526.33	706	1.34	0.51
		48	0.299	0.569	0.610	538.81	667	1.30	0.47
TSDDI (With three TRRs)	13.3%	48	0.261	0.440	0.665	552.55	1004	1.32	0.54
	23.3%		0.231	0.442	0.835	592.15	908	1.13	0.28
	33.3%		0.201	0.458	0.972	621.71	889	0.95	0.23

For the ring specimens cut from the cups under TSDDI tests, in combination with the data in Table 2, it can be found that with the increase of the TRR, the deformation and damage value of the cup both increased. The more significant work hardening caused the ring specimens requiring greater loading forces at the same displacement. Meanwhile, the subsequent loading capacity of the cup is weaker, leading the decrease of F_{max}, x and dx. When combining the results under the OSDD, TSDD and TSDDI three different deep drawing forming methods, it can be observed that overall, the lower the damage value, the higher the F_{max}, i.e., the better the subsequent tensile capacity of the part. However, the change in dx is not regular, and it may be necessary to consider the effect of factors such as the deformation degree on the subsequent ductility.

4. Summary and Outlook

In this paper, three types of deep drawing tests including OSDD, TSDD, and TSDDI were conducted and the damage distribution of the cups were determined based on simulations. Through comparing the performance of ring specimens cut from different drawn cups in the tension test. It is concluded that for the blank with same diameter the deformation of the cup side wall under TSDD was slightly larger than that under OSDD, while the damage value under TSDD was less. Under the same forming process, the equivalent plastic strain and damage value of the sidewall of the cup both increased with the increase in blank diameter. For TSDDI, as the TRR increased, the maximum damage value of the cup increased; the increased hydrostatic compressive stress resulted in a more homogeneous deformation of the sidewall. The ring tensile tests on deep-drawing formed parts revealed that, in general, under the same deep drawing method, the lower the damage value, the higher the subsequent tensile strength and ductile properties of the part. After comparing the result of the different drawing methods, the lower the damage value, the higher its maximum load, while the displacement difference between the maximum load point and the fracture onset point is not necessarily higher, which may be related to factors such as the degree of deformation and so on.

For the multi-stage forming process where deformation is under a nonlinear strain path, the material may need to be characterized using a more suitable constitutive model. Therefore, follow-up work could investigate the changes in the mechanical behavior of materials under complex loading paths and the effect on damage accumulation to improve the accuracy of the model in predicting the fracture behavior.

5. Acknowledge

This research was supported by Shanghai Outstanding Academic Leaders Plan (21XD1422000).

References

[1] Nezamaev A V and Karzhavin V V 2020 Deep drawing failure modes *IOP Conf. Ser. Mater. Sci. Eng.* **966** 012126

[2] Zhang W, Yang H, Zhuang X, Wu H and Zhao Z 2022 Comparative study of various hardening models for the prediction of plastic responses under strain path change conditions *Int. J. Mater. Form.* **15** 38

[3] Gorji M B and Mohr D 2018 Predicting shear fracture of aluminum 6016-T4 during deep drawing: Combining Yld-2000 plasticity with Hosford–Coulomb fracture model *Int. J. Mech. Sci.* **137** pp 105–20

[4] Kim M and Hong S 2019 Prediction of shear fracture in multi-stage deep drawing process including ironing process for Al3003 sheet *Int. J. Adv. Manuf. Technol.* **101** pp 747–55

[5] Basak S and Panda S K 2023 Use of uncoupled ductile damage models for fracture forming limit prediction during two-stage forming of aluminum sheet material *J. Manuf. Process.* **97** pp 185–99

[6] Müller M, Herrig T and Bergs T 2023 Numerical analysis of the load paths and the resulting damage evolution during deep drawing of dual-phase steel *IOP Conf. Ser. Mater. Sci. Eng.* **1284** 012006

[7] Tekkaya A E, Bouchard P O, Bruschi S and Tasan C C 2020 Damage in metal forming *CIRP Ann.* **69** pp 600–23

[8] Tekkaya A E, Allwood J M, Bariani P F, Bruschi S, Cao J, Gramlich S, Groche P, Hirt G, Ishikawa T, Löbbe C, Lueg-Althoffa J, Merklein M, Misiolek W Z, Pietrzyk M, Shivpuri R and Yanagimoto J 2015 Metal forming beyond shaping: Predicting and setting product properties *CIRP Ann. - Manuf. Technol.* **64** pp 629–53

[9] Besserer H B, Hildenbrand P, Gerstein G, Rodman D, Nürnberger F, Merklein M and Maier H J 2016 Ductile damage and fatigue behavior of semi-finished tailored blanks for sheet-bulk metal forming processes *J. Mater. Eng. Perform.* **25** pp 1136–42

[10] Tekkaya A E, Ben Khalifa N, Hering O, Meya R, Myslicki S, and Walther F 2017 Forming-induced damage and its effects on product properties *CIRP Ann. - Manuf. Technol.* **66** pp 281–4

[11] Fayolle S, Bouchard P O, and Mocellin K *Modelling the strength of self-piercing riveted joints*, (Chergui 2004). Woodhead Publishing Limited, 2014.
[12] Zhang J, Gu N, Yang H and Zhuang X 2022 Study on ductile fracture and forming limit of battery case steel *J. Plast. Eng.* **29** pp 173–82.

IOP Conf. Series: Materials Science and Engineering **1307** (2024) 012009 doi:10.1088/1757-899X/1307/1/012009

Influence of local heat treatment of rivets on the joint formation of a versatile joining process

F. Kappe[1*], M. Bobbert[1], G. Meschut[1]

[1] Laboratory for material and joining technology, Paderborn University, Paderborn, Germany

*fabian.kappe@lwf.upb.de

Abstract. An efficient lightweight construction method is the combination of different materials in order to adapt the structure to the applied load. To join these multi-material structures mechanical joining technologies are applied. However, the rigid tooling systems cannot be adjusted to changing boundary conditions which is why new, versatile joining technologies are required. In the versatile self-piercing riveting (V-SPR) process presented in [1] different material combination are joined by using a multi-range capable rivet. The rivet head is formed onto the respective thickness of the joint by an outer punch. In order to punch thru the upper sheet a great rivet hardness is required whereas a lower hardness is required for the subsequent forming of the rivet head. To achieve a combination of these requirements, this study investigates a local heat treatment of the rivet. The aim is to determine the feasibility of such a heat treatment as well as to investigate the influence on the joint formation.

1. Introduction

Due to the ecological and social developments of the recent years as well as the climate change, greater focus is being placed on resource-saving technologies to reduce CO_2 emissions significantly. Also, increasing efficiency is an objective in many areas [2]. A far-reaching change is also occurring in the mobility sector. As this sector is a major contributor to carbon dioxide, various strategies, such as the use of lightweight constructions are being pursued to reduce these emissions. Multi-material design, in which materials with different properties are combined, also belongs to this type of construction technology. However, due to their properties, the different materials can no longer be joined using conventional thermal joining methods, For this reason, mechanical joining processes such as self-piercing riveting are increasingly being used.

1.1. Self-piercing riveting

For joining two or more sheet with auxiliary joining element, self-piercing riveting (SPR) is of particular interest. The process sequence can by divided into four stages (see Figure 1) and begins by fixing the sheets on the die. Subsequently, the rivet penetrates the punch-sided sheet. The resulting slug is stored in the rivet shank. The further feed leads to a plastic deformation of the rivet, which expands radially into the die-sided sheet. Hereby, form-fit and force-fit joint is created [3]. Several works have focused on the extension of the process limits of the SPR process [4]. These focus either on the rivet geometry [5] or the process kinematics [6]. Also the combination of rivet geometry and kinematics were investigated as shown in [7]. However, the strategies for increasing the process limits described above often lead to increased process complexity and a great degree of specialisation of the process due to the rigid tools used.

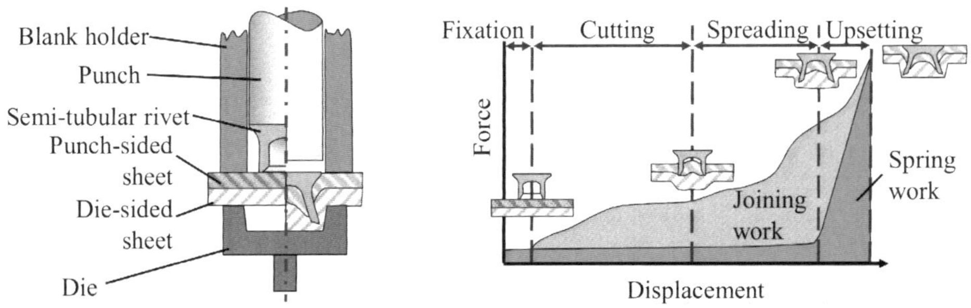

Figure 1. Tooling for the self-piercing riveting process (left) and process sequence for self-piercing riveting and exemplary joining force-displacement curve (right) [3]

1.2. Versatile self-piercing riveting

With regard to versatile process chains joining processes have to be able to react to changing boundary conditions. The V-SPR process presented in [1] allows these changes by combining an extended tool actuator technology with a multi-range capable rivet. To form the joint, a two-piece punch is used. The process itself relies on the process principle of the conventional SPR process (see Figure 2). Using the inner punch the rivet is pushed into the parts to be joined whereby an form and force-fit joint is created. The rivet head is formed after setting using the outer punch. Subsequent forming allows flexible reaction to changing sheet thicknesses.

The multi-range capable rivets are currently produced by machining. As with the production of conventional SPR rivets, they are subsequently heat treated and tempered to a hardness of 480 HV. The friction and corrosion resistance between the rivet and the joining parts is improved by using an ALMAC coating.

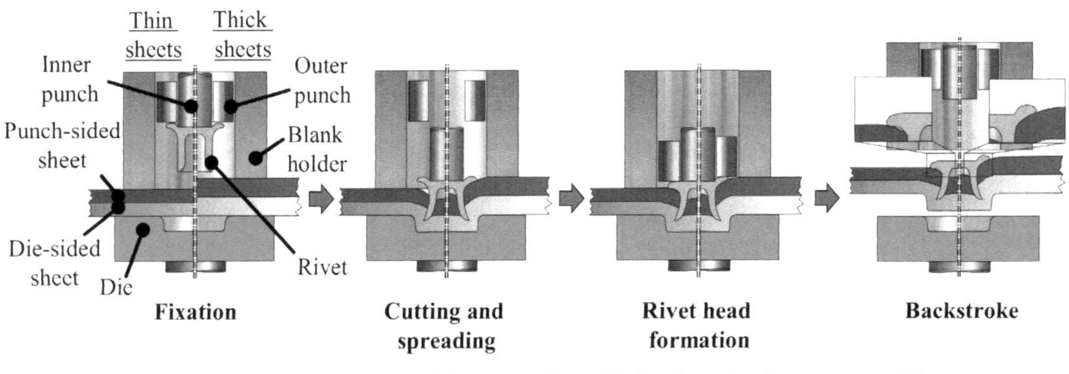

Figure 2. Process sequence of the versatile self-piercing riveting process [1]

Previous studies on the V-SPR process have mainly focused on the process development and joint formation analysis. Especially for aluminium joints, large material thickness ranges could be joined with a single rivet-matrix combination [1]. In addition, the preferred joining direction thin to thick was identified with the help of a numerical analysis [8]. The joint formation when using multi-material joints could also be proven [9]. However, it was shown that the use of steel as the die-sided joining part requires a large deformation in the head area, as the rivet cannot penetrate deep into the joining parts. This increases the risk of cracking in the setting head during the formation of the rivet head. For this reason, the use of a local heat treatment instead of the current heating and tempering process is to be investigated in this study. A ductile head could also improve the forming behaviour of the rivet head to the punch-sided joining part.

IOP Conf. Series: Materials Science and Engineering **1307** (2024) 012009 doi:10.1088/1757-899X/1307/1/012009

2. Experimental details

For the combination of a high shaft hardness with a ductile and formable rivet head, the setting of graded rivet properties is required. A possible method to achieve this is the local heat treatment of the rivet by induction. In this study, the feasibility of such a local heat treatment is investigated. The following section discusses the applied material and the general experimental setup in more detail.

2.1. Test materials

The multi-range-capable rivet was manufactured using C35 B+Cr steel. Due to its good ductility and strength properties even in the hardened state this material is well-suited for use as rivet material. The base material is a cold-drawn wire with a thickness of 10 mm. The rivet geometry is produced from this material by means of machining processes. Joining tests are carried out to prove the feasibility of local heat treatment of the rivet. For this purpose, joining parts of the aluminium material EN AW-6014 in temper T4 are used. This aluminium alloy is often used in crash-relevant components in the automotive industry due to their good formability and high energy absorption. The chemical composition of the rivet as well as the sheet material is shown in Table 1.

Table 1. Chemical compositions and mechanical properties of investigated material EN AW-6014 temper T4 [10] and C35 B+Cr

Chemical composition EN AW-6014, T4 (weight %)									
Elements	Si	Fe	Cu	Mn	Mg	Cr	Zn	Ti	V
Min.	0.30			0.05	0.40				
Max.	0.60	0.35	0.25	0.20	0.80	0.20	0.10	0.10	0.10
Chemical composition C35 B+Cr (weight %)									
Elements	Si	C	Cu	Mn	P	S	Cr	Mo	B
Min.	0.044	0.359	0.010	0.737	0.010	0.007	0.227	0.005	0.0031

2.2. Experimental set-up

Multi-range capable rivets are used for the investigations of local heat treatment. The geometry of these rivets is shown in Figure 3, a). In addition, Figure 3, b) shows the experimental set-up for the heat treatment of the rivets. For the experiments, the rivets are positioned below the induction coil using a rivet holder. Both copper (Figure 3, c) and chamotte (Figure 3, d) were used as rivet holder material to influence the heat flow into the rivet head. The distance between the rivet and the coil is adjusted by means of a spacer. In order to be able to analyse the influence of the coil position, different series of tests were carried out with varying coil positions. Both a positioning in the area of the rivet foot (top) and in the area of the rivet shank (middle) were selected (see Figure 3 e).

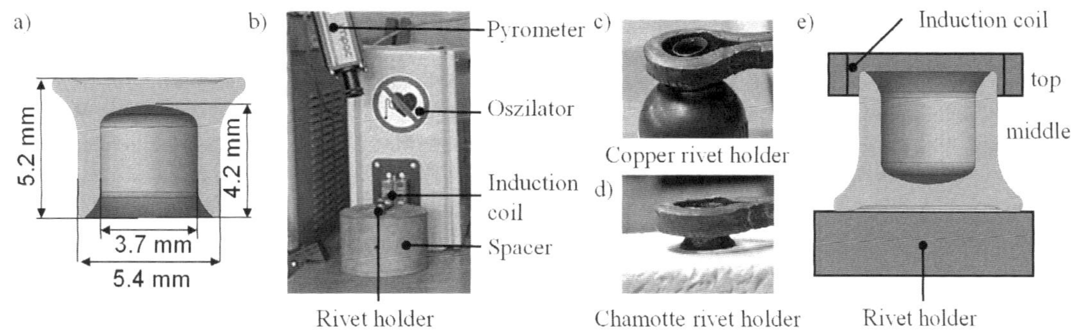

Figure 3. Illustration of the rivet geometry used (a), experimental set-up for local heat treatment by means of induction (b), and rivet holder made of copper (c) and chamotte (d) and position of the coil investigated (e)

During the entire hardening process, the temperature of the rivet element is measured using the pyrometer positioned above the induction coil. A heating time of 0.2 seconds was specified for all tests carried out. The heating time is followed by a holding time at the specified temperature. During the holding phase, the temperature is regulated using the pyrometer. The temperature is varied in the different test series. After the holding time, the riveting element is rapidly quenched to room temperature within a few seconds using a defined air volume flow.

The feasibility of the local heat treated rivets is analysed using a joining system with an extended tool-actuator technology. It was presented in detail in [1]. In order to create the joint, the inner punch sets the rivet. Subsequently the outer punch forms the rivet head onto the respective thickness of the punch-sided joining part. To produce these specimens the joining speed was set to 5 mm/s. The defined displacements for the inner and outer punch were kept constant for all joining tests. In order to define the process parameters specimens with the dimensions 45 mm x 45 mm are used. The joint is set in the geometric centre of the joint.

The specimens are examined with regard to their joint formation, the resulting microstructure and the ensuing hardness after joint formation. Vickers hardness testing was conducted using a hardness testing machine KS 30 S and according to DIN EN ISO 6507-1.

3. Results and Discussion

In order to prove the feasibility of the local heat treatment of the versatile rivet geometry and to be able to determine the influencing parameters on the hardening process, first experimental tests were carried out. The results of the hardening tests are shown in Figure 4 compared to a conventionally hardened rivet which consists of a hardness of 480 HV (Reference). Generally, it is shown that the hardening of the rivet is possible by means of induction heating. The local heat treatment including the following quenching process leads to an increased hardness in the shank area of the rivet. In the rivet foot area a hardness of 600 HV to 700 HV is reached. It can be assumed that in this area, due to the high hardness and the rapid cooling rate as a result of the subsequent quenching process, a completely martensitic structure has been formed. The hardness decreases with increasing distance from the rivet foot towards the rivet head. The transition range between the hardness levels is relatively short. The initial hardness of the base material is present in the rivet head.

The position of the induction coil (middle and top, Figure 3, e) and the heating temperature used (860 °C, 900 °C, 950 °C; measured using the pyrometer) have a decisive influence on the hardened area of the rivet shank. The examined temperatures do not correspond to the austenitization temperature (850 °C) of the rivet material due to the short heating rate (0.2 seconds). If the coil is positioned at the top position, less heat is applied to the rivet and a smaller area is hardened. Increasing the temperature while remaining at the top coil position can increase the hardened area. If the coil is positioned in the middle, the hardening temperature can be lowered to obtain the same hardening result (Figure 4 II and III). The increase in temperature also leads to an enlarged hardened area for a middle coil position. However, the differences in hardened areas are no longer significant when comparing the hardening temperature of 900 °C and 950 °C. This effect can be attributed to the use of copper as a rivet holder material during this test series. The good heat conduction properties of copper ensure rapid heat dissipation from the rivet head, which is placed onto the rivet holder, into the copper material. The heat introduced into the rivet shank, where the coil is positioned is therefore not sufficient to heat the transition area between the rivet shank and the rivet head as well as the rivet head sufficiently. For this reason, the following quenching process after the heat treatment does not result in a microstructural transformation.

IOP Conf. Series: Materials Science and Engineering **1307** (2024) 012009 doi:10.1088/1757-899X/1307/1/012009

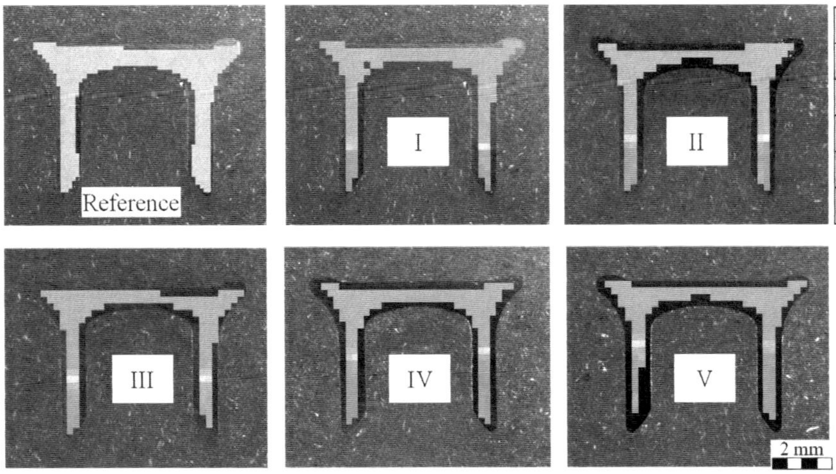

	T in °C	t in s	Position
I	860	1.5	top
II	900	1.5	top
III	860	1.5	middle
IV	900	1.5	middle
V	950	1.7	middle

Hardness in HV0.3

150	370	590
205	425	645
260	480	700
315	535	

Figure 4. Influence of the induction parameters induction coil position and temperature on the resulting rivet hardness when using a copper rivet holder compared to a conventionally hardened rivet

All the hardness parameters examined show that the rivet head does not undergo any change in hardness. The basic ductility of the material is maintained. At the same time, this can lead to deformation of the rivet head as a result of the high joining forces when setting the joint. In order to be able to exclude this, the joint formation was investigated experimentally. The result is shown as an example for a rivet with hardness parameters IV in Figure 5. It can be seen, that the hardness of the rivet shank is sufficient to punch through the punch-sided joining part. However, the shank collapses at the point where the hardened area transitions to the non-hardened area. The strength of the material is not great enough to absorb the joining forces, which increase exponentially during the joint formation. As a result, the inner punch, which introduces the load into the rivet, embosses into the rivet head. The good ductility of the material improves the forming of the rivet head onto the punch-sided joining part using the outer punch.

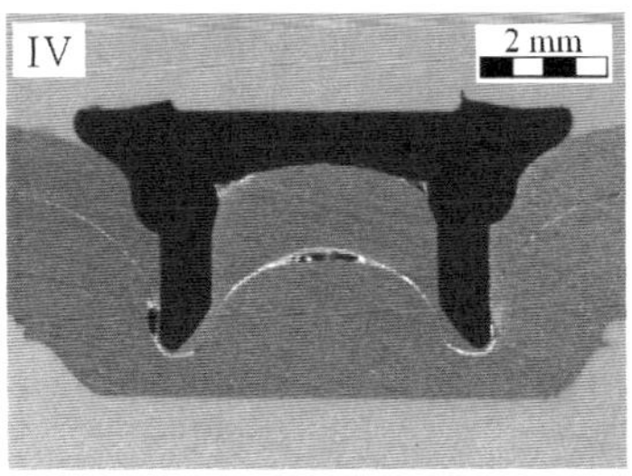

Figure 5. Joint formation using a local heat treated rivet; middle coil position, copper as rivet holder, 900°C, 1.5 s

The aim of the local heat treatment is thus to harden the shank along its entire length to prevent collapsing during the formation of the joint. Since the rivet holder material copper prevents an increased hardening range in the shaft area due to its good thermal conductivity, chamotte was used as the rivet holder material in further tests. The lower thermal conductivity aims to prevent the heat from flowing out of the rivet head in order to achieve continuous hardening of the shank.

In order to identify the effects of the local heat treatment, the joints made with chamotte as rivet holder material were examined metallographically. An exemplary result is shown in Figure 6 for a hardening temperature of 860°C.

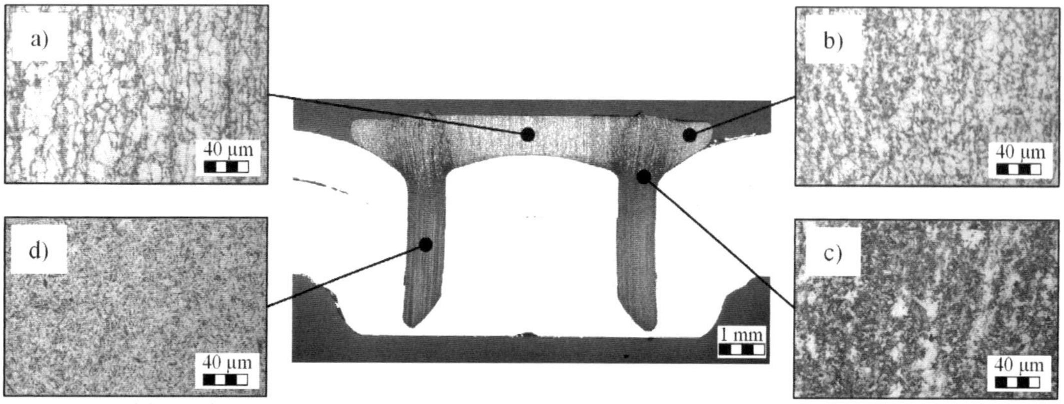

Figure 6. Resulting microstructure using inductive local heat treatment at 860°C for 3.0 seconds and a middle coil position with chamotte as rivet holder

The ferritic basic structure of the material is visible in the central area of the rivet head (Figure 6, a). The ferrite appears as white round grains in the micrograph. The individual grains are delimited from each other by the black grain boundaries. In addition, the grooves resulting from the manufacturing process by means of drawing are visible. In this area, the microstructure was not affected by the heat treatment.

In the rivet head protrusion (Figure 6, b), which is formed onto the joining partner on the punch side, a mixed microstructure of ferrite (white) and pearlite (black) is found. The microstructural transformation here shows that the induced heat has penetrated into the rivet head. Since only pearlite is present here in addition to the ferrite, the hardness has not increased too much. Sufficient ductility, which is required for the forming process, is present.

In the transition area between the rivet head and the rivet shank, seen in Figure 6, c), a mixed structure of ferrite, pearlite and bainite is present. In parts, individual martensite needles may also be present. Due to the mixed structure, a high hardness is already present in this shank area, which prevents the shank from collapsing when the rivet is set.

In the rivet foot (Figure 6, d), a martensitic structure is found exclusively. The individual martensite needles, which spread from one side of the grain to the other, are clearly visible. This is also the area with the greatest heat input since the induction coil is placed here. As a result of the martensite, the highest hardness values are found in the area of the rivet foot and ascending to the rivet head.

In addition to the analysis of the microstructure, joining tests were carried out for the locally heat treated rivets with chamotte as rivet holder material for different temperatures. Again, after the heating (0.2 seconds) and holding time (3.0 seconds) a quenching process is carried out using a defined air volume flow. The results are shown in Figure 7. In addition, the hardness distribution is shown next to the micrograph. Despite the use of chamotte, a heating temperature of 800 °C is not sufficient to form a proper joint. Again, collapsing of the shank occurs. However, if the temperature is increased at the same induction coil position (middle), the shaft collapse is prevented. The prevented heat dissipation also provides a softer transition between the hardness regions in the shank. The resulting maximum hardness does not change.

The evaluation of the hardness reveals that the use of chamotte leads to an increased hardness in the rivet head area. This is caused by the accumulating heat as a result of the rivet holder material used. At a hardening temperature of 900 °C, a hardness of 520 HV to 600 HV is obtained in the head. The high hardness of the entire rivet suggests that a pure martensitic structure has been produced in large areas. The great hardness has in turn a negative effect on the formation of the rivet head. Equal punch displacements result in a small gap between the rivet head and the punch-sided joining part. If a hardening temperature of 860°C is used, the result is a head hardness of 320 HV to 420 HV, which ensures a good forming behaviour of the rivet head. At this hardening temperature, the hardness of the base material is still present in the centre of the rivet head, which suggests that no microstructure transformation has occurred here.

In general, the hardness of the shank prevents the spreading of the rivet in the pure aluminium joint. At both 860°C and 900°C, only a small interlock is achieved. In addition, a low minimum die-side material thickness is formed in both joints.

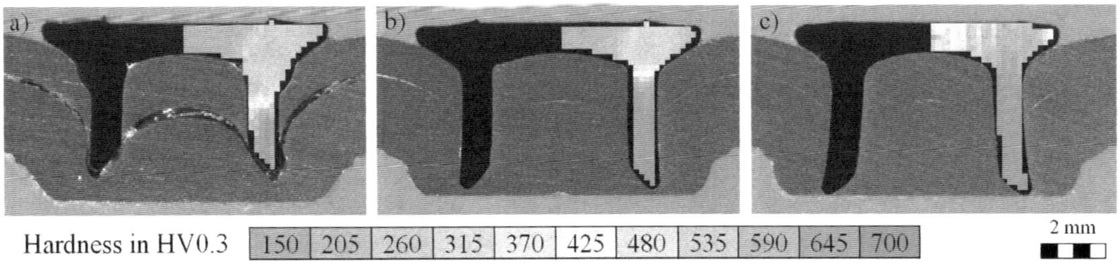

Figure 7. Joint formation and resulting rivet hardness using local heat treated rivets; middle coil position, chamotte as rivet holder, 800°C (a), 860°C (b) and 900°C (c)

In principle, the results showed that local heat treatment of the rivet element can be used to achieve a joint formation. The advantage of this procedure is that little energy is required to harden the rivet element. Complex hardening and tempering processes in large furnaces can therefore be partially avoided, which can have a positive effect on ecological aspects and the overall energy consumption. In addition, the local heat treatment of the rivets can increase the application range of the multi-range capable rivets. The increased design freedom therefore allows the increase in the degree of lightweight construction, as different rivet elements can be avoided and shorter rivets can be used due to the multi-range capability and the subsequent forming of the rivet head.

4. Conclusion

In the study presented the local heat treatment of multi-range capable self-piercing rivets was investigated. In the process sequence, the rivet head is formed to the respective thickness of the joining partners after the joining process. This subsequent forming process requires a certain ductility. However, a corresponding hardness is necessary for the punching of the punch-sided joining partner. The local heat treatment aims to influence the properties of the rivet in such a way that a hard rivet shank is combined with a ductile rivet head. The heat treatment was carried out inductively with the induction coil placed around the rivet shank. After reaching the hardening temperature and the expiry of the holding time, a quenching process using a defined air volume flow was carried out. Basically, it was shown that both the position of the induction coil as well as the hardening temperature used influence the hardness distribution in the rivet.

During heat input, the rivet head was placed on a rivet holder. Copper and chamotte were tested as holder materials. It was found that copper as material prevents the complete hardening of the shank. When using chamotte, the heat can rise higher in the rivet, which prevented the shank from collapsing during joint formation. However, the high hardness due to induction heating inhibits the spreading of the rivet and thus the joint formation in pure aluminium joints. The reduced hardness in the rivet head area could improve the forming of the rivet head. Future work aims to include a tempering process in the process sequence in addition to the inductive hardening process. Furthermore, the use of induction-hardened rivets will be investigated for joining ultra-high strength materials.

Acknowledgment

Funded by the Deutsche Forschungsgemeinschaft (DFG, German Research Foundation) – TRR 285/2 – 418701707. The authors thank the German Research Foundation for their organisational and financial support.

References

[1] Kappe F, Wituschek S, Bobbert M, Meschut G. Determining the properties of multi-range semi-tubular self-piercing riveted joints. Prod Eng 2022;16:363–78. https://doi.org/10.1007/s11740-022-01105-2.
[2] Allwood J, Cullen JM. Sustainable materials: With both eyes open. UIT Cambridge 2012:373.
[3] DVS/EFB-Merkblatt 3410. Self-pierce riveting - Overview. Düsseldorf: Europäische Forschungsgesellschaft für Blechverarbeitung e.V. (EFB); 2020.
[4] Meschut G, Merklein M, Brosius A, Drummer D, Fratini L, Füssel U, et al. Review on mechanical joining by plastic deformation. J Adv Join Process 2022;5:100113. https://doi.org/10.1016/j.jajp.2022.100113.
[5] Uhe B, Kuball C-M, Merklein M, Meschut G. Improvement of a rivet geometry for the self-piercing riveting of high-strength steel and multi-material joints. Prod Eng 2020;14:417–23. https://doi.org/10.1007/s11740-020-00973-w.
[6] Li Y, Wei Z, Li Y, Wang Z, Zhu X. Friction Self-Piercing Riveting (F-SPR) of AA6061-T6 to AZ31B. Vol. 2B Adv. Manuf., American Society of Mechanical Engineers; 2013, p. 1–7. https://doi.org/10.1115/IMECE2013-64212.
[7] Alves LM, Afonso RM, Martins PAF. Double-sided self-pierce riveting. Int J Adv Manuf Technol 2020;108:1541–9. https://doi.org/10.1007/s00170-020-05503-7.
[8] Kappe F, Zirngibl C, Schleich B, Bobbert M, Wartzack S, Meschut G. Determining the influence of different process parameters on the versatile self-piercing riveting process using numerical methods. J Manuf Process 2022;84:1438–48. https://doi.org/10.1016/j.jmapro.2022.11.019.
[9] Kappe F, Wituschek S, Bobbert M, Lechner M, Meschut G. Joining of multi-material structures using a versatile self-piercing riveting process. Prod Eng 2022. https://doi.org/10.1007/s11740-022-01151-w.
[10] Novelis Global Automotive. Datasheet: Novelis Advanz™ 6F - e170; Aluminium Sheet For Exterior Applications, 2019.

Experimental investigation on process parameters of laser-assisted robotic roller bending of a thin-walled structure

Junpeng Qiu[1], Chenyang Xu[2,3], Yi Liu[1], Fei Han[2,3] and Junying Min[1,4]

[1] School of Mechanical Engineering, Tongji University, Shanghai, China
[2] Baosteel Research Institute, Baoshan Iron & Steel Co., Ltd., Shanghai, China
[3] State Key Laboratory of Development and Application Technology of Automotive Steels, Baosteel, Shanghai, China

[4] Junying.min@tongji.edu.cn

Abstract. Laser-assisted robotic roller forming (LRRF) combines the process capabilities of robot-based manufacturing and laser-assisted forming. In this work, the LRRF process was applied to bending DP1180 steel sheets to thin-walled structures designed for seat trails. Comprehensive experimental investigations were conducted to explore the influences of laser power, forming passes and scanning speed on the forming forces, springback and bending radii of final parts. Experimental results show that the effects of process parameters on the springback and bending radii are similar to those on the forming forces, while forming passes make an insignificant difference to the springback. The optimized process window was subsequently determined out of the balance between geometrical accuracy and experimental efficiency. By applying the optimized process parameters (laser power of 750 W, 6 forming passes, scanning speed of 5 mm/s), the peak force during LRRF was reduced to ~2.1kN. Meanwhile, a thin-walled profile with higher precision was achieved. Specifically, the springback angle was reduced to ~4.1° and a compact profile with a radius-to-thickness ratio of ~1.0 was obtained.

1. Introduction

Ultra-high strength steels have been widely applied in the manufacturing of thin-walled structures to achieve better performance in various working environments. Due to their difficult-to-form characteristic at room temperature, ultra-high strength steels are usually processed through the traditional multi-pass roll forming process to achieve complex cross-sectional profiles [1]. However, increasing requirements for compact geometrical profiles with small springback and sharp bending radius usually bring out additional roll configurations; the tendency towards customized geometry design also demands the redesign of forming dies. Both issues lead to additional process costs, thus limiting the industrial application of this forming process. Therefore, the development of novel sheet metal forming methods for the flexible production of complex thin-walled structures is essential.

Robot-based sheet metal forming has been considered an alternative to traditional metal forming methods, which responds promptly to customer-specific requirements and avoids additional costs for die manufacturing [2]. Robot-based incremental forming is one typical process that has been widely applied in the manufacturing of complex thin-walled structures [3]. To achieve better geometrical accuracy of the products, various studies have been focusing on parameter optimization for this forming process. Dai et al. [4] developed a multi-pass incremental forming process and determined the optimal process parameters by orthogonal experimental design. Subsequent tool path compensation for the height of the stepped plane significantly decreased the maximum deviation at the stepped feature by 70%. Störkle et al. [5] improved the geometrical accuracy in double-sided incremental forming of DC04 steel sheets by varying the workpiece orientation and forming sequence. Mohanty et al [6] added two degrees of freedom (DOF) manipulators to the conventional incremental forming process to improve process flexibility. The effects of additional process parameters (part inclination and rotation) on forming force and formability were analytically investigated to improve the formability of aluminum sheets with steep wall angle.

Heat-assisted forming technology has also been widely applied and studied to achieve higher forming accuracy for difficult-to-form materials. Hao et al. [7] adopted the pre-heat treatment and heat-assisted forming strategy in the incremental forming of TA1 titanium sheets, which achieved better maximum forming angle and surface quality. Wei et al. [8] proposed a novel heat-assisted incremental bending process using the induction heating system to fabricate large curvature metal plates. The developed forming process was proved to be effective in improving forming accuracy. Ortiz et al. [9] applied an intelligent process model that compensates for springback deviations to improve the geometrical accuracy in hot single point incremental forming (SPIF) of one typical aerospace Ti-6Al-4V part, achieving an improvement in dimensional accuracy of up to 49%. Specifically, laser-assisted forming has attracted attention due to softening effects in limited areas without mechanical properties transition of entire steel sheets. Mohammadi et al. [10] introduced laser heating to SPIF and achieved a 42% reduction of bulge height by appropriate selection of laser positioning strategy. Gisario et al. [11] developed an external force laser-assisted bending process to bend Titanium Grade-2 sheets. The application of optimal laser parameters achieved sharp bending angles of nearly 140° and small bending radii of ~2 mm.

Based on the process capabilities of laser-assisted forming and robot-based manufacturing, one laser-assisted robotic roller forming (LRRF) process has been proposed and applied to bending ultra-high strength steel sheets. With the introduction of synchronized laser heating and incremental roller forming, the generation of cracks in traditional cold forming can be prevented [12] and compact thin-walled structures with smaller springback and sharper bending radii can be achieved [13]. Prior investigations were mostly focused on the process capability [12] as well as the microstructure evolution during LRRF [16], while the process window for this process was not well understood. Thus, full-factor experiments were performed in this study to investigate the effects of multiple process parameters on the forming forces, springback and bending radii of DP1180 thin-walled structures manufactured by LRRF. The optimized process window was subsequently determined out of the balance between geometrical accuracy and forming efficiency, which facilitated the manufacturing of compact DP1180 thin-walled structures designed for seat trails.

2. Experimental details

The aforementioned LRRF process is applied to bending steel strips in this study, which is schematically illustrated in figure 1. The corresponding experimental setup is presented in figure 2. The as-received DP1180 steel sheets with a thickness of 1.6 mm were cut into 100 mm × 50 mm strips, and then fixed by the clamp onto the fixture with a fillet radius of 1mm. The forming roller with a diameter of 50 mm and a width of 25 mm is installed at the end flange of the industrial robot through the connecting shaft to bend the steel strips by multiple passes. A high-power laser source is installed beside the end flange to heat steel strips synchronously during the forming process.

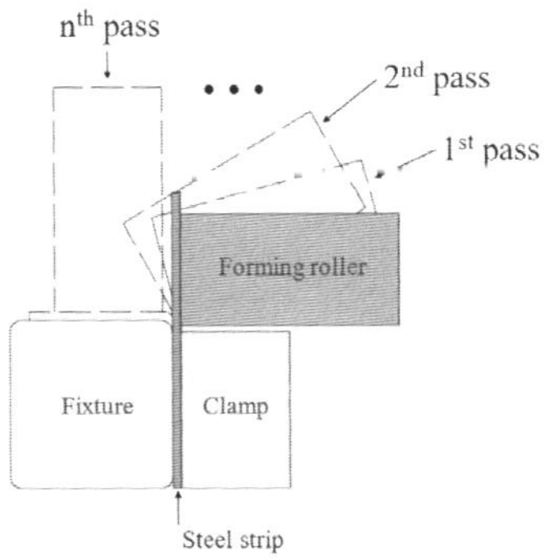

Figure 1. Schematic diagram of LRRF process.

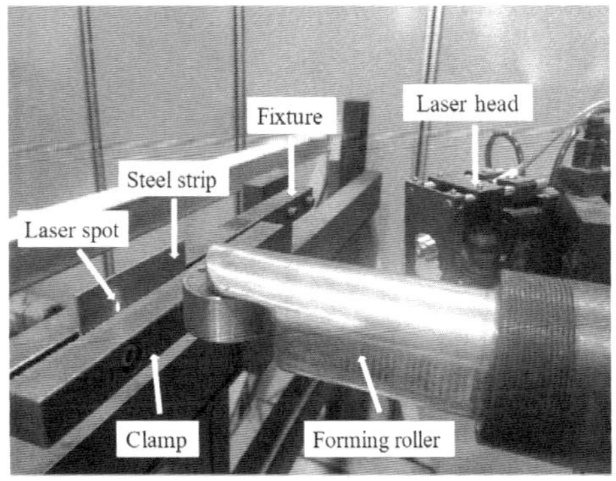

Figure 2. Experimental setup of LRRF process.

Full-factor experimental design was devised to investigate the effects of laser power, forming passes and scanning speed on the forming forces and dimensional accuracy. Due to the significant temperature gradient of laser heating, lower scanning speeds were selected for thicker steel strips compared with the previous study [13] to achieve more pronounced softening effects through the thickness direction of irradiated areas. The highest laser power was also determined by preliminary experiments to avoid melting on the irradiated surface. The detailed process parameters are summarized in table 1, with a total number of 32 tests performed. The forming forces during LRRF were measured by the load sensor installed at the end flange of the industrial robot. The springback angles were measured by a digital dipmeter, and the bending radii were obtained from cross-sectional profiles of final parts through tangent circle fitting.

Table 1. LRRF process parameters.

Laser spot size	Scanning speed	Laser power	Forming passes	Angle increment per pass
4 mm × 2 mm	5 mm/s	500 W	3	30°
		600 W	4	22.5°
	10 mm/s	700 W	6	15°
		800 W	9	10°

3. Results and discussion

3.1. Effects of process parameters on forming forces

For robot-based manufacturing, higher demands for allowable load and robot stiffness of industrial robots usually stand for additional facility costs, which emphasizes the significance of controlling the magnitude of forming forces. The forming forces during 6-pass LRRF with different scanning speeds are presented in figure 3. The peak force of each pass fluctuates with the increase of pass numbers due to the dual influences from the accumulation of work hardening and softening effects by laser heating. With the increase of laser power, forming forces are significantly decreased because of the lower yield strength of steel sheets at elevated temperatures. The scanning speed of 5 mm/s further decreases the forming forces by increased laser power density, and the forming forces come to a stable level with the laser power of over 700 W. The relationship between forming forces and laser process parameters of 3, 4 and 9 passes is similar to that of 6 passes, and those similar figures are not repeated here.

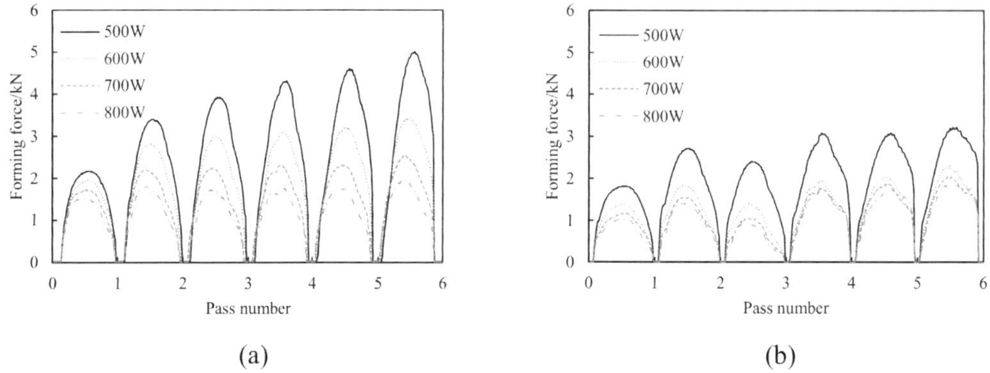

Figure 3. Forming forces during 6-pass LRRF with scanning speeds of (a) 10 mm/s, (b) 5 mm/s.

To clarify the effect of forming passes on forming forces, the peak forces during LRRF at different passes are extracted and presented in figure 4. With the increase of forming passes, the peak force is reduced due to minor plastic deformation in each pass. The peak force can be significantly reduced by 6 passes with the scanning speed of 10 mm/s; for the scanning speed of 5 mm/s, 4-pass forming is sufficient because of more significant softening effects by higher laser power density. Also, a plot of average experimental responses for each experimental factor, i.e., the main effect plot (MEP) of peak force is presented in figure 5 to reveal the significance of multiple process parameters on peak force. According to MEP, all three parameters exert significant changes to the peak force. Besides, the degree of variation decreases with the increase of laser power and forming passes, which points out the possibility of parameter optimization for the balance between resource consumption and facility costs.

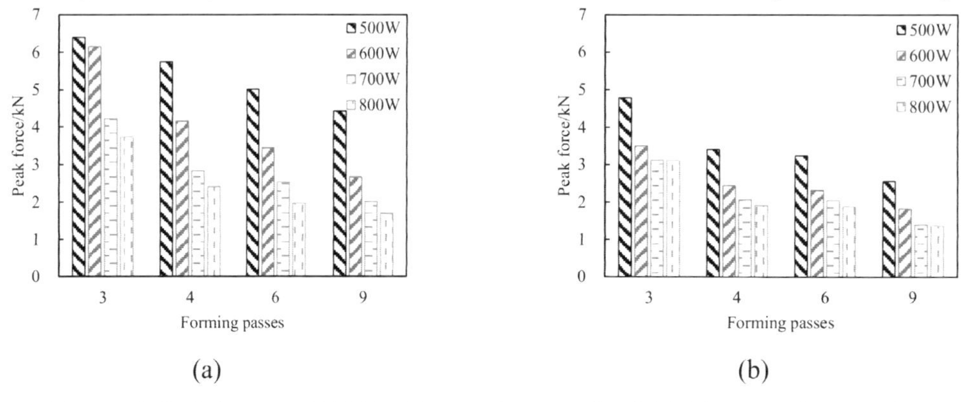

Figure 4. Peak forces in LRRF with scanning speeds of (a) 10 mm/s, (b) 5 mm/s.

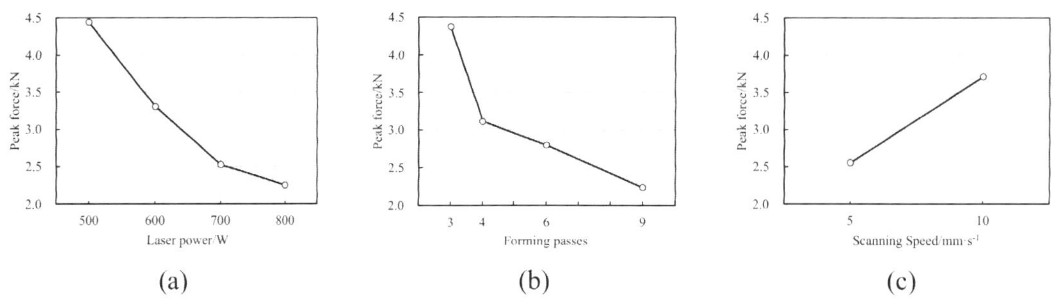

Figure 5. Main effect plot (MEP) of peak force vs. process parameters.

3.2. Effects of process parameters on springback and bending radii

As mentioned before, advanced industrial design calls for the manufacturing of thin-walled structures with small springback and sharp bending radii. Thus, the effects of process parameters on springback and bending radii in LRRF are investigated here to assist in the determination of a proper process window for subsequent application.

The springback angles of final parts with different scanning speeds are presented in figure 6. According to the previous study [13], the springback in LRRF is usually related to stress at loading and Young's modulus of materials at unloading. The springback based on the release of elastic deformation is positively correlated to forming forces, and thus can be significantly reduced by the increase of laser power and decrease of scanning speed; while the increase of forming passes exerts insignificant changes to the springback. Although LRRF with more forming passes can significantly decrease the forming forces, the lower Young's modulus of steel strips at unloading due to the heat accumulation still triggers higher elastic recovery, which alleviates the decrease of springback. Similarly, the MEP of the springback angle is presented in figure 7, which demonstrates obvious influences of laser power and scanning speed, while insignificant influence of forming passes on springback.

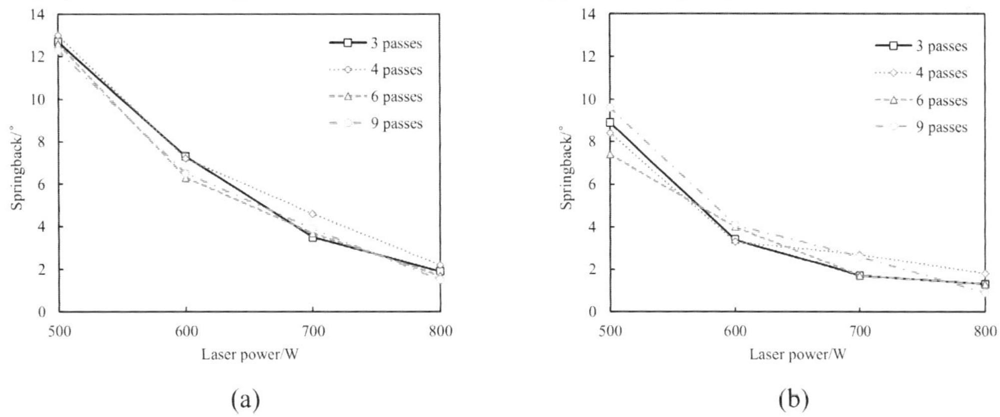

Figure 6. Springback angles of final parts by LRRF with scanning speeds of (a) 10 mm/s, (b) 5 mm/s.

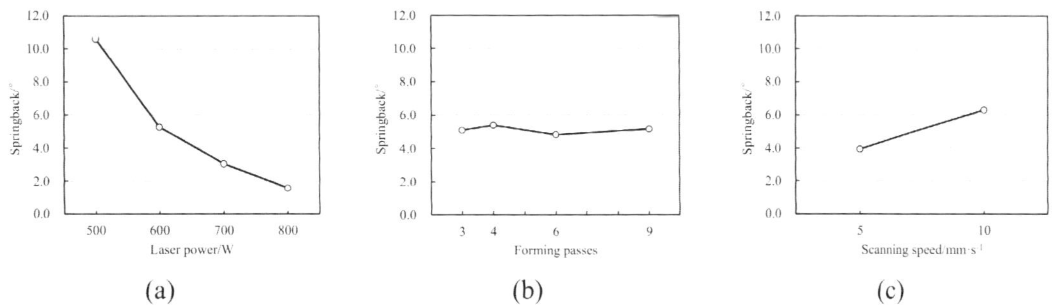

Figure 7. Main effect plot (MEP) of springback angle vs. process parameters.

Radius-to-thickness (*R/t*) ratio was adopted here to evaluate the effects of process parameters on the bending radii. The *R/t* ratios of final parts with different scanning speeds are presented in figure 8. With the increase of laser power and forming passes, the *R/t* ratio is decreased significantly because of lower resistance to deformation of steel strips at elevated temperatures. The decrease in scanning speed from 10 mm/s to 5 mm/s also strengthens the softening effects of materials, which leads to smaller *R/t* ratios. With the low springback angle of ~2°, the smallest bending radius of ~1.6 mm still has a gap from the fillet radius of 1 mm possessed by the fixture, which can be attributed to the prominent impact of robot stiffness deformation on bending radii in LRRF [17]. The MEP presented in figure 9 also reveals the

IOP Conf. Series: Materials Science and Engineering **1307** (2024) 012010 doi:10.1088/1757-899X/1307/1/012010

effects of process parameters on the *R/t* ratio. The laser power obviously influences the *R/t* ratio, followed by the forming passes; the scanning speed exerts the least apparent changes to the bending radii.

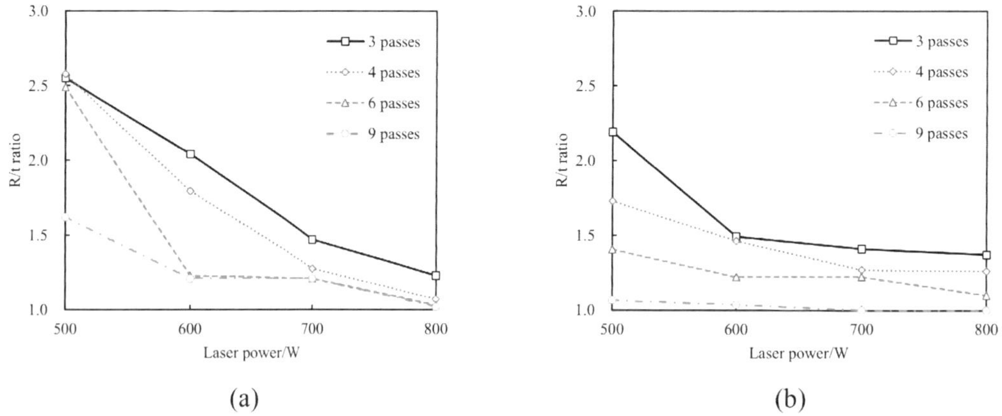

Figure 8. R/t ratios of final parts by LRRF with scanning speeds of (a) 10 mm/s and (b) 5 mm/s.

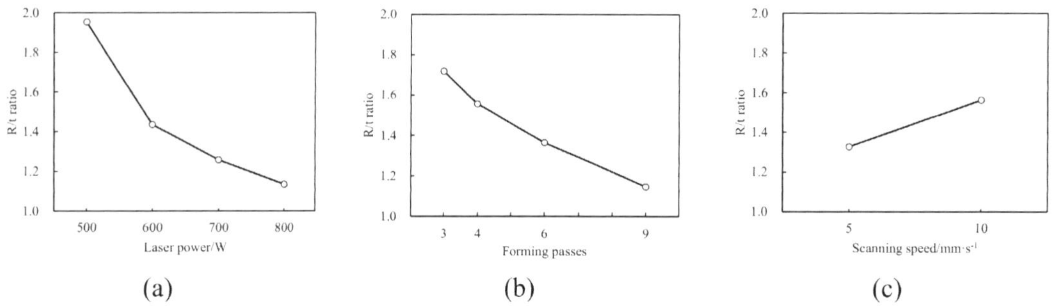

Figure 9. Main effect plot (MEP) of *R/t* ratio vs. process parameters.

3.3. Determination and application of optimized process window

According to the above experimental results and analyses, the optimized process window with laser power of 700~800 W, 6 forming passes and scanning speed of 5 mm/s was determined to achieve better geometrical accuracy with controlled heat input and process time. One thin-walled structure designed for seat trails was subsequently manufactured by optimized process parameters, aiming at the validation of process availability in the manufacturing of complex geometrical profiles, with the detailed procedure illustrated in figure 10. The applied laser power was chosen as 750 W to achieve more distinct softening effects of materials while avoiding melting on the irradiated surface. The final complex thin-walled structure fabricated via LRRF is also presented in Figure 11.

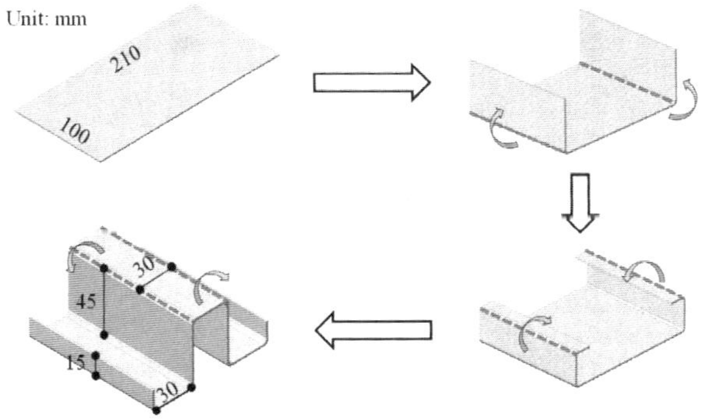

Figure 10. Fabrication procedure for a thin-walled structure designed for seat trails.

Figure 11. Thin-walled structure fabricated via LRRF process.

Table 2 presents the peak forces, springback angles and *R/t* ratios of final components obtained via robotic roller forming (RRF) and LRRF. Compared with RRF at room temperature, the optimized process parameters reduced the peak force to ~2.1kN, which facilitates the reduction of facility costs in LRRF. The manufacturing of a thin-walled structure with higher dimensional accuracy was also validated: the springback angle was reduced from ~11.7° to ~4.1°, and a compact profile with an *R/t* ratio of ~1.0 was achieved.

Table 2. Peak forces and geometrical accuracy of thin-walled structures formed by RRF and LRRF.

	Peak force (kN)	Springback angle (°)	*R/t* ratio
RRF	5.20	11.73	2.91
LRRF	1.97	4.07	1.02

4. Conclusions

In this study, LRRF experiments with multiple process parameters including laser power, forming passes and scanning speed were primarily performed on DP1180 steel sheets to obtain an optimized process window, which balances the dimensional accuracy and forming efficiency. One complex thin-walled structure designed for seat trails was subsequently manufactured by optimized process parameters to validate the availability of this forming process. The main conclusions are summarized below:

(1) Laser power and scanning speed are closely correlated to the softening effects of materials during the LRRF process, and forming passes affect the magnitude of plastic deformation in each pass. All three parameters exert obvious influences on forming forces.

(2) Springback angles and bending radii of final parts in LRRF are distinctly affected by laser power and scanning speed. The increase of forming passes can remarkably decrease the bending radii while exerting insignificant influence on springback.

(3) Optimized process parameters with laser power of 750 W, 6 forming passes and scanning speed of 5 mm/s were determined to fabricate DP1180 thin-walled structures, which significantly reduced the peak force to ~2.1kN compared with RRF. A cross-sectional profile with higher precision was also achieved, with the springback angle reduced from ~11.7° (RRF) to ~4.1° (LRRF) and radius-to-thickness ratio reduced from ~2.9 (RRF) to ~1.0 (LRRF).

Acknowledgments

J. Qiu, Y. Liu, and J. Min would like to acknowledge the financial support from the Science and Technology Commission of Shanghai Municipality (Grant number: 21170711200).

References

[1] Wang, J., Liu, H. M., Li, S. F., & Chen, W. J. (2022). Cold Roll Forming Process Design for Complex Stainless-Steel Section Based on COPRA and Orthogonal Experiment. *Materials*, *15*(22), 8023.

[2] Sazonnikova, N. A., Ilyuhin, V. N., Surudin, S. V., & Svinaryov, N. N. (2020, September). Increasing of the Industrial Robot Movement Accuracy at the Incremental Forming Process. In *2020 International Conference on Dynamics and Vibroacoustics of Machines (DVM)* (pp. 1-8). IEEE.

[3] Ullah, S., Xu, P., Li, X., Li, Y., Han, K., & Li, D. (2022). A Review on Part Geometric Precision Improvement Strategies in Double-Sided Incremental Forming. *Metals, 12*(1).

[4] Dai, P., Chang, Z., Li, M., & Chen, J. (2019). Reduction of geometric deviation by multi-pass incremental forming combined with tool path compensation for non-axisymmetric aluminum alloy component with stepped feature. *Int. J. Adv. Manuf. Technol.*, *102*, 809-817.

[5] Störkle, D. D., Seim, P., Thyssen, L., & Kuhlenkötter, B. (2016, October). Influence of part orientation on the geometric accuracy in robot-based incremental sheet metal forming. In *AIP Conference Proceedings* (Vol. 1769, No. 1). AIP Publishing.

[6] Mohanty, S., Regalla, S. P., & Daseswara Rao, Y. V. (2019). Robot-assisted incremental sheet metal forming under the different forming condition. *J. Braz. Soc. Mech. Sci. Eng.*, *41*, 1-12.

[7] Yuan, H., Li, Y. L., Liu, Y. Y., Zhao, G. L., & Li, F. Y. (2023). Improving the forming performance of incrementally formed sheet parts with customized heat treatment strategies. *Adv. Manuf.*, 1-16.

[8] Wei, B., Wei, Y., Zhang, F., He, K., Dang, X., & Du, R. (2021). Springback control and plastic deformation of metal plates with large curvature in heat-assisted incremental bending process. *Int. J. Adv. Manuf. Technol.*, *112*, 1483-1500.

[9] Ortiz, M., Penalva, M., Iriondo, E., & López de Lacalle, L. N. (2019). Accuracy and surface quality improvements in the manufacturing of Ti-6Al-4V parts using hot single point incremental forming. *Metals*, *9*(6), 697.

[10] Mohammadi, A., Vanhove, H., Van Bael, A., & Duflou, J. R. (2016). Towards accuracy improvement in single point incremental forming of shallow parts formed under laser assisted conditions. *Int. J. Mater. Form.*, *9*, 339-351.

[11] Gisario, A., Mehrpouya, M., Venettacci, S., & Barletta, M. (2017). Laser-assisted bending of Titanium Grade-2 sheets: Experimental analysis and numerical simulation. *Opt. Lasers Eng.*, *92*, 110-119.

[12] Liu, Y., Min, J., Zhang, J., Cai, W., Carlson, B. E., Bobel, A. C., Hector, L.G., Sachdev, A. K. (2022). Laser-assisted robotic roller forming of an ultrahigh strength martensitic steel. *J. Manuf. Processes, 82*, 192-202.

[13] Min, J., Wang, J., Lian, J., Liu, Y., & Hou, Z. (2023). Laser-Assisted Robotic Roller Forming of Ultrahigh-Strength Steel QP1180 with High Precision. *Materials*, *16*(3), 1026.

[14] Liu, Y., Qiu, J., Wang, J., Bambach, M., & Min, J. (2023). Fabrication of thin-walled hat-shaped beams from ultrahigh strength steel by laser-assisted robotic roller forming. *Smart Manuf.*, 1-8.

[15] Liu, Y., Min, J., & Lin, J. (2021). Robotic Roller Forming Process and Strategies to Eliminate Geometrical Defect of Edge Waves. In *Forming the Future* (pp. 479-491).

[16] Liu, Y., Wang, J., Cai, W., Carlson, B. E., Lian, J., & Min, J. (2023). A thermo-metallurgical-mechanical model for microstructure evolution in laser-assisted robotic roller forming of ultrahigh strength martensitic steel. *J. Mater. Res. Technol.*, *25*, 451-464.

[17] Liu, Y., Qiu, J., Wang, J., Lian, J., Hou, Z., & Min, J. (2024). An iterative path compensation method for double-sided robotic roller forming of compact thin-walled profiles. *Rob. Comput. Integr. Manuf.*, 86, 102689.

IOP Conf. Series: Materials Science and Engineering **1307** (2024) 012011 doi:10.1088/1757-899X/1307/1/012011

Clarification of appropriate conditions for forming complex shapes with multiple steps in two-stage deep drawing process

Shuki Nakamura[1], Takeshi Ichikawa[1], Masato Ishihara[2], Haruyuki Konishi[2]

[1]Application Technology Centre, Technical Development Group, Kobe Steel, LTD.1-5-5 Takatsukadai, Nishi-ku, Kobe, HYOGO 651-2271 Japan
[2]Aluminum Sheets & Coils Development Department, Moka Works, Steel & Aluminum Business, Kobe Steel, Ltd. 15, Kinugaoka, Moka-city, Tochigi 321-4367, Japan

nakamura.shuki@kobelco.com

Abstract. In this study, appropriate conditions for forming complex shapes with multiple steps, such as automotive side door inner in a two-stage deep drawing process are investigated, and the effectiveness of the process is verified in a test. In this forming process, the inner area of the part is formed in the first stage and the outer area is formed in the second stage. And in the second stage, a forming pad is used to reduce strain in the inner area that is formed in the first stage. The results of the forming test showed that it is necessary to control the blank holding force (BHF) and the pad force (PF) in the second stage to avoid blank ruptures in all areas. Therefore, the concept of the BHF and PF for blank rupture risk at each site was studied using the finite element method (FEM). An appropriate range of BHF and PF was estimated from the results of FEM and the validity of the estimated values was confirmed in an actual test. In addition, it was confirmed that a two-stage deep drawing process using the above appropriate values can form a more severe shape than a one-stage deep drawing process.

1. Introduction

In some case, lightweight aluminum alloy side doors are used in automobiles to improve fuel efficiency [1]. However, since the formability of aluminum alloys is inferior to that of conventional materials (mild steel), multiple deep drawing processes are sometimes performed to compensate for the formability when forming automobile side doors [2],[3]. A major example of a multiple deep drawing process is the re-drawing process of beverage cans. This is done by forming a large shape in the first stage and then forming a smaller and smaller shape as the process proceeds [4]. However, the re-drawing process is not suitable for forming an automobile side door inner. This is because the outer area of the part is formed in the first stage, so when the inner area is formed, an overhang forming is performed, and deep shapes cannot be formed. Therefore, an automobile side door inner should be formed from the inner area [2],[3]. And in the second stage, the inner area of the part is pressed down with a forming pad to suppress deformation to prevent increased strain and blank rupture in the inner area. The blank holding force and the pad force in the second stage are very important to avoid blank rupture, but the concept and the appropriate conditions for them are not yet clear.

The aim of this study is to form a complex and deep shape with multiple steps, such as automotive side door inner, with aluminum alloy sheet. For this purpose, a process where deep drawing was performed twice was tested. The test results indicate that the blank holding force and pad force in the second stage must be controlled to obtain the effect of the two-stage deep drawing process. Therefore, the concept of the blank holding force and pad force for blank rupture risk at each site was studied using the finite element method (FEM) and an appropriate range was estimated. The appropriate range and the effectiveness of the two-stage deep drawing process were confirmed in actual tests.

2. Experimental procedure and test results

2.1. Test Procedure

In this study, forming test was performed using a 3-steps square cylinder shape simulating the shape of a front side door inner of a commercially available automobile to confirm the effectiveness of process that deep drawing is performed twice. The die shape of the 3-steps square cylinder is shown in Figure 1. The upper part of the die simulates the inner section of the side door inner, the middle part simulates the outer section, and the lower part simulates flange section and remnants of die. And the height of the upper part (H in Figure 1) can be changed to 20 mm or 25 mm.

The schematic diagram of testing process is shown in Figure 2. The forming tests in one-stage deep drawing process were performed under the condition that flange pressure (blank holding force divided by the flange area of the initial blank) was 1.8 MPa, because preliminary tests using alloys with similar properties (equivalent yield stress levels) shown that setting flange pressure too low would result in wrinkling and surface defects. (Figure 2(a)). In the two-stage deep drawing process, the upper part, in which little material flows in, was formed in the first stage of the process, and the remaining middle part and lower part were formed in the second stage. (Figure 2(b)). The first deep drawing was performed under the condition that Flange pressure = 1.8 MPa. The second deep drawing was performed under the same flange pressure as the first stage, flange pressure = 1.8 MPa. In addition, the part formed in the first stage was held down by a forming pad during the second stage because it would deform again and rupture if formed without a forming pad. And the pressure applied to the blank by the forming pad (called inner pressure, the value obtained by dividing the pad force by the area restrained by the forming pad) was 5.9 MPa. In order to prevent deformation of the blank by the pad during die-release, the die and pad were raised sufficiently faster than the speed at which the holder is raised.

The properties of the 1.0 mm thick aluminum alloy sheet A5182-O used in the test are shown in Table 1. Test specimens are 600 mm x 500 mm rectangle. The rust prevention oil R303P manufactured by Sugimura Chemical Industrial Co., Ltd. was used as the lubricant for the tests.

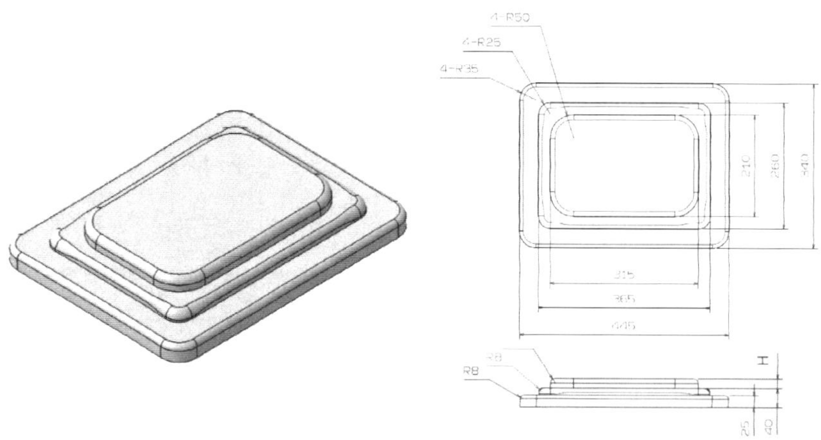

Figure 1. Die shape

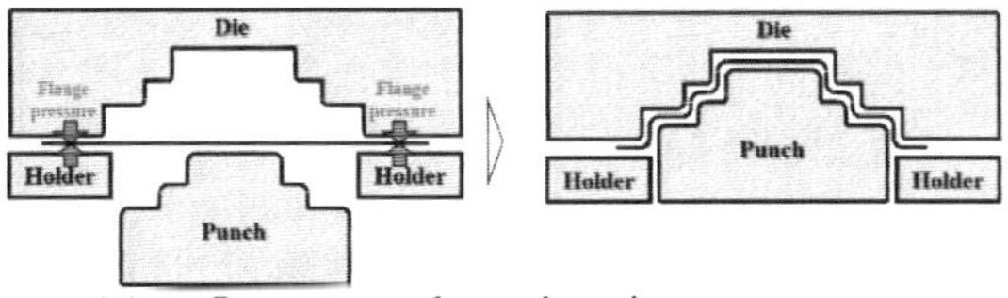

(a) One-stage deep drawing process

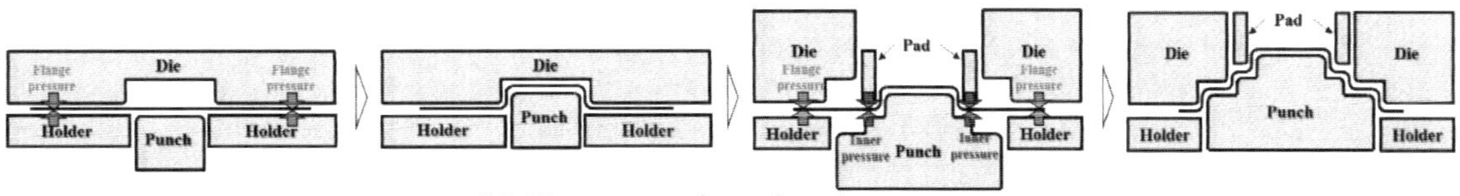

(b) Two-stage deep drawing process

Figure 2. Schematic diagram of the testing process

Table 1. Material properties of A5182-O.

Thickness / mm	Tensile Direction / DEG.	YS / MPa	TS / MPa	EL / %	*r-value (at 15%)*
	0	116	270	31.8	0.74
1.0	45	115	269	31.4	0.68
	90	115	266	30.7	0.60

2.2. Test Results

Test results are shown in Figure 3. When the blanks were formed in a one-stage deep drawing process, the blank cracked in the upper part under the H=25mm condition. On the other hand, when the blanks were formed in a two-stage deep drawing process, no cracking occurred in the upper part of the blank. But necking occurred in the middle part even under the condition H= 20 mm where no cracking occurred in the one-stage deep drawing process. The application of the two-stage deep drawing process was expected to improve formability, but no clear effect was obtained under the aforementioned conditions. In the second stage of the process, the area formed in the first stage were pressed down by the forming pad, which reduced the outflow of material, resulting in insufficient material supply to the middle part of blank, which may have caused necking. Therefore, optimizing the conditions of inner pressure and flange pressure may avoid necking.

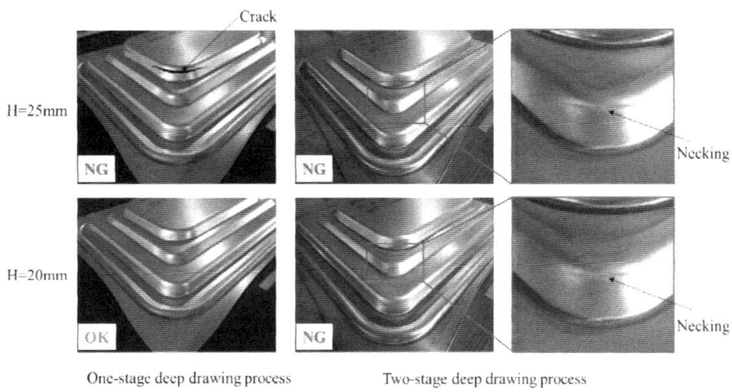

Figure 3. Test results

3. Investigation of the influence of the Flange pressure and Inner pressure by CAE

3.1. CAE Procedure

In order to maximize the effect of a two-stage deep drawing process, the effects of inner pressure and flange pressure in the second stages of the process on the formability were investigated using CAE. A quarter-symmetric model was used in CAE to reduce computation time. In order to reproduce the actual test process in CAE, the upper part was formed in the first stage and remaining middle part and lower part in the second process as in the actual test. Flange pressure in the first stage was set at 1.8MPa. And in the second stage, flange pressure was set at three levels (0.7, 1.0 and 1.8 MPa) and inner pressure was set at six levels (1.6, 2.7, 3.8, 4.9, 5.9 and 7.0 MPa). In order to simulate the material, the Barlat '91 yield function [5], calculated based on the material properties described above (see Table 1), was used. A constant value of 0.14 was used for the friction coefficient, assuming coulomb friction. The calculations were performed using PAM-STAMP.

3.2. CAE Results

The rupture risk values of the punch shoulders at the upper, middle, and lower corners were measured from the blanks after forming, and the results are shown in Figure 4. Rupture risk is one of the functions of the PAM-STAMP and means the distance in the FLD between the point representing the element and the FLC. Overall, the lower the value of flange pressure, the lower the rupture risk. However, the effect of inner pressure on the rupture risk varied depending on each part of the blank. First, inner pressure had little effect on the rupture risk of the lower part of the blank. Next, in the upper part, there was a tendency for the rupture risk to increase rapidly as inner pressure fell below 3.8 MPa. On the other hand, in the middle part, inner pressure, at which the rupture risk was minimized, changed with flange pressure.

The amount of material outflow in the diagonal direction was measured in the plane directly above the punch shoulder in the upper, middle and lower part from the blank after forming. The amount of material inflow at the flange edge of the blank was also measured. The results of these measurements are shown in Figures 5. The amount of material outflow measured near the punch shoulder at the upper, middle, and lower part of the blank increased with a decrease of inner pressure at all locations. Material inflow increased with higher inner pressure and lower flange pressure.

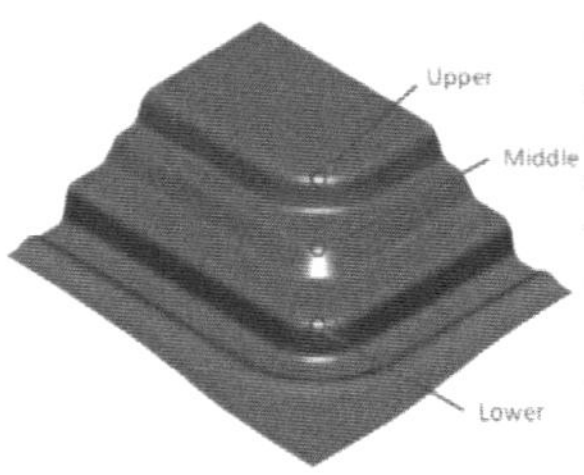

(a)Measurement position

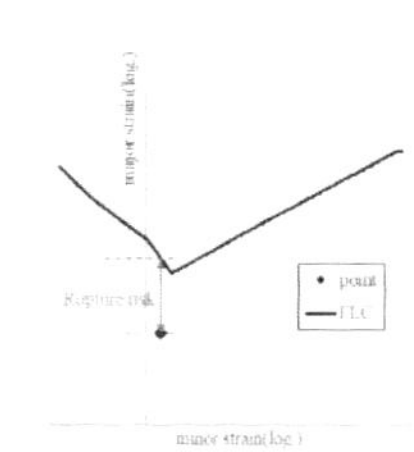

(b)schematic diagram of rupture risk

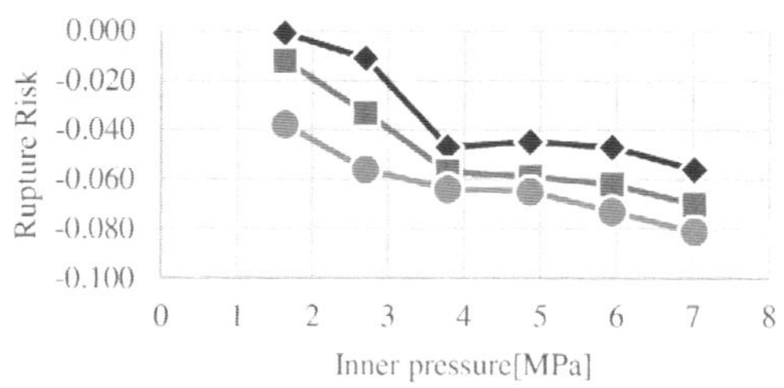

(c)Upper part

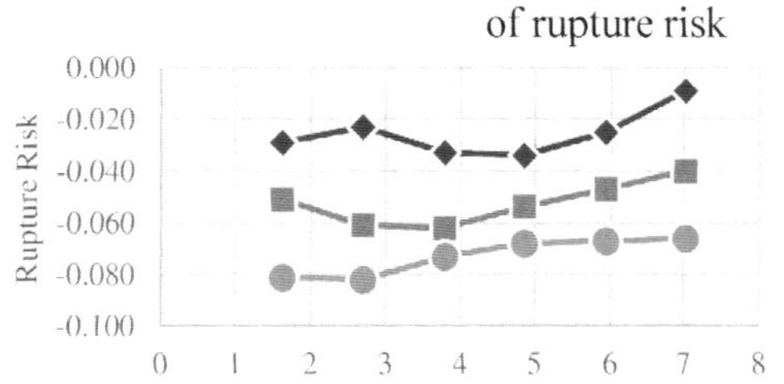

(d)Middle part

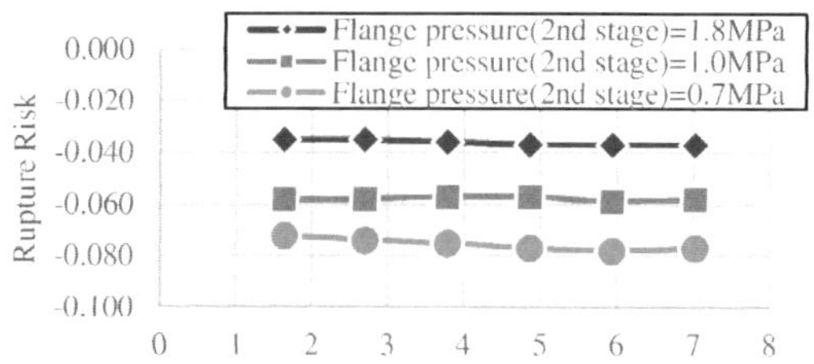

(e)Lower part

Figure 4. Rupture risk values of the punch shoulders at the upper, middle, and lower corners

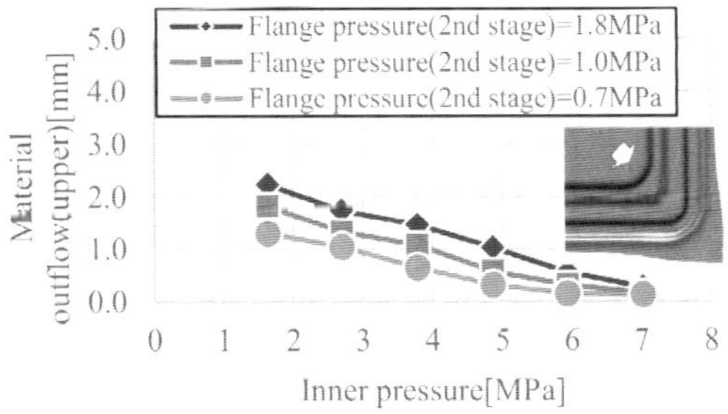

(a) Material outflow at upper part

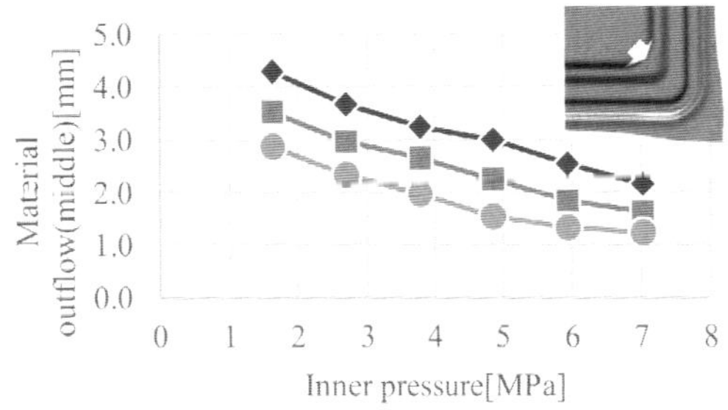

(b) Material outflow at middle part

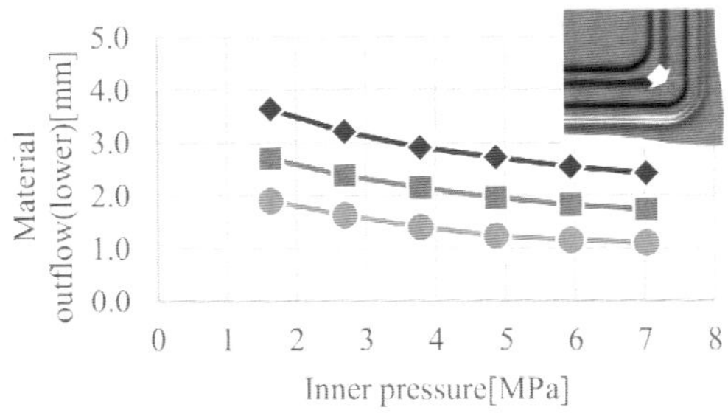

(c) Material outflow at lower part

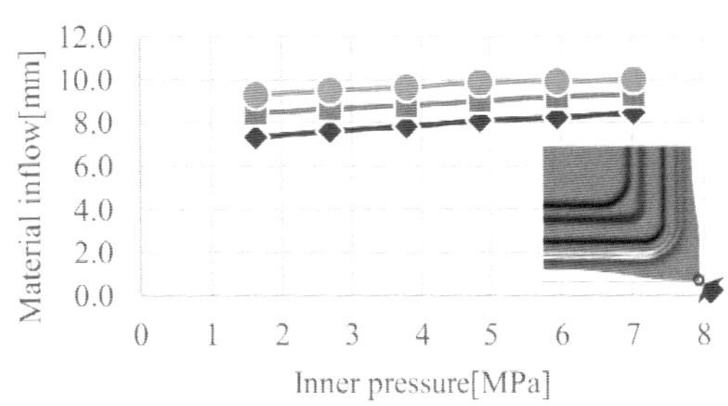

(d) Material inflow at blank edge

Figure 5. Material outflow and inflow

3.3. Concept of the flange pressure and the inner pressure for the blank rupture

A schematic diagram of the corner section cut out in the diagonal direction is shown in Figure 6. In the second stage of the process, the material at the upper part of the blank only moves in outflow direction during forming. In addition, the strain in the upper part increases as the material flows out. From the above figure 5, a larger inner pressure suppresses material outflow at the upper part, while a larger flange pressure promotes material outflow. Therefore, it can be said that the strain near the upper punch is affected by both inner pressure and flange pressure. Therefore, the rupture risk of the upper part of the blank may be organized by the ratio of inner pressure to flange pressure (= inner pressure divided by flange pressure). A comparison graph is shown in Figure 7. A correlation was observed between the ratio of inner pressure and flange pressure and the rupture risk of the upper part.

Next, the rupture risk of the lower part of the blank is examined. The aforementioned analysis results show that the rupture risk of the lower part is hardly affected by inner pressure. (Figure 4 above). Since the lower part of the blank is the closest to the blank holder among the three locations, the influence of the blank holder may be very large. The results of the comparison of rupture risk of the lower part and flange pressure are shown in Figure 8. A correlation was observed between them.

Finally, the rupture risk of the middle part of the blank is examined. Since this area is located between the forming pad and the blank holder, the strain generated is considered to be affected by both. In addition, since the corner radius of the middle part of the blank is smaller than that of the others, minor strain varies greatly under the influence of material outflow. Therefore, the major strain and minor strain at the site where the rupture risk was measured are examined. The stronger inner pressure and flange pressure are, the greater the major strain at this location should be. Figure 9 compares the major strain with the sum of inner pressure and flange pressure. The graph shows that there is a correlation between them. In the case of minor strain, on the other hand, inward movement of the blank causes compressive strain, while outward movement of the blank causes tensile strain. Therefore, a large inner pressure results in circumferential compression and a large flange pressure results in circumferential tension. Indeed, a correlation was obtained by comparing the minor strain to the ratio of inner pressure and flange pressure. (Figure 10). Plotting the major and minor strains of the middle part of the blank on the FLD, a trend of increasing major and minor strains for higher flange pressure and increasing major strain but decreasing minor strain for higher inner pressure appeared. (Figure 11). In this study, Rupture risk increased as Inner pressure became stronger, but it may decrease if the shape of the FLC of the material is different. Therefore, it is important to predict the rupture risk in advance using CAE.

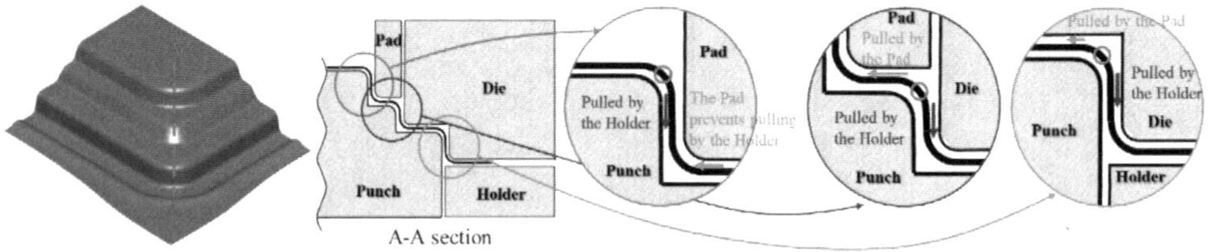

Figure 6. Schematic diagram of the corner section cut out in the diagonal direction

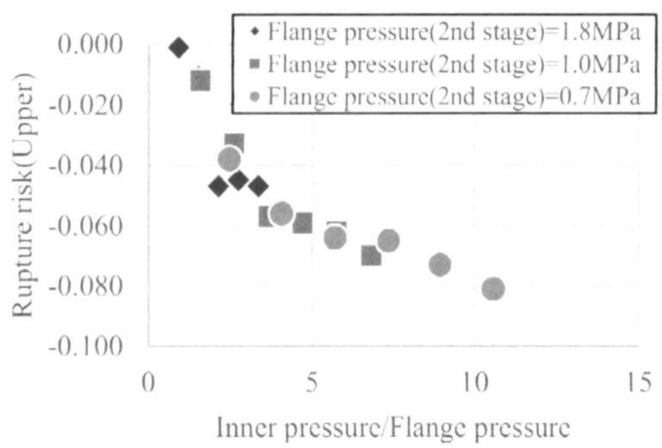

Figure 7. Graph comparing rupture risk of upper part with ratio of inner pressure to flange pressure

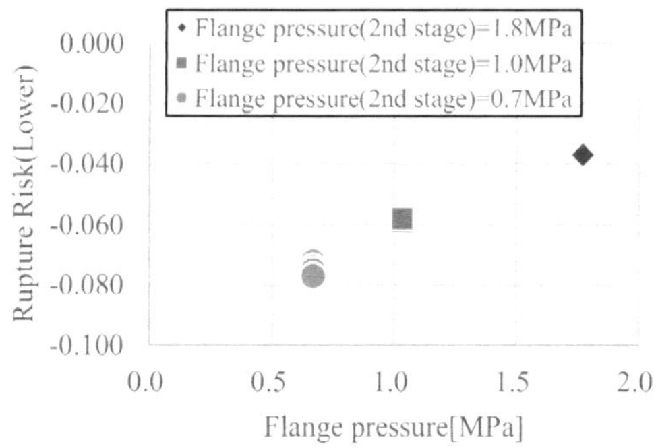

Figure 8. Graph comparing rupture risk of lower part with flange pressure

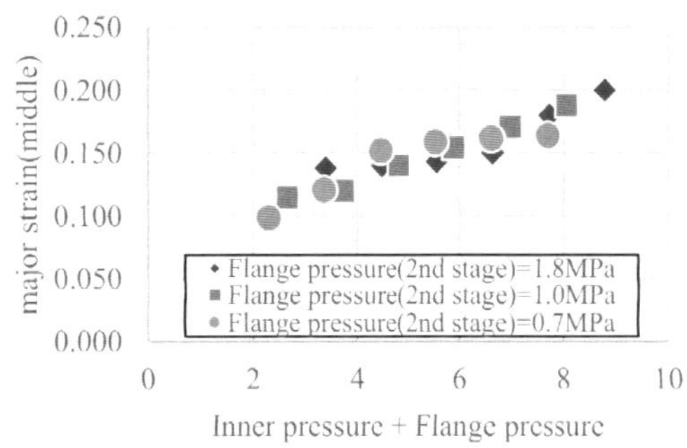

Figure 9. Graph comparing major strain in middle part with sum inner pressure and flange pressure

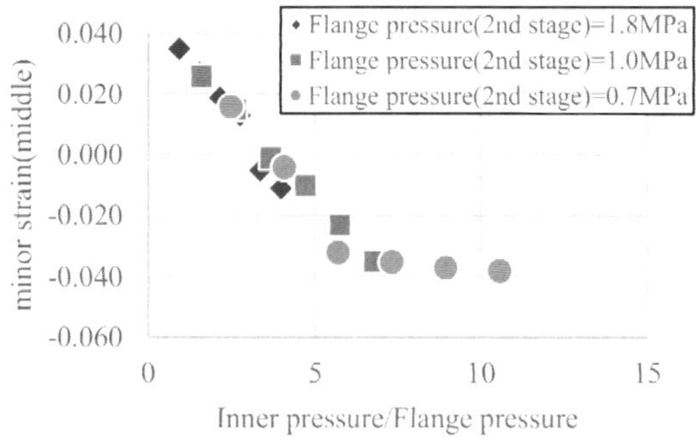

Figure 10. Graph comparing minor strain in middle part with ratio of inner pressure to flange pressure

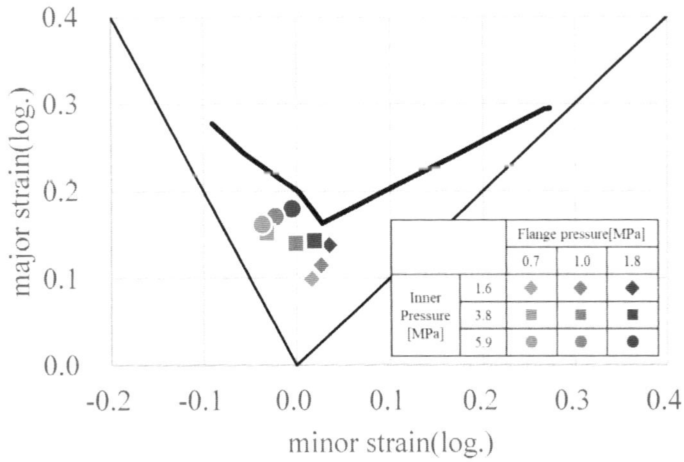

Figure 11. The major and minor strains of the middle part of the blank plotted on the FLD

3.4. Estimation of an appropriate range of the flange pressure and inner pressure

It was assumed that necking could be avoided if the rupture risk was less than -0.050. Under the conditions of this study, inner pressure must be at least 3.5 times larger than flange pressure to avoid necking in the upper part, based on Figure 7 above. In order to avoid necking in the lower part, flange pressure must be less than or equal to 1.2 MPa, based on Figure 8 above. To avoid necking in the middle part, it is necessary to set:
flange pressure ≤ 1.0 MPa and 2.7 MPa ≤ inner pressure ≤ 5.5 MPa
or
flange pressure ≤ 0.7 MPa

based on the CAE calculation results shown in Figure 4 above.
Based on the above, the appropriate forming conditions for the die shape studied in this study are:
flange pressure = 1.0 MPa, 3.5 MPa ≤ inner pressure ≤ 5.5 MPa
or
flange pressure = 0.7 MPa, 2.7 MPa ≤ inner pressure ≤ 7.0 MPa.

4. Verification of the appropriate values and effectiveness of the two-stage deep drawing process

4.1. Test Procedure

Another forming test was performed by the two-stage deep drawing process to confirm the validity of the values derived in the previous section. In the first stage of the process, the blank periphery was held at flange pressure = 1.8 MPa for deep drawing. In the second stage, the blank periphery was held at two conditions, flange pressure = 1.0 MPa and 1.8 MPa, and the area formed in the first stage was held at two conditions inner pressure = 3.8 MPa and 5.9 MPa, for deep drawing.

In addition, to confirm the effectiveness of the two-stage deep drawing process, the height of the upper part (H in Figure 1) was tested as 20 mm and 25 mm, with flange pressure = 1.0 MPa and inner pressure = 3.8 MPa. The results obtained were compared with the results of the aforementioned forming test by the one-stage drawing process (see Figure 2).

4.2. Test Results

The test results for different conditions of flange pressure and inner pressure for forming in two-stage deep drawing process using H = 20mm die are shown in the Figure 12. It can be seen that necking can be avoided under conditions where appropriate values are applied. It was found that the aforementioned approach was able to derive the appropriate conditions for flange pressure and inner pressure.

Comparison of results formed by one-stage deep drawing process and two-stage deep drawing process are shown in Figure 13. The blank could only be formed up to a height of H = 20 mm at the upper part in one-stage deep drawing process, but it could be formed up to H = 25 mm in two-stage deep drawing process. Therefore, this result confirms the effectiveness of the two-stage deep drawing process.

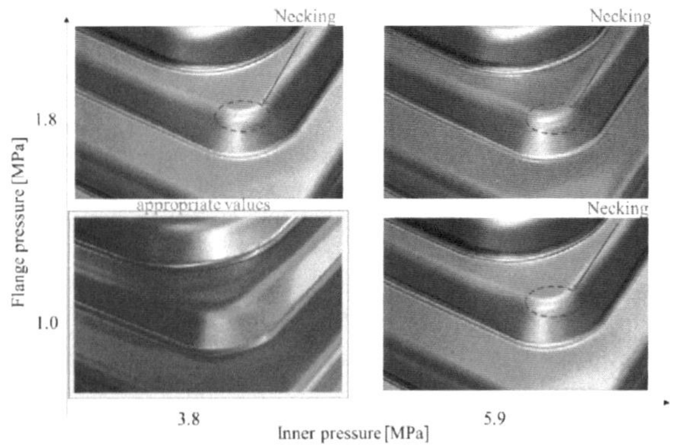

Figure 12. The validation of the appropriate values for the two-stage deep drawing process

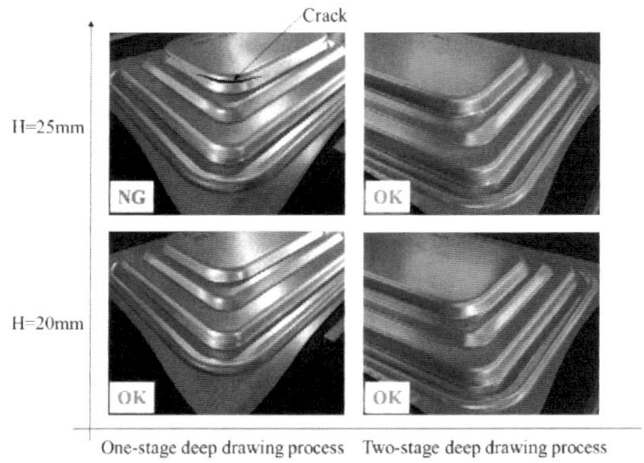

Figure 13. Comparison of results formed by one-stage deep drawing process and two-stage deep drawing process

5. Conclusion

1) Since the pad pressure acts in the direction of inhibiting cracking in the upper part and promoting cracking in the middle part, it is necessary to set the pad pressure so that both can be achieved. If inner pressure is set lower than flange pressure, too much material will flow out, so the value of Inner pressure should basically be set higher than flange pressure.

2) Assuming that rupture risk ≤ -0.050 can avoid rupture and necking of the blank, the appropriate values of flange pressure and inner pressure in the second stage of the process in this study are "Flange pressure = 1.0 MPa and 3.5 MPa ≤ inner pressure ≤ 5.5 MPa" or "flange pressure = 0.7 MPa and 2.7 MPa ≤ Inner pressure ≤ 7.0 MPa". The actual tests confirmed that cracking and necking could be avoided within the aforementioned appropriate range.

3) The results of test by the one-stage deep drawing process and the two-stage deep drawing process were compared. As a result, it was confirmed that the shape formed by the two-stage deep drawing process was 5 mm deeper than that formed by the one-stage deep drawing process.

References

[1] Keisuke AKASAKI. Solution Technology for Applying Aluminum Alloy Sheet to Automobile Bodies: KOBE STEEL ENGINEERING REPORTS. 2019;69(1):50-54.

[2] Martin Halliwell, Robin Warmington. JLR Future Lightweight Doors. In: Automotive-Circle. DOORS & CLOSURES IN CAR BODY ENGINEERING 2010, Nov. 16-17 2010; Bad Nauheim DEU.

[3] Naonori Sakamoto, Takahiro Kawase. Doors and Closures of the New Honda CLARITY FUEL CELL. In: Automotive-Circle. DOORS & CLOSURES IN CAR BODY ENGINEERING 2016, Nov. 15-16 2016; Bad Nauheim DEU.

[4] Masatoshi Koide, Kiyohito Tsuruda. Developments and Future Trends in Aluminum Can Stock for Beverage Containers: KOBE STEEL ENGINEERING REPORTS. 2005;55(2):75-80.

[5] Frédéric Barlat, Daniel J. Lege, John C. Brem. A six-component yield function for anisotropic materials. International Journal of Plasticity. 1991;7(7):693-712.

IOP Conf. Series: Materials Science and Engineering **1307** (2024) 012012 doi:10.1088/1757-899X/1307/1/012012

High speed impact cutting (HSIC) of advanced high strength steel 42SiCr under exploitation of adiabatic shear bands

P. Birnbaum[1], A. Kunke[1] and V. Kräusel[2]

[1] Institute for Machine Tools and Production Processes (IWP), Chemnitz University of Technology, Chemnitz, Germany

[2] Fraunhofer Institute for Machine Tools and Forming Technology (IWU), Reichenhainer Straße 88, 09126 Chemnitz, Germany

E-mail: peter.birnbaum@mb.tu-chemnitz.de

Abstract. Shear Cutting of Advanced High-Strength Steels poses technological challenges due to the substantial mechanical loads imposed on cutting tools, leading to elevated wear rates. A strategy for cutting high-strength materials involves the utilization of high-speed impact cutting (HSIC), wherein component separation occurs along a locally adiabatically heated shear band, resulting in reduced cutting forces. The steel alloy 42SiCr undergoes heat treatments involving Quenching+Tempering (Q+T) as well as Quenching+Partitioning (Q+P) for two sheet thicknesses. This results in the formation of martensitic microstructures with varying retained austenite content, as determined through X-ray Diffraction (XRD). Subsequently, the heat-treated steel samples are subjected to tensile testing for mechanical property evaluation, revealing ultimate tensile strengths exceeding 1500 MPa and fracture elongations ranging from 2 % to 12 %. Following this, the material is subjected to HSIC using the AdiaPress Adia 7 machine, employing predefined cutting energies. It is observed that both Q+T and Q+P-treated materials can be successfully cut using HSIC, although distinct cutting edge morphologies are evident. Optical examinations of the cut edges, conducted through top-view and cross-sectional analysis using Scanning Electron Microscopy and 3D laser scanning microscopy, confirm the presence of adiabatic shear bands and discrete zones.

1. Introduction and state of the art

Quenching and partitioning (Q+P) steels have been widely investigated in terms of mechanical properties, fabrication routes and processing, since Speer et al. [1] first reported the partitioning of carbon from supersaturated quenched martensite into retained austenite (RA) in 2003. As a result of their high strength and ductility, Q+P steels represent the third generation of advanced high strength steels (AHSS) for the automotive industry [2, 3]. The reason for their outstanding properties is the initial thermal treatment of the steel, resulting in a fully martensitic structure with increased RA content [4–7]. Q+P steels gain their high strength properties from the martensitic content and the enhanced ductility by RA of up to 20 % of phase content. Plastic deformation can be absorbed by the RA which transforms simultaneously to martensite and strengthens the material. The mechanical processing of Q+P steels is strongly limited by its high strength. When considering shear cutting, the tools are subjected to increased wear, by which the tool life quantity is significantly reduced. As an enhancement of the conventional shear cutting process, high-speed impact cutting (HSIC) emerged. With the help of a defined energy input which accelerates the punch up to 10 m/s, the blanking material is deformed very fast along the cutting line which heats the material up by several hundred Kelvin within 100 µs [8]. Adiabatic heating of the shear zone occurs in a range of less than 100 µm [9–12]. About 95 % of the cutting energy is converted into heat [13] which softens the material and enables the cutting process. Winter et. al. [9] investigated the temperature rise during adiabatic cutting of a press hardening steel based on a thermomechanical model approach. Yang's studies in orthogonal cutting [14] show enhanced shear band formation when low thermal conductivity, low heat capacity, low strain rate sensitivity, and high yield stress occur in the material. Adiabatic shear bands begin when a critical, material-specific deformation rate is reached and sufficient energy is available for plastic dissipation [15]. In order to activate the adiabatic shear bands for the cutting process the punch has to penetrate the component less than 10 % of the blank thickness. As a result of the low penetration of the punch, spring back aspects are negligible in comparison to cutting with lower punch velocities [13]. The HSIC process thus results in improved cut surfaces, nearly free of burrs, high contouring accuracy and the reduction of shear cutting forces [16, 17].

2. Experimental procedure

2.1. Chemical composition

The HSIC tests were carried out on the experimental cast of 42SiCr steel, that was investigated for the sheet thickness of 1.8 mm and 3.6 mm after rolling. The chemical composition of the material was analyzed by OES Bruker Tasman Q4 (see Table 1).

Table 1: Chemical composition of 42SiCr base material [wt %], Fe-balanced

C	Si	Mn	Cr	Mo	Ni	Cu	Al	Co	Sn	V	P	S
0.426	2.151	0.679	1.329	0.037	0.051	0.039	0.0078	0.013	0.008	0.0057	0.01	0.0036

2.2. Initial heat treatment

The base material was exposed to three different variants of heat treatment in order to set various mechanical properties (Figure 1). The thermal treatment is based on preliminary investigations and literature values on Q+P [4–7]. After the austenitization subsequent quenching with $>$ 30 K/s is performed in a heated salt bath. The temperature of the salt bath is set on a temperature level between the martensite start temperature (M_s) and martensite finish temperature (M_f) of the material. The partitioning process takes place in the heated salt bath, at which the dissolved carbon atoms diffuse out of the supersaturated martensite into the intermediate space between the martensite needles and stabilize RA. According to the literature [4–7], the diffusion-controlled process takes between 15 and 20 min. The quenched and tempered (Q+T) material state is quenched to temperatures below M_f and subsequently tempered, as in conventional tempering and annealing processes. The routes Q+P–1 and Q+P–2 are quenched and partitioned with divergent temperatures and intervals (Table 2).

2.3. Tensile testing of heat treated base material

In order to evaluate the mechanical properties of the base material after heat treatment, all tests were performed by the universal testing machine H&P Inspekt 150 according to DIN EN ISO 6892-1. The tensile tests were carried out for all heat treated conditions and both sheet thicknesses at room temperature and a testing velocity of 0.5 mm/s. The initial gauge length was L_0 = 40 mm.

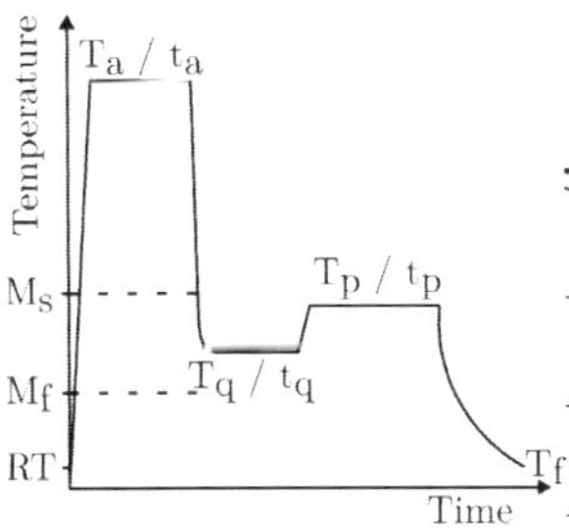

Figure 1: Schematic route of heat treatment

Table 2: Heat treatment routes for sheet thickness 1.8 mm and 3.6 mm

Process \ Thickness		1.8 mm Q+T	1.8 mm Q+P-1	1.8 mm Q+P-2	3.6 mm Q+T	3.6 mm Q+P-1	3.6 mm Q+P-2
Austenizing	T_a	950 °C	950 °C	950 °C	950 °C	950 °C	950 °C
	t_a	6 min	6 min	6 min	6 min	6 min	6 min
Quenching	T_q	60 °C	200 °C	150 °C	60 °C	200 °C	150 °C
	t_q	1 min	5 min	5 min	1 min	5 min	5 min
Partitioning	T_p	250 °C	300 °C	250 °C	250 °C	300 °C	250 °C
	t_p	10 min	10 min	10 min	10 min	10 min	10 min
Chilling	T_f	20 °C	20 °C	20 °C	20 °C	20 °C	20 °C

2.4. XRD analysis of heat treated base material

After the as-received condition had been heat treated, a phase analysis was performed by means of X-ray diffraction to identify the RA content. The samples were measured on the cross-sectional area. Quantitative analyses were performed according to Rietveld using the Siemens D5000 X-ray diffractometer.

2.5. High speed impact cutting experiments

The HSIC process was carried out on an AdiaPress Adia 7 machine, which is driven by a hydraulic cylinder and can provide cutting energies stepwise between 1.7 and 7 kJ. A tool for shearing with a circular cutting punch was used. The diameter of the active tool parts were 30.0 mm for the punch and 30.2 mm for the die. This resulted in a cutting clearance related to the sheet thickness of 5.6 % for a base material of 1.8 mm and 2.8 % for a base material of 3.6 mm sheet thickness. Prior publications indicates a suitable clearance for HSIC between 2 to 4 % of the sheet thickness [8]. This clearance is significantly higher compared to fine blanking (~ 0.5 %), but lower compared to conventional blanking (5 – 15 %) [18]. For the sheet thickness of 1.8 mm, cutting energies of 1.7, 2.5 and 3.5 kJ were performed. The test of 42SiCr steel with the sheet thickness of 3.6 mm was carried out with cutting energies of 2.5 to 4.5 kJ, whereby the impact energy was increased by 1 kJ stepwise. In each possible parameter combination, a minimum of 3 experiments were conducted.

2.6. Optical evaluation of cut material

After the HSIC tests, the quality of the cut surfaces at the cut out part and at the sheet metal strip was investigated, using optical microscope OLYMPUS GX51, SEM LEO 1455VP and ZEISS NEON 40EsB as well as 3D laser scanning microscope Keyence VK-9700. The cut surfaces were analyzed in front view and in cross-section (Figure 2). For the cross-sectional analysis, the material was polished and etched with picric acid for 50 seconds.

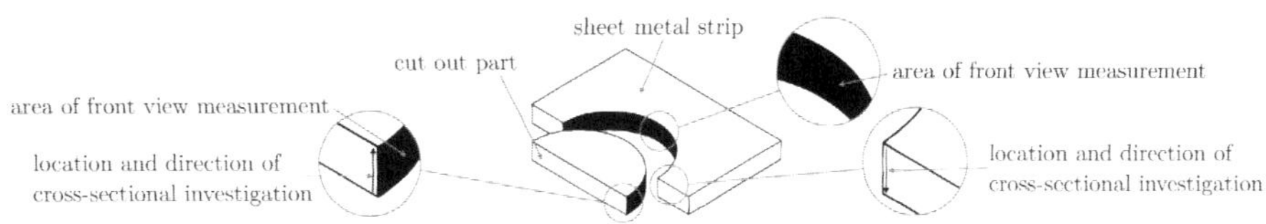

Figure 2: Cross-sectional view on cut sheet metal strip and cut out part

3. Results

3.1. Mechanical properties of heat treated basic material

With the help of tensile tests of heat-treated sheet materials of 1.8 mm thickness, the highest tensile strength (R_m = 2040 MPa) and the smallest ultimate strain (A_{40mm} = 7.5%) were measured with Q+T treated specimens (Figure 3 left). Compared to Q+T, the Q+P treatment slightly reduces the tensile strength but increases the ultimate strain almost by the factor of 2. Thus, the maximum measured tensile strength and ultimate strain for Q+P–1 is $R_m = 1720$ MPa and $A_{40mm} = 13.5$ % and for Q+P–2 $R_m = 1950$ MPa and $A_{40mm} = 10.5$ %. Compared to sheets of 3.6 mm thickness (Figure 3 right), 1.8 mm sheets tend to have an increased yield stress by up to 200 MPa but an about $\Delta A_{40mm} = 2$ % reduced ultimate strain. For Q+P–2, the treatment of 3.6 mm thick material provides a maximum tensile strength of $R_m = 1680$ MPa and an ultimate strain of $A_{40mm} = 12.5$ %. Q+P–1 exhibits a tensile strength and an ultimate strain of $R_m = 1560$ MPa, respectively $A_{40mm} = 15$ %. For all three individual heat treatments, there is a deviation in the stress-strain curve during consecutive tensile tests, despite undergoing an identical heat treatment. This can be attributed to the use of the experimental cast 42SiCr and the discontinuous aspect of the heat treatment. Additionally, it is possible that the mechanical properties are highly sensitive to the heat treatment and RA content.

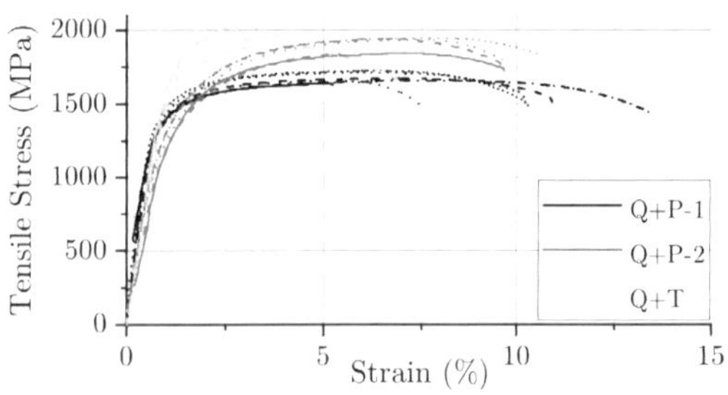

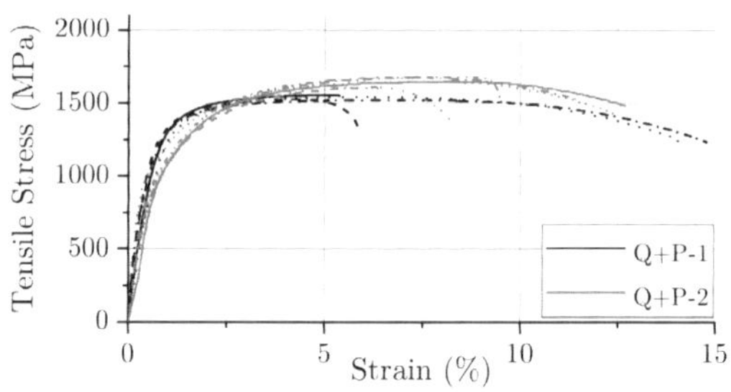

Figure 3: Stress-strain curve after Q+T and Q+P for sheets with 1.8 mm (left) and 3.6 mm (right) thickness

3.2. Phase composition and microstructure after Q+P and Q+T treatment

The phase analysis of the material provides information on the RA content. Measurements were performed on surfaces that were cut by saws or water-jet, ground and polished. It can be assumed that the RA on the surface turns into martensite due to the mechanical processing. As given in Figure 4, there is a significant difference between Q+T and between Q+P treated samples. For example, the proportion of RA is less than 2 % for 1.8 mm thick sheet metal after Q+T treatment and less than 3 % for 3.6 mm thick material. In contrast, the fraction of RA of Q+P–1 heat-treated sample is much more pronounced. A content of 9 % can be determined for 1.8 mm thick material and 9.5 % for 3.6 mm thick material. The highest RA content was found in Q+P–2 treated samples ranging from 9.5 % for 1.8 mm thick material to 12.2 % for 3.6 mm thick material. So, the RA content increase for 1.8 mm material is between ΔRA = 0.5 – 2.7 % lower than for 3.6 mm thick material. The influence of the preparation is also verifiable, since all sawn samples have lower RA content than samples that were ground and polished after water-jet cutting. As illustrated in Table 3, after the respective heat treatments the specimen reveal different martensitic microstructures in the inner material. The images were taken by optical microscope. After a Q+P–1 treatment of the 1.8 mm thick material another phase is evident in the form of light grains that do not react with picric acid. According to the Q+P treatment and the analyzed austenite content, it is supposed to be grains of RA.

3.3. Cutting experiments with HSIC

The HSIC experiments for all 3 heat treatment conditions and both sheet thicknesses were carried out. The smallest cutting energy input was high enough to cut the samples. Q+P and Q+T-treated 1.8 mm sheets were cut with 1.7 kJ successfully just like 3.6 mm thick material at 2.5 kJ. No evident differences were found at the optical evaluation of the cut surfaces in regard to the heat treatment.

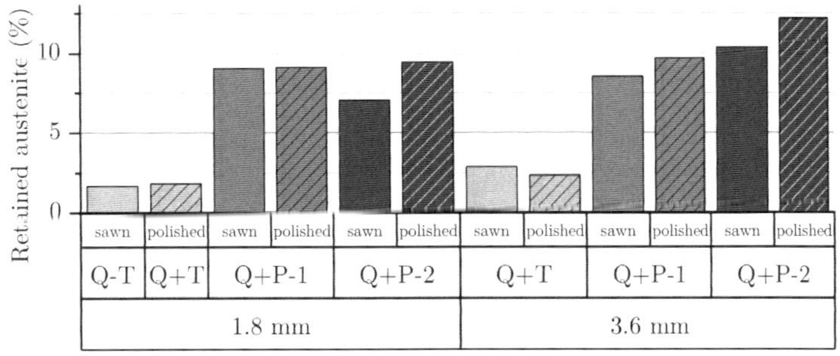

Figure 4: RA content after Q+T, Q+P–1 and Q+P–2 treatment of 1.8 mm and 3.6 mm sheet

Table 3: Microstructure of inner layer after heat treatments of 42SiCr, 100 x, optical microscope

	Q+T	Q+P–1	Q+P–2
1.8 mm	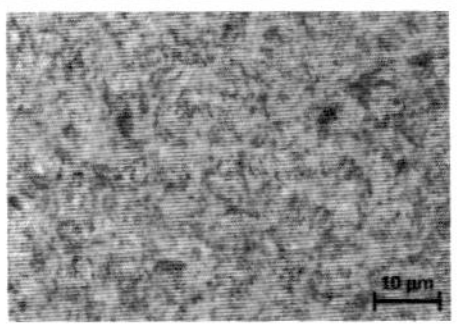	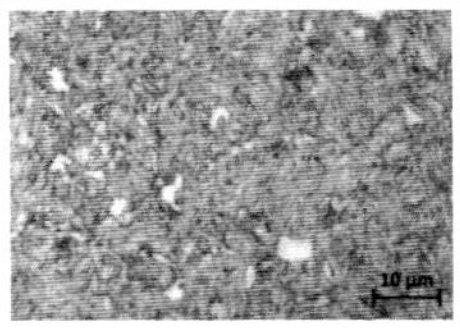	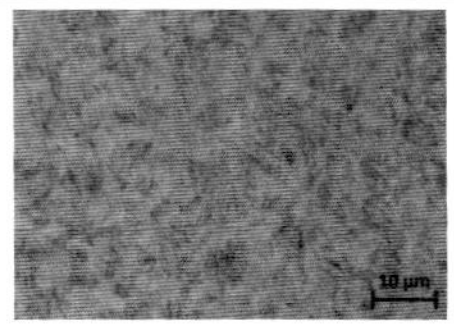
3.6 mm	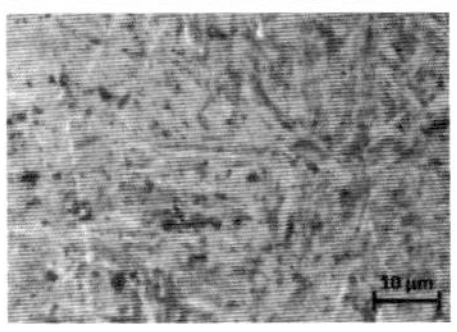	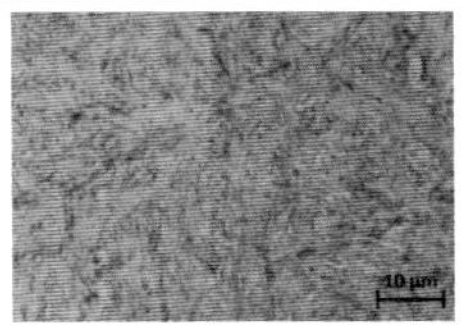	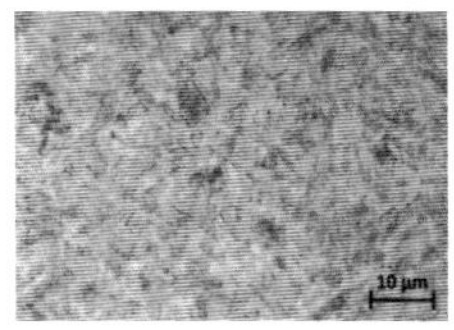

3.4. Microscopic analysis of the cut surface in cross section

The investigated cross section from the cut surface into the middle layer of the material can be divided into three distinct zones according to the distance from surface. Figure 5 illustrates the cross section of 3.6 mm Q+P–1 treated cut-out material. The undeformed martensitic base material in zone 1 does not show any obvious deformation. In contrast, severely deformed martensite needles were formed in zone 2. The curvature of the martensite needles amounts to 90 °. The average diameter of the zone is 10 to 15 µm. Bordering to this layer of strongly curved martensite needles, the highly deformed zone 3 with a thickness of 8 to 15 µm is present. It differs significantly from the underlying structure. Between zones 2 and 3, a sharp transition of the structure can be detected. The curved martensite needles from zone 2 in Figure 5 are shown in detail in Figure 6 for similar HSIC cut material. Obviously, the needles do not break due to the shear process (6 a). The deformation is endured by the martensite without obvious fracture. Zone 3-grains of similar HSIC cut material are spherical and slightly stretched in the direction of deformation (Figure 6 b), whereby a length up to about 200 nm and a height between 100 and 150 nm was determined. Thus, the area is ultrafine-grained. Delamination of zone 3 is illustrated in Figure 6 c.

3.5. Microscopic analysis of the cut surface in front view

SEM images demonstrate the characteristic formation reported previously [8]. The fracture zones in the front view are exemplified in Figure 7 for a 3.6 mm sheet after Q+P–1 treatment and HSIC with 2.5 kJ. The upper edge of the cut surface shows a rollover height ranging from 80 to 150 µm, followed by the burnish. Consistent with prior HSIC research [8], the burnish height accounts for 4 % of the sheet thickness. An adjacent dimple fracture area, 50 to 150 µm thick,

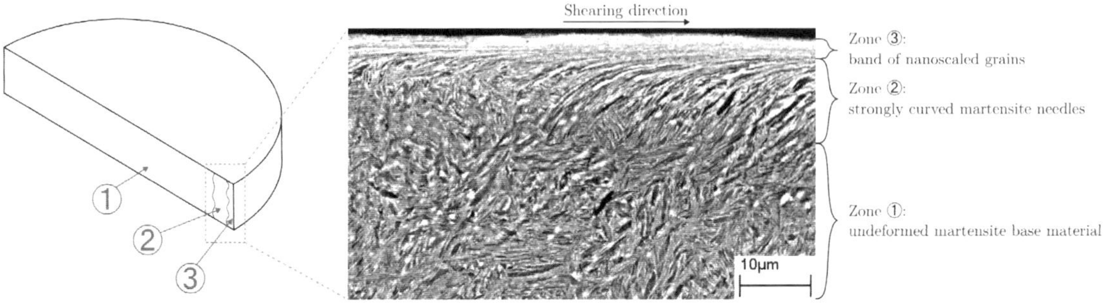

Figure 5: Cross section of Q+P–1 cut-out part with 3.6 mm sheet thickness after HSIC with 2.7 kJ, SEM 1000 x

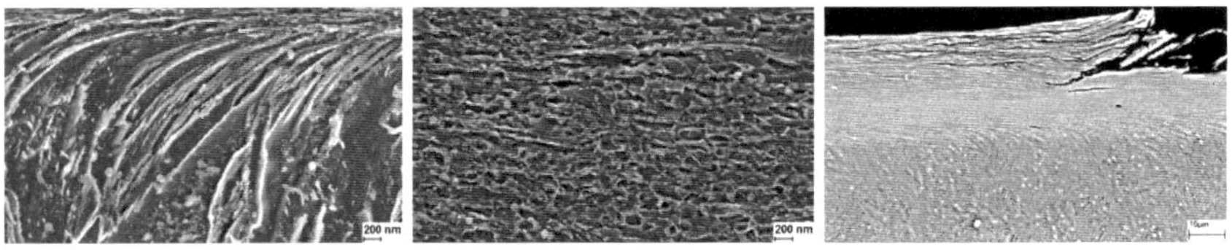

(a) (b) (c)

Figure 6: (a) Highly deformed martensite in zone 2 (Q+P–2, 3.6 mm thickness, 4.5 kJ cutting energy, SEM 25K x), (b) Nanoscaled, equiaxed grains in Zone 3 (Q+T, 1.8 mm thickness, 1.7 kJ cutting energy, SEM 25K x), (c) Fracture through a shear band (Q+T, 3.6 mm thickness, 2.5 kJ cutting energy, SEM 10K x)

is visible in the cutting direction (Figure 7). Below the dimple fracture zone, a broad smooth area ascends towards the opposite sheet surface, accumulating in a plate-like manner up to the tear-off at the burr. The cut surface morphology is similar for the 1.8 mm sheet thickness, with a less pronounced extent of the plate-like structure and a smaller increase of the even surface towards the opposite edge. The rollover height is lower in comparison to 3.6 mm sheet thickness, with the burnish cut accounting for 5 % of the sheet thickness. The dimple fracture below the burnish is smaller than in the 3.6 mm thick sample. Once again, a region of reduced roughness is evident below the burnish with a subsequent dimple fracture.

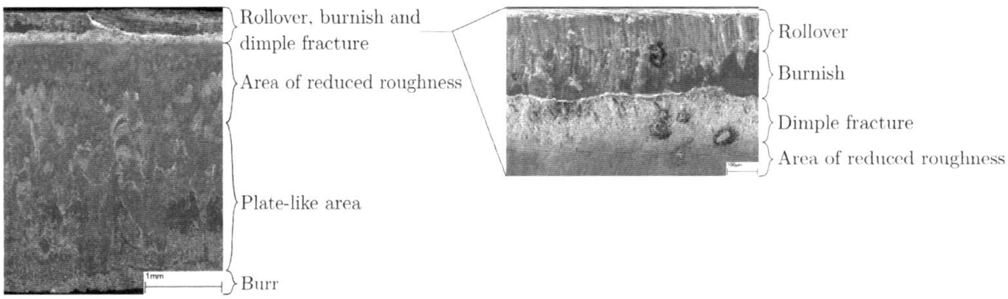

Figure 7: Front view on HSIC cut surface 3.6 mm thickness after Q+P–1 and 2.5 kJ cutting energy SEM 20 x (left) and detailed view 140 x (right)

3.6. Topographic measurement of the cut surface

For both Q+T and Q+P treatments, no significant differences in the cut surface morphology are observed for their respective sheet thicknesses. To assess the shape of the cut surface of cut-out parts, topographical measurements of the 3.6 mm thick sample after Q+P–1 treatment and HSIC with 2.5 kJ were performed using a Keyence laser microscope, displaying morphology similar to optical microscopy and SEM. The height profile, enlarged by 200 %, highlights key areas. In Figure 8, the burnish cut portion forms a flat surface on the left. The subsequent dimple fracture is followed by a rapid fall-off of the cut surface, with the adjacent area exhibiting an increased profile. The strong plate-like elevation, around 25 µm relative to the surrounding material, is visible in the cross-sectional view in Figure 8. In comparison to the 3.6 mm thick material, the 1.8 mm thick material exhibits a similar but less pronounced cut morphology. Underneath the burnish, the cut surface displays a less distinctive dimple fracture, followed by a flat and even area. A local elevation of approximately 10 µm forms a structure akin to the plate-like structure observed in the 3.6 mm thick material (Figure 7). This shape has been previously observed for AHSS and described as an S-shaped curve [12].

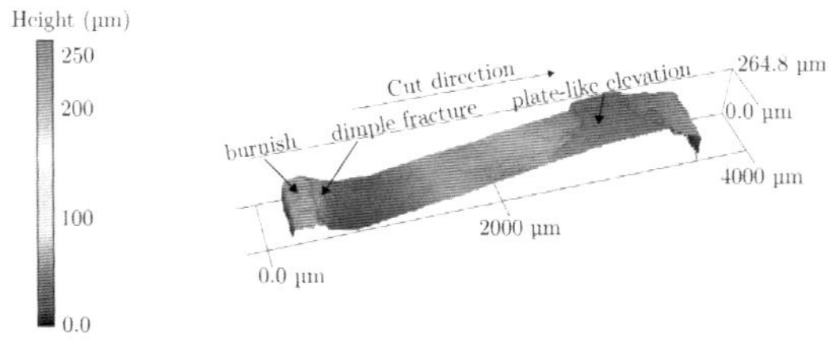

(a)

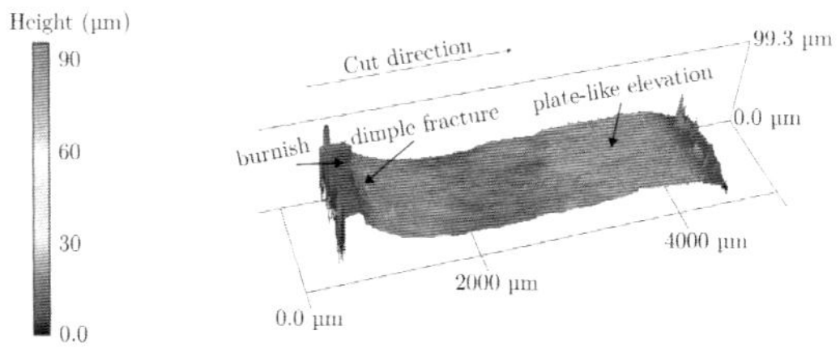

(b)

Figure 8: Topography (200 % enlarged) of HSIC cut surface (cut-out part) with (a) 2.5 kJ of 3.6 mm thickness Q+P–1 treated, (b) 2.5 kJ 1.8 mm thickness Q+P–1 treated

4. Discussion

Investigations show that 42SiCr sheet metal achieves desired properties via applied heat treatments. Interrupting the cooling process and partitioning between M_s and M_f leads to higher ductility compared to quenching below M_f with subsequent tempering. Q+T-treated material exhibits higher tensile strength than Q+P-treated 42SiCr. Despite different properties observed through testing and energy input in HSIC, the cut surface morphology remains similar. Significant differences arise when comparing sheet thicknesses, possibly due to the different sheet thickness related cutting clearance. Metallographic analysis of cut surfaces reveals a three-zone structure, including undeformed martensite (zone 1), strongly curved martensite in shear direction (zone 2), and ultrafine-grained equiaxial microstructure towards the cut surface (zone 3). Zone 3 likely contains adiabatic shear bands for deformation during the shear process. Examining the front view, a multi-level structure is evident, with a small proportion of rollover and burnish heights. A dimple fracture surface adjacent to the burnish suggests ductile fracture behavior. Below it, the cut surface deepens towards the base material and increases towards the burr on the opposite sheet surface. A zone of reduced roughness followed by plate-like structures aligns with the adiabatic shear bands in the cross section. SEM images confirm separation through the shear band and at the boundary between highly curved martensite (zone 2) and the adjacent shear band (zone 3). See Figure 9 for a schematic representation of the cut surface morphology based on metallographic analysis.

5. Conclusion

From the investigations it could be determined that the performed heat treatments lead to Q+P properties. The examined steel can be cut by HSIC with low energy. The examination of the surface showed a multi-zone structure in front view and in cross-section. The cross-section of the cut surface offers a 3-zone structure consisting of a zone of undeformed martensite, a second zone of strongly deformed martensite needles and a third zone of ultrafine equiaxial microstructure. The fine-grained structure is only visible as a narrow band in certain areas of the surface and is defined as adiabatic shear band. In front view of the cut surface, the shear bands are plate-like structures in the direction of the burr. Topography measurements support

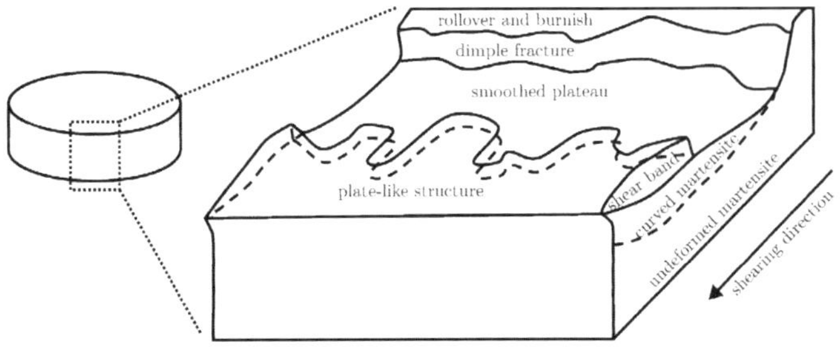

Figure 9: Schematic sketch of cut surface morphology and microstructure after HSIC

this statement, as plate-like structures of up to 25 µm stand out from the surface. From this a basic cut surface morphology model for adiabatic cutting can be developed. There is a need for further investigations on how the adiabatic cutting process will be influenced by increasing RA contents and what effects this will have on the microstructure at the cut surface. Furthermore, it is necessary to demonstrate how a cut surface would appear in conventional shear cutting under the same material and die clearance conditions. Wear studies on cutting tools during HSIC and conventional cutting of Q+P steel are crucial to quantify the advantages of HSIC.

References

[1] Speer J, Matlock D K, De Cooman B C and Schroth J G 2003 *Acta Mater.* **51** 2611–2622 ISSN 13596454

[2] Sun J and Yu H 2013 *Mater. Sci. Eng. A* **586** 100–1107 ISSN 09215093

[3] Matlock D K, Speer J G, De Moor E and Gibbs P J 2012 *JESTECH* **15** 1–12

[4] Jirková H and Mašek et al B 2015 *Mater. Today Proc.* **2** S627–S630 ISSN 2214-7853

[5] Černý I and Mašek et al B 2011 *Procedia Eng.* **10** 3310–3315 ISSN 1877-7058

[6] Jirková H, Mašek B, Wagner M F, Langmajerová D, Kučerová L, Treml R and Kiener D 2014 *J. Alloys Compd.* **615** S163–S168 ISSN 09258388

[7] Jenicek S, Vorel I, Kana J, Opatova K, Rubesova K, Kotesovec V and Masek B 2017 *IOP Conf. Ser. Mater. Sci. Eng.* **181** ISSN 1757899X

[8] Kräusel V 2012 *The design and evaluation of a single stroke cutting procedure, using rigid tools, with particular emphasis on the quality of sheet metal cutting surface* Habilitation Chemnitz, Techn. Universität

[9] Winter S, Schmitz F, Clausmeyer T, Tekkaya A E and Wagner M F 2017 *IOP Conf. Ser. Mater. Sci. Eng.* **181** ISSN 1757899X

[10] Neugebauer R, Kräusel V and Weigel P 2010 *Stahl und Eisen* **130** 39–42 ISSN 03404803

[11] Lazzarotto L and Michon R 2008 *5th Car Body Colloq. 2008 - Chemnitz*

[12] Schmitz F, Winter S, Clausmeyer T, Wagner M F and Tekkaya A E 2020 *CIRP Ann.* **69** 269–272 ISSN 17260604

[13] Grünbaum M, Breitling J and Altan T 1996 *ERC Rep.* **19** 5–96

[14] Yang Q, Liu Z and Shi Z 2012 *Mater. Sci. Forum* **723** 41–49 ISSN 16629752

[15] Dodd B and Bai Y 1989 *Mater. Sci. Technol. (United Kingdom)* **5** 557–559 ISSN 17432847

[16] Neugebauer R, Kräusel V, Weigel P, Silbermann K and Sterzing A 2009 *ICIT&MPT 2009*

[17] Krutz P, Leonhardt A, Graf A, Winter S, Galiev E, Rehm M, Kräusel V and Dix M 2023 *J. Manuf. Mater. Process.* **7** ISSN 2504-4494

[18] Zheng Q, Zhuang X, Gao Z, Guan M, Ding Z, Hong Y and Zhao Z 2019 *Int. J. Adv. Manuf. Technol.* **104** 4129–4141 ISSN 14333015

Biaxial Tensile Testing of Cruciform Samples to Determine the Impact of Pre-Strain on the Forming Limit Diagram of Aluminum Alloy 5052 through Finite Element Simulation

Shrutee Pradeep Pawar, Pavan Kumar and K. Narasimhan

Indian Institute of Technology Bombay, India

nara@iitb.ac.in

Abstract. The present work focuses on the effect of pre-strain on the forming limit curves (FLCs) of aluminium alloy 5052 sheet of 2.5 mm thickness using cruciform samples. To identify the effect of pre-strain on the FLC, the finite element simulations are performed on cruciform samples. The cruciform samples are deformed to a small level of pre-strain in different strain paths. Thereafter, the further deformation under the various strain paths to measure the effect of pre-strain on the forming limits are carried out. Pre-strain in uniaxial, plane strain and biaxial loading conditions are considered in this work. For each case, i.e., uniaxial, plane strain and biaxial conditions two pre-strain conditions are considered. The cruciform samples are carefully thinned in the central region to promote development of large plastic strains there and eventually to neck and fail. The output obtained through simulations are presented in terms of forming limit curves for various strain paths.

1. Introduction

One of the earliest methods for predicting the necking or failure of sheet metal components is the forming limit diagram (FLD) [1]. In the conventional FLD, Major and minor strains are taken into account on the ordinate and abscissa respectively [2]. Typically, the strain-based FLD is divided into two sections; Goodwin created the left side, while Keeler and Backofen introduced the right side [3,4]. Numerous techniques are described in the literature to ascertain the forming limits of metallic sheet [5]. These techniques can be divided into two categories: out of plane deformation and in-plane deformation methods [6,7]. Most of the literature reports the determination of the FLD using out-of-plane deformation [8]. The cruciform sample method of FLD determination falls under the category of in-plane deformation methods [9]. The sample shape developed by the Kuwabara et al. [10,11] was adopted as the international standard for the cruciform sample in 2014. Determination of FLD using cruciform samples is very promising due to their frictionless character, access to a pure two-dimensional stress area and in-situ measurement of strain on both side of the gauge region [12]. Nikhare [13] suggested that conventional forming limit test limits the possibilities of applying multiaxial loading. To overcome this limitation, he used the biaxial cruciform specimen which had provided the capability to test in multiaxial stress states by varying the independent load or displacement on two independent axes. Siddiqui et al. [14] designed a biaxial cruciform sample to evaluate the material properties under biaxial loading. They suggested that the identification of biaxial mechanical properties using the planar biaxial testing technique is more accurate compare to the other out-of-plane biaxial testing techniques. A novel cruciform biaxial tensile testing technique was utilized by Xiao et al. [15] to assess the limit strains of a TA1 titanium alloy. The FLD measured at various angles relative to the direction of material rolling indicated that the limit strain was anisotropic. Wang et al. [16] had proposed the two different types of cruciform geometries for the prediction of forming limits, one for the left side and one for its right side. Yu et al. [17] designed a cruciform sample to study the sheet forming limit under complex strain loading paths. They optimized the chamfer on the arms and the central region by FE simulation and reported that by altering the velocity boundary conditions, the various strain paths in the central region could be achieved. Jocham et al. [18] optimized the cruciform geometry for the evaluation of strain limits in the center of the of the specimen. They reported that the developed specimen allows the testing of materials under multi-axial strain states with a reduced testing effort. Song et al. [19] evaluated the four in-plane cross specimen shapes through finite element simulations for getting the large strain and fracture in their central point. Based on these numerical findings, a new form with optimized dimensions is suggested for figuring out the forming limits at fracture of 2 mm thick DP600 sheet metal. Yuan et al. [20] explored the effect of pre-strain on forming limit curves of 5754-O aluminum alloy sheet through biaxial tensile technique. Through experiments and simulations, they reported the pre-strained uniaxial tension followed by strain tension or equibiaxial tension can improve the formability of sheet metals.

In the present work the biaxial cruciform samples are used to explore the impact of pre-strain on the forming limit diagram of aluminum alloy 5052 through finite element simulation. In this work the biaxial tensile testing with the cruciform sample is opted due to several advantageous features. This method offers the flexibility to test different strain paths using the same sample. Additionally, the absence of friction between the sample and the tool is a notable benefit, as it eliminates potential interference that could otherwise impact the accuracy of the results. This approach not only allows for a comprehensive examination of strain paths but also ensures the reliability of the obtained data by minimizing external factors that might influence the outcomes.

2. Material and methods

In this study, the influence of pre-strain on the FLD using cruciform sample through finite element (FE) simulations is explored. Initially individual samples are subjected to each condition i.e., uniaxial, plane strain and biaxial to obtain the FLD of as-received material. Based on the maximum strains, 3% and 5% strain values are decided as the pre-strain values in each case. During the simulation once the 3% or 5% strain is reached for a certain condition additional condition is imposed for strain path change. For example, initially the sample is stretched in uniaxial condition up to 3% strain, thereafter, strain path change is imposed by applying the biaxial condition until necking is observed. The similar procedure was adopted for each case to identify the effect of strain path change or the effect of pre-strain on FLD using cruciform sample.

Aluminium alloy AA 5052 sheet of 2.5 mm thickness is used in this study to get the desired outputs. The chemical composition of the AA 5052 material is depicted in Table 1.

Table 1. Chemical composition of AA 5052 sheet

Element:	Mg	Cr	Co	Fe	Mn	Si	Zn	Al
Et. percent	2.4	0.32	0.09	0.38	0.08	0.23	0.1	Balance

The tensile tests were performed using an MTS Landmark 100 kN servo-hydraulic uniaxial testing machine with a displacement control rate of 1 mm/min. The tensile specimens were prepared according to the ASTM E8 standard, and a minimum of three tests was conducted along the rolling direction (RD), diagonal direction (DD) and transverse direction (TD) to evaluate the material properties. Strain data was measured in the 25 mm gauge length using extensometer which is mounted on the specimen during the tests. The stress strain curve along the rolling direction and the material parameters obtained from the tensile tests are depicted in Figure 1 and Table 2 respectively.

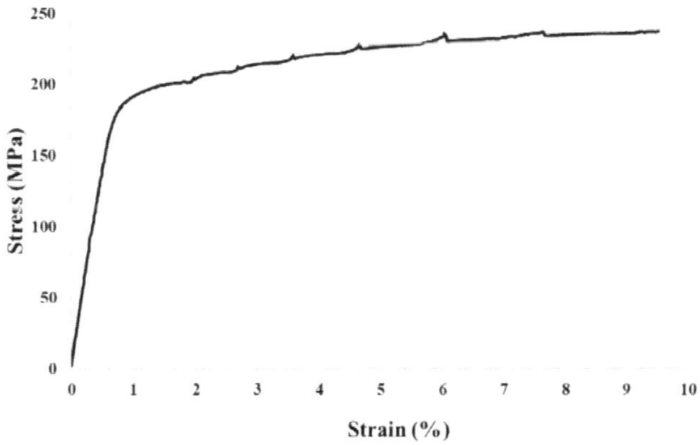

Figure 1. Stress-strain curve along the rolling direction

Table 2. Material properties of AA 5052 sheet

S. No.	Parameters		Values
1.	Youngs modulus		70×10^9 Pa
2.	Poisson's ratio		0.33
3.	Density		2.68×10^{-6} Kg/mm^3
4.		R_0	0.66
5.	Lankford coefficient	R_{45}	0.64
		R_{90}	0.68

3. Finite element modelling and simulation

To perform the simulation on the cruciform samples initially the CAD model of the sample was prepared in CATIA V5 R 20. The cruciform sample geometry was adopted from the work of Siddiqui et al. [14] as shown in Figure 2.

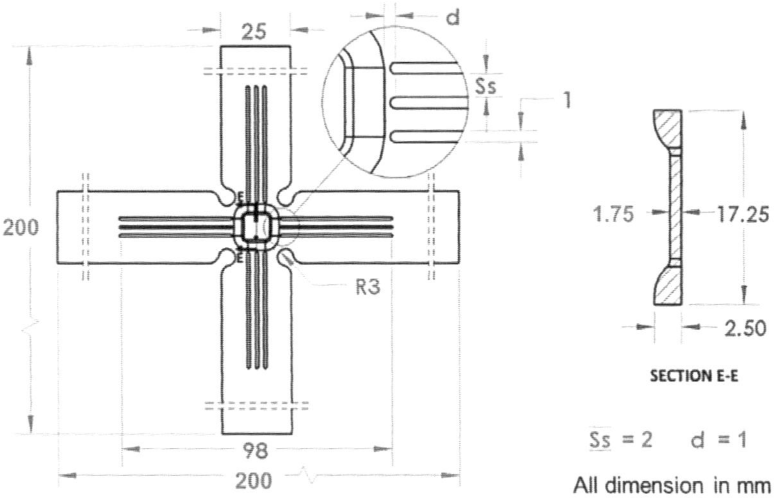

Figure 2. Biaxial cruciform specimen with its dimensions

The model of the cruciform sample was imported as the IGES file in PAMSTAMP. After importing the sample in the simulation software, the material properties were assigned and the sample was meshed with the quad element (four node shell element). Afterwards, different displacement boundary conditions were applied on the arms of the samples to obtain the desired outputs (Figure 3). The boundary conditions applied on the samples for getting the desired FLDs are mentioned in Table 3.

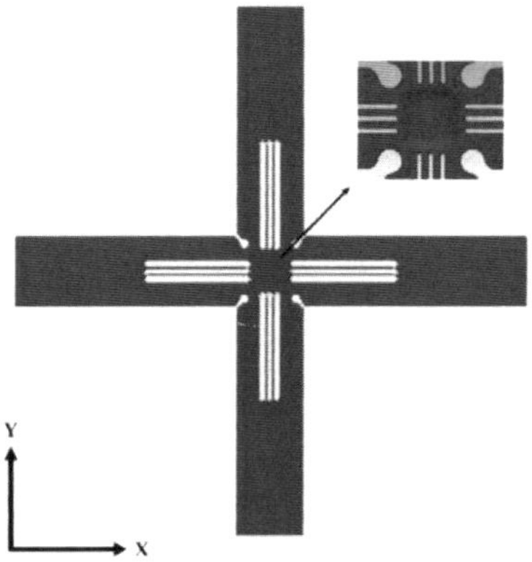

Figure 3. Finite element model of cruciform sample

Table 3. Primary boundary conditions

S. No.	Strain path conditions	Displacement control	
		X-direction	**Y-direction**
1.	Uniaxial	Displacement	Unconstrained
2.	Plane Strain	Displacement	Constrained
3.	Biaxial	Displacement	Displacement

Once the simulations with primary boundary conditions were performed the pre-strain values of 3% and 5% were selected on the basis of the maximum major strain achieved in each case. Based on the pre-strain values multiple simulations were performed by changing the boundary condition until necking occurred during the simulations. These boundary conditions are briefly explained with the help of Figure 4a. Figure 4b shows the schematic of strain path change.

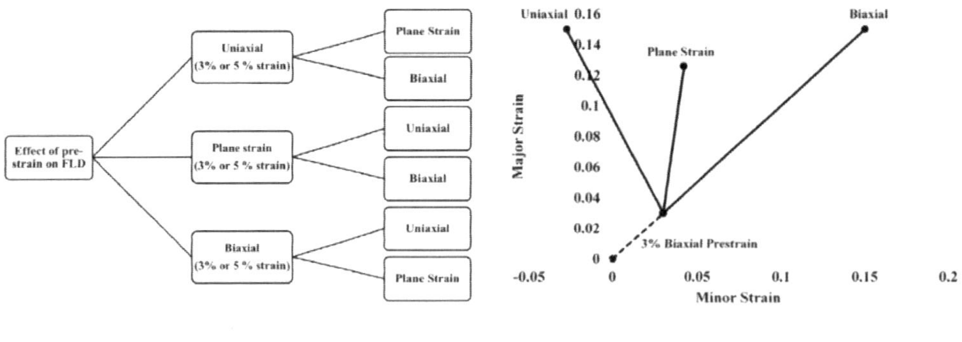

Figure 4. (a) Flowchart of varying boundary conditions, and (b) Schematic of strain path change

4. Result and discussion

The findings of this work are derived from the post-analysis of FE simulations. The main goal of this work is to uncover the effect of pre-strain on the FLD, while performing the biaxial in-plane deformation using cruciform samples. Typical in sheet metal forming processes, localized necking occurs before failure. Therefore, the thickness gradient criterion (TGC) is employed as the failure criteria to determine the limiting strain at necking during the simulation.

The TGC criterion used for this work has been proposed in 1994 [21]. It has been validated experimentally by different researchers and further used extensively to find forming limits [22–24].

Experimental research revealed that the existence of a critical thickness gradient in the sheet is subjected to the presence of a localized neck [21]. When a metal sheet begins to deform, a thickness gradient known as "R_{tg}" forms in the sheet and is expressed as follows:

$$R_{tg} = \frac{t_{neck}}{t_{neighbour}} \tag{1}$$

In the above equation R_{tg} represent the thickness gradient, t_{neck} corresponds to the current thickness of a neck element and $t_{neighbour}$ indicates the current thickness of neighboring element. As forming operations progress, the thickness gradient values fall from their initial value of 1.0 over time. The thickness gradient hits a crucial value at the point where diffuse necking transitions to localized necking, and this can be stated as follows:

$$R_{tg} = R_c \tag{2}$$

A critical thickness gradient value of 0.92 was reported experimentally in the literature[24]. Necking of sheet metal components is considered to occur, if the thickness gradient (R_{tg}) is less than 0.92. Throughout the simulations, we consistently measured the thickness gradient values for critical elements at various time frames. This observation was conducted for each scenario, i.e., uniaxial, plane strain, and biaxial loading condition. Examining the thickness gradient involved considering four neighboring elements from the simulation's initiation to the point of necking for each case. Figure 5 presents a visual representation of the thickness ratio to the neighboring elements (1 to 4) at different time frames for the biaxial case of as-received material. Notably, Element number 2 emerged as the first to meet the thickness gradient criterion, as illustrated in Figure 5.

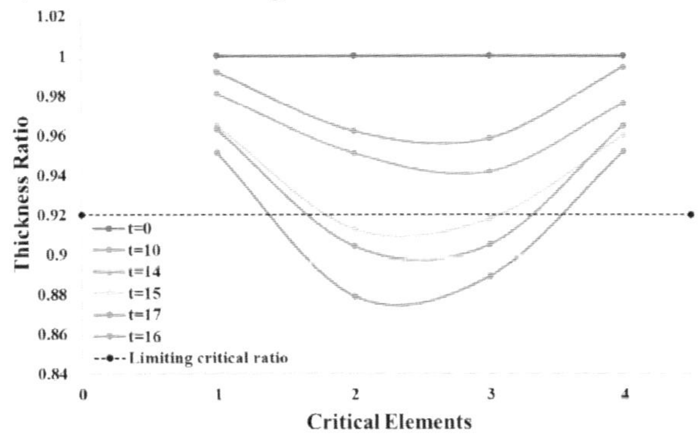

Figure 5. Thickness ratio vs critical elements for linear biaxial strain path

This critical element's (element no. 2) major and minor strains are regarded as the limit strain under biaxial conditions. Following this approach, we obtained major and minor strain values at critical element for the uniaxial and plane strain which ultimately provided us with the FLD of as-received material (Figure 6).

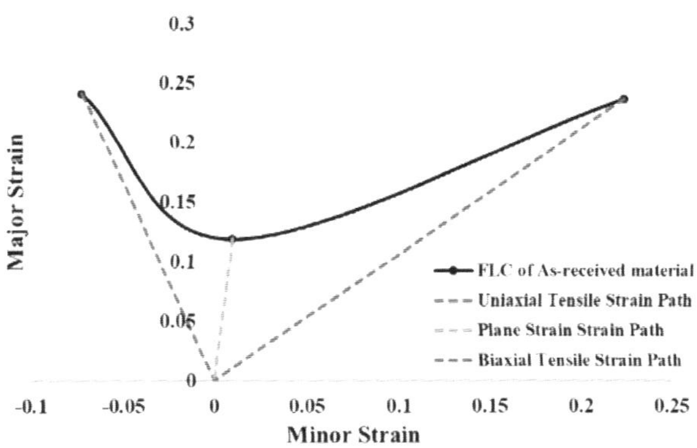

Figure 6. FLD of as-received material

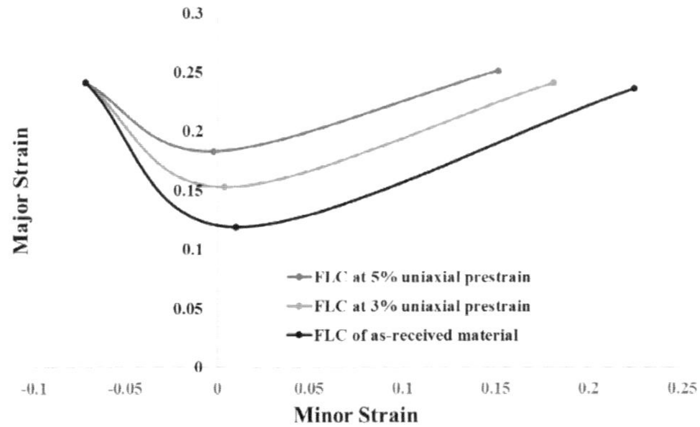

Figure 7. As-received material's FLD with uniaxial pre-straining

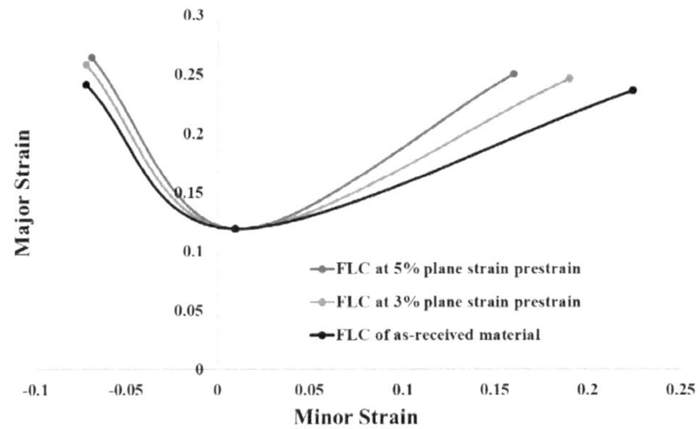

Figure 8. As-received material's FLD with plane strain pre-straining

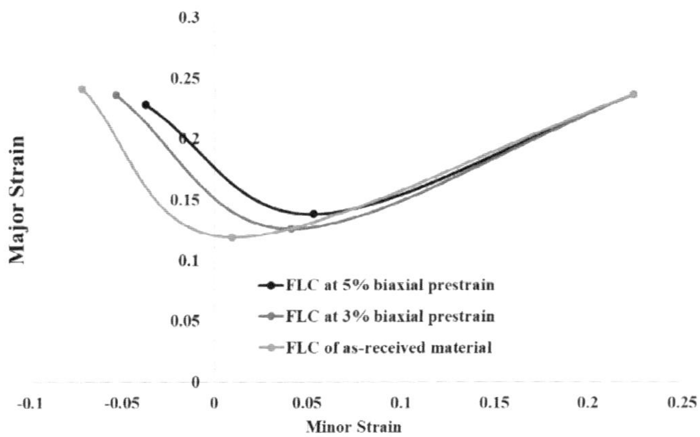

Figure 9. As-received material's FLD with biaxial pre-straining

As mentioned earlier, we determined two pre-strain values (0.03 and 0.05) based on the maximum major strain obtained from the FLD of the as-received material under uniaxial, plane strain, and biaxial conditions. Based on these pre-strain values, we conducted additional simulations by introducing changes in the strain path during the simulation process. The methodology previously employed to select the critical element and identify forming limits was consistently applied in these subsequent simulations. Figure 7 and Figure 8 and Figure 9 shows the effect of pre-strain on forming limit diagrams for uniaxial, plane strain and biaxial condition.

As illustrated in Figure 7, in case of uniaxial pre-straining (both 3% and 5%), we observe higher forming limits in the plane strain and biaxial domains when compared to the forming limit curve (FLC) of the as-received material. Notably, the influence of plane strain pre-straining leads to a growth of forming limits on both the left and right sides of the FLD, as depicted in Figure 8. Moving on to biaxial pre-straining (Figure 9), it becomes evident that both 3% and 5% pre-straining result in increased forming limits within the uniaxial zone. Simultaneously, there is a notable shift of the FLD_0 from its original plane strain position in an upward direction. Interestingly, in all cases, the new forming limit point coincides with the original point, adding a layer of consistency to these observations. This analysis provides valuable insights into how different pre-strain conditions impact the material's forming behavior across various stress scenarios.

5. Conclusion

This study delves into the influence of pre-strain on the FLD of aluminium alloy 5052, employing a biaxial cruciform sample through FE simulation. The preliminary phase involved determining pre-strain levels of 3% and 5% through FE simulations conducted under uniaxial, plane strain, and biaxial conditions to obtain the as-received material's FLD. Subsequently, based on the pre-strained values additional simulations were performed. The key highlights of this investigation are summarized below:

- Uniaxial pre-straining led to the enhancement of forming limits in both the plane strain and biaxial zones.
- Plane strain pre-straining showcased improved forming limits on both sides, with FLD_0 maintaining its original position when compared to the FLD of the as-received material.
- Similarly, biaxial pre-straining resulted in enhanced forming limits.

These findings highlight how different pre-strain conditions can positively impact the material's behaviour.

Acknowledgement
The authors would like to thank Mr. Adwait Pandey from ESI Group for supporting in the simulation activities.

References

[1] Hosford W F and Caddell R M 2011 Metal Forming Mechanics and Metallurgy, 4th Edition - William F Hosford & M. Cadell ISBN 978-1-107-00452-8 331
[2] Banabic D, Carleer B, Comsa D S, Kam E, Krasovskyy A, Mattiasson K, Sester M, Sigvant M and Zhang X 2010 Sheet metal forming processes: Constitutive modelling and numerical simulation Sheet Metal Forming Processes: Constitutive Modelling and Numerical Simulation 1–301
[3] S. P. Keeler and W. A. Backhofen 1963 Plastic Instability and Fracture in Sheet Stretched over Rigid Punches ASM Transactions Quarterly 56 25–48
[4] Goodwin G M 1968 Application of Strain Analysis to Sheet Metal Forming Problems in the Press Shop SAE Technical Papers
[5] Paul S K 2021 Controlling factors of forming limit curve: A review Advances in Industrial and Manufacturing Engineering 2 100033
[6] Kasaei M M and Oliveira M C 2022 Influence of the contact with friction on the deformation behavior of advanced high strength steels in the Nakajima test Journal of Strain Analysis for Engineering Design 57 193–207
[7] Achineethongkham K and Uthaisangsuk V 2020 Analysis of forming limit behaviour of high strength steels under non-linear strain paths using a micromechanics damage modelling International Journal of Mechanical Sciences 183
[8] Basak S, Katiyar B S, Orozco-Gonzalez P, Baltazar-Hernandez V H, Arora K S and Panda S K 2019 Microstructure, forming limit diagram, and strain distribution of pre-strained DP-IF steel tailor–welded blank for auto body application International Journal of Advanced Manufacturing Technology 104 1749–67
[9] Zhang R, Shao Z, Shi Z, Dean T A and Lin J 2021 Effect of cruciform specimen design on strain

paths and fracture location in equi-biaxial tension Journal of Materials Processing Technology 289 116932

[10] Kuwabara T, Kuroda M, Tvergaard V and Nomura K 2000 Use of abrupt strain path change for determining subsequent yield surface: experimental study with metal sheets Acta Materialia 48 2071–9

[11] Anon ISO 16842:2014(en), Metallic materials — Sheet and strip — Biaxial tensile testing method using a cruciform test piece

[12] Mitukiewicz G, Głogowski M, Stelmach J, Leyko J, Dimitrova Z and Batory D 2019 Strengthening of cruciform sample arms for large strains during biaxial stretching Materials Today Communications 21 100692

[13] Nikhare C P 2018 Experimental and numerical investigation of forming limit differences in biaxial and dome test Journal of Manufacturing Science and Engineering, Transactions of the ASME 140 1–12

[14] Siddiqui A H, Patil J P and Mishra S 2023 Design of a Biaxial Cruciform Specimen with a High Degree of Plastic Deformation and Yield Locus Evolution Experimental Mechanics 63 853–69

[15] Xiao R, Li X X, Lang L H, Song Q and Liu K N 2017 Forming limit in thermal cruciform biaxial tensile testing of titanium alloy Journal of Materials Processing Technology 240 354–61

[16] Wang Z, Guines D, Chu X and Leotoing L 2022 Characterization of forming limits at fracture from shear to plane strain with a dedicated cruciform specimen International Journal of Material Forming 15 1–12

[17] Yu Y, Wan M, Wu X D and Zhou X Bin 2002 Design of a cruciform biaxial tensile specimen for limit strain analysis by FEM Journal of Materials Processing Technology 123 67–70

[18] Jocham D, Gaber C, Böttcher O, Wiedemann P and Volk W 2017 Experimental prediction of sheet metal formability of AW-5754 for non-linear strain paths by using a cruciform specimen and a blank holder with adjustable draw beads on a sheet metal testing machine International Journal of Material Forming 10 597–605

[19] Song X, Leotoing L, Guines D and Ragneau E 2017 Characterization of forming limits at fracture with an optimized cruciform specimen: Application to DP600 steel sheets International Journal of Mechanical Sciences 126 35–43

[20] YUAN W, WAN M, WU X, MA B, LU X and YANG B 2020 Influence of uniaxial tensile pre-strain on forming limit curve by using biaxial tensile test Chinese Journal of Aeronautics 33 2268.

[21] Kumar S, Date P P and Narasimhan K 1994 A new criterion to predict necking failure under biaxial stretching Journal of Materials Processing Tech. 45 583–8

[22] Zhang R T, Chen X F, Su H B and Chen Z Y 2013 Numerical and Experimental Study of Forming Limit Diagrams in Sheet Metal Forming and Seamed Tube Hydroforming Applied Mechanics and Materials 395–396 914–9

[23] Singh A K and Narasimhan K 2019 Prediction of necking & thinning behaviour during hot stamping conditions of 22MnB5 steel IOP Conference Series: Materials Science and Engineering 651

[24] Ganguly S, Raj A and Narasimhan K 2021 A Novel Approach for Process and Tool Design During Manufacturing of Industrial Sheet Metal Components Transactions of the Indian Institute of Metals 74 1203–12

Applicability of the formability evaluation method for advanced high strength steels

Changwei LIAN[1,2], Chao NIU[1,2] and Fei HAN[1,2]

[1]R&D Center, Baoshan Iron & Steel Co., Ltd, Shanghai 201900, China
[2]State Key Laboratory of Development and Application Technology of Automotive Steels (Baosteel), Shanghai 201900, China

E-mail: lianchangwei@baosteel.com

Abstract. The use of high strength and ultra-high strength steels has become the main technical solution for reducing vehicle weight and improving safety. It is more complicated to evaluate the formability of advanced high strength steels with different microstructure and deformation characteristics. In this paper, the applicability of existing formability evaluation methods for advanced high strength steels has been verified by experiment and theoretical analysis. The experiment and data analysis of conventional formability evaluation methods, such as strain hardening index, forming limit curve and hole expansion ratio, were carried out on advanced high strength steels, such as dual-phase steel, complex-phase steel, quenched and tempered steel and dual-phase steel with high formability, which are widely used in automotive industry. It is found that due to the complex work hardening characteristics of advanced high strength steels, the work hardening homogenization and forming limit cannot be characterized by a single work hardening index or forming limit diagram, and the standard hole expansion method cannot reflect the quality sensitivity of the formed edge. A new formability evaluation index is proposed and discussed, which can be used to more accurately compare the formability of advanced high strength steels and provide a reference for material evaluation and part selection.

1. Introduction

Steel is still the most widely used material in automotive bodies and components because of its combined strength and cost advantages [1]. In response to the demand for environmental protection and safety in the automotive industry, automotive lightweighting has become an industry hot spot. The application of high strength steels and ultra-high strength steels has gradually become the main technical solution to reduce weight and improve safety, and more and more high strength steel materials have been developed and applied to the automotive industry [2]. Steel companies are constantly developing higher performance high strength steel products to meet the needs of automotive industry users [3]. The application of these materials to the vehicle body requires a combination of formability and performance, which requires an accurate understanding of the mechanical properties and forming characteristics of the material at the design stage.

Conventional steels include mild steel, interstitial-free atomic steel (IF), baking hardening steel (BH), carbon-manganese steel (CMn), and low alloy high strength steel (HSLA), and so on. Advanced high strength steels usually have complex-phase organization, achieve the required chemical composition and microstructure by strictly controlling the heat treatment process, and adopt various strengthening mechanisms to achieve different strength, ductility, and fatigue properties. The first generation of advanced high strength steels includes duplex steel (DP), transformation induced plasticity steel (TRIP), complex phase steel (CP), press hardening steel (PHS), and martensitic steel (MS) [3]. The second generation of advanced high strength steels are fully austenitic, including austenitic stainless steels and twinning induced plasticity steel (TWIP) [4], which have excellent strength and plasticity balance. By introducing metastable austenite in ultra-fine grain ferrite, bainite or martensite organization [5], the elongation of the new generation of advanced high strength steels is improved compared to the first generation steels at the same strength level through certain strengthening mechanisms and plasticity growth mechanisms, such as quenching & partitioning steel (QP), dual-phase steels with enhanced formability (DH), complex-phase steels with enhanced formability (CH), and medium Mn-TRIP Steel [6-8]. The formability properties of steel sheets generally decrease with increasing strength, and the complex deformation mechanisms caused by the complex phase microstructure, as well as the inherently low total elongation, present difficulties and challenges in the evaluation of formability properties and forming applications.

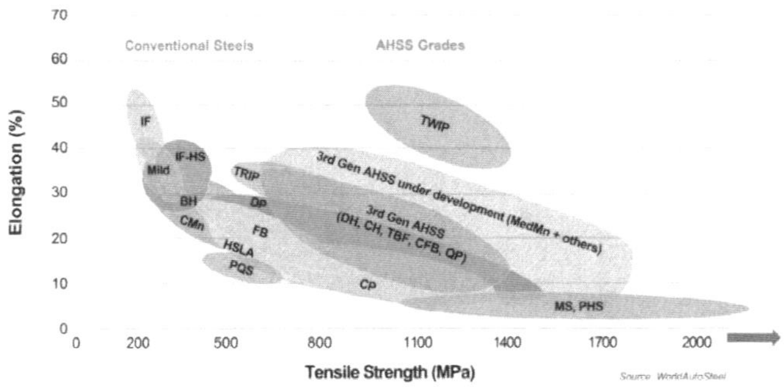

Figure 1. The Global Formability Diagram of automotive steel [1]

In this paper, advanced high strength steels commonly used in automotive bodies and traditional automotive steel sheets are selected to evaluate the applicability of conventional formability evaluation methods and indices for high strength steels through the analysis of experimental formability evaluation data, and new formability evaluation methods and indices are discussed through theoretical analysis.

2. Evaluation method for formability

Formability generally refers to the ability of a material to adapt to different forming processes such as stamping and bending. Formability indices are used to select and evaluate the forming properties of sheet materials. When designing forming tools and analyzing forming processes, it is necessary to refer to the data in the formability indices. For steels used in cold forming, the strain hardening index (n-value), forming limit curve and hole expansion ratio are commonly used indicators for testing and evaluating forming properties.

The strain hardening index is a performance index to characterize the work hardening behavior of metallic materials. It is usually calculated from the stress-strain curve obtained from a uniaxial tensile test. This index is often used to evaluate the forming properties of metallic materials, the higher the value of n, the greater the strain homogenizing ability of the material. Ludwick first proposed the concept of strain hardening index n [9], Hollomon established the power exponent equation (1) [10].

$$\sigma = K\varepsilon^n \quad (1)$$

Where σ is the true stress, ε is the true strain, and K is the strength factor. The calculation of the strain hardening index is related to the selected strain range. For conventional steel materials, the strain hardening index remains basically constant after the strain exceeds the initial yield point, such as mild steel with uniform elongation over 20%, and the value of n is usually obtained by taking the engineering strain in the range of 10 to 20%.

Forming limit is a criterion to analyze whether the process can develop stably in the forming process, and it characterizes the ability of the sheet to resist local necking or fracture under different strain states in the forming process. Sheet in different strain paths can be achieved in different ultimate strain, these ultimate strain values in the coordinate system constitutes the ultimate strain distribution area, as well as based on the ultimate strain points into a curve known as the forming limit diagram (FLD), from the ultimate strain points into a curve known as the forming limit curve (FLC). The forming limit is measured in the laboratory by using a hemispherical or flat-bottomed cylindrical die, and by varying the width of the specimen to obtain different strain paths and strain states [11-12]. In this paper, a hemispherical die was used for stamping tests on a forming test machine, and the strain measurements were performed using a digital image correlation (DIC) method, in which random scattering spots were sprayed on the specimen surface, and two sets of high-definition cameras were used to record the forming process and to collect the calculated strain information. Ultimate strain points were obtained using the cross-section line method or the time-dependent method recommended by ISO 12004. And FLC_{min} refers to the minimum principal strain value in the forming limit curve.

Hole expansion ratio index refers to the ability of sheet material to resist edge cracking during the forming process, which is an effective method for evaluating the characteristics of material flanging and forming. According to ISO 16630, the hole expansion test includes two steps of punching the hole and pressing the hole using a cone-head convex tool. The tool is loaded by the testing machine until the hole is cracked through the thickness of the edge under the action of the convex tool to stop the test. The ultimate expansion ratio of the hole can be calculated by taking all the repetitions of the test.

3. Materials and results

Several steel grades are selected to investigate the applicability of the formability evaluation method for advanced high strength steels. They are DP980, CP980, MS1180, QP980, QP1180, DH980, Mn-TRIP980, and TWIP980. Also 3 conventional steels of DC06, HC220B, and HC340LA are selected for comparison.

The results of mechanical properties, n value, FLC_{min}, and hole expansion ratio of the selected grades are shown in Table 1.

Table 1. Testing results of selected grades.

Steel grades	Thickness (mm)	Yield strength (MPa)	Tensile strength (MPa)	Total elongation (%)	n	FLC_{min}	hole expansion ratio(%)
DC06	0.7	140	290	49	0.24	0.34	130
HC220B	0.8	245	360	38	0.19	0.28	108
HC340LA	1.2	398	493	29	0.17	0.26	66
DP980	1.4	660	1051	12	0.10	0.12	34
CP980	1.4	881	1005	11	0.06	0.07	55
MS1180	1.4	1029	1289	7	0.05	0.10	33
QP980	1.6	687	1068	22	0.15	0.18	27
QP1180	1.2	976	1239	17	0.10	0.16	29
DH980	1.4	715	992	17	0.12	0.13	29
Mn-TRIP980	1.4	668	1030	42	0.51	0.21	25
TWIP980	1.4	511	982	60	0.35	0.39	40

3.1. *Strain hardening index*

In order to apply and describe the forming properties of materials, it is necessary to analyze the obtained test stress and strain data to obtain the evolution rule of the hardening characteristics. For conventional steels, the work hardening index remains essentially constant after the strain exceeds the initial yield point. For advanced high strength steels, the strain is distributed among several phases, especially for austenite-containing steel grades, and the work-hardening index changes with the change in strain. Therefore, the instantaneous work hardening index is used for analysis. The instantaneous work hardening index during the tensile process can be expressed by the following equation.

IOP Conf. Series: Materials Science and Engineering **1307** (2024) 012014 doi:10.1088/1757-899X/1307/1/012014

$$n' = \frac{\partial ln\sigma}{\partial ln\varepsilon} \tag{2}$$

According to the calculation method of Song Y [13], the difference is used instead of differentiation, based on the test data arbitrary strain. The instantaneous hardening index at any strain of the test data can be approximated by equation (3).

$$n'_m \approx \frac{\Delta ln\sigma_m}{\Delta ln\varepsilon_m} = \frac{ln\sigma_{m+1} - ln\sigma_m}{ln\varepsilon_{m+1} - ln\varepsilon_m} \tag{3}$$

In the formula, ε_m andσ_m are the true strain, and the true stress at the place respectively. Directly calculated results from data obtained by conventional tensile test methods usually have vibration, and in the actual calculation, it is necessary to set a reasonable strain step or moving average method to deal with the microscopic vibration caused by the measurement in the test process, or firstly, use the preset hardening model to fit the test data to get the functional relationship between stress and strain and then obtain the derivation by equation (3). The application of digital image correlation (DIC) technology and more accurate load cell can get better quality of test data.

Different microstructural properties and plastic deformation mechanisms cause different patterns of change in strain hardening rate. In order to analyze the changes of strain hardening index of different types of microstructures, the instantaneous strain hardening index of selected steels was calculated by using equation (3). At the same time, in order to facilitate the comparison of steel grades with different uniform elongation, the relative plastic strain, which is expressed as the ratio of the plastic strain to the plastic strain corresponding to the single-phase tensile uniform elongation of the steel grades, is used to determine the instantaneous strain hardening index. The calculated results are shown in Figure 2.

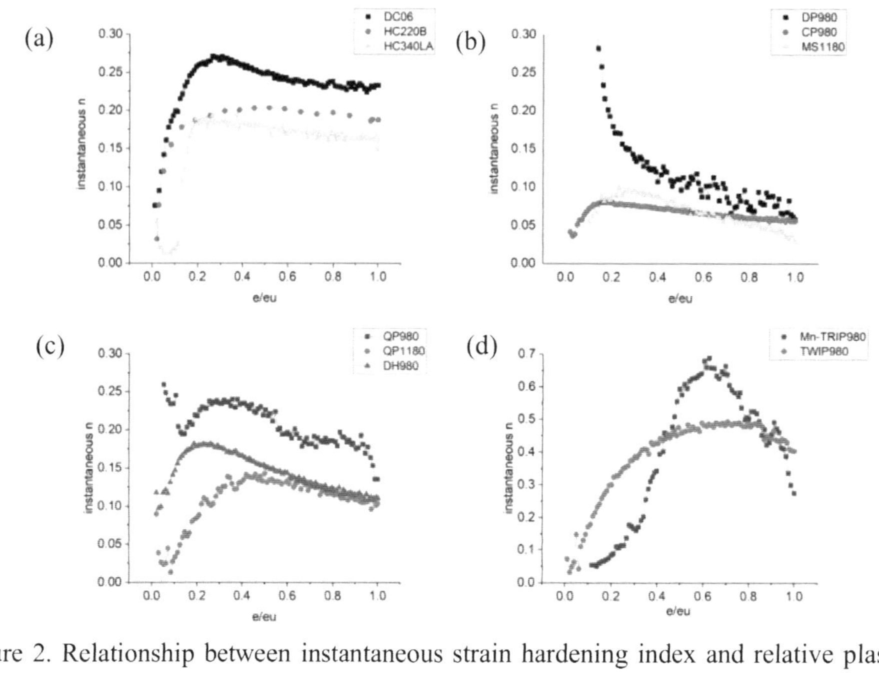

Figure 2. Relationship between instantaneous strain hardening index and relative plastic strain of different types of steel

For the conventional steel grades DC06, HC220B and HC340LA, the n values are almost constant after stabilization as shown in Figure 2(a), so the n values can be used to compare the formability and strain homogenizing ability of different steel grades. DP980, CP980 and MS1180, the first generation of advanced high strength steels, show a strong decreasing trend of n value with increasing strain as shown in Figure 2(b). The continuous decreasing trend of n value for the 3rd generation advanced high strength steels of QP980, QP1180 and DH980 is significantly changed by the introduction of the TRIP effect with respect to the first generation of advanced high strength steels, as shown in Figure 2(c). The TRIP effect is more pronounced for Mn-TRIP980 due to the presence of a large amount of austenite, which exhibits a sharp increase followed by a sharp decrease in the n value, while the n value of TWIP steels increases slowly to a very high value as shown in Figure 2(d). Analyzed in terms of strain mechanisms, unidirectional ferrite steels, which are dominated by dislocation mechanisms, have very stable n values, and multiphase materials show an overall decreasing trend in n values, while advanced high strength steels, which are introduced by phase transitions and twinning mechanisms, show a more complex change in their n values.

As can be seen from the results, the instantaneous strain hardening indices of all steel grades, except for the conventional steels, showed significant changes at the later stage. Therefore, the traditional method of selecting the average n value of a certain strain range to represent the overall performance can no longer be applied to advanced high strength steels as an evaluation index of forming performance.

3.2. Forming limit curve

Forming limit experiments were performed for different steels. The forming limit curves were obtained using the polynomial segment fitting method, and the results are shown in Figure 3.

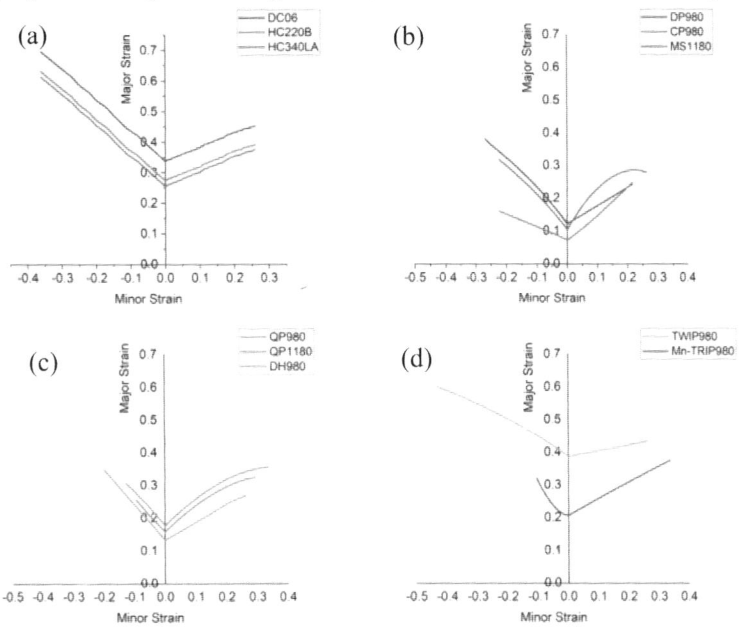

Figure 3. Forming limit curve of different types of steel

For conventional steel grades, it is common to compare the forming capability of different materials laterally using FLC_{min}, which is defined as the lowest point of principal strain in the forming limit diagram, i.e., the forming limit value of the material in the plane strain state. From the above results, it can be seen that the FLC_{min} of advanced high strength steels in the strength class of 780 MPa~1180 MPa are all in the range of 0.1~0.2. However, from the analysis in the previous section 3.1, it is found that the strain hardening indices of different steel grades are very different, so further comparison of their forming limits is needed. DP980 and QP980 with thicknesses of 1.8 mm and 1.6 mm were selected, and the FLC_{min} values and limit points of the two grades were approximately the same, as shown in Figure 4(a), but there was a significant difference in the limiting dome heights (LDH) of the specimens of different widths from 20mm (specimen number 1) to 180mm (specimen number 9), as shown in Figure 4(b) [14]. Teflon film was used as a lubricant and the test parameters were set to be the same for all forming limit tests of both materials, so that the FLC and LDH could be directly compared using the same strain paths. This indicates that, when reaching the same forming height, QP980, due to higher n values, obtains a more homogeneous strain distribution on the formed parts, and has a better part forming capability, although it reaches an approximate ultimate strain at the point of instability. Therefore, comparison of FLC_{min} or FLC does not fully reflect the forming performance of different advanced high strength steels. Uniform deformation can delay the onset of failure, so that the overall forming capability of the material is greatly improved. The ability to homogenize deformation is also an important index for evaluating the forming performance.

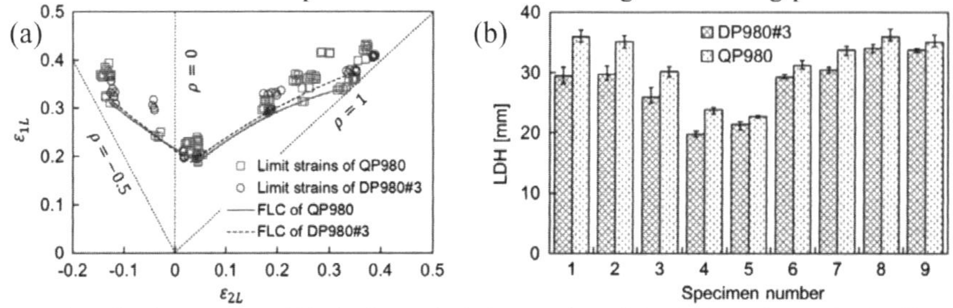

Figure 4. Forming limit curve and limit dome height results of QP980 and DP980 [14]

3.3. *Hole expansion ratio*

Hole expansion test was carried out in order to obtain the hole expansion ratio data of each steel grade. The results are shown in Figure 5(a), which shows that for conventional high strength steels, the hole expansion ratio index decreases linearly with the tensile strength. For advanced high strength steels with a tensile strength of 780 MPa or more, the hole expansion ratio is in the range of 24-42%, which does not correlate well with the tensile strength.

Hole expansion ratio was also compared using different edge processing methods for DP980, QP980, MS1180 and Mn-TRIP980. The edge quality obtained from punched holes as specified in the hole expansion ratio test standard is poor, usually with a small amount of bright banding with burrs, and the hole expansion ratio obtained under these conditions do not effectively discriminate the crack susceptibility of the edges of advanced high strength steels. In contrast, wire EDM and edge polishing can be used to achieve near flawless edge quality. From the results in Figure 5(b), it can be seen that the hole expansion ratio of DP980 and MS1180 are relatively high under punching conditions, and DP980 is higher than MS1180, while DP980 is lower than MS1180 under wire-cut machining conditions. If the edge quality sensitivity coefficient is defined as the ratio of the decrease in the hole expansion ratio under punching conditions to that of under wire-cut conditions, the MS steel is the most sensitive with a decrease of 67%. Therefore, the addition of two parameters, hole expansion ratio with wire-cut edge and sensitivity coefficient, provides a more comprehensive characterization of the edge forming performance of advanced high strength steels.

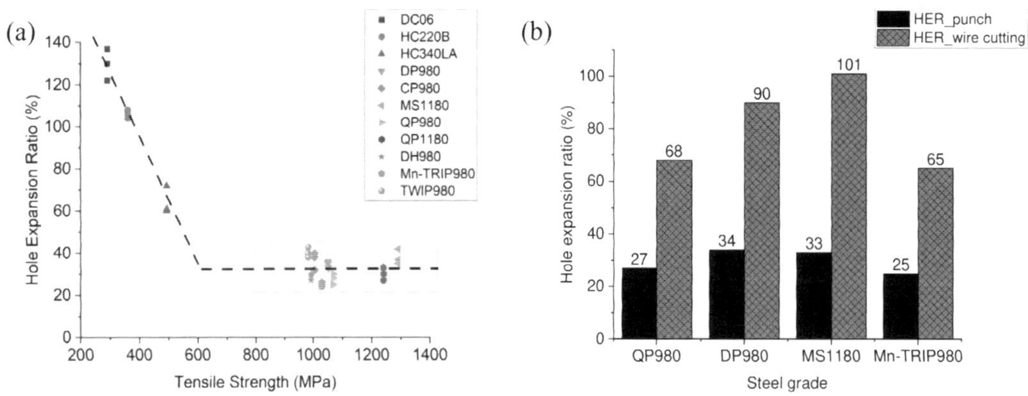

Figure 5. Testing results of hole expansion with different steel grades and different hole processing method

3.4. *New evaluation index for forming strain homogenization capability*

From the above analysis, it is found that the existing n value, forming limit curve and hole expansion ratio indexes can be used to evaluate the forming properties of high strength steels, but they cannot reflect the forming properties of advanced high strength steels well and comprehensively. Therefore, more accurate evaluation indices for advanced high strength steels need to be developed, and the feasibility of testing methods should be considered at the same time.

In terms of the overall forming capacity, considering that neither the change of n value with strain nor the forming limit can fully reflect this capacity, this paper proposes a new index of forming strain homogenization capacity (SHC). It is defined as the ratio of the integral value of the instantaneous strain hardening index (n' value) over the plastic strain (ε_p) to the plastic strain interval from ε_p^{min} to ε_p^{max}, which can be expressed by equations (4).

$$SHC = \frac{\int_{\varepsilon_p^{min}}^{\varepsilon_p^{max}} n'(\varepsilon_p)\, d\,\varepsilon_p}{\varepsilon_p^{max} - \varepsilon_p^{min}} \quad (4.a)$$

$$\varepsilon_p^{min} = 0.1\varepsilon_u \quad (4.b)$$

$$\varepsilon_p^{max} = \varepsilon_i^{ps} = FLD_{min} \quad (4.c)$$

$n'(\varepsilon_p)$ is the strain hardening index of the material at the specified strain. ε_p^{min} and ε_p^{max} are the upper and lower limits of the calculated strain interval. ε_u is the true plastic strain value corresponding to the uniform elongation. FLD_{min} is the instability limit of the material in the plane strain state, i.e. the value of the principal strain at the lowest point of the forming limit curve of the material, which can be obtained by forming limit experiments.

The index generalizes the calculation of the n value. As seen in equation (4.a), the SHC can be interpreted as a generalized average n value for a specific strain interval. When ε_p^{min} and ε_p^{max} are the plastic strains corresponding to engineering strains at 10% and 20%, respectively, the SHC simplifies to the conventional definition of the n-value. Moreover, SHC extends the upper limit of the strain interval to FLD_{min}, which covers the overall forming strain interval of the material. In addition, the reason for defining the value of ε_p^{min} as $0.1\varepsilon_u$ is to take into account the influence of the initial yielding stage on the calculation of the n-value by neglecting the data in the interval 0 to $0.1\varepsilon_u$. The uniaxial tensile test cannot provide the n-value data beyond the uniform strain. Alternatively, the n-value data can be obtained by hydraulic bulge test, in-plane torsion test, and other tests to obtain stain hardening data under large strain interval, or by extrapolation of the predictive hardening model. Combined with the instantaneous strain hardening index data and forming limit data, the SHC values of DP980 and QP980 were calculated according to equation (4), respectively, and DC06 and HC220B were also selected as reference, as shown in Figure 6(a). It can be seen that the difference between QP980 and DP980 is reflected by this index, which is consistent with the experimental results in Figure 4. Under the same die and stamping process conditions, QP980 and DP980 were used to produce the reinforcing beam parts in the actual stamping process. After the drawing process, QP980 did not crack when punched to the final position, while DP980 cracked at positions A and B, as shown in Figure 6(b). And the principal strains measured by the mesh method at different positions after the drawing process showed that QP980 was lower than DP980, as shown in Figure 6(c). This indicates that QP980 has better strain homogenizing ability and higher formability, and the SHC index can be used as one of the indices to be considered when selecting materials for parts and components.

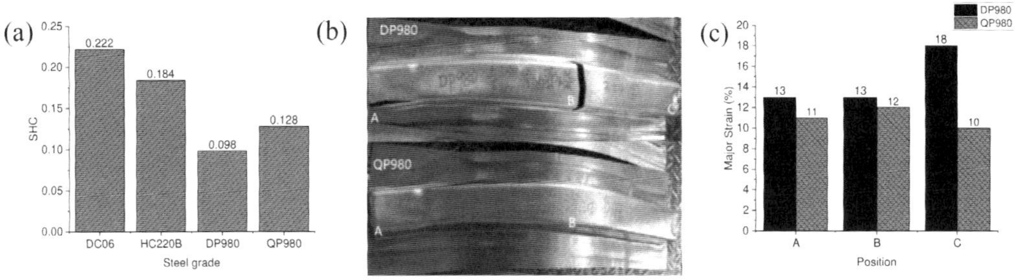

Figure 6. Calculated results of SHC for different steel grades and Forming test results of QP980 and DP980

IOP Conf. Series: Materials Science and Engineering **1307** (2024) 012014 doi:10.1088/1757-899X/1307/1/012014

Therefore, SHC provides the possibility of comparing the forming strain homogenizing ability of different steel grades. Combined with the conventional indexes, the forming properties of advanced high strength steels can be evaluated more comprehensively.

4. Conclusion

From the above test results and analysis, the following conclusions can be drawn.

(1) Advanced high strength steel due to its complex microstructure and containing sub-stable austenite phase caused by the complexity of the strain hardening behavior, strain hardening index with the strain changes significantly, cannot be used to select a certain strain range of the average n value on behalf of the performance of the whole. And forming limit diagram cannot comprehensively reflect the advanced high strength steels of the overall forming ability, must take into account the effect of deformation homogenization ability at the same time.

(2) The traditional hole expansion test method cannot effectively distinguish the flanging and edge forming performance of advanced high strength steels, especially ultra-high strength steels. The addition of two parameters, namely, the hole expansion ratio of the wire cutting edge and the sensitivity factor of the punching edge quality, can more comprehensively characterize the edge forming performance of advanced high strength steels.

(3) A new index describing the overall forming capability, the strain homogenization capability SHC, is proposed. As a generalized average n value, it provides a way to compare the overall forming performance and strain homogenization capability of different steel grades. Combined with the traditional evaluation indices, the forming performance of advanced high strength steels can be more comprehensively evaluated, providing an accurate reference for the design of the stamping process.

References

[1] WorldAutoSteel 2018 *Advanced high strength steels application guidelines* (Beijing: Metallurgical Industry Press)

[2] Chen X, Niu C, Lian C and Lin J 2017 *Procedia Engineering* **207** 556-561.

[3] Zhong Y, Wang L and Zhang Y 2018 *Baosteel Technical Research* **12** 9-20.

[4] Zhang L, Song R, Zhao C, Yang F, Qin S and Yang XU 2014 *Materials guide* **28** 111-118

[5] Schmitt J H and Iung T 2018 *Comptes Rendus. Physique* **19** 641-656

[6] Matlock D K, Speer J G, Moor E D and Gibbs P J 2012 *Engineering Science and Technology an International Journal* **15** 1-12

[7] Han Q, Zhang Y and Wang L 2015 *Baosteel Technology* **4** 9-17

[8] Matlock D K and Speer J G 2008 *Proc. Int. Conf. on Microstructure and Texture in Steels and Other Materials* (Jamshedpur: India)

[9] Ludwik P 1909 *Elemente der Technologischen Mechanik* (Heidelberg: Springer Berlin)

[10] Hollomon J H and Member J 1945 *Metals Technology* **12** 268-90

[11] UNE-EN ISO 2008 *Metallic Materials-Sheets and Strip-Determination of Forming-Limit Curves-Part 2: Determination of Forming-Limit Curves in the Laboratory* (ISO 12004-2: 2008)

[12] Panich S, Barlat F, Uthaisangsuk V, Suranuntchai S and Jirathearanat S 2013 *Materials & Design* **51** 756-766

[13] Song Y, Cheng Y and Liu Y 2001 *Science in China (Series E)* **44** 10

[14] Zhang L, Lin J, Min J, You Y and Kang L 2016 *Journal of Materials Engineering & Performance* **25** 2296-2306

IOP Conf. Series: Materials Science and Engineering **1307** (2024) 012015 doi:10.1088/1757-899X/1307/1/012015

The influence of intrinsic formability parameters on the sheared-edge ductility of 120K-class AHSS

Gihak Yim, Jiyoung Kim, Jinhwa Jeon, Joosik Hyun and Brandon M Hance

Applications Engineering Group, Hyundai Steel, Dangjin, South Korea

gyim88@hyundai-steel.com

Abstract. With enduring versatility, advanced high-strength steels (AHSS) play an increasingly important role in securing automotive safety, performance and sustainability targets in future automotive architectures. However, local formability (resistance to cracking) is of specific relevance to AHSS, and various intrinsic fracture-related concepts have emerged with the goal of understanding and improving the local formability of AHSS (sheared-edge ductility, bendability). This analysis is focused on the relationships between intrinsic formability parameters and other laboratory-scale performance indicators—i.e. the ISO 16630 hole expansion ratio (HER). Included is the AHSS material with minimum tensile strength designation above 1.2GPa: 120K-class 3rd generation AHSS. Additionally, the impact of material property uniformity on HER, and local toughness concept based on the true fracture strain is explored.

1. Introduction

Concerns related to lightweighting in the automotive industry are increasing due to the need for improving automotive fuel efficiency, meeting stringent CO2 emission standards, and complying with safety regulations. To address these demands, advanced high strength steel (AHSS) is widely applied in automotive components due to its favorable strength-to-density ratio. Despite its advantages AHSS faces challenges related to local formability failure modes, such as edge cracking, which are difficult to predict using conventional forming limit curves due to the absence of observable thinning or necking before fracture.[1] To evaluate sheared-edge ductility, it is common practice to perform hole expansion testing per ISO 16630. However, the hole expansion ratio (HER) is regarded as a relative "local formability" parameter due to the intra-laboratory scatter controversy, and the ratio itself has a limited meaning as a practical failure criterion. Thus, to evaluate the performance of AHSS grades, a practical performance classification system considering both global and local formability based on true fracture strain measurement was proposed and explored continuously by steelmakers, automakers and international industry consortiums.[2]

The paper aims to evaluate the performance of AHSS materials using the intrinsic global and local formability parameters derived from standard uniaxial tension tests.[3] Additionally, a toughness concept based on the true fracture strain value was investigated to explore the impact of material property uniformity on HER.

IOP Conf. Series: Materials Science and Engineering **1307** (2024) 012015 doi:10.1088/1757-899X/1307/1/012015

2. Materials and procedures

2.1. Material
This analysis includes a prototype AHSS material with minimum tensile strength designation above 1.2GPa. The basic material description is given in Table 1.

Table 1. Basic material description

Grade	Thickness	Description
1180MPa	0.9mm	Third generation steel (prototype) Uncoated cold rolled steel

2.2. Procedures

2.2.1. Tensile test. Two types of standardized tension test specimen were prepared for validation of fracture surface measurement method for fracture strain derivation. All specimens were aligned in the transverse direction (90° to the sheet roll direction). Specimen dimensions are given in Table 2.

Table 2. Nominal tensile test specimen dimensions

Specimen type	Overall length	Gage length	Gage Width
ASTM E8 subsize	120mm	25mm	6mm
EN ISO 6892-1	250mm	95mm	20mm

2.2.2. Hole expansion test. Table 3 shows the testing parameter for the hole expansion test in accordance to the ISO 16630 standard. All testing was carried out in the same lab with a universal sheet metal testing machine. First, the specimen was pierced by using the correct tool for even clearance for different thickness. Then, the specimen was expanded with a conical punch until through-the-thickness crack was observed. Note that 0.9mm thickness is not in the range of the thickness the standard states.

Table 3. Hole expansion test specimen specification

Specimen size	Hole size	Clearance	Punch type	Punch angle
100mm x 100 mm	10mm	12±2%	Conical	60 degree

2.3. Fracture area measurement method. Various fracture area measurement methods described by Hance were used to derive the true fracture strain value and the intrinsic formability parameters for this study.[3] Figure 1 shows the locations of the thickness and the width measurement. For the thickness, two points at the edges, one point at the center and two points at the quarter width positions were measured. For the fracture width, a single measurement at mid-thickness was done with an optical microscope. Finally, fracture area (A_f) was determined by the methods defined in Table 4. All measurements for the fracture surface were made with respect to the projected plane normal to the tensile axis.

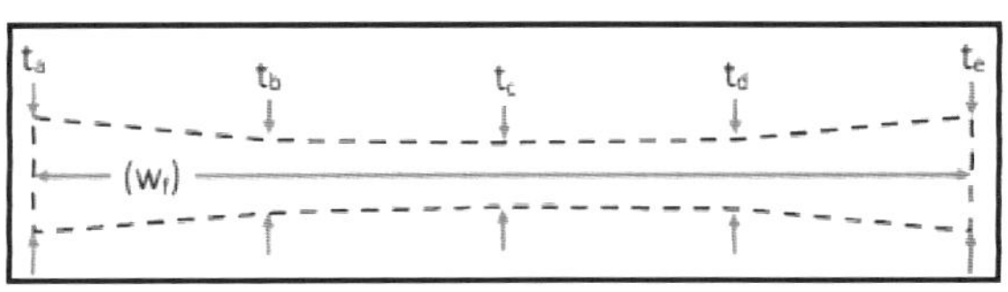

Figure 1. Schematic representation of a tension test specimen fracture surface. Five thickness and one width measurement locations are as shown [3]

Table 4. Fracture area measurement methods[3]

Methods	Fracture thickness (t_f)	Fracture area (A_f)
A	$t_f = t_c$	
B (ASTM E8)	$t_f = (1/6) \cdot (t_a + 4t_c + t_e)$	$A_f = t_f \cdot w_f$
C	$t_f = (1/5) \cdot (t_a + t_b + t_c + t_d + t_e)$	

3. Results and discussion

3.1. Validation of measurement method. In order to define which fracture area measurement method to use, two types of tensile test specimen, ASTM E8 subsize and EN ISO 6892-1 were prepared from a 120K grade steel sheet regarding the sheet rolling direction. The considered orientation of the specimen are 0°,45° and 90° respectively. After tensile testing, the fracture area was measured for each according to the mentioned method. Then, the true fracture strain(TFS) was derived with the acquired measurement for comparison with the method described by Hance.[3]

$$TFS = \varepsilon_f = \ln\left(\frac{A_0}{A_f}\right) \quad (1)$$

Where A_0 is the tension test specimen cross-section area before testing and A_f is the projected fracture surface area after fracture.

Table 5. True fracture strain measurement (n=5 for each)

		EN				ASTM subsize			
		0°	45°	90°	Avg.	0°	45°	90°	Avg.
Width [mm]	w_0	20.11	20.13	20.11	20.12	5.99	6.01	6.01	6.01
	w_f	17.47	17.37	17.49	17.44	4.68	4.62	4.65	4.65
Thickness [mm]	t_0	0.90	0.91	0.92	0.91	0.91	0.90	0.92	0.91
	Method A	0.48	0.45	0.50	0.48	0.55	0.53	0.59	0.56
	Method B	0.53	0.49	0.55	0.52	0.58	0.54	0.61	0.58
	Method C	0.54	0.50	0.55	0.53	0.61	0.55	0.64	0.60
Area [mm^2]	A_0	18.09	18.35	18.46	18.30	5.44	5.40	5.52	5.45
	Method A	8.39	9.32	9.39	9.04	2.59	2.72	2.85	2.72
	Method B	9.32	8.55	9.54	9.14	2.72	2.50	2.84	2.68
	Method C	9.39	8.71	9.66	9.25	2.85	2.56	2.95	2.79
TFS	Method A	0.77	0.86	0.74	0.79	0.74	0.80	0.70	0.75
	Method B	0.66	0.76	0.66	0.70	0.70	0.77	0.67	0.71
	Method C	0.66	0.75	0.65	0.68	0.65	0.75	0.63	0.67

Table 5 is the comparison result for the effect of test specimen type and fracture measurement method on true fracture strain values for 120K grade steel. Overall, Method B showed similar variation as Method C, but note that Method C is conceptually more accurate and conservative method.

3.2. Impact of material property uniformity. One of the crucial assessments at the development stage is to evaluate the material property uniformity for quality control and problem identification. ASTM

IOP Conf. Series: Materials Science and Engineering **1307** (2024) 012015 doi:10.1088/1757-899X/1307/1/012015

E8 subsize samples in the transvers orientation were acquired from various width locations for assessment, as shown in Figure 2. As presented in Table 6, material property variability across the coil-width blank was observed for the 120K prototype grade steel, particularly evident for yield and tensile strength.

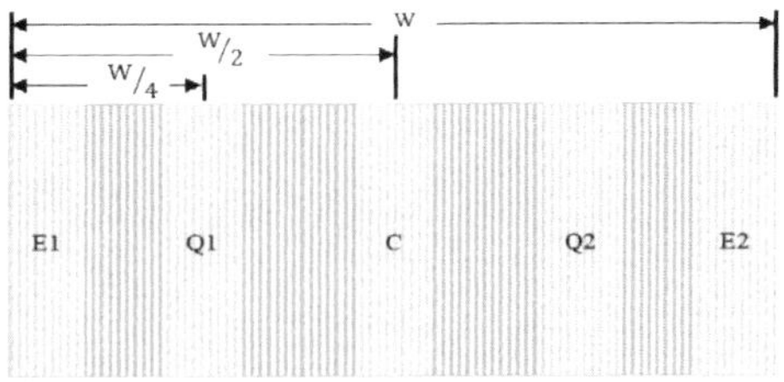

Figure 2. Schematic representation of a coil-width blank for assessment of material property uniformity. Samples were acquired from each highlighted zone, as indicated.

Table 6. Mechanical Properties (n=5 for each)

	E1	Q1	C	Q2	E2	Average
Uniform-El [%]	11.3	11.1	11.0	10.6	10.6	10.9
Total El [%]	16.2	15.0	14.9	15.1	14.8	15.2
YS [Mpa]	882	854	885	925	963	902
UTS [MPa]	1284	1294	1294	1286	1279	1287

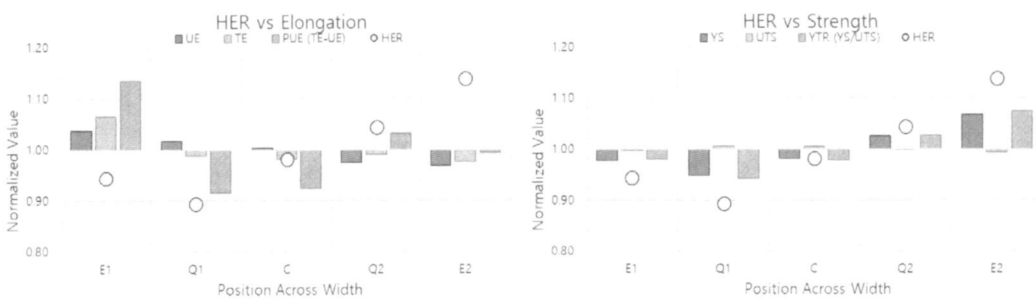

Figure 3. Test result comparison cross the width.

Thus, the impact of cross-width property uniformity on the intrinsic formability parameters and the correlation between the local formability performance indicator, HER was investigated. The intrinsic formability parameters described by Hance were used, as in Table 7[3]. Note that local/global strain ratio and formability index are strain-based parameters, and the performance index is based on both strain and the ultimate tensile strength. The calculated intrinsic parameter results, along with the measured hole expansion ratio is summarized in Table 8.

Table 7. Intrinsic parameters based on true fracture strain value

Parameters	Equation
Local/Global Strain ratio (L/GSR)	TFS/ε_u
Formability Index (F.I.)	$\mathrm{F.I.} = \sqrt{\varepsilon_u \cdot TFS}$
Performance Index (P.I.)	$\mathrm{P.I.} = \mathrm{UTS} \cdot \mathrm{F.I.} = \mathrm{UTS} \cdot \sqrt{\varepsilon_u \cdot TFS}$

Table 8. Formability Assessment (n=5 for each)

		Position across width					
		E1	Q1	C	Q2	E2	Average
Intrinsic Formability Parameters	TFS	0.63	0.54	0.57	0.61	0.67	0.60
	L/GSR	5.87	5.14	5.47	6.08	6.63	5.84
	F.I.	0.26	0.24	0.24	0.25	0.26	0.25
	P.I .(MPa)	334	309	315	320	331	322
Hole Expansion	HER (%)	30.0	28.4	31.2	33.2	36.2	31.8

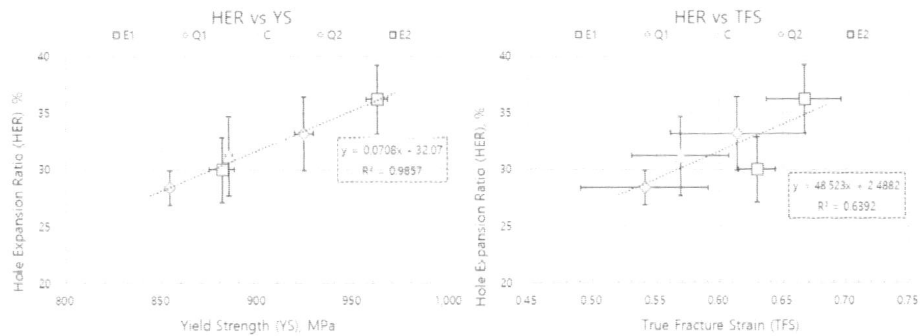

Figure 4. Linear relationship comparison for hole expansion ratio

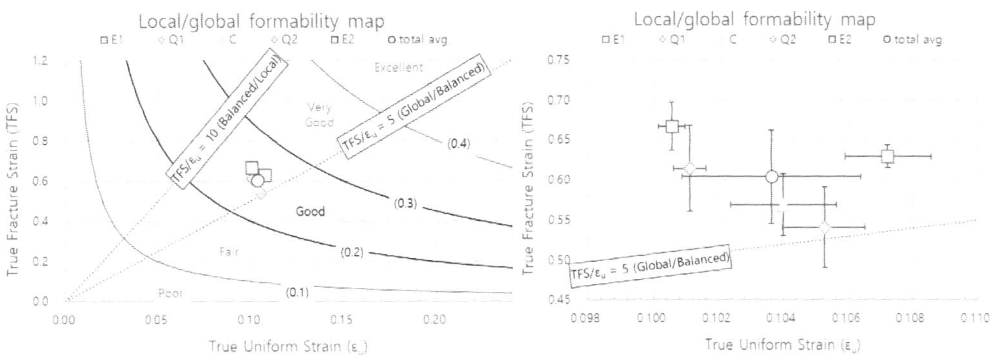

Figure 5. Local/Global Formability map (n=25 for total average, n=5 for each position)

3.3 Local Toughness concept. From the above results, the HER results had high linear correlation with the global property, yield strength as shown in Figure 4. For a better interpretation of the formability performance, simplified toughness approach regarding the true fracture strain was considered, as significant correlation between HER and fracture toughness was verified by previous studies.[5,6] Table 9 is the summary of the formability assessment results with the toughness results included.

Table 9. Overall summary (n=5 for each)

		120K				
		E1	Q1	C1	Q2	E2
Total El	e_t	0.162	0.150	0.149	0.151	0.148
YS (MPa)		882	854	885	925	963
UTS (MPa)		1284	1294	1294	1286	1279
TFS		0.63	0.54	0.57	0.61	0.67
TFS (eng. form)	e_0	0.88	0.72	0.77	0.85	0.95
L/GSR		5.87	5.14	5.47	6.08	6.63
F.I.		0.26	0.24	0.24	0.25	0.26
P.I. (MPa)		334	309	315	320	331
Global toughness	$^1/_2\,(\mathrm{YS}+\mathrm{UTS})e_t$	175	161	163	167	166
Local concept	$^1/_2\,(\mathrm{YS}+\mathrm{UTS})e_0$	949	773	836	940	1063
HER (%)		30.0	28.4	31.2	33.2	36.2

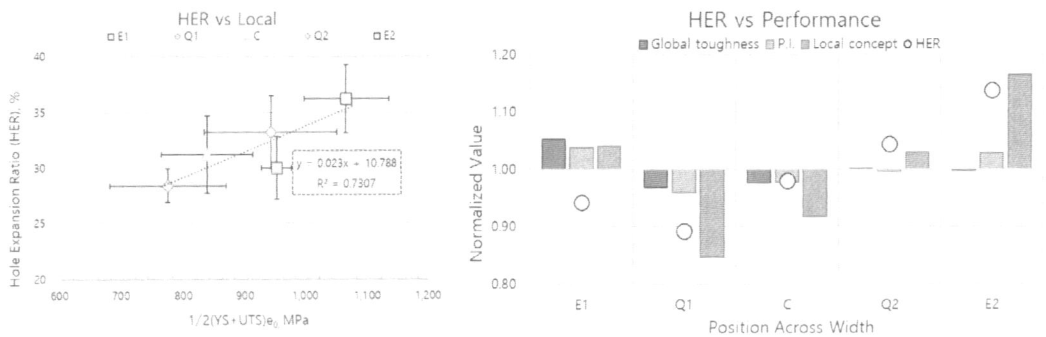

Figure 6. Various measure of AHSS performance

From the result, when yield strength and TFS are considered together, the correlation with HER is substantially higher than for TFS alone. Note that these data represent various positions across a single coil-width blank. In addition, relatively significant linear relationship was found between the local toughness concept and the hole expansion ratio than for yield strength or TFS alone when different range of grades were considered together(not detailed due to confidentiality reasons).

4. Conclusion

In conclusion, this study examined the intrinsic formability parameters and their influence on the sheared-edge ductility of 120K grade steel regarding the material property uniformity. The analysis focus on the investigation of the relationship between the formability parameter and laboratory-scale performance indicators such as the ISO 16630 hole expansion ratio. Additionally, toughness concept was considered for further investigation. The following conclusion was drawn from this analysis:

- The observed cross width variability observed in global material property influenced both intrinsic parameters and the laboratory-scale performance indicators.
- Through the formability assessments, local toughness and the hole expansion ratio portrayed significant linear relationship than for true fracture strain alone.
- The above investigation and method will be expanded for different AHSS grades for future work.

References

[1] Shen F, Wang H, Liu Z, Liu W, Könemann M, Yuan G, Wang G, Münstermann S and Lian J 2022 *J. Mater. Process. Technol.* **299** 117368
[2] Hance B 2017 *SAE Int. J. Engines* **10** Issues 2
[3] Hance B 2018 *SAE Int. J. Mater. Manuf.* **11(4)** pp. 505-516
[4] Yoon J, Jung J, Joo S, Song T, Chin K, Seo M, Kim S, Lee S and Kim H 2016 *Mater. Letters* **180** pp. 322-326
[5] Frómeta D, Lara A, Grifé L, Dieudonné T, Dietsch P, Rehrl J, Suppan C, Casellas D and Calvo J 2021 *Metall. and Mater.Trans. A* **52(6)**
[6] Heibel S, Dettinger T, Nester W, Clausmeyer T and Tekkaya A.E 2018 *Materials* **11(5)** 761
[7] Larour P, Freudenthaler J, Kerschbaum M and Dolzer D 2020 Mater. Sci. Eng. **967** 012080
[8] Kim J and Hance B 2022 *IABC* (Detroit)
[9] Hance B and Link T M 2019 *IOP Conf. Ser.: Mater. Sci. Eng.* **651** 012061
[10] ASTM International Standard Designation: E8/E8M
[11] ISO Technical Specification: ISO/TS 16630
[12] Verband der Automobilindustrie Test Standard: VDA 238-100

Comparison of hole expansion ratios in a dual phase steel for holes prepared by conventional punching and fine piercing

Abrar Salam Ebrahim+, U Aravind and Uday Chakkingal*

Dept of Metallurgical and Materials Engineering, IIT Madras, Chennai 600 036, INDIA
+ Currently at University of New Hampshire, USA

udaychak@iitm.ac.in

Abstract. Dual phase steels generally exhibit poorer stretch flangeability compared to lower strength steels. Stretch flangeability is normally evaluated using hole expansion testing, wherein a specimen with a 10 mm central hole is expanded using a conical punch till onset of cracking at the edge of the hole. The percentage increase in hole diameter is referred to as the hole expansion ratio (HER). In this study, hole expansion ratios of a DP 600 steel were evaluated. Holes were prepared on a standard 90 mm x 90 mm specimen using two different hole preparation conditions: i) conventional punching where the hole edge exhibits both fracture and shear regions and ii) fine piercing where the hole edge exhibits almost completely smooth shear. Hole expansion testing was carried out on these specimens as per standard ISO 16630. The specimens with holes produced by fine piercing exhibited lower values of HER in spite of having a hole edge with a smooth surface. The specimens with holes prepared by conventional punching exhibited higher HER values. The influence of strain hardening of the cut edge during hole preparation and the surface roughness of the cut edge on the HER are discussed.

1. Introduction

Dual Phase steels (DP Steels) have gained widespread use in automobile manufacturing due to their favorable characteristics, including a low yield-to-ultimate strength ratio, enhanced strain hardening, and continuous yielding [1]. However, DP steels are susceptible to failure during edge stretching operations, primarily owing to their limited stretch flangeability. Stretch flangeability refers to a material's capacity to resist through-thickness crack propagation when subjected to edge stretching operations. The Hole Expansion Ratio (HER) serves as a quantifiable measure of edge stretchability, determined through the hole expansion test (HET).

Numerous studies have been conducted to establish relationship between stretch flangeability and material properties, including parameters such as yield strength, ultimate tensile strength, post-uniform elongation, fracture toughness etc. [2–5]. Additionally, extrinsic factors, such as hole-making techniques and hole edge quality, have been identified as influencing HER. Prior research has employed various methods such as punching, drilling, laser machining, and water jet machining for HET [6,7]. Upon investigating the stretch flangeability of punched hole edges, surface roughness, extent of sheared and fracture zones, etc. have been recognized as important parameters impacting HER, as concluded by [8]. Researchers have also reported the influence of expanding punch geometry [9], and the tooling configuration used in the punching process [10] on HER.

In the current study, the stretch flangeability of DP600 steel is examined by employing two different hole preparation methods: conventional punching and fine piercing. Fine piercing stands out as a precision metal shearing operation that delivers products with near-finished quality and tight tolerances [11]. The hole edge quality is characterized by parameters including surface roughness and hardness and the influence of burr orientation and hole edge quality on the HER is investigated.

2. Experimental Procedure

The Hot Rolled Pickled and Oiled (HRPO) Dual Phase (DP) 600 steel grade with thickness 2.6 mm was used for the studies. The material composition of the obtained DP 600 steel revealed by the Optical Emission Spectroscopy is shown in Table 1.

Table 1. Chemical composition of the as-received DP600 steel

Element	C	Si	Mn	Cr	Ni	Al	S	P	Fe
Weight percentage	0.08	0.13	0.94	0.57	0.017	0.03	0.005	0.039	98.1

The microstructure of Dual Phase 600 steel was analyzed using an optical microscope. After etching with Nital reagent (3 ml HNO3 and 97 ml ethanol), the sample revealed a ferrite-martensite microstructure, where dark areas represented martensite colonies, and the bright background indicated the ferrite matrix as shown in Figure 1. Approximately 22% of martensite was uniformly distributed within the microstructure. Grain size analysis, conducted with ImageJ software, determined an average ferrite grain size of 4.1 μm and a martensite colony size of 2.5 μm.

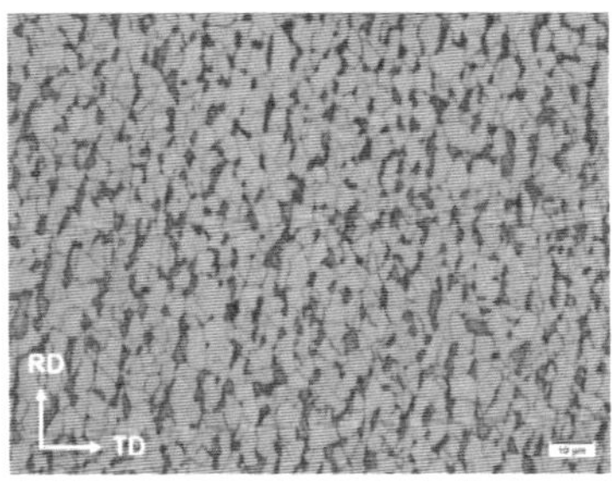

Figure 1. Microstructure of the as received DP600 steel.

The Hole Expansion Test (HET) was conducted in accordance with ISO 16630 standard geometry, as illustrated in Figure 2. This standard specifies the use of 90 mm X 90 mm square samples, which were obtained by shear cutting from the bulk sheet. In the center of each sample, a 10 mm diameter hole was prepared. To create these holes in this study, punching and fine piercing methods were employed. The punched holes were generated using a tooling setup featuring a punch-die clearance of 8% of the sheet thickness.

A custom fine piercing tooling setup was designed and developed specifically for this study. Figure 3 illustrates the key components involved in the fine piercing setup. Typical tool-die clearance of fine piercing tooling lie between 0.5% to 1% of the sheet thickness. The present work chose a clearance of 10 μm, which is 0.5% of the 2.6 mm thickness of Dual Phase (DP) steel sheet. The selection of tooling parameters was based on the sheet's thickness and the desired cut geometry. The fine piercing process encompasses shearing force, blank holder force, and counterforce. To determine the magnitude of these forces, empirical relations from standard guidelines for fine blanking process design, as outlined by [12], were employed. The sum of these forces—shearing, counter, and blank holding—determines the maximum total force required for fine piercing. Consequently, a hydraulic press with a capacity of 600

IOP Conf. Series: Materials Science and Engineering **1307** (2024) 012016 doi:10.1088/1757-899X/1307/1/012016

kN was chosen and the custom tooling was installed on it. Subsequently, the square specimens were subjected to the fine piercing process to produce 10 mm diameter holes at the center.

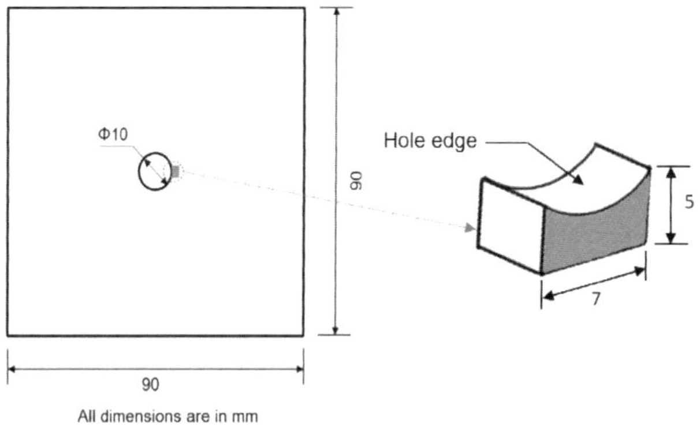

Figure 2. Specimen dimensions for hole expansion testing. The enlarged view of material extracted for hole edge characterization (grey shaded portion encircled in red) is shown on the right.

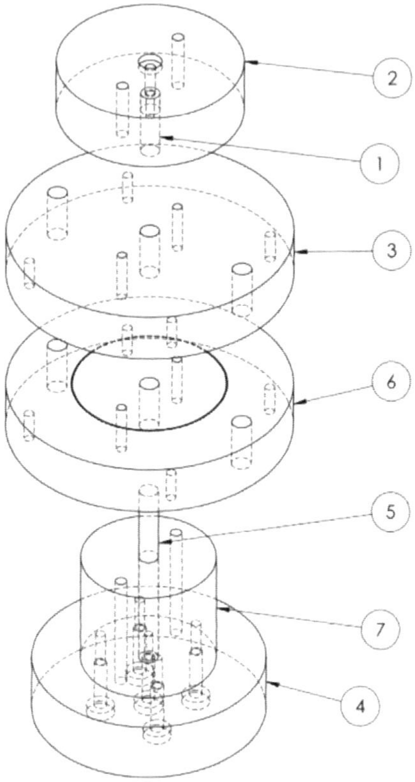

Figure 3. Fine piercing tool setup 1) Counter Punch 2) Counter punch retainer 3) Die 4) Punch Retainer 5) Punch 6) Vee-ring block and 7) Supporting ring for punch.

To evaluate surface roughness and strain hardening at the hole edge, a 5 mm X 7 mm specimen, as shown in Figure 2, was extracted from the hole periphery using electrical discharge machining (EDM). Macroscopic images of the hole edges were acquired through stereo microscopy, while high-magnification images of the hole edge surface were captured using a scanning electron microscope (SEM) in secondary electron mode. The hardness assessment of the hole edge was conducted along its thickness using a Vickers hardness testing apparatus, applying an indentation load of 1 kg and dwell time of 10 s as per ASTM standard E-92-17.

To acquire a detailed surface profile of the hole edge, a non-contact optical profiler was utilized for optical scanning and determination of the mean surface roughness (S_a). Three measurements were taken along the thickness direction, each positioned at the centre of the cut sample.

The setup for the hole expansion test (HET), which was mounted on a 300 kN double action press, is illustrated in Figure 4. This configuration comprises a 60° conical punch employed for hole expansion. To secure the sample from drawing-in, it was clamped between the blank holder and the die with a clamping force of 36.8 kN. During the test, the ram descended at a velocity of approximately 20 mm/min. For real-time monitoring of the hole deformation, a portable camera equipped with an additional light source was attached to the ram and linked to a computer. The operation of the ram was halted, and the punch was retracted when a through-thickness crack was detected. Subsequently, the diameter of the ruptured hole was measured along two perpendicular directions, avoiding the crack. The resulting mean value of the final hole diameter is denoted as D_f.

Hole expansion ratio (HER) λ is given by,

$$\lambda = \frac{D_f - D_i}{D_i} X\ 100 \tag{1}$$

Where, D_i (= 10 mm) is the hole diameter before the expansion.

To examine the impact of burr orientation on the stretch flangeability, the Hole Expansion Test (HET) was conducted using two specific orientations of the test specimen in relation to the burr's position. In the first case, the specimen was set up so that the burr was situated at the top, while in the other case, the burr was placed at the bottom, facing the expanding punch.

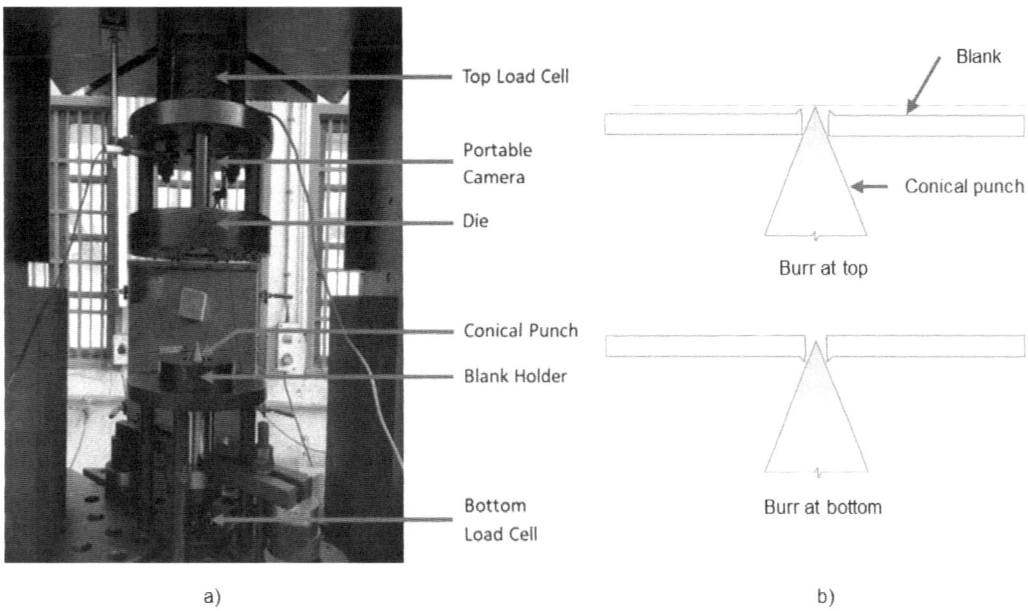

Figure 4 a) Hole expansion test setup and b) schematic of blank orientation.

3. Results and Discussion

The surface features of the hole edges prepared by conventional punching and fine piercing were observed using a stereo microscope and are shown in Figure 5. Noticeable difference between punched and fine pierced surfaces are observed. The punched hole surface has two distinct regions: shear and fracture zones. The shear zone presents a smooth and finely finished surface, while the fracture zone is distinguishable by its high surface roughness and the presence of microcracks, as detailed by [13]. Notably, approximately 40% of the punched hole edge was found to be attributed to the shear zone.

On the other hand, the fine piercing process created holes with over 95% of their edges composed of the shear zone, accompanied by small tears dispersed over the edge. It is essential to highlight that a lower clearance ratio serves to expand the shear zone region, ultimately enhancing the quality of the edge surface, as underscored by [14]. At the end of punching stroke, a small burr region is formed in both the hole edges. A close look of the hole edges is provided by SEM images in Figure 5.

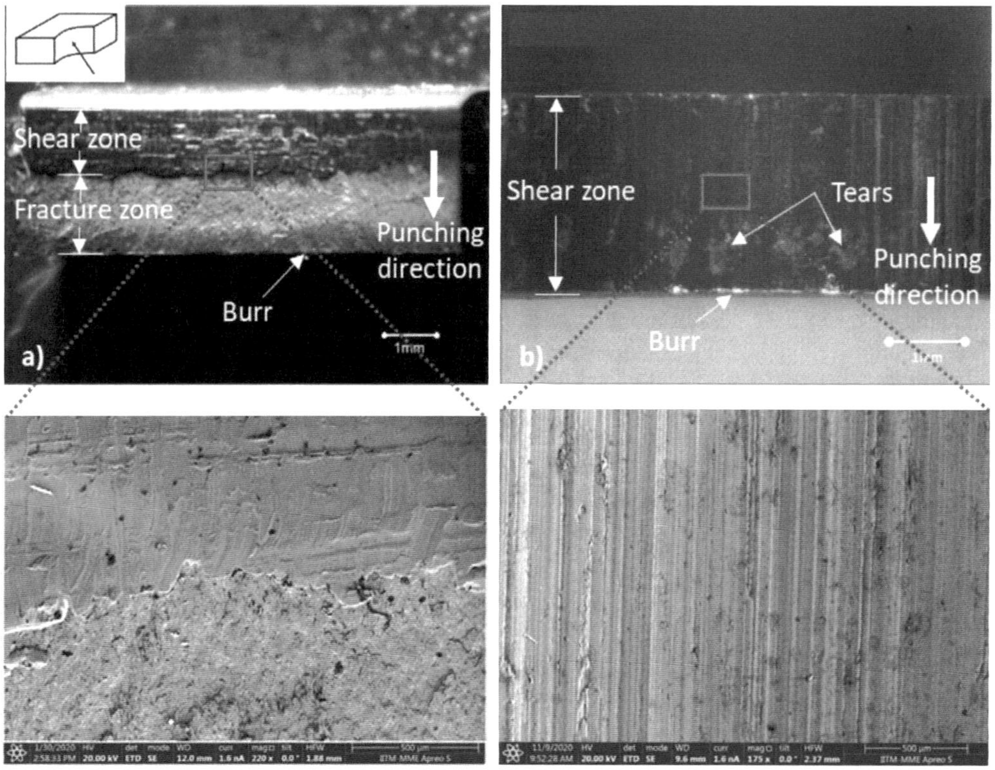

Figure 5 Hole edges captured using stereo microscope a) Punched b) Fine pierced.

The optical profilometry measurements of hole edge surface roughness revealed S_a values of 2145 nm for the punched edge and 1602 nm for the fine-pierced edge. The surface topography is visually depicted in Figure 6. The hardness of the hole edges was characterized by Vickers hardness test. The fine-pierced edge exhibited a hardness value of 437 HV at the point where the punching tool initiates shearing and reached its peak value of 489 HV near the burr, where the material underwent maximum strain hardening just prior to the separation of the blank from the sheet. In contrast, the punched hole edges displayed a lower hardness value. Specifically, the punched hole edge's hardness ranged from its lowest value of 359 HV at the beginning of the shear zone to its highest value of 419 HV at the end of the shear zone. The fine pierced hole edge exhibited higher Vickers hardness when compared to the

IOP Conf. Series: Materials Science and Engineering **1307** (2024) 012016 doi:10.1088/1757-899X/1307/1/012016

punched edge. This can be attributed to the small punch-die clearance used in the fine piercing operation, which results in delayed fracture and allows for high strain hardening of the material prior to rupture [8]

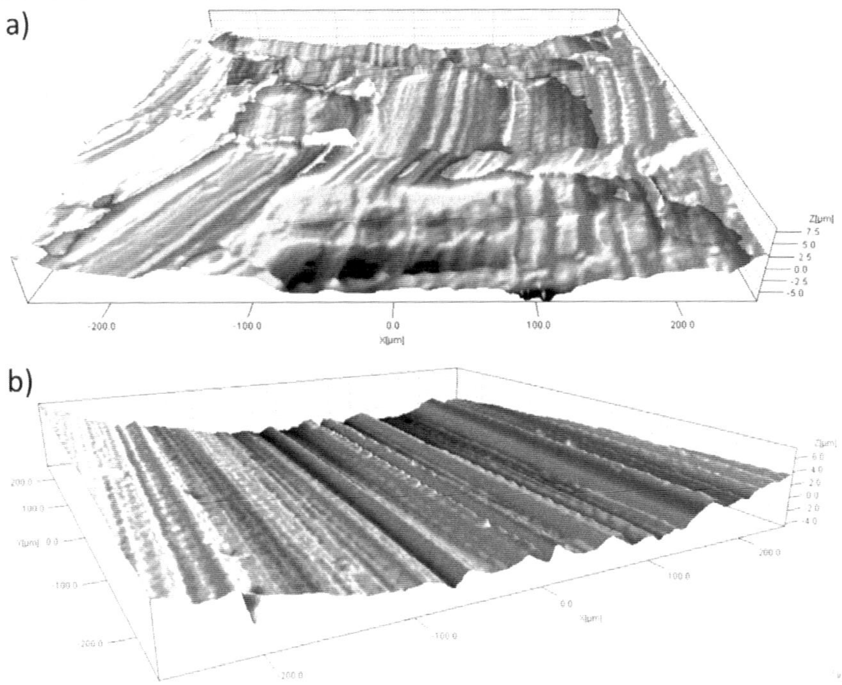

Figure 6. Surface topology of a) punched edge b) fine pierced edge captured using optical profiler.

A comparison of the hole expansion ratios (HER) values for punched and fine-pierced holes in two different burr orientations is presented as a bar chart in Figure 7. Specimens with conventionally punched holes had HER values of 34 and 48, whereas specimens with fine-pierced holes had lower HER values of only 31 and 42 for burr on top and burr on bottom orientations respectively. When assessing stretch flangeability, it becomes evident that samples with punched holes exhibit a higher HER compared to those with fine-pierced holes. This result is noteworthy, considering that punched holes feature a larger fracture zone containing numerous micro-cracks and exhibit higher surface roughness at the hole edge. This indicates that large strain hardening that occurs at the vicinity of the hole edge is a critical factor in influencing stretch flangeability. Fine piercing typically induces large plastic strains at the cut edge. Therefore, a smooth surface produced by a fine blanking operation is not conducive for good stretch flangeability. Furthermore, both the hole edges show remarkable sensitivity to the orientation of the burr. The HER reaches its maximum when the burr is oriented downward.

The lower HER observed when the burr is oriented towards the top can be attributed to the strain hardening of the hole edge. During the hole expansion test (HET), the upper edge of the hole undergoes the most substantial strain [15]. When the blank is positioned such that the burr is at the top during the HET, the portion of the edge that has already experienced significant strain hardening is positioned at the top. As a result, the upper edge can only withstand a limited additional amount of strain. This limitation consequently restricts the stretch flangeability of both punched and fine-pierced edges when the burr is oriented upwards.

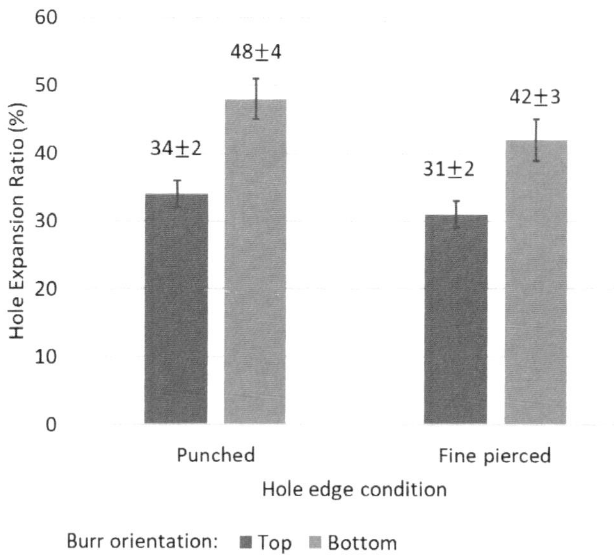

Figure 7. Comparison of HERs of punched and fine pierced holes with respect two different burr orientations (Values represent mean and standard deviation of five measurements in each condition).

4. Conclusions

A comparative analysis of stretch flangeability in DP600 steel was conducted using two distinct hole creation methods: conventional punching and fine piercing. The fine-pierced holes exhibited a smooth surface free of fracture zones, while the punched holes displayed both shear region and a substantial fracture region. Despite the fine-pierced holes' superior surface finish, they yielded lower HERs in comparison to the punched holes. This difference in HERs is attributed to the elevated level of strain hardening induced by the fine piercing process along the hole edge. Furthermore, the orientation of the burr exhibited a notable impact on stretch flangeability. The material displayed enhanced resistance to edge cracking when the burr was oriented downward. The lower HER observed with fine-pierced holes and a burr-downward orientation underscores the critical significance of strain hardening at the hole edges. It can be inferred that the degree of strain hardening imparted by the hole-making technique plays a pivotal role in determining the material's HER compared to the surface roughness of the hole edge.

References

[1] Hilditch T, Souza T and Hodgson P 2015 *Welding and Joining of Advanced High Strength Steels* Ed. M. Shome and N. Tumluru (Woodhead Publishing) 9–28
[2] Chatterjee S and Bhadeshia H K D H 2007 *Materials Science and Technology* **23** 606.
[3] Paul S K 2014 *J. of Materi Eng and Perform* **23** 3610.
[4] Yoon J I, Jung J, Lee H H, Kim G-S and Kim H S 2016 *Met. Mater. Int.* **22** 1009.
[5] Prasad K, Ebrahim A S, Krishnaswamy H, Chakkingal U and Banerjee D K 2022 *IOP Conf. Ser: Mater. Sci. Eng.* **1238** 012038
[6] Karelova A, Krempaszky C, Werner E, Tsipouridis P, Hebesberger T and Pichler A 2009 *Steel Research International* **80** 71-77.
[7] Wang K, Luo M and Wierzbicki T 2014 *Int J Fract* **187** 245.
[8] Mori K, Abe Y and Suzui Y 2010 *Journal of Materials Processing Technology* **210** 653.
[9] Pathak N, Butcher C and Worswick M 2016 *J. of Materi Eng and Perform* **25** 4919–32
[10] Shih H-C, Zhou D and Konopinski B 2017 *SAE Int. J. Engines* **10** 2051.
[11] Aravind U, Gopalakrishnan C K, Uday C and Venugopal P 2017 *Procedia Engineering* **207** 1523.

[12] Haack J and Bircer F 1984 *Fine-blanking Practical Handbook* (Lyss, Switzerland: Feintool AG).
[13] Wu X, Bahmanpour H and Schmid K 2012 *Journal of Materials Processing Technology* **212** 1209
[14] Mori K, Abe Y, Kidoma Y and Kadarno P 2013 *International Journal of Machine Tools and Manufacture* **65** 41.
[15] Sartkulvanich P, Kroenauer B, Golle R, Konieczny A and Altan T 2010 *CIRP Annals* **59** 279.

IOP Conf. Series: Materials Science and Engineering **1307** (2024) 012017 doi:10.1088/1757-899X/1307/1/012017

Tool surface texturing by shot peening: initial results and lessons learned

Eneko Sáenz de Argandoña[1]*, Markel Mendez[1], Joseba Mendiguren[1] and Alaitz Zabala[1]

[1] Mondragon Unibertsitatea, Faculty of Engineering, Mechanics and Industrial Production, Loramendi 4, Mondragon 20500, Gipuzkoa, Spain

esaenzdeargan@mondragon.edu

Abstract. Abrasive and adhesive wear is a major issue in sheet metal forming processes. Although different solutions exist to reduce its impact, there is still room for progress. Surface texturing presents a big potential for tribological behaviour improvement by introducing oil retention voids on the surface. However, the economic efficiency of the surface functionalization is paramount for its implementation in industrial processes. In the present study, a widely known technology, shot peening, is applied for the creation of the voids. A grinding+polishing process is then applied to ensure a flat load bearing area. The textures are generated in 1.2379 tool steel blocks hardened at 60 HRc that emulate the tool. Different shot peening strategies are evaluated by carrying out strip drawing tests with AA5754H111 aluminium alloy sheets. Contact pressures up to 25 MPa are applied and the adhesive wear reduction potential is evaluated. First results point out the potential of this solution although the applied shot peening strategy needs to be carefully defined.

1. Introduction

All the metal forming processes, as sheet metal forming, have in common that a relative movement must exist between the workpiece and the tool in order to be able to deform the material. This leads to the generation of a friction force that opposes to this relative movement. The generated frictional force causes wear on tools and increases the total energy required to complete the process. It is estimated that around 30% of the energy generated globally is lost due to friction, so developed countries allocate approximately 5% of their GDP to develop strategies to minimize friction [1].

The most common method for reducing friction in processes involving relative velocity between the workpiece and the tool is the proper use of lubricant agents. Lubrication not only helps to reduce tool wear and the force required to complete the process, but also contributes to achieving better geometrical and surface tolerances [2].

It has been observed that liquid lubricant in conjunction with surface textures helps to reduce the coefficient of friction even more compared to the use of lubricant in the absence of textures [3, 4, 5]. However, these textures need to be within certain parameters to prevent friction from increasing instead of decreasing.

In terms of the geometry of the textures, it has been observed that the longitudinal channels perpendicular to the sheet's direction of movement have shown the greater reduction in the coefficient of friction, while longitudinal channels parallel to the sheet's direction of movement have demonstrated the greater increase in the coefficient of friction, as can be seen in the studies conducted by Costa et al. [6] and Sulaiman et al. [5]. However, the most versatile texture is the one with the pattern of dimples, as its geometry doesn't favour one direction over the other in the reduction of the coefficient of friction.

Regarding to the geometric dimensions, based on the studies conducted by Vorhold et al. [7], Costa et al. [6], and Saeidi et al. [8], it can be concluded that the diameter of the texture should fall within the range of 20 to 100 μm. If the diameter is too small, it leads to an increase of the coefficient of friction, and if it's too big, it doesn't reduce the coefficient of friction as much as the optimal diameter [7, 8]. So, a diameter of 50 μm is the most suitable choice.

For the depth of the texture, on the works realized by Godi et al. [1], Saeidi et al. [8], Sulaiman et al. [5], and Hazrati et al. [9], it has been seen that depth less than 7 μm have shown a coefficient of friction similar to the absence of texture, and greater than 10 μm have not shown significant improvement compared to 10 μm. Therefore, a depth of 10 μm has been considered the most appropriate one.

With respect to the textured area fraction, it has been observed in the studies conducted by Hazrati et al. [9] end Sulaiman et al. [5] that the greatest reduction on the coefficient of friction is achieved with a textured area fraction around 20%. It is also worth mentioning that high textured area fraction (higher than 25% for the study conducted by Hazrati et al. and 40 % for the study conducted by Sulaiman et al.) can lead to an increase of the coefficient of friction. Furthermore, Hazrati et al. have observed that very low textured area fractions (lower than 15%) can also lead to an increase of the coefficient of friction compared to the untextured tools.

Traditionally technologies such as laser texturing, micro-machining or chemical machining have been applied to the texturing of surfaces. Even though good results have been reported, the texturing or large areas is not industrially feasible nowadays. In the presence study, a more industrial approach has been carried out with the purpose of evaluating textures made by shot peening strategies. In order to ensure an appropriate load bearing area, surfaces have been polished after the texturing. The main driver of the analysis is the industrial readiness of shot peening technologies and the possibility of directly apply them in tool texturing. This way, dimples of different sizes have been created by shot peening and the response of the different textures has been evaluated by strip drawing tests with AA5754H111 aluminium alloy sheets. Contact pressures up to 25 MPa are applied and the adhesive wear reduction potential is evaluated.

2. Materials and methods

2.1. Materials and lubricants

Commercial-grade aluminium AA5754H111 alloy sheets with mill finish surfaces and 1.5 mm thickness have been used for the study. Mechanical properties were tested, and the results are shown in Table 1. Table 1 also shows the surface roughness of the tested aluminium sheets that has been measured with a 2D tactile Mitutoyo roughness meter following the DIN EN ISO 4288 standard. The chemical composition of the aluminium sheets is given in Table 2.

Table 1. Mechanical properties of tested material AA 5754H111.

Property	Symbol	Unit	Value
Yield stress	Rp0.2	MPa	124.9
Yield strength	Rm	MPa	216.3
Total elongation (A50%)	A50	%	23.96
Average roughness	Ra	μm	0.292
Peak roughness	Rz	μm	1.3
Root mean square roughness	RSM	μm	33.26

Table 2. Chemical composition of tested material AA 5754H111.

A5754	Mg	Fe	Mn	Si	Cu	Ti	Cr	Zn	Pb
H111	2.800	0.330	0.270	0.120	0.020	0.100	0.009	0.008	0.004

The strip drawing test blocks that emulate the drawing tools are made of 1.2379 steel with a contact area of 30 x 50 mm and a hardness of 60 HRC. Table 3 shows the chemical composition of the 1.2379 tool steel blocks.

Table 3. Chemical composition of tool steel 1.2379.

1.2379	C	Cr	Mn	Mo	P	S	Si	V
	1.550	11.530	0.320	0.780	0.025	0.004	0.470	0.780

And the lubricant used in the strip drawing test is Walter Zepf Schmierungstechnik AL-200WL; with a kinematic viscosity of 90 mm2/s. The used lubricant amount is 3 g/m^2.

2.2. *Blocks surface texturing*

First, the blocks have been grinded using a tangential griding machine to ensure that the surface is completely flat. As a result of the grinding, at the borders of the flat surfaces sharp edges have been created. To remove these sharp edges, the radius of the edge of the tools has been restored by the use of different abrasive stones.

Once the contact surface of the blocks is completely flat, and the sharp edges have been eliminated, the tools have been textured by shot peening. This texturing process allows the generation of dimple shaped texture, which consist of a bunch of pronounced peaks and valleys and is easily applicable in an industrial environment.

The desired texture should consist of a plane sustaining surface with dimples in it, so in order to turn the peaks into plateaus generating a bigger sustentation surface, the textured tools have been manually grinded with P1000 grain silicon carbide abrasive paper after their texturing. After the grinding, the tools have been polished with 6 µm size diamond particles. This way, the roughness of the sustentation surface is reduced to a mirror like roughness to eliminate the effect of the sustentation surface roughness on the test.

As a result of the previous procedure, three different surface topographies have been generated. A reference topography, with no texture, that has only been polished achieving a surface roughness Ra of 0.01 µm. Texture A, that has been textured and manually grinded during 3 minutes with a P1000 grain silicon carbide abrasive paper after the texture application, and texture B that has been textured and manually grinded during 12 minutes with a P1000 grain silicon carbide abrasive paper after the texture application. The three surfaces have suffered a final polishing with 6 µm size diamond particles.

During all this procedure the 2D roughness values and the 3D topography evolution of the surfaces have been traced. For measuring the roughness, a 2D tactile Mitutoyo roughness meter has been used, following the DIN EN ISO 4288 standard. For the area 3D surface topography measurement, a Sensofar Senox profilometer has been used, and the data obtained has been processed with the SensoMAP software.

2.3. *Strip drawing test*

Strip Drawing Test (SDT) has been used to analyse the response of the textures with respect adhesion phenomena. This method allows to emulate the contact between the sheet metal forming tool and the workpiece material under controlled conditions of pressure and velocity. This way, the movement of the strip generates a tangential fiction force that combined with the applied normal force allows the calculation of the coefficient of friction using Coulomb's friction force equation.

As mentioned before, the tests have been carried out in order to compare the response of the three different surface finishes. The conditions at which the strip drawing tests have been carried out are shown in Table 4. During the set up of the testing, Fujifilm's Prescale Films have been used in order to verify an homogeneous pressure distribution in all the contact area between the sheet and the blocks.

Table 4. Conditions at which Strip Drawing Tests have been carried out.

Replicates	Contact pressure	Lubricant amount	Drawing speed	Sliding distance	Tool surface finish	Sheet rolling direction
6	25 MPa	5 g/s2	10 mm/s	150 mm	Mirror polish (Ra 0.01 μm) and shot peened (textures A and B)	Parallel to the strip movement

2.4. Post test surface analysis
After the realization of the strip drawing tests, an analysis of the surfaces of the blocks has been carried out. For the surface composition analysis, a SEM CI Nova Nanosrm 450 has been used, and the data obtained has been processed with the Aztec3.3 software.

3. Results

3.1. Surface texture
Figure 1 shows the evolution of the surface of the textured blocks. Figure 1a shows the surface after the texturing treatment. It can be observed that the surface is a combination of peaks and valleys what correlates to a surface textured by shot peening. As mentioned before, and in order to generate flat sustentation areas, Figure 1b shows the surface after a manual grinding process where the peaks have been eroded generating flat areas. And finally Figure 1c shows a surface that has followed the same approach as the one in Figure 1b but with the application of a longer manual griding process as explained before. It can be observed that the application of the manual griding process generates a sustentation area in both surfaces being this area greater in the surface that has been manually grinded for a longer time where larger areas colored in white are observed.

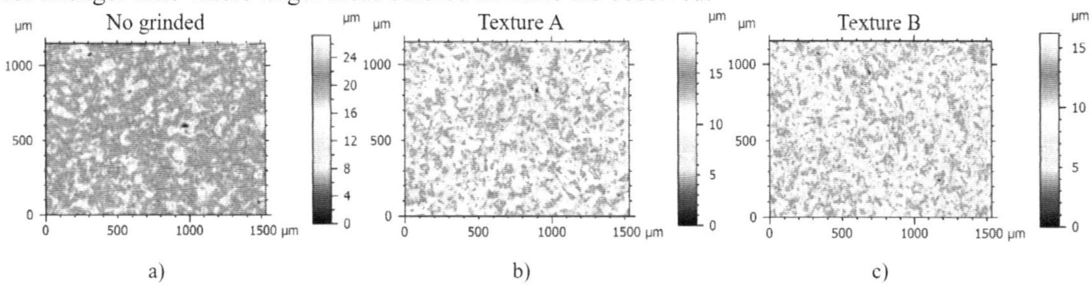

Figure 1. Surface topography: after texturing (a), after texturing and manually grinding for three minutes (b) and after texturing and manually grinding during twelve minutes (c)

In order to make a more quantitative evaluation of the textured surfaces, Table 5 presents the most representative 3D roughness values for each of the surfaces. It must be stated that during the strip drawing tests the sheet is clamped between two blocks, and for this reason the table presents the results of both blocks for each kind of texture: texture A, texture B and reference blocks that were polished achieving a surface roughness of 0.01 μm.

Table 5. Topographical parameters of the textured surfaces.

		Vvv (0=80)	Sdr (%)	Sa [μm]	Sq [μm]
Textured but not grinded	Mean	0.427	25.606	2.177	2.855
	Deviation	0.020	0.272	0.133	0.168
Texture A	Mean	0.407	9.557	1.318	1.833
	Deviation	0.037	0.063	0.008	0.034

Texture B top	Mean	0.377	6.347	1.047	1.538
	Deviation	0.040	2.099	0.206	0.239
Mirror polish	Mean	0.006	0.002	0.038	0.048
	Deviation	0.0006	0.0009	0.0012	0.0021

3.2. Coefficient of friction

For each couple of blocks a total of six strip drawing tests were carried out. As mentioned in Table 4, the tests were carried out at a contact pressure of 25 MPa, at a sliding velocity of 10mm/s and the sliding distance at each test was 150mm. Figure 2 shows the results achieved when computing the coefficient of friction (COF from now on). First of all, it can be observed the high repeatability of the tests. The surface finish that offers the lowest COF is the mirror polish surface with values about 0.02. On the other hand, in the case of the texture surfaces, both of them offer higher values of COF being texture A the one offering the highest COF. Furthermore, the signal obtained during the testing of texture A surface presents a high scattering which is typical of stick-slip contact mechanism. However, and as it will be shown later, no galling or stick-slip phenomena was observed in texture A blocks after the testing. The reason for this noisy signal is the topography of the sustentation surface in texture A compared to texture B. A sustentation surface deeper analysis was carried out and it was observed that due to the shorter polishing time texture A had more and severe peaks in the sustentation area.

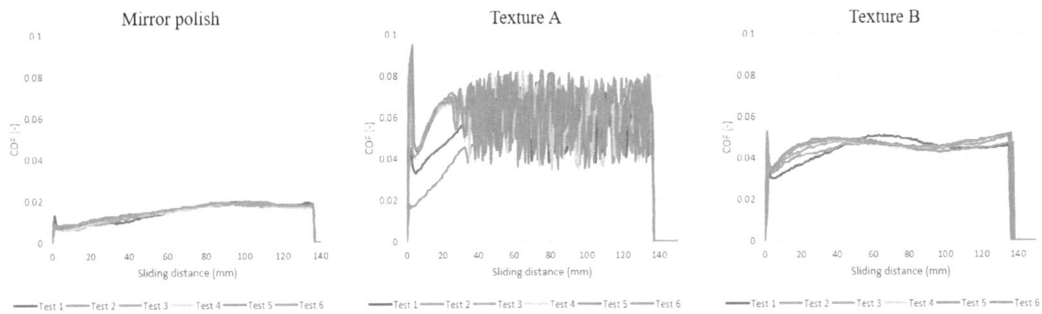

Figure 2. Coefficient of friction obtained for the different blocks surface condition.

3.3. Tool wear

After the realization of the strip drawing tests, an analysis of the surfaces of the blocks was carried out. The final situation of the blocks after the realization of the previous mentioned 6 tests for each block can be observed in Figure 3. For the sake of simplicity, only lower blocks are shown.

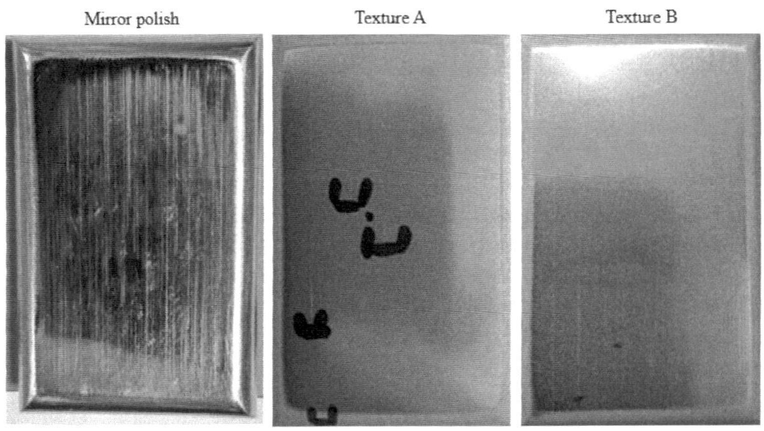

Figure 3. Blocks surfaces after the realization of the strip drawing test.

It can be observed that the mirror polished block is the one that has shown the greatest wear. It must also be stated that even there are many longitudinal traces all over the surface, the depth of these traces is not higher than 0.2 μm. However, and since the initial polish of the surface makes its roughness to be very low, the longitudinal traces are very easily observed.

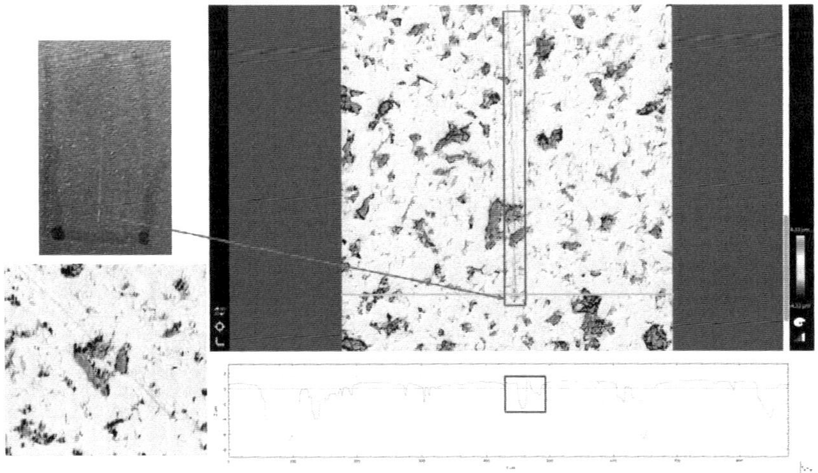

Figure 4. Detail of a longitudinal wear trace observed in a textured surface. Measurement carried out at surface with texture A.

On the other hand, the textured blocks do no present many traces or signs of wear at their surfaces. After analyzing the origin of the longitudinal wear traces, it has been observed that the traces start in areas where ceramic grains used during the texturing of the surface were initially embedded (see Figure 4). So the hypothesis is that the ceramic grains are embedded during the texturing, and when realizing the strip drawing tests, they are detached from the surface creating a wear trace during all the travel until leaving the surface of the block.

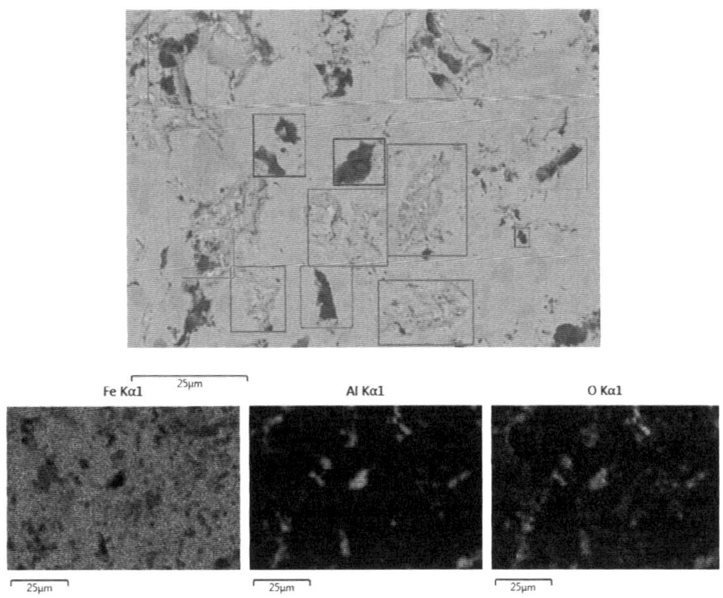

Figure 5. SEM analysis of textured surface after the texturing and manual grinding process. Squares in red represent ceramic grains flushed in the sustentation surface. Squares in green represent ceramic grains at the bottom of the texture valleys. Squares in blue represent valleys without the presence of ceramic grains. Measurement carried out at surface with texture B.

In order to verify this hypothesis, the composition of one of the textured surfaces has been obtained in a SEM CI Nova Nanosrm 450 before the realization of the strip drawing tests. It must be noted at this point that the material used during the texturing of the surfaces has been alumina (Al2O3). This way, by looking to the presence of these elements, the next conclusions have been extracted as shown Figure 5. Traces of alumina have been found at the textured surface. Some of the alumina traces are found at the bottom of the valleys, where they are not detrimental. However, some other traces of alumina have been found flushed at the sustentation surface. These traces are assumed to be the ones that during the strip drawing test realization detach from the sustentation surface and generate longitudinal wear traces in the textured surfaces. Different longitudinal traces have been measured and their height is lower than 2 μm in all the cases.

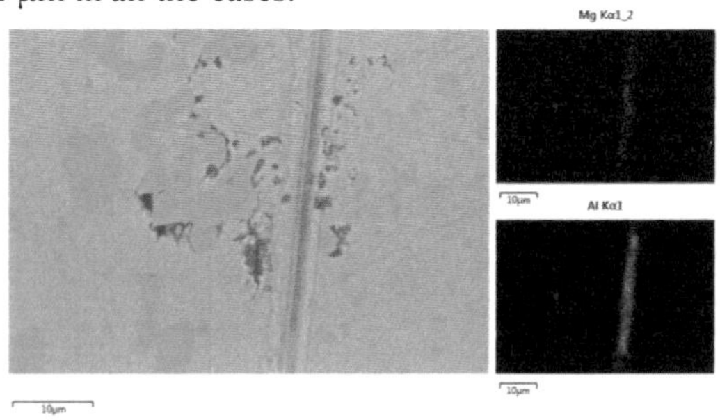

a)

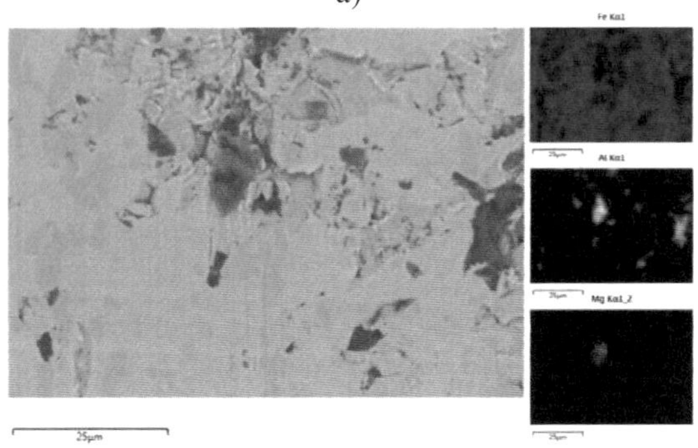

b)

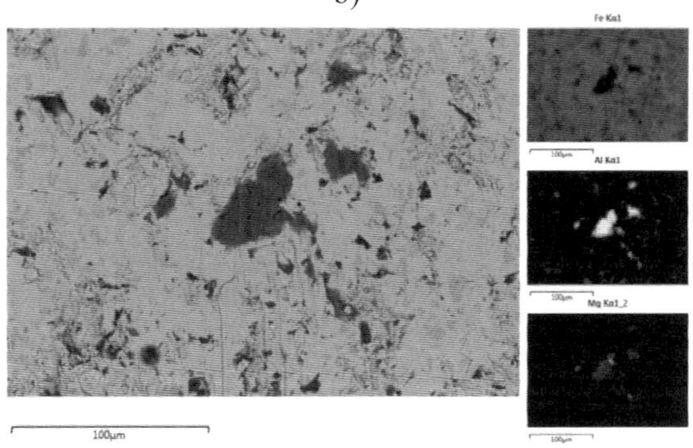

c)

Figure 6. Surface composition analysis of the different blocks by SEM analysis. Results from the mirror polished surface block a), results from the textura A surface block b) and results from the textura B surface block c).

And finally it has also been analyzed if there has been any transfer of aluminum from the sheets to the blocks. For doing so, a surface composition analysis of the different blocks has been carried out using a SEM CI Nova Nanosrm 450 after the realization of the strip drawing tests. Figure 6 shows the results of this analysis. It can be observed in Figure 6a that aluminum and magnesium traces have been found in the mirror polish surface after the realization of the tests. This means that aluminum transfer phenomena has taken place from the aluminum sheets to the blocks. On the other hand, in the case of the texture blocks, it has been observed that although aluminum and magnesium transfer has also been taken place from the aluminum sheets to the steel blocks, the aluminum has concentrated in the valleys of the texture not finding any aluminum and magnesium traces in the sustentation area. Anyway one question arises here at this stage and is the how the transfer of the material would affect the overall performance of the textured surfaces for longer periods of contact time.

4. Conclusions

In the present work a combination of shot peening technologies and post-polishing techniques has been carried out in order to generate functional anti-galling surfaces. The main objective was to develop surface with valleys able to retain lubricants combined with flat sustentation areas able to support the contact between the surfaces. Next conclusions have been observed:

- Even if no aluminum transfer to the sustentation area of the textured surfaces has been observed, traces of aluminum have been detected in the valleys of the textured surfaces.
- A deeper analysis of the transfer mechanism and the impact in the long term contact periods is necessary to understand the severity of such phenomena.
- Furthermore, wear traces generated by the detachment of alumina particles used during the shot peening process have been observed after the strip drawing tests.
- As a result, it is proved that ceramic particles can not be used for the generation of the textures by shot peening and new analysis will be focused in the texturing of the surfaces using metallic particles.

References

[1] Godi A, Grønbæk J and De Chiffre L 2015 *CIRP J Manuf Sci Technology*, vol. 11, pp. 28–35
[2] Velkavrh I 2017 *Tribology International*, vol. 114, pp. 418–428
[3] Wang Z, Ye R and Xiang J 2023 *Tribology International*, vol. 177
[4] Bech J, Bay N and Eriksen M 1998 *CIRP Annals* Vol74 pp. 221–226
[5] Sulaiman M H, Christiansen P and Bay N 2017 *Procedia Engineering* vol. 207 pp. 2263–2268
[6] Costa H L and Hutchings I M 2009 *J. Mater Process Technol*, vol. 209, no. 3, pp. 1175–1180
[7] Vorholt J 2017 Procedia Engineering, 2017 *Procedia Engineering* vol. 207 pp. 2209–2214
[8] Saeidi F, Meylan B, Hoffmann P and Wasmer K 2015 *Wear* vol. 348–349, pp. 17–26,
[9] Hazrati J, Stein P, Kramer P and Van Den Boogaard A H 2018 *IOP Conf. Ser.: Mater. Sci. Eng.* 418 012095

IOP Conf. Series: Materials Science and Engineering **1307** (2024) 012018 doi:10.1088/1757-899X/1307/1/012018

Sustainable tool technology: Wood-based forming tools for deep drawing of sheet metal

Michael Geueke[1]*, Rainer Steinheimer[1], Peter Frohn-Sörensen[1], Michael Schiller[1] and Bernd Engel[1]

[1] *Forming Technology, Institute of Production Technologies, University of Siegen, 57076 Siegen, Germany*

* michael.geueke@uni-siegen.de

Abstract. In order to improve adaptability to evolving manufacturing processes in an era marked by mass customization and individualization, while also meeting geopolitical climate objectives, it is necessary to employ more sustainable methods in sheet metal forming to enhance flexibility. Metal based rigid tooling systems require a lot of material and energy resources, when aiming for high flexibility and scalability with respect to small batch size production. To reduce the overall resource consumption and to meet climate restrictions, non-metallic tooling approaches can counteract these issues when aiming for individualization in sheet metal forming. Compared to conventional metal-based tooling approaches, wood-based forming tools may enable a potential resource reduction (e.g., carbon footprint) and offer an alternative tooling approach for sheet metal forming. Although the mechanical performance of wood is lower compared to conventional tooling steel, certain wood-based forming tools can potentially be implemented for sheet metal forming. In this work a feasibility study of wood-based tooling materials is executed to investigate the mechanical performance on drawn sheet metal parts in a deep drawing operation. Different composite structures are investigated for wooden forming tools of black locust (robinia pseudoacacia).

1. Introduction

If production is aimed at small batch sizes, economic and ecological success depends heavily on whether the desired results can be scaled in the respective production areas. On the one hand, increasing automation, adaptability and digitization approaches are powerful tools when it comes to agile and scalable production systems. However, as these methods and techniques are still cost-intensive, they are mostly used in production systems with large batch sizes. On the other hand, the paradigm shift from mass production to flexibility in all production and consumer industries is continuously increasing the demand for highly individualized mass customization and small batch sizes. From an ecological point of view, the industry is no longer focusing solely on mass reduction, e.g. when striving for lightweight construction applications. There are various approaches dealing with sustainable, future-orientated production technologies such as circular economy, intelligent material replacement strategies or alternative drive technologies. To achieve a sustainable paradigm shift to high flexible and agile production systems, technological combined with ecological alternatives are inevitable. Increasing changeability in manufacturing is one of the key levers responding to growing derivatives and the associated smaller batch size up to individualization of one part only [1].

According to today's state of the art, sheet metal parts are almost exclusively manufactured in mass production in high quantities [2]. New approaches are required to reconcile both objectives of enabling flexibility in metal forming under geopolitical restrictions within an environment of increasing globalization and price competition. [3]. Sheet metal forming is a widely used manufacturing process that produces finished products of good quality and high precision, meaning that deep-drawn components can be used almost immediately after production. While flexibility is inherent in generic forming tools (e.g. freeform bending) such as freeform bending, it can be further enhanced through alternative, agile, or smart production systems. [4]. In 2017, Potting et al. introduced the concept of the circular economy and outlined 9R strategies aimed at establishing a more sustainable production model transitioning from a linear to a circular economy. [5]. The 9R strategies encompass various methods to decrease natural resource consumption and environmental impact. Implementing these strategies in traditional manufacturing contexts holds promise for fostering sustainable manufacturing technologies and systems.

This study reveals a wood-based method following alternative tooling approaches for sheet metal forming. The performance of composite structures from black locust (robinia pseudoacacia) punch tools for sheet metal forming is experimentally evaluated based on the deep drawing process.

2. Literature Review

In deep drawing, a sheet metal blank is inserted between the die and the blank holder. By bringing these two tool parts together and applying force on the blank holder, the sheet is clamped. The force of the blank holder on the sheet metal blank must be adjusted, to avoid wrinkles (not enough blank holder force) or to avoid cracks (blank holder force to high) in the sheet metal. The force adjustment of the blankholder may either be performed via external springs or with an external hydraulic unit. By moving the punch relative to the blank holder and the die, the part is formed (see Figure 1). Compared to stretch forming, the material flow of the sheet metal is allowed in deep drawing to ensure a uniform wall thickness of the deep drawn parts [6]. The drawing gap between the die and the punch is usually bigger than the sheet metal thickness to avoid stretching.

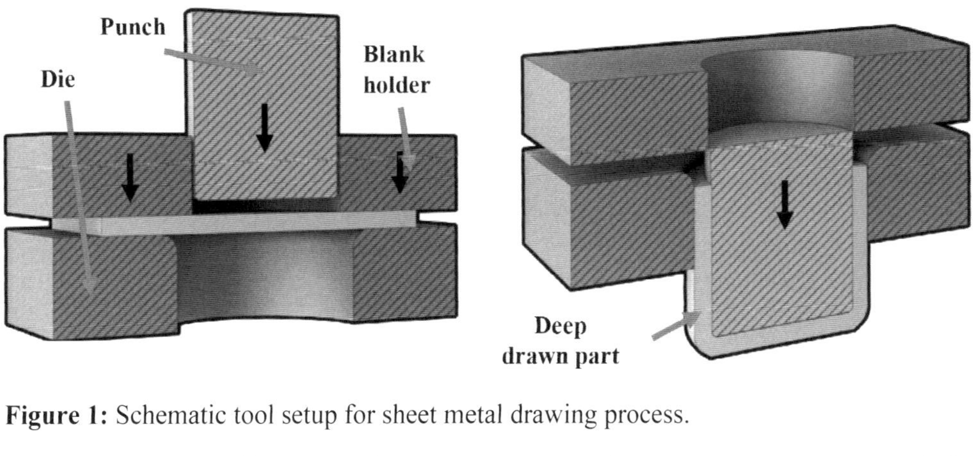

Figure 1: Schematic tool setup for sheet metal drawing process.

Deep drawing is a tensile-compression forming process because, in contrast to stretch forming, localized tensile and compressive stresses are located on the workpiece (e.g., flange on the cup). The interface properties in the forming zone determine the process limits in deep drawing [7]. However, the process parameters, such as the drawing speed or the drawing ratio, also limit the deep drawing process. The workpiece material used is decisive for the load conditions during the deep drawing process. This is determined based on the required component properties of the product to be produced. As a system component, the workpiece material specifies the restrictions for a suitable overall system. The drawing limit ratio β is the ratio of the largest blank diameter in relation to the punch diameter that can still be drawn without tearing (d_{blank}/d_{punch}). It is reached when the largest drawing force that occurs can no longer be transferred to the forming zone (limiting drawing ratio $\beta_{limit} \sim 2$) [6].

3. Related Research

While there have been numerous research projects about deep drawing with conventional tools for large-scale production, including reinforcement or hybrid construction, there are only a few results for alternative tool approaches for small batch size production and high product individualization.

Holzer et al. and Guilleaume et al. investigated concrete based deep drawing tools for small batch size applications [8, 9], whereas Witulski et al. studied the performance of hard material coated polymer deep drawing tools with respect to spatial deviation after usage. Giorleo and Ceretti, Bergweiler et al. and Holzer et al. examined the performance and mechanical stability of additive manufactured deep drawing tools from polymers for small batch applications [8, 10, 11]. Geueke et al. investigated the viability of using wooden forming tools for sheet metal forming [12, 13]. In a hydromechanical drawing tool configuration, the punch was fabricated from laminated black locust wood and tested in a small batch series production involving 64 parts made of conventional drawing steel (DC04). Kolleck et al. conducted a feasibility study to assess the effectiveness of wood-based tooling material in a deep drawing process. [14]. The wood-based forming tools were created by bonding together 4 mm sheets of black locust (BL). They demonstrated effective performance in low batch size production scenarios (<500 parts), particularly for contact normal stresses on corner radii exceeding 10 mm. Pinto et al. assessed the efficiency of both polymer and wood-based tools in sheet metal forming, with batches comprising 500 parts each [15]. They found that both types of tools were generally feasible for radii of 7 mm, although there was slightly greater deformation observed with the densified wood-based tools compared to the polymer tools. Geueke et al. investigated the performance of wood-based forming tools for profile bending. The BL laminated composite structures were investigated for two different wood direction alignment (0° and 45°), were they achieved good performance of the wooden tools for both orientations [16].

4. Methodology

The objective of this study is to explore the possible replacement of traditional tooling materials with black locust (robinia pseudoacacia) for deep drawing applications.

4.1. Mechanical characterization of black locust

Black locust (BL) is a hardwood with favorable mechanical properties that can thrive in nutrient-poor and arid environments, outperforming other wood or biobased materials. As noted by Green et al. and Risse and Richter, black locust exhibits superior mechanical properties compared to all other native wood species in Europe [17, 18].

Before conducting the feasibility study on wood-based deep drawing tools, a mechanical assessment of untreated black locust (tensile, compression, bending, hardness) was carried out using a Zwick Roell Z250 universal testing machine, following applicable standards such as DIN EN ISO 527-1, DIN 52192, and DIN 52186, as shown Figure 2.

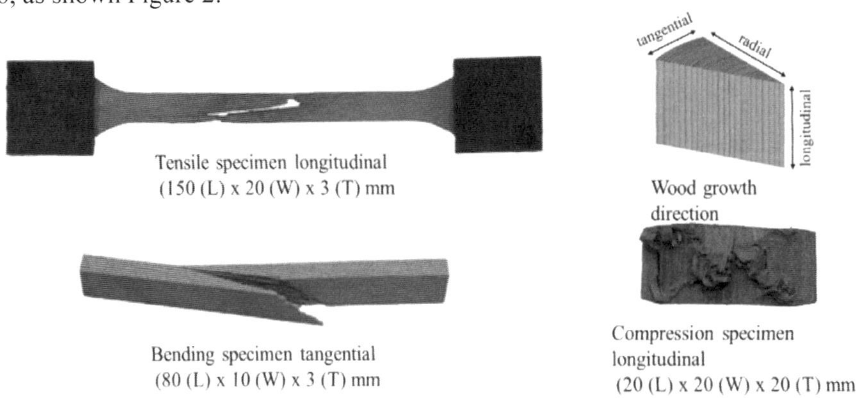

Figure 2: Black locust test specimen (dimension prior testing) after mechanical characterization.

The mechanical properties are condensed in Table 1.

Table 1: Mechanical characterization results for black locust.

Tensile properties		Characterization
Tensile strength parallel to fiber	[N/mm^2]	153.14 (± 23.31)
Tensile strength perpendicular to fiber	[N/mm^2]	7.59 (± 1.36)
Tensile modulus parallel to fiber	[N/mm^2]	9325.95 (± 308.10)
Tensile modulus perpendicular to fiber	[N/mm^2]	1394.06(± 409.82)
Compression properties		
Compressive strength longitudinal	[N/mm^2]	75.63 (± 2.01)
Compressive strength tangential	[N/mm^2]	20.86 (± 3.63)
Compressive strength radial	[N/mm^2]	28.02 (± 2.36)
Compressive modulus longitudinal	[N/mm^2]	8720.43 (± 622.56)
Compressive modulus tangential	[N/mm^2]	603.71 (± 28.03)
Compressive modulus radial	[N/mm^2]	750.24 (± 48.18)
Bending properties		
Flexural strength tangential	[N/mm^2]	141.54 (± 11.83)
Flexural strength radial	[N/mm^2]	145.95 (± 14.04)
Flexural modulus tangential	[N/mm^2]	11107.98(± 1327.36)
Flexural modulus radial	[N/mm^2]	11334.11(± 1198.52)
Hardness properties		
Brinell Hardness longitudinal	[N/mm^2]	79.15 (± 5.80)
Brinell hardness tangential	[N/mm^2]	41.94 (± 4.95)
Brinell hardness radial	[N/mm^2]	39.93 (± 3.38)

4.2. Wood-based deep drawing tools manufactured of black locust

To evaluate the effectiveness of wood-based deep drawing tools, a composite structure was created using black locust (BL) materials. This structure consisted of rectangular pieces made from 3 mm sheets of BL, which were aligned and bonded together using polyurethane adhesive according to [12]. The process involved cutting solid black locust planks into 3 mm thin veneer sheets, sanding them with coarse grit (P80) using a surface sander, and then stacking and bonding 34 layers in total. Each stack comprised BL sheets with widths ranging from 85-130 mm, ensuring alignment and proper adhesion between layers. Before final assembly, the contact surfaces were bonded and then sanded to achieve a uniform 3 mm thickness. To guarantee a stable and strong composite structure, the bonding process followed the guidelines specified in the data sheet for adhesive 501.0 from the manufacturer KLEIBERIT.

After creating the composite BL stack, various orientations for the layer arrangement were examined for the tool punches of the deep drawing tool (see Figure 3, left). To investigate the anisotropic mechanical performance of the wood-based structures, the composite block was cut in horizontal (H1) and vertical (V1) direction and then shaped into the desired tool form using a CNC milling machine, according to Habraken et al. [19]:

- Tool punch horizontal alignment (H1): 60 mm (height), 60 mm (diameter). 5 mm radius
- Tool punch vertical alignment (V1): 60 mm (height), 60 mm (diameter). 5 mm radius

IOP Conf. Series: Materials Science and Engineering **1307** (2024) 012018 doi:10.1088/1757-899X/1307/1/012018

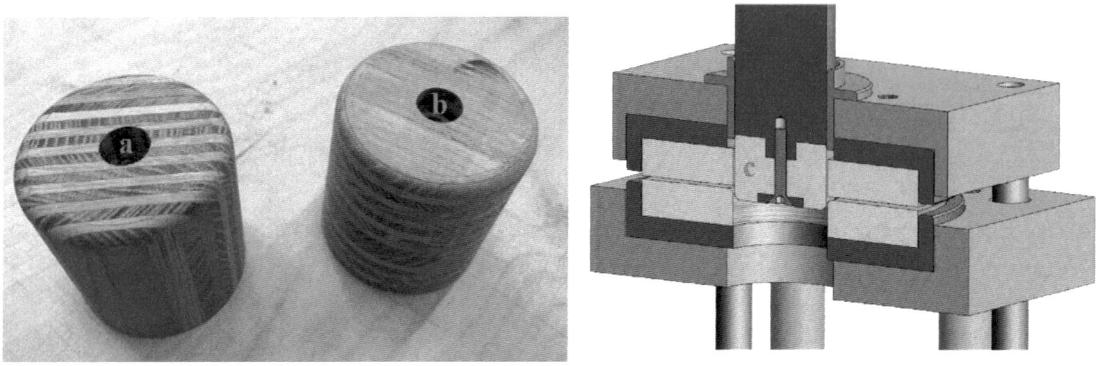

Figure 3: Black locust composite punch alignment of 3 mm sheets rectangular glued together with polyurethane (vertical laminate orientation (V1=a), horizontal laminate orientation (H1=b)). After milling on a CNC machine, the tools were implied in the deep draw test bed as tool punches (c).

4.3. Mechanical performance and optical quality determination of wooden drawing tools

Following the manufacture of both punch variants, the mechanical stability and performance of the tools was investigated on 1 mm DC04 sheet metal. According to Habraken et al. the sheet metal diameter was chosen as 120mm aiming for a drawing ratio of 2 [19]. The drawing operations were path-controlled performed on a Zwick Roell Z250 universal testing machine, applying 18 bar on the blank holder with an external hydraulic unit and a stamping speed of 1mm/s.

The drawing tests were repeated until the stress strain curve stabilized to a consistent performance. This was first obtained for the V1 punch, therefore both variants were investigated until H1 stabilized after the 6th stroke. Without any further deflection in the curve no additional densification was assumed. After the small batch series of 6 parts each, the optical evaluation revealed a rounded drawn cup with a greater radius compared to the pristine H1 punch. On the other hand, the V1 variant showed a remaining flat bottom section and smaller radius deviation compared to the H1 variant. Slight wrinkles (c.f. Figure 4, right drawn cup along the radius) along the radius were detected on the area were the rectangular alignment of the punch and therefore a transition between longitudinal and radial/tangential load was in use.

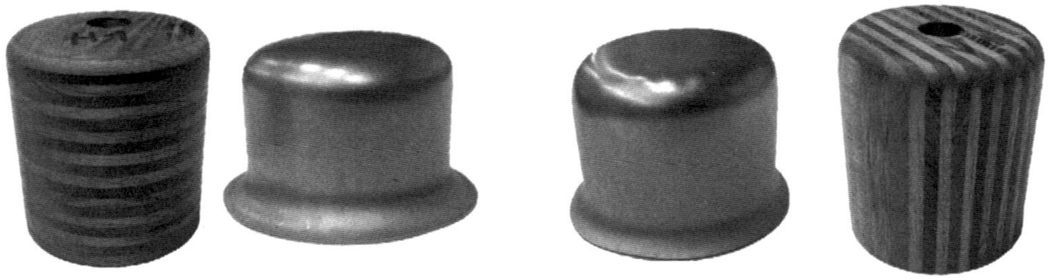

Figure 4: Optical correlation of both punch variants (H1 left, V1 right) next to their according deep drawn cups.

Following the optical correlation of the drawn cups the tools were inspected using a GOM ATOS scanner. Due to the anisotropic behaviour of wood and the resulting mechanical properties (c.f. Table 1), the horizontal stacked punch (H1) densified over the whole contact surface along the punch base increasing to the radius of the punch up to 3.5 mm. The vertical aligned punch was not condensed on the base surface and the maximum radius deviation of 1.16 mm is significantly lower than the permanent plastic densification of variant H1. Figure 5 depicts the spatial deviation of both punch variants (H1 left, V1 right).

Figure 5: Optical correlation of the spatial deviation with GOM ATOS for both punch variants (H1 left, V1 right) and a section graph with corresponding deviation between the unused and the punch after the 6th stroke.

5. Conclusion

Wood-based deep drawing tools provide avenues for enhancing inherent flexibility and enabling mass customization in sheet metal forming processes. Overall, a feasibility was demonstrated, even though the spatial deviation of the horizontal stacked punch tool was inadequate due to high radius deformation and height reduction of more than 1.5 mm. This is due to the bad mechanical strength of the wood layers in radial and tangential direction which can be correlated to the mechanical performance and durability of the anisotropic structure of wood. The vertical aligned punch tool showed a better performance, since the overall height of the punch was still the same as it was before and the remaining plastic deformation along the radius was roughly 1.2 mm smaller than the pristine punch radius of 5 mm.

To sum up, the vertically aligned wood-based deep drawing tools showed satisfactory performance for small-scale batch production. Future research should delve into the high-stress areas of the bending tools and the arrangement and orientation of the layers. Additionally, there should be further analysis of the deformation behavior of wooden forming tools through numerical simulations and practical tests to understand how local mechanical tool properties, like local tool stiffness, affect product geometry. Next to this, the high stressed areas along the radii could be milled in a two-step approach, with an initial dimension bigger than the desired radii and a locally densification after the first punch gaining higher mechanical strength. Further tribological investigations could help to identify a suitable process window for wooden deep drawing tools.

Acknowledgements

The authors acknowledge the financial support by the European Union and the Federal Ministry for Economic Affairs and Climate Action in the project SkaLaB (project number 13IK025B).

Funded by the European Union
NextGenerationEU

Supported by:
Federal Ministry for Economic Affairs and Climate Action
on the basis of a decision by the German Bundestag

References

[1] M. Gude *et al., FOREL-Wegweiser: Handlungsempfehlungen für den ressourceneffizienten Leichtbau*. Dresden: Technische Uni Dresden, 2020. [Online]. Verfügbar unter: https://ris.uni-paderborn.de/record/20318

[2] R. Sousa, "Incremental Sheet Forming Technologies," in *Reference module in materials science and materials engineering*, Elsevier, uuuu-uuuu. [Online]. Verfügbar unter: https://www.sciencedirect.com/science/article/pii/B9780128035818040558

[3] G. Herrigel, "Globalization and the German industrial production model," *J Labour Market Res*, Jg. 48, Nr. 2, S. 133–149, 2015, doi: 10.1007/s12651-014-0170-5.

[4] D. Y. Yang *et al.,* "Flexibility in metal forming," *CIRP Annals*, Jg. 67, Nr. 2, S. 743–765, 2018, doi: 10.1016/j.cirp.2018.05.004.

[5] Potting, José & Hekkert, M.P., E. Worrell und A. Hanemaaijer, *Circular Economy: Measuring innovation in the product chain.: Policy Report* (issue 2544). PBL Netherlands Environmental Assessment Agency, 2017.

[6] K. Lange, *Umformtechnik Handbuch für Industrie und Wissenschaft: Band 4: Sonderverfahren, Prozeßsimulation, Werkzeugtechnik, Produktion*. Berlin, Heidelberg: Springer, 1993.

[7] J Filzek und P Groche, "Assessment of the Tribological Function of Lubricants for Sheet Metal Forming," in S. 97–108, doi: 10.1520/STP10502S.

[8] K. Holzer, H. Schaal-Mulacek, E. Marker und W. Volk, "Kostenreduktion in der Prototypenfertigung," *ZWF*, Jg. 118, 7-8, S. 511–515, 2023. doi: 10.1515/zwf-2023-1097. [Online]. Verfügbar unter: https://www.degruyter.com/document/doi/10.1515/zwf-2023-1097/html

[9] C. Guilleaume, A. Mousavi und A. Brosius, "Hybrid deep drawing tool for lubricant free deep drawing," *AIP Conf. Proc.*, Jg. 1896, Nr. 1, S. 140003, 2017. doi: 10.1063/1.5008159. [Online]. Verfügbar unter: https://pubs.aip.org/aip/acp/article/1896/1/140003/998831/Hybrid-deep-drawing-tool-for-lubricant-free-deep

[10] L. Giorleo und E. Ceretti, "Deep drawing punches produced using fused filament fabrication technology: Performance evaluation," *Journal of Manufacturing Processes*, Jg. 84, S. 1–9, 2022. doi: 10.1016/j.jmapro.2022.09.054. [Online]. Verfügbar unter: https://www.sciencedirect.com/science/article/pii/S1526612522006752

[11] G. Bergweiler, F. Fiedler, A. Shaukat und B. Löffler, "Experimental Investigation of Dimensional Precision of Deep Drawn Cups Using Direct Polymer Additive Tooling," *JMMP*, Jg. 5, Nr. 1, S. 3, 2021. doi: 10.3390/jmmp5010003. [Online]. Verfügbar unter: https://www.researchgate.net/publication/348001796_Experimental_Investigation_of_Dimensional_Precision_of_Deep_Drawn_Cups_Using_Direct_Polymer_Additive_Tooling

[12] Geueke M., Steinheimer R., Lutz M. und Engel B., "Sustainable tool technology: Wood-based forming tools," *Materials Research Proceedings*, Jg. 28, S. 1967–1976, 2023, doi: 10.21741/9781644902479-212.

[13] M. Geueke, D. Nebeling und B. Engel, "Alternative Werkzeugtechnik – holzbasierte Umformwerkzeuge," *Transformation in der Biegetechnik : Tagungsband zum 6. Biegeforum Siegen*, 2023, doi: 10.25819/ubsi/10389.

[14] R. Kolleck, C. Koroschetz, G. Schickofer und Agustin M., "Alternativer Werkstoff Holz spart Werkzeugkosten für die Blechumformung," *Maschinenmarkt*, 2008 2008. [Online.] Verfügbar: https://www.maschinenmarkt.vogel.de/alternativer-werkstoff-holz-spart-werkzeugkosten-fuer-die-blechumformung-a-142940/?p=3

[15] M. Pinto, A. D. Santos, P. Teixeira und P. J. Bolt, "Study on the usability and robustness of polymer and wood materials for tooling in sheet metal forming," *Journal of Materials Processing Technology*, Jg. 202, 1-3, S. 47–53, 2008, doi: 10.1016/j.jmatprotec.2007.08.082.

[16] M. Geueke, D. Nebeling und B. Engel, "Sustainable tool technology: Wood-based forming tools," *Materials Research Proceedings*, 2023, doi: 10.1016/j.procir.2023.09.093.

[17] M. Risse und Richter K., *Das Holz der Robinie – Eigenschaften und Verwendung: Beiträge zur Robinie* (LWF Wissen), 2020.

[18] D. W. Green, J. E. Winandy und D. E. Kretschmann, *The Mechanical Properties of Wood, 1999* (1999).

[19] A. M. Habraken *et al.,* "Analysis of ESAFORM 2021 cup drawing benchmark of an Al alloy, critical factors for accuracy and efficiency of FE simulations," *Int J Mater Form*, Early Access. doi: 10.1007/s12289-022-01672-w.

IOP Conf. Series: Materials Science and Engineering **1307** (2024) 012019 doi:10.1088/1757-899X/1307/1/012019

Cryogenic Deep Drawing Tool Design for Reduced Heat Transfer

M Tulke[1] and A Brosius[1]

[1] Chair of Forming and Machining Processes, TUD Dresden University of Technology, Germany

marc.tulke@tu-dresden.de

Abstract. Subject of this study is the usage of macro-structured tools for deep drawing of aluminum alloys at cryogenic temperatures in order to increase the drawing depth. The advantage of the macro-structure is the reduced contact area, which primarily leads to a reduction in frictional forces in the flange area and also minimizes the heat transfer from the tool to the cooled sheet metal. In addition to the geometry of the tools, the heat transfer is also influenced by the thermal properties of the tool materials. Therefore, the tool must be designed to reduce heat flux and thermal diffusivity. In this regard, substitute materials for the tools, specifically polymer, and steel-shielded polymer, were investigated. In this context, the heat transfer coefficients between the tool material and the aluminum blank are determined experimentally. The influence of thermal conditions on the cryogenic deep drawing process are investigated numerically and experimentally. The result of the study is an improved process understanding to ensure the blank temperature in a cryogenic range and increase the resistance against bottom cracks.

1. Introduction

This paper demonstrates the deep drawing process of aluminum alloys at cryogenic temperature in order to expand the process windows. The main defects in deep drawing are bottom cracks and wrinkling. For stabilizing the process, macro-structured tools are used which enable lubricant free deep drawing by reducing the contact surface and friction force in the flange area, resulting in increased drawing ratio. In addition, the wavy structure induces an alternate bending that increases the resistance against wrinkling. Overall, the macro-structured tools stabilize the cryogenic deep drawing process, especially with minimal lubrication due to the low temperature [1].

As lightweight material, aluminum alloys can play a significant role in resource efficiency and CO_2 reduction. Despite the relatively low formability of aluminum alloy sheets at room temperature, their ductility can be significantly increased in cryogenic range whereas uniform elongation is reduced at higher than room temperature [2]. Yuan et al. [3] showed that improved deep drawability can be achieved at lower temperature, while Grabner et al. [4] demonstrated the suitability of the process for complex semi-industrial parts. This was achieved using an isothermal forming process with cooled tools. In contrast, Brosius and Tulke used adapted, macro-structured tools, aiming at reducing the heat flux and friction force in cryogenic deep drawing [5], therefore removing the need for expensive and complex tool cooling. Additionally, it was demonstrated that the thermal properties of the punch material influence the blank temperature and consequently the deep drawing process.

While the influence of the macro-structured tool design on wrinkling is shown in [6], the focus here is on the temperature profile in the aluminum blank depending on the tool material. In this work, the influence of the tool materials for the die and blank holder as well as the punch are investigated regarding the development of the blank temperature and the resulting drawing depth.

2. Cryogenic deep drawing process

In the presented cryogenic deep drawing process, the aluminum blank is cooled in liquid nitrogen (LN_2) down to $\vartheta_{\mathrm{init}}$ = -196 °C. To ensure that the blank remains in the cryogenic range for as long as possible, the heat flux from the uncooled tools ($\vartheta_{\mathrm{tool}}$ = 20 °C) to the blank needs be reduced in order to maintain the improved formability and strength of aluminum alloys. For this reason, the contact surface between the tools and the blank is reduced by using macro-structured tools. To achieve this, the full contact tools (depicted in Figure 1 (a)) are replaced by a blank holder and die with a wavy structure, see Figure 1 (b). Furthermore, the punch is modified so that the front surface is hollow for a reduced contact area, see detail I in Figure 1, in order to prevent the contact heat conduction and heat radiation into the aluminum blank. Another way to reduce the heat flux from the tools to blank is to substitute steel as tool material with a plastic material that has a lower thermal diffusivity, see Table 1.

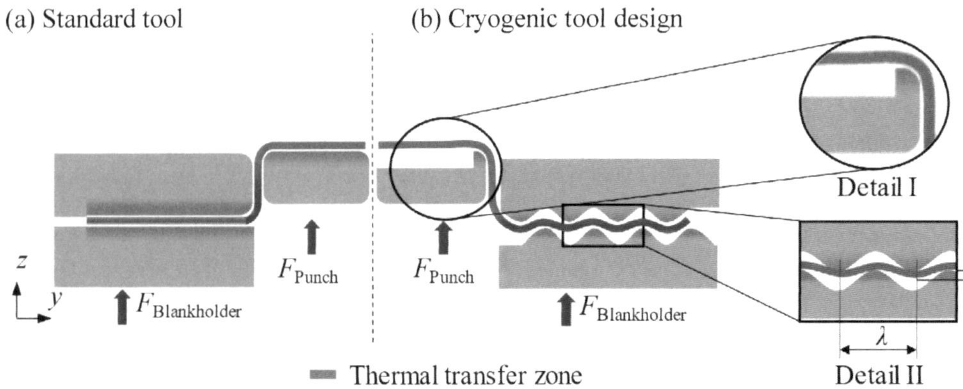

Figure 1. Approach for cryogenic deep drawing with adapted tools according to [7]

2.1. Approach

The objective here is to analyze how the tool material influences the temperature profile in the blank and how the temperature profile affects the formability and strength of the aluminum alloy in the deep drawing process. The aim is to maintain the blank temperature in a cryogenic range in the critical area for an increased resistance against bottom cracks.

3. Investigated tool systems

In general, a tool material with a low thermal conductivity λ_{th} and thermal capacity c_{p} is advantageous regarding the heat transfer to the blank. With low thermal conductivity, the heat flow in direction of the negative temperature gradient (to the cooled blank) is lower than with high thermal conductivity. Moreover, when using a material with low thermal capacity, the amount of energy that can be released is reduced compared to materials with high c_{p}. This means that low values for λ_{th} and c_{p} are advantageous regarding a reduced heat transfer from the tool to the blank. The tool materials examined here are a steel (1.7225) and a polyurethane-based engineering plastic (SikaBlock LAB850). The low thermal conductivity of the plastic compared to steel is relevant in order to reduce the heat transfer from the tool to the sheet metal. Due to the very different thermal properties of the two materials, the thermal diffusivity α_{th} is used here for comparison, as this parameter depends on both the thermal conductivity and the heat capacity in accordance with equation (1):

$$\alpha_{\mathrm{th}} = \frac{\lambda_{\mathrm{th}}}{\rho \cdot c_{\mathrm{p}}} \tag{1}$$

with λ_{th} the thermal conductivity, ρ the density and c_p the thermal capacity. When comparing the thermal diffusivity of the tool materials, it is evident that the heat transfer in the plastic material is significantly lower than in the steel tool, meaning that a lower heat transfer from the plastic tool to the sheet metal can be assumed. The thermal properties of the used tool materials and investigated blank material AA6014 are summarized in Table 1.

Table 1. Thermal properties of the tool material and investigated aluminum alloy AA6014

			Tool material		Blank material
			1.7225	Block LAB850	AA6014
Density	ρ	in $\mathrm{kg \cdot m^{-3}}$	$7.72 \cdot 10^3$	$1.18 \cdot 10^3$	$2.7 \cdot 10^3$
Thermal conductivity	λ_{th}	in $\mathrm{W \cdot m^{-1} K^{-1}}$	45	0.2	120
Thermal capacity	C_p	in $\mathrm{J \cdot kg^{-1} K^{-1}}$	465	1875	600
Thermal diffusivity	α_{th}	**in** $\mathrm{mm^2 \cdot s^{-1}}$	**12.54**	**0.09**	**74.07**
Heat transfer coefficient	h_c*	**in** $\mathrm{kW \cdot m^{-2} \cdot K^{-1}}$	**5 – 25**	**1000 - 5000**	-

*(Pressure dependent value for contact between AA6014 and tool material, given for a contact pressure of 5 MPa and 30 MPa)

The heat transfer coefficient h_c, used in FEM to consider the contact heat conduction, is determined experimentally with an experimental setup according to [6]. The experimental setup consists of two punches and an aluminum plate in the middle, see Figure 2. According to the cryogenic deep drawing process, the aluminum plate is cooled in liquid nitrogen (ϑ = -196 °C) prior the test, while the punches remain at room temperature (ϑ = 20 °C). Corresponding to the deep drawing tools the punches here are made of steel on the one hand and plastic material on the other hand. Since it is a symmetrical model, both punches are made of the same material, so that a total of four punches are required; two punches made of steel and two made of plastic material. Each punch and the aluminum plate are equipped with several thermocouples to measure the local and temporal temperature development while in contact. The contact pressure varies from 0.1 MPa up to 30 MPa to examine the pressure dependency of h_c. Subsequently, the heat transfer coefficient is determined by a parameter identification based on the experimental data of the temperature development in the blank and punches. Here the Fourier's law of heat conduction, see equation (2), is used for calculating the thermal conduction within the material, and equation (3) is used for the contact heat conduction with h_c as a proportional factor. The hypothesis is that c_p and λ_{th} are constant.

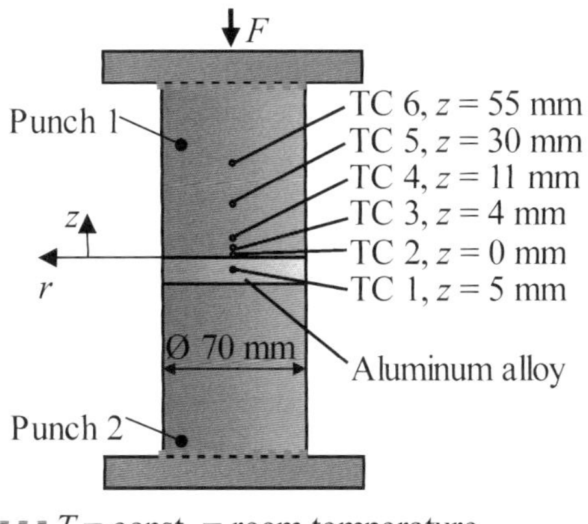

Fourier's law of heat conduction

$$\frac{\partial T}{\partial t} - \frac{\lambda_{\mathrm{th}}}{\rho \cdot c_{\mathrm{p}}} \frac{\partial^2 T}{\partial z^2} = 0 \qquad (2)$$

Contact heat conduction

$$q = -h_c \cdot \Delta T \qquad (3)$$

▪▪▪ T = const. = room temperature
TC: Thermocouple

q: Thermal energy
z: Position in z -direction

Figure 2. Experimental setup for determining the contact heat transfer coefficient h_c according to [6]

In this investigation, macro-structured tools with a wave length λ = 10 mm are used (see Figure 1, detail II), with one tool set made of 1.7225 and a second made of plastic material SikaBlock LAB850.

Due to the low hardness of the plastic material, a steel sheet with a thickness of 0.1 mm is bonded to the die and blank holder as wear protection. An additional advantage of the thin steel sheet is the lower heat capacity in direct contact with the sheet and reduces the energy which can be transferred to the aluminum blank. In case of the punch made of plastic material, no steel-shield is needed because there is no relative movement between punch and blank at high contact pressure. The principal setup and the different tool variants are shown in Figure 3. The investigated tool combinations are summarized in Table 2.

Table 2. Combination of the different tool systems

	Die material	Blank holder material	Punch material / design
Variant 1	Steel	Steel	Steel / hollow
Variant 2	Steel	Steel	Plastic / hollow
Variant 3	Shielded plastic	Shielded plastic	Plastic / hollow
Variant 4	Shielded plastic	Shielded plastic	Plastic / full contact

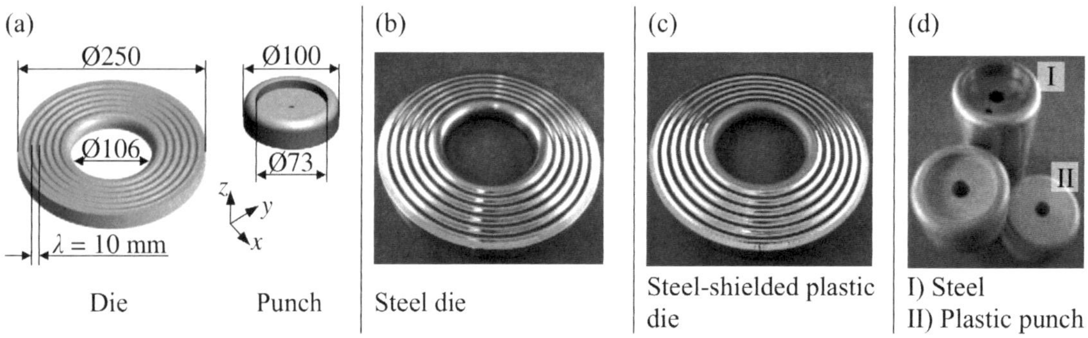

Figure 3. Various tool geometries and materials for the experimental investigations

4. Determination of the temperature development

In order to determine the local and time-related development of the blank temperature, the blank is equipped with thermocouples at the middle and outer edge. In addition to [5], the temperature development is measured for the different tool systems made of steel and SikaBlock. It was measured during positioning of the blank and closing of the die and blank holder with a resulting immersion depth $\delta = 0.1$ mm. No measurement is possible during the deep drawing process, since the tool contact would destroy the thermocouples. For this reason, the punch is not varied while the hollow punch made of plastic material is used to complete the tool system. The die edge radius and punch edge radius are $r_{DIE} = r_{PUNCH} = 10$ mm. Here the initial blank diameter is $D_0 = 200$ mm which results in a drawing ratio $\beta = 2.0$. The results of the experimental investigation are depicted in Figure 4. Here, the time frame between $-10 < t < 0$ s is used for positioning the blank on the die and at $t = 0$ s the die and blank holder are closed with $\delta = 0.1$ mm. In the remaining time frame up to $t = 30$ s the blank temperature is measured while the punch is in contact, but without deep drawing. It shows a larger temperature gradient in the blank for the steel blank holder and die compared to plastic tools. This results in a large temperature difference $\Delta\vartheta_{steel} = 155$ K between the flange area and the middle of the blank for the steel tools, especially in the first five seconds after closing the tools. In contrast, the temperature difference in the blank for plastic tools $\Delta\vartheta_{plastic} = 55$ K is significantly lower. The slight temperature difference at the beginning of the measurement is due to the manual test procedure, so that the temperature is not exactly the same when the tools are closed at time $t = 0$ s. Nevertheless, it is clearly visible that the temperature difference between the two tool designs quickly increases when closing the tools. This preliminary examination serves to validate the following numerical investigations. The experimental results enable a realistic evaluation of the numerically determined temperature development in the blank. In principle, it is already shown here that there is a temperature gradient from the flange to the bottom area that is affected by the tool material, but a calculation is required for a determination while deep drawing.

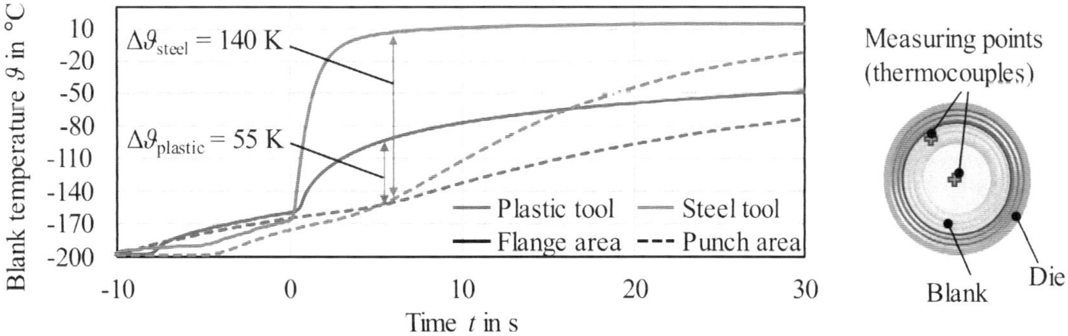

Figure 4. Experimental temperature development in the flange area for different tool materials

Due to the direct contact of the tools with the sheet metal and the stretching of the blank surface, a spatially resolved measurement from the edge to the center of the blank is not possible during the deep drawing process. Therefore, an additional numerical analysis of the temperature development is carried out using LS-Dyna 12.0.0. The blank is meshed with a fully integrated 3D-shell element type with a mesh size of 1.0 mm while the tools are modeled with solid elements for a correct consideration of the thermal capacity. The material behavior is described by temperature dependent flow curves, see Figure 5 (a), using MAT133 (Barlat_YLD2000) with isotropic hardening behavior. The deep drawing tools are modelled as rigid bodies using MAT020. According to the experimental tests, the hollow punch made of plastic is used with a punch diameter d_{punch} = 100 mm and the blank diameter is D_0 = 200 mm. The initial blank temperature is $\vartheta_{blank.0}$ = -150 °C while the initial tool temperature is $\vartheta_{tool.0}$ = 20 °C.

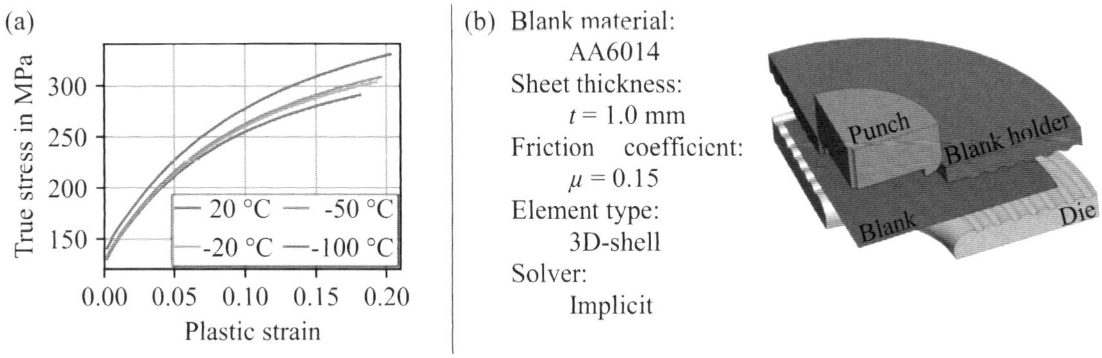

Figure 5. FE-model: (a) Temperature dependent flow curves; (b) Characteristics of the FE-model

The numerical results of the temperature development along the blank radial direction are summarized in Figure 6. The temperature profile is presented for punch displacement values of 0 mm, 5 mm, 15 mm and 35 mm. At the beginning of the deep drawing process, a temperature profile in the flange corresponding to the design of the macro-structure is recognizable and the temperature distribution is almost the same for the three tool combinations, see Figure 6 (a). A clear difference in the temperature profile can be seen after a punch displacement s_p = 5 mm (Figure 6 (b)). In case of the steel tools, the blank temperature in the flange area rises significantly faster than at the punch area with increasing drawing depth. In the further course of the process, the blank temperature in the flange increases significantly compared to the bottom area. This results in a negative temperature gradient from the flange to the bottom of the cup leading to an increasing strength in the bottom area without work hardening.

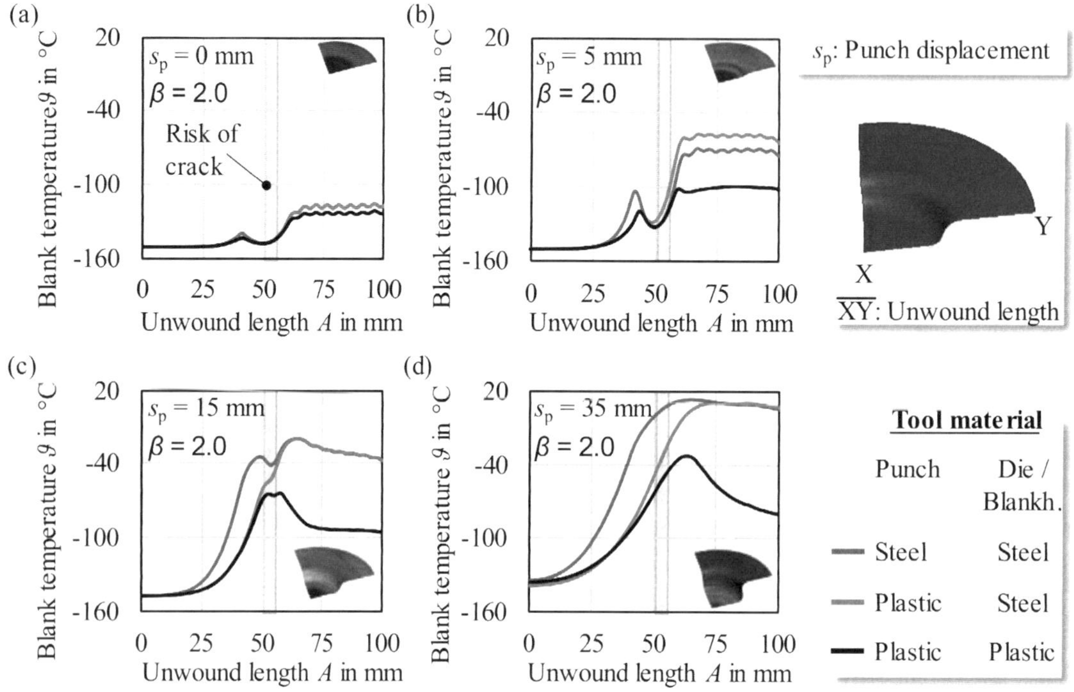

Figure 6. Development of the blank temperature for three tools systems at different punch displacements

In contrast, the temperature in the blank is significantly lower when forming with a plastic blank holder and die compared to steel. But the temperature in the critical zone for cracking, area above the punch edge, rises faster compared to the flange area. When the punch displacement reaches $s_p = 15$ mm (Figure 6 (c)) the temperature at the die corner is higher as in the flange area. Thus, the strength in the flange or the frame of the cup exceeds the strength at the punch edge radius due to work hardening and lower blank temperature in the flange. Consequently, bottom cracks will occur sooner than with a negative temperature gradient towards the punch. The experimental results in 5.1. will provide evidence for this assumption.

In summary, an increase in the drawing ratio or drawing depth compared to conventional deep drawing can only be achieved if the blank temperature in the bottom area is significantly lower than in the flange area. Thus, it can be assumed that the strength in the bottom area is greater than in the flange area due to the low temperature, which has a positive effect on the crack resistance.

5. Cryogenic deep drawing: Influence of the tool material

For evaluating the influence of the tool material on the drawing depth experimentally, the tools at the flange area, blank holder and die, and the punch, mainly responsible for heat transfer at the bottom area, are evaluated separately. The circular cups are deep drawn till bottom cracks occur; a fully deep drawn cups were not reached with this high drawing ratio $\beta \geq 2.2$.

5.1. Tool material of blank holder and die

For analyzing the influence of the tool material of die and blank holder on the deep drawing process, experimental tests were performed. The blank diameter is $D_0 = 220$ mm that results in a drawing ratio $\beta = 2.2$ mm and the initial sheet thickness is t = 1.0 mm. Here, all tests are performed with the hollow punch made of plastic. Figure 7 shows the resultant drawing depth for the tool system with a die and blank holder made of steel compared with the tool system made of plastic material. It is clearly visible that the reached drawing depth with the plastic tools is much lower compared to the steel tools.

Furthermore, the gradient of the force increase is greater with the plastic tool compared to the steel tool. Taking into account the temperature distribution determined in chapter 4. , the lower sheet temperature in the plastic tool results in a higher material strength in the aluminum sheet, so that a greater forming force is required. This also means that a higher temperature in the flange area than in the bottom area is favorable for the deep drawing process. As the material in the flange area undergoes work hardening by the forming process, the strength increases here compared to the base area, as no forming occurs here. As a result of the lower temperature in the base area, the strength increases there due to the material properties of aluminum alloys at cryogenic temperatures. This increases the strength at the bottom without work hardening, thus increasing the resistance to bottom crack. However, if there is no temperature gradient from the flange to the bottom, this effect cannot be utilized and bottom crack occurs at a lower drawing depth.
As shown in Figure 7, an increase in the drawing depth of up to 60 % is achieved while using steel for the die and blank holder for a drawing ratio $\beta = 2.2$ due to the high temperature difference between the flange and bottom area, although the increase in material strength is approximately 15 % at -100 °C compared to room temperature (see Figure 5 (a)). The result is explained by the different areas of work hardening in the blank. Basically, the bottom area does not undergo work hardening, but instability at the punch edge must be avoided in order to prevent bottom cracks. The temperature difference between the flange and the bottom area shown in Figure 6 is sufficient to increase the strength at the bottom of the cup and allows greater drawing depths. Therefore, the relatively small change in strength due to very low temperatures is the key to expand the drawing depth.

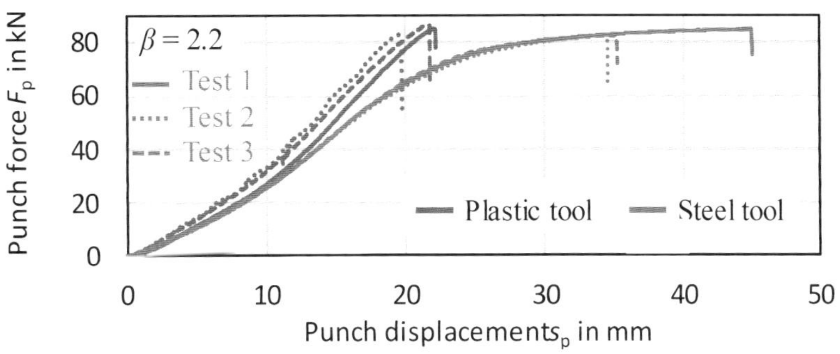

Figure 7. Experimental drawing depth depending on the tool material for blank holder and die

5.2. Design and material of the punch
For evaluating the influence of the punch design on the drawing depth, three different designs are investigated: a steel punch with reduced contact area; a plastic punch with reduced contact area; a plastic punch with full contact, see Figure 3 (d). Since the plastic material has a significantly lower thermal diffusivity than steel, the punch with full contact is used to investigate whether the change of material is sufficient to increase the drawing depth or whether the reduction in contact area also leads to a noticeable increase in the drawing depth for the plastic material. According to the results in 4. and 5.1. here the steel blank holder and steel die are used because of the higher reached drawing depth.

The result in Figure 8 shows that the use of a full-surface punch made of plastic already enables an increase in the drawing depth compared to a steel material. However, the additional reduction of the contact surface on the front side leads to a significant increase in drawing depth. Consequently, a large bottom surface of the cup, without tool contact, seems to be advantageous in order to utilize the increased strength of aluminum alloys under cryogenic conditions.

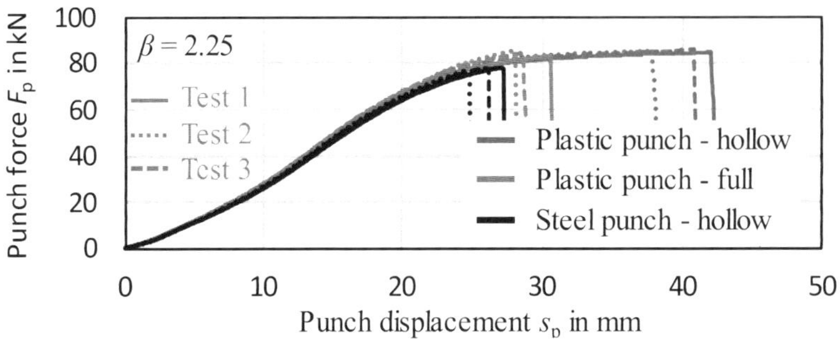

Figure 8. Experimental drawing depth depending on the punch design and material

6. Conclusion

The presented work shows the influence of the temperature profile in the blank on the deep drawing process. A negative temperature gradient must be set from the flange to the bottom area, which results in an increase in strength in the critical zone for bottom crack at the punch radius. This enables higher drawing depths to be achieved compared to deep drawing without an adapted temperature profile in the aluminum sheet metal. Here, the temperature profile was set by using a plastic material for the punch, as this has a lower thermal diffusivity than steel, while steel tools were used in the flange area. The macro-structured tools in the flange, on the other hand, are mainly relevant for stabilizing the deep drawing process. By reducing friction in the flange, the required drawing force is reduced. The combination of both effects; reduced punch force and increased strength in the area of the punch edge radius; results in an increase in achievable drawing depth compared to conventional deep drawing.

Acknowledgement

The authors are grateful to Outokumpu Nirosta GmbH, especially Mr. Jörn Teipel, for providing thin sheet steel for wear protection of the plastic tools and the Center for Information Services and High Performance Computing [Zentrum für Informationsdienste und Hochleistungsrechnen (ZIH)] at TU Dresden for providing its facilities for high throughput calculations.

References

[1] Brosius A and Mousavi A 2016 Lubricant free deep drawing process by macro structured tools. *CIRP Annals* pp 253–6, DOI: 10.1016/j.cirp.2016.04.060.
[2] Bruschi S, Bertolini R, Simonetto E and Ghiotti A 2019 Analysis of mechanical behaviour of AA6082-T6 sheets deformed at low temperatures. *Procedia Manufacturing* pp 435–42, DOI: 10.1016/j.promfg.2019.02.159.
[3] Yuan S, Cheng W, Liu W and Xu Y 2020 A novel deep drawing process for aluminum alloy sheets at cryogenic temperatures. *Journal of Materials Processing Technology*, DOI: 10.1016/j.jmatprotec.2020.116743.
[4] Grabner F, Gruber B, Schlögl C and Chimani C 2018 Cryogenic Sheet Metal Forming - An Overview. *MSF* pp 1397–403, DOI: 10.4028/www.scientific.net/MSF.941.1397.
[5] Brosius A and Tulke M 2023 Forming of aluminium alloys with macro-structured tools at cryogenic temperature. *CIRP Annals* pp 225–8, DOI: 10.1016/j.cirp.2023.03.018.
[6] Tulke M, Lafarge R, Wolf A and Brosius A 2024 Contact Conditions and Temperature Distribution During Cryogenic Deep Drawing with Macro-structured Tools. *Proceedings of the 14th ICTP* pp 233–43, DOI: 10.1007/978-3-031-40920-2_25.
[7] Tulke M, Wolf A, Lafarge R, Grunow F and Brosius A 2022 Influence of Macro-Structured Tools on the Formability of Aluminum Alloys in the Cryogenic Temperature Range. *KEM* pp 760–7, DOI: 10.4028/p-nhu9o5.

IOP Conf. Series: Materials Science and Engineering **1307** (2024) 012020 doi:10.1088/1757-899X/1307/1/012020

Development of a novel forming tool for the production of high-strength aluminum components with tailored properties

N Rigas[1], M Merklein[1]

[1] Institute of Manufacturing Technology, Friedrich Alexander-Universität Erlangen-Nürnberg, Egerlandstr. 13, 91058 Erlangen, Germany

nikolaos.rigas@fau.de

Abstract. Because of growing environmental and technical requirements, the efficient manufacturing of components with tailored properties using recyclable materials is mandatory. The combination of high-strength aluminum alloys and thermal-assisted forming operations is an innovative method for the production of components with varying mechanical properties. By locally adjusting the cooling rates during a thermo-mechanical forming operation, it is possible to modify the microstructure and precipitation conditions. A subsequent aging operation causes different microstructural and mechanical characteristics. This allows both deep-drawing and adjustment of the component properties in one single step. For this reason, this contribution focuses on the production of components with tailored properties. With the aid of a non-contact measuring system, previous investigations have proved that mechanical material properties can be adapted during a forming process by adjusting the cooling rates. In this contribution, this knowledge is transferred to a novel temperature-controlled rectangular cup tool. A variation of the local tool temperature and holding duration in the tool will be carried out. Then, the mechanical and microstructural properties of the manufactured components will be characterized through hardness and DSC investigations. In addition, the influence of different tool temperatures on the component temperature, the sheet thickness distribution and process forces will be investigated. As a result, it has been demonstrated that mechanical, process-related and geometric advantages can be achieved by using locally temperature-controlled forming tools.

1. Introduction

Changing mobility and lightweight requirements need the use of alternative production methods in combination with alternative materials. High-strength aluminum alloys of the 6xxx and 7xxx series permit the realization of lightweight design and fulfill these demands. Due to their limited formability, high-strength aluminum alloys can only be formed without failure using thermally supported forming methods [1]. In particular, the so-called Hot Form and Quench (HFQ®) process first introduced by Garret et al. offers beneficial forming and component properties [2]. It consists of the process steps of solution annealing, simultaneous quench forming and subsequent artificial aging. The elevated forming temperatures permit reduced forming forces, decrease springback while increasing the formability. Rapid quenching during forming causes a **s**uper**s**aturated **s**olid **s**olution (SSSS) and enable the subsequent formation of strengthening precipitates during artificial aging. In this way, the failure-free production of components with high final material strength can be achieved. For the 6xxx alloy type, different precipitation phases can be generated during the aging operation, which varies in both the morphology as well as the strength potential. At room temperature, globular and spherical clusters and

GP-zones occur, which only cause a slight increase in mechanical strengthening. With increasing artificial aging temperature, needle-shaped β''-precipitants are present, which cause a maximum increase in mechanical performance [3]. A further increases of the artificial aging temperature or duration lead to the formation of semi-coherent, rod-shaped β'-phases or platelet-shaped β-phases, which result in lower material strength because of over-aging [4]. For this reason, precipitation-hardenable aluminum alloys have different mechanical material properties depending on the thermal treatment.

2. Production of aluminum components with tailored mechanical properties

Manufacturing aluminum components with tailored properties has become an important topic in recent years to fulfill the growing requirements in the field of mobility. It has already been proven that steel components with tailored characteristics can be produced using modified joining processes, heat treatment strategies or rolling operations [5]. These methods for local adjustment of material properties can also be transferred to aluminum alloys. Due to the high quench sensitivity of precipitation-hardenable aluminum alloys, it is possible to grade the solid solution during the forming operation [6]. By varying the cooling rate, specific areas can be created that do not have a supersaturated solid solution. With decreasing cooling rates, however, different precipitates can already be generated during the quenching operation [7]. As a result, the mechanical properties are affected both, after quenching and after artificial aging. By this, the production of components with locally adapted mechanical properties is possible. Investigations by Gao et al. revealed, that the alloy AA6082 has a critical cooling range between 240 °C and 440 °C [8]. Studies by Sajadifar et al. demonstrated that various cooling rates can be obtained by adjusting the tool temperature [9]. In addition to the die temperature, recent studies have shown that the holding time in the tool also has a significant influence on the mechanical properties and the precipitation behavior. Using a non-contact in-situ measuring method, it was proven that even short holding times at elevated tool temperatures lead to a significant increase in coarse precipitates [10]. Thus different mechanical properties can be combined in one component by adjustment of the cooling rate during the simultaneous forming and quench process (Tailor Quenched Forming).

3. Material, developed forming tool and measurement technique

The commercial alloy AA6082 in the initial condition T6 with a sheet thickness of t_0 = 2 mm was used for the investigations. The blanks for deep-drawing operations were laser cut.

3.1. Developed tool for Tailor Quenched Forming (TQF) and process steps

To manufacture components with tailored local properties, a forming tool with different local temperatures was developed. It is designed so that both the drawing depths and the temperature of the actively heated tool side can be freely adjusted. The tool is installed in the hydraulic press type Lasco TSP 1000. In Figure 1, both the forming tool as well as the blank with its relevant dimensions are shown schematically. The punch, blank holder and die of the tool each consist of an actively cooled and a temperature-controlled tool zone. To prevent uncontrolled sheet thinning during forming, the deep-drawing operation is carried out with a distanced blank holder. The tool has a constant drawing gap of 3 mm and a uniform drawing depth of 30 mm is realized for all forming operations. To determine the required process forces during forming the punch is equipped with a 1200 kN load cell.
First, the sheets are coated with a graphite-based lubricant type Graphitex®CR 25-9/14 in order to prevent undesired wear during forming. Subsequently, the components are solution heat treated in a furnace of model Rohde ME17 at 540 °C for 15 minutes each. Then, the heated aluminum blanks are then transferred in the locally tempered tool and simultaneously formed and quenched. Using a thermal camera type Flir SC7600, the temperature distribution of the aluminum blanks and the formed components is analyzed contact-free before and after forming. For the forming operations, the die temperature of the temperature-controlled tool zone is varied between 20 °C and 400 °C. In addition, the duration of the closed tool differs between 5, 20 and 60 seconds. After forming, the components cool

in ambient air and are then artificially aged at 155 °C for 16 hours in a convection-furnace type Nabertherm N120. The components are then analyzed in terms of sheet thickness distribution, hardness and precipitation morphology.

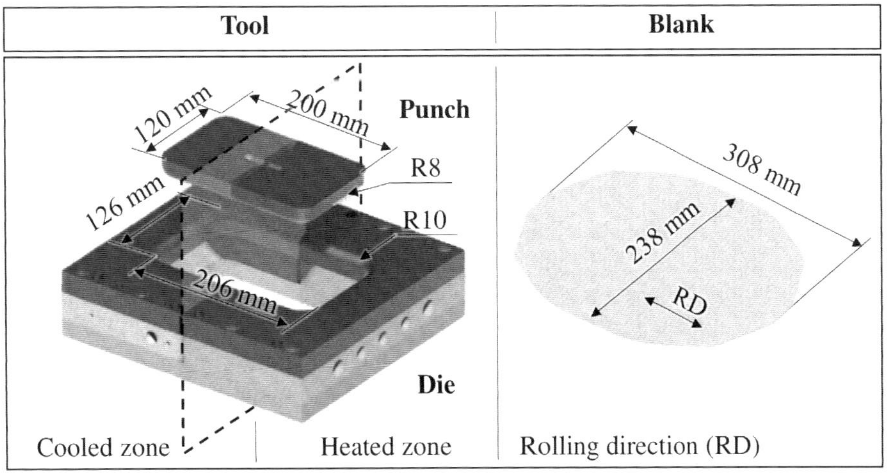

Figure 1. *Developed tool and blank geometry for the production of rectangular cups with tailored properties*

3.2. Analysis of the components after forming

After the forming operations, the components were cleaned and subsequently coated with a reflection-reducing paint for optical component measurement. Then, the rectangular cups are optically scanned by using an optical 3D measurement system type ATOS. Using the software GOM-Inspect the sheet-thickness distribution of the digitized components is determined.

Using a hardness tester of type Instron Wolpert 930/250, the Brinell hardness of the manufactured components was determined at seven different test positions according to DIN EN ISO 6506-1. In this way, both the flange and the component bottom were examined. DSC samples were extracted from the formed components to record the precipitation condition. In order to prevent the thermal influence of a machining or laser based cutting operation, the samples were removed from the cups at three different positions using wire erosion in a water bath. For this purpose, rotation-symmetrical discs with a diameter of 5 mm have been cut out. The samples were then analyzed in a MettlerToledo DSC822 measuring device. In Figure 2, the positions of the hardness measurements and the extraction position of the DSC samples are illustrated.

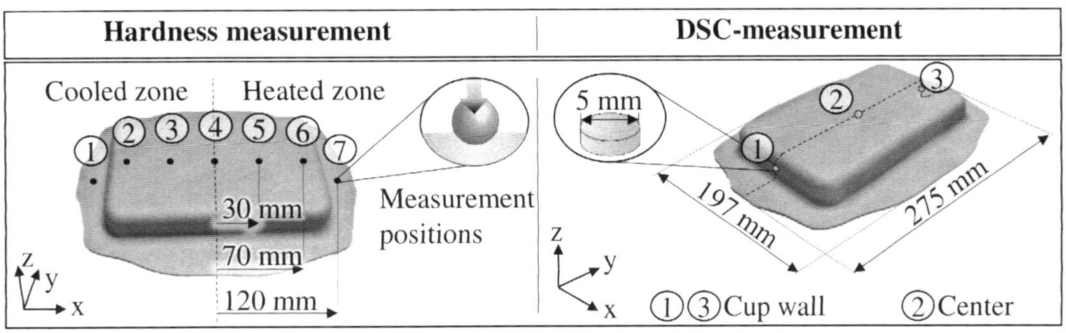

Figure 2. *Measurement positions for hardness and DSC analyses of the formed cups after aging*

4. Results and discussion

To determine the influence of different tool temperatures on the cooling behavior of components, the temperature distribution before and after forming was measured (Figure 3). The results reveal that a homogeneous temperature distribution of 540 °C is achieved before the tailor quench forming operation. All components remain for 20 seconds in the local tempered tool. For the untempered tool, the results show that an almost homogeneous temperature distribution occurs. However, a slight temperature increase of approximately 70 °C can be seen in the bottom of the component. This is because no counter-punch is currently installed in the novel tool. As a result, no contact between the component and the mold is ensured at the bottom of the rectangular cup, which leads to a lower cooling rate and therefore to a higher removal temperature in the centre. At a forming tool temperature combination of 20 °C and 200 °C, a clear temperature gradation can be seen between the actively cooled die area and the temperature-controlled die zone. The results illustrate that a distinct gradation zone results in the flange bottom and wall of the part. In contrast, with a tool temperature combination of 20 °C and 400 °C, a more recognizable transition zone can be seen between the different component areas. This is due to the high differences in destination temperatures. As soon as the forming tool opens, a temperature transfer takes place from the actively tempered zone to the cooled area. In addition, a significant transition zone can be seen at the bottom of the component. This is due to the combination of no counterholder and the heat transfer of the aluminum components. This leads to the conclusion that different component temperatures can be realized with the help of tempered tool cavities. However, the results show that as the temperature difference increases, a less narrow temperature gradation results and larger transition zones are created. It can also be noted that after opening the tool, locally cooled component areas heat up in an uncontrolled manner due to thermal conduction. This can lead to undesirable precipitate formation and thus to an influence on the mechanical component properties.

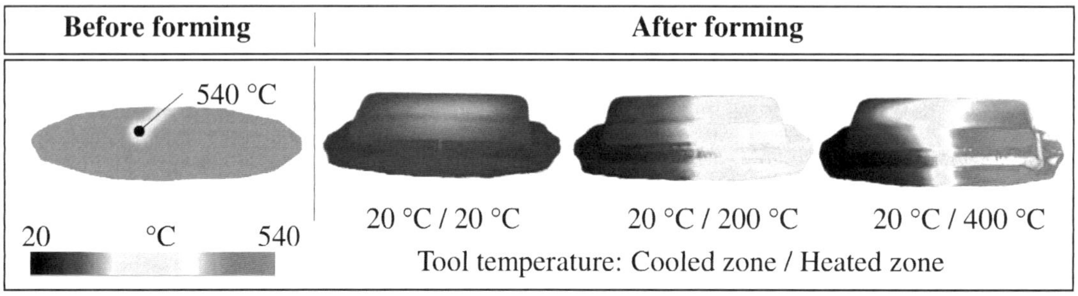

Figure 3. *Temperature distribution before and after forming at different tool temperatures*

Using the load cell, the influence of different tool temperatures on the process forces during tailor quench forming was determined. For this purpose, the load curve was identified over the forming duration and the maximum force for the different tool temperature combinations is compared, which is presented in Figure 4. In addition to the tool temperature, the holding time in the tool also varied between 5, 20 and 60 seconds. The results reveal that the maximum punch forces decrease with increasing tool temperature. This is due to the varying cooling behavior during the forming operation. At elevated temperatures, aluminum exhibits reduced flow resistance, which leads to lower forming forces. At high temperatures, the components cool down less quickly during forming in the actively tempered tool area, especially the flange and cup wall area. This reduces the hold-back force and results in lower maximum punch forces. For this reason, the maximum punch forces decrease by 53 % from 46.2 ± 1.5 kN at non-tempered quench forming to 21.7 ± 0.6 kN at a tool temperature combination of 20 °C and 400 °C. The variation of the holding time of the closed forming tool does not influence the resulting maximum punch force. It can therefore be concluded that increased tool temperatures lead to reduced process forces during forming.

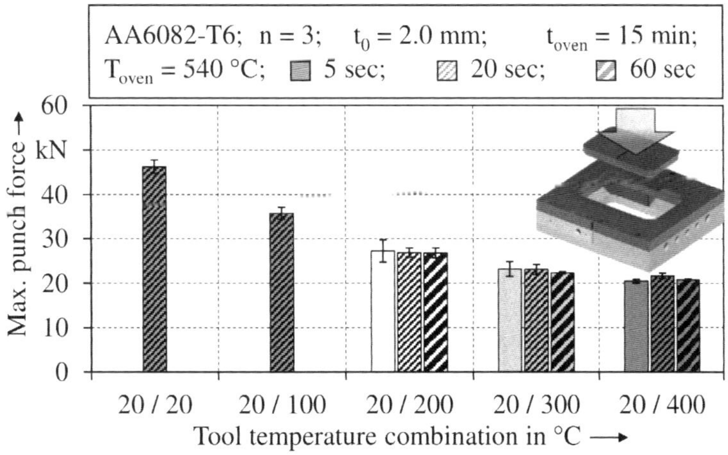

Figure 4. *Forming force for tailor quench forming in dependence of the tool temperature*

To evaluate if the material flow behavior is influenced during a tailored quenching operation, the sheet thickness distribution of the formed components is investigated. In Figure 5, the sheet thickness distribution of five differing tool temperature combinations is compared. The results illustrate that in an untempered quenching tool, symmetrical sheet thinning in the corner areas of the bottom of the component takes place. This can be explained by the rapid cooling of the flange and cup wall region due to the low blankholder and die temperatures. Because these areas cool down faster than the bottom, increased restaining forces occur and lead to a material flow from the bottom area into the cup wall zone. As a result, the thickness of the components in the corner area decreases. In the flange area, however, the formation of wrinkles can be seen. The reason for this is that the forming operations are carried out with a distanced blank holder. As a result, tangential compressive stresses occur in the flange area, leading to wrinkling. For a tool temperature combination of 20 °C and 200 °C, the sheet thinning on the tempered tool side decreases noticeably around the bottom area. This is due to the improved flow behavior of the aluminum material at higher temperatures and the reduced withdrawing forces. At a temperature combination of 20 °C and 400 °C, the sheet thinning in the component bottom drops to a minimum and is only still visible in the transition area between the bottom and cup wall. Furthermore, the results show that increased tool temperatures do not lead to local wrinkling in the flange area, but to a homogeneous accumulation of material in the areas where compressive stresses are superimposed. This leads to improved visual surface properties of the formed components. Therefore, it can be summarized that increased tool temperatures lead to more homogeneous sheet thickness distributions and reduced wrinkling at formed components.

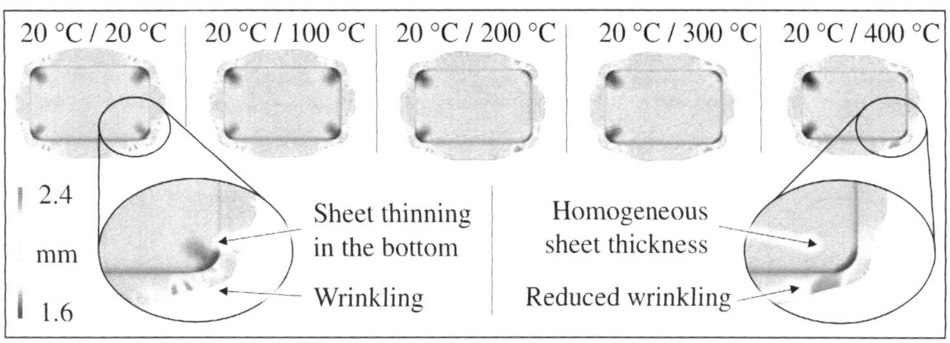

Figure 5. *Sheet thickness distribution of the formed components*

In order to investigate the influence of different tool temperatures and holding times on the mechanical component properties, Brinell hardness measurements were carried out. Figure 6 compares the material hardness for the tool temperature combinations of 20 °C with 200 °C, 300 °C or 400 °C in each case. In addition to the temperature combination, the holding time in the closed tool is also varied between five, 20 and 60 seconds. The material hardness was measured at seven positions for every component. Measurement position one and seven correspond to the flange area of the cups. Positions two to six are in the bottom section of the cups. Position four corresponds to the center of the component. The results illustrate that a tool temperature combination of 20 °C / 200 °C results in the highest average material hardness without a clear transition zone. It can also be seen that the material hardness at the edge positions one and seven is slightly higher than in the middle positions of the component base. This is due to the higher cooling rate in the edge zones and, as a result, the quicker crossing of the critical temperature range. In addition, the outcomes show that the maximum material hardness results from a 20-second holding time, independent of the respective measuring position. This can be explained by two reasons. On the one hand, with a five-second holding duration, it can be assumed that the component bottom has not reached the desired target temperatures of 20 °C and 200 °C. Therefore no sufficiently rapid cooling takes place in this component area to generate a supersaturated solid solution. On the other hand, with a 60-second holding duration in the forming tool, a first material overaging can be assumed. Because the target temperature of 200 °C is already higher than the usual artificial aging temperature, the first coarse precipitates are generated, which reduces the subsequent potential for the formation of fine-distributed needle-shaped β''-precipitates during artificial aging.

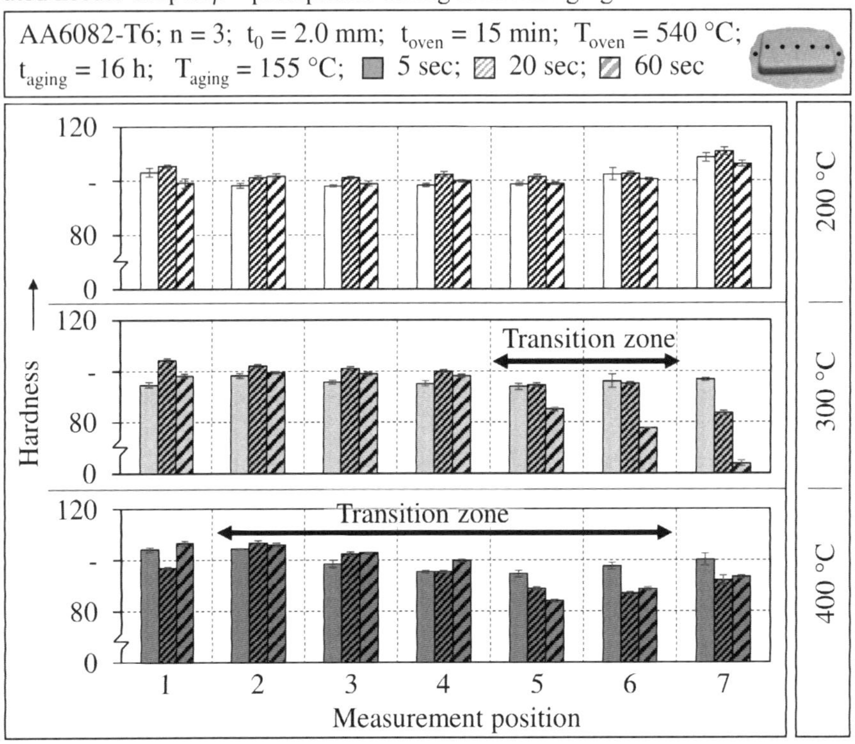

Figure 6. *Hardness distribution of the formed aluminum components*

With a tool temperature combination of 20 °C / 300 °C, a comparable hardness distribution can be seen between measuring positions one to four. Starting from measuring positions five to seven, however, a transition zone caused by a decrease in hardness can be seen towards the edge of the component. This is due to the increased tool temperature of 300 °C, which is reached more and more towards the component flange. Furthermore, it can be seen that a minimum material hardness is achieved, especially with a holding duration of 60 seconds. This can also be attributed to the increasing overaging, which occurs faster at 300 °C than at 200 °C. At a tool temperature combination of 20 °C / 400 °C, a comparable hardening behavior to 300 °C can be seen. However, it is less distinct. In addition, a larger transition area can be seen, which confirm the results of Figure 3. This can be attributed to a combination of the high tool temperatures and the subsequent cooling of the component in ambient air. At such high tool temperatures, coarse β'- and β- phases are generated after a short holding time, which leads to a reduction in strength [10]. However, the die temperature of 400 °C is in the upper range of the critical cooling rate. For this reason, it can be assumed that the subsequent cooling in ambient air will nevertheless result in a partially supersaturated solid solution, which ensures a moderate increase in strength during artificial aging.

DSC measurements were used to determine the precipitation distribution at three different locations of two components after tailor quenched forming and subsequent artificial aging. For this purpose, components manufactured and held for 20 second each in the forming tool at temperature combinations of 20 °C / 20 °C and 20 °C / 375 °C were compared. Based on Fröck's methodological approach, the formation and dissolution of different precipitation phases during a continuous heating operation with a heating rate of 30 K/min is identified [11]. In Figure 7 the results are illustrated. For the components produced by forming in the unheated tool, it can be seen that the heat flow of the β'' and β'-precipitant peaks are approximately the same high in the cup wall zones for the first and third extraction positions. In the bottom area of the component, however, a significantly increased formation of β'-phases can be seen. As explained before, this behavior can be attributed to the lower cooling rate in the middle of the component. With a tool temperature combination of 20 °C / 375 °C, however, the formation phase of β'' precipitates is significantly lower compared to the β' precipitation phase and is at a lower curve level. Therefore, it can be concluded that due to the thermally assisted forming operation at high tool temperatures and the subsequent aging, a supersaturated material state is present, which only proves the formation of coarse β' and β precipitates significantly during the heating operation in the DSC measurement system. As a result, it can be summarized that components produced in a locally tempered forming tool have different precipitate distributions.

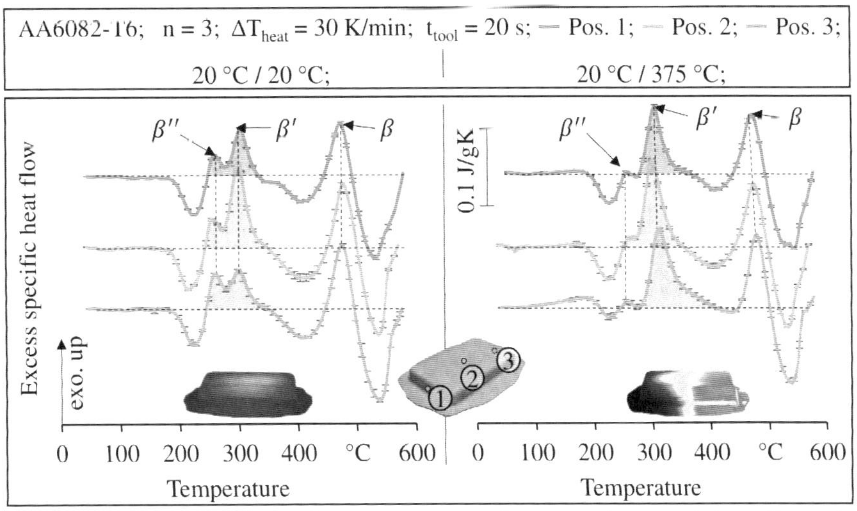

***Figure** 7. Influence of the tool temperature on the occurring precipitate formation*

5. Summary and outlook

In this study, a new tool design for adjustment of mechanical component properties by using tailor quenched forming was presented. The results show that failure-free forming and local adjustment of component properties is possible. It was shown that the temperature distribution of the manufactured components is highly dependent on the tool temperature and the contact conditions between the tool and

the workpiece. The subsequent sheet thickness distribution and component quality is also affected by the tool temperature. At tool temperatures of 200 °C and above, the components reveal a more homogeneous sheet thickness distribution and less wrinkling. Tool temperatures of 300 °C in particular cause a significant reduction in hardness in the tempered tool area. It was also shown that the holding time in the closed tool influences the resulting material hardness. Furthermore, it was shown that the process forces for component forming decrease with increasing tool temperature.
In the future, the analysis of further process variables and the use of additional precipitation-hardenable materials is planned to further expand the know-how regarding the potential of tailor quench forming.

Acknowledgments
The authors would like to thank the German Research Foundation (DFG) for supporting the present investigations, which were performed within the scope of the research project ''Manufacturing of tailored aluminum parts by controlling the local cooling rates in a combined forming, quenching and hardening process (Tailor Quenched Forming)'' (DFG 456415464). The authors would also like to thank the student Mahmoud Attia for his support during his project thesis.

References
[1] Behrens, B.-A.; Hübner, S.; Vogt, H.: Formability of 7000 aluminum alloys in warm and hot forming condition: Materials Science and Engineering, S. 12–27
[2] Garrett, R. P.; Lin, J.; Dean, T. A.: An investigation of the effects of solution heat treatment on mechanical properties for AA 6xxx alloys: experimentation and modelling. International Journal of Plasticity 21(2005)8, S. 1640–1657
[3] Hidalgo-Manrique, P.; Cao, S.; Shercliff, H. R.; Hunt, R. D.; Robson, J. D.: Microstructure and properties of aluminium alloy 6082 formed by the Hot Form Quench process. Materials Science and Engineering: A 804(2021)
[4] Teichmann, K.; Marioara, C. D.; Pedersen, K. O.; Marthinsen, K.: The effect of simultaneous deformation and annealing on the precipitation behaviour and mechanical properties of an Al–Mg–Si alloy. Materials Science and Engineering: A 565(2013), S. 228–235
[5] Merklein, M.; Johannes, M.; Lechner, M.; Kuppert, A.: A review on tailored blanks—Production, applications and evaluation. Journal of Materials Processing Technology 214(2014)2, S. 151–164
[6] Dolan, G. P.; Flynn, R. J.; Tanner, D. A.; Robinson, J. S.: Quench factor analysis of aluminium alloys using the Jominy end quench technique. Materials Science and Technology 21(2005)6, S. 687–692
[7] Milkereit, B.; Starink, M. J.; Rometsch, P. A.; Schick, C.; Kessler, O.: Review of the Quench Sensitivity of Aluminium Alloys: Analysis of the Kinetics and Nature of Quench-Induced Precipitation. Materials (Basel, Switzerland) 12(2019)24
[8] Gao, M.; Wang, M.; Wen, L.; Li, G.: TTP Curves and Microstructural Evolution Mechanism After Quenching in Aluminum Alloy 6082. Metallogr. Microstruct. Anal. 1(2012)3-4, S. 165–169
[9] Sajadifar, S. V.; Scharifi, E.; Weidig, U.; Steinhoff, K.; Niendorf, T.: Effect of Tool Temperature on Mechanical Properties and Microstructure of Thermo-Mechanically Processed AA6082 and AA7075 Aluminum Alloys. HTM Journal of Heat Treatment and Materials 75(2020)3, S. 177–191
[10] Lypchanskyi, O.; Rigas, N.; Korpała, G.; Merklein, M.; Prahl, U.: Ex-situ and in-situ investigations of the microstructural evolution of AA6082 aluminum alloy during heat treatment. Materials Science and Engineering: A, (2023)870, S. 144828
[11] Fröck, H.: Kurzzeit-Wärmebehandlung von AlMgSi-Legierungen - Phasenumwandlungen und mechanische Eigenschaften - Dissertation (Hrsg.): Forschungsberichte des Lehrstuhls für Werkstofftechnik der Universität Rostock. Bd. Band 9. Universität Rostock, 2020

IOP Conf. Series: Materials Science and Engineering **1307** (2024) 012021 doi:10.1088/1757-899X/1307/1/012021

Robustness of macro-structured deep drawing process

A Wolf[1,*], M Tulke[1] and A Brosius[1]

[1] Chair of Forming Processes, TUD Technical University of Dresden, 01069 Germany

* *alexander.wolf3@tu-dresden.de*

Abstract. Reaching robustness of forming processes is one of the main challenges in manufacturing technologies. Volatile material properties [1] and fluctuations in process parameters often result in insufficient part quality and rising production costs, especially in production chains [2]. As a common solution for this problem, numerical simulations are used to consider scattering of material parameters and define a useable process window with a specified safety margin. A new approach is the use of macro-structured tools. Result shown here indicate, in addition to the general enlargement of the useable process window [3], a high tolerance toward volatile material properties and variations of the blank. The produced parts show consistent quality with a reduction in residual stresses [4]. To examine the robustness in particular, virtual studies for steel alloy DX54 with varying alignment of the blank were performed. The resulting properties, such as the blank draw-in, thinning and springback, were examined in predefined positions on the blank to compare the experiments. The enlarged process window will be analysed and a strategy for a deeper understanding of the forming process will be generated, which will sustain the design of robust tools and processes.

1. Introduction

Deep drawing parts are used in a various range of manufactured parts across all industries. Herein sheet metal blanks were formed in various and complex three-dimensional shapes. Deep drawing tools, which usually comprise a combination of a shape-defining die, a forming punch, and a blank holder, were designed with the goal of producing a constant part quality over the lifetime of the tool. Reaching this process capability is a major aim as it reduces waste and improves tool durability. Considering the life cycle cost of a deep drawing tool, resources for failed parts and reworking due to process abnormalities have a significant impact on sustainability. By reaching a robust process window, a contribution to resource efficiency is made, not only by reducing the waste of failed parts but also by minimizing or even eliminating costs for prolonged try-out phases, reworking of the tool produced parts through reduced variability. However, more complex shapes, tight tolerances and newly developed materials like high-strength steels or improved lightweight materials complicate the design process. Nevertheless, archiving a robust forming process is crucial, which is why there are numerous approaches regarding forming processes and deep drawing to overcome uncertainty [5]. The finite element method (FEM) is a commonly used procedure to significantly reduce the material resources required to design a new tool and predict a process window. However, a safety margin always has to be included to handle process variations such as material scatter or changing environmental terms. A deep drawing variant to mitigate the quality-influencing factors was presented by Brosius et al. [6] with the macro-structured deep drawing. In the current research robustness of the process will be investigated by reenacting some influencing factors using FEM.

IOP Conf. Series: Materials Science and Engineering **1307** (2024) 012021 doi:10.1088/1757-899X/1307/1/012021

2. State of the art

2.1. Deep drawing process

The deep drawing process is the most commonly used procedure for manufacturing complex three-dimensional shapes such as kitchen sinks, fuel tanks and many other parts in our daily lives. Therefore, research in this area is carried out to improve the process by analyzing the parameters [7], using them to reduce springback [8], wrinkling [9] and excessive thinning [10] for the manufactured parts. Beside the tool design, friction plays a major role regarding the influencing factors. Due to the large contact area of the blank between the die and the blank holder, friction is a research topic, e. g. talking about analyzing the contact properties [11], improving the determination of friction coefficients [12] or eliminating the lubricant and perform dry forming processes [13].

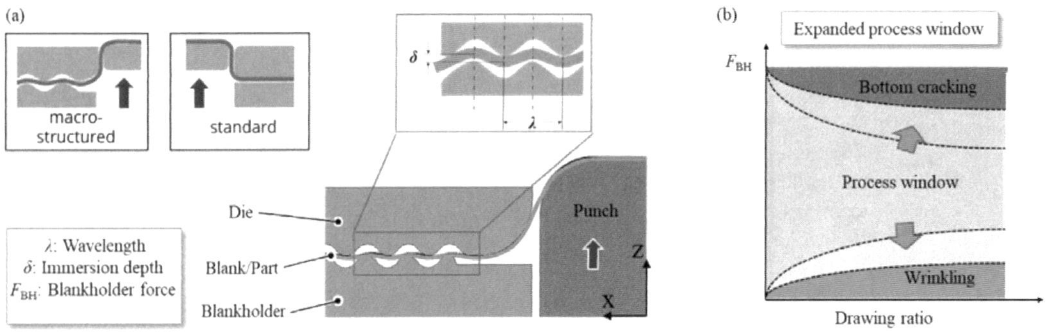

Figure 1. a) Comparison of deep drawing with standard and macro-structured tools and b) Expansion of the process window by macro-structured deep drawing [3]

What all have in common is that the basic design of the tools, consisting of a die and blank holder combination with flat surfaces, is mostly untouched and set as state of the art (see figure 1a). This concept was enhanced by Brosius et al. [6], where the structure of the tools changed to a wave-like structure. This macro-structure showed benefits in current research, such as improved process window (figure 1b) [3], less residual stresses and therefore reduced springback [4] next to the point that it can be used without lubrication. Therefore, good process robustness can be assumed which is further substantiated in the present study.

2.2. Process robustness and influencing parameters

Due to the wide usage of the deep drawing in manufacturing processes, several 3D-shaped parts are produced on daily base. Consistent quality inside the tolerances must be ensured for the process to call it robust. Any variation can lead to discrepancies. Grossenbacher divides the main influencing factors into three groups: material scatter, process fluctuations and modelling uncertainties [14]. Material scatter is quite important, whereas the mechanical properties of the material fluctuate between batches and even in one batch, e. g. up to ±50 MPa for the tensile strength [1]. The dependency on strain-rates regarding the forming velocity, forming temperature and possible preform operations [15] was also considered.

Talking about process fluctuations offers a wide range of parameters leading to possible scrap or rework of the parts. First of all, the tribological conditions depending on the surface of tools [16], blanks and chosen lubricant [17] are an main areas of research. The forming press used in combination with the tool and their combined stiffness during the process is also an area of interest because of their influence on the part quality [18]. Another not negligible parameter are the blank or tool misalignment. The definition of the blank dimensions is the initial step to determine the perfect blank geometry for minimal or no waste after trimming. Hereby, the shape and its prediction can vary by the chosen algorithm, e.g. by slip line field analysis. [19]. Furthermore, wrong placement of the blank can occur for various reasons and have an influence of the resulting shape [20]. Nowadays, robots place the blank with a certain repeatability on the same position to mitigate this uncertainty but often it is done regardless of the surrounding conditions. In this case, tool wear or a more simple misalignment of the punch [21] can affect the resulting shape and the induced strains. Even with all these variables that could cause instability in the process, FEM provides the ability to test and assess various configurations before determining the optimal design configuration for the component. However, the FEM is also a potential source of uncertainty because of assumptions made for the material modelling, contact modelling, meshing, and additional replacement models that need to be defined [22].

Despite all this uncertainty factors strategies were implied to reach a robust process. A capable process is defined by [23] as a process ranging in a defined and controlled process window. There are different approaches to reach this robustness (without claiming completeness): Statistical Process Control (SPC), design of experiments (DoE), Failure Mode and Effects Analysis (FMEA), sensitivity analysis, continuous monitoring and feedback and the well-known process capability analysis [24].

To conclude the state of the art, there are well or less well-known influencing factors which affect the product quality. Most of them are research topics and more or less understood regarding various possible different options for solving issues. Beside there are methods and algorithms to gain robustness and cover uncertainties in the process, like implementation of a closed loop control [25]. Nevertheless, uncertainties remain that can influence the manufacturing process and result in out-of-tolerance parts. Deep drawing with macro-structured tools can offer a solution to this challenge, as the process is very tolerant of known influencing factors such as friction conditions in the flange area, shape of the blank, and the resulting thickness [26] and offers a robust process despite alternating conditions.

3. Approach and methods

To prove the robustness of macro-structured tools, virtual studies using FEA (LS-Dyna R14) were conducted for steel alloy DX54. A T-shaped cup geometry will be used. For the dimensions refer to figure 2b. Robustness will be evaluated by comparing a deep-drawn component drawn with standard tools and a component manufactured with macro-structured tool (figure 2a) for altering conditions, testing different friction coefficients as well as lateral offset of the blank. The macro-structured tool set used has a wavelength of λ = 10 mm and the immersion depth δ is varied. The punch radius r_{punch} and die radius r_{die} were both 10 mm. The drawing depth is set to 55 mm, the die has no bottom. The tool set for standard deep drawing is the same size and, in contrast, has a flat surface for the blank holder and die.

Due to the stresses occurring in the process, the shape of the cup offers some areas for typical defects in deep drawing (see figure 3a). The top side of the T is through the length of side vulnerable for enhanced draw-in and wrinkling (4) which is usually countered by the use of draw beads. The concave area is prone to bulging in the bottom area above (2). The compression stresses accumulate and push the material in the bottom of the cup without proper material flow. Thinning and resulting bottom cracks occur around area (3). Around the convex areas, wrinkling is likely (1). Therefore, the geometry is well suited to test different frictions conditions and offset of the blank to check their influence on the resulting geometry.

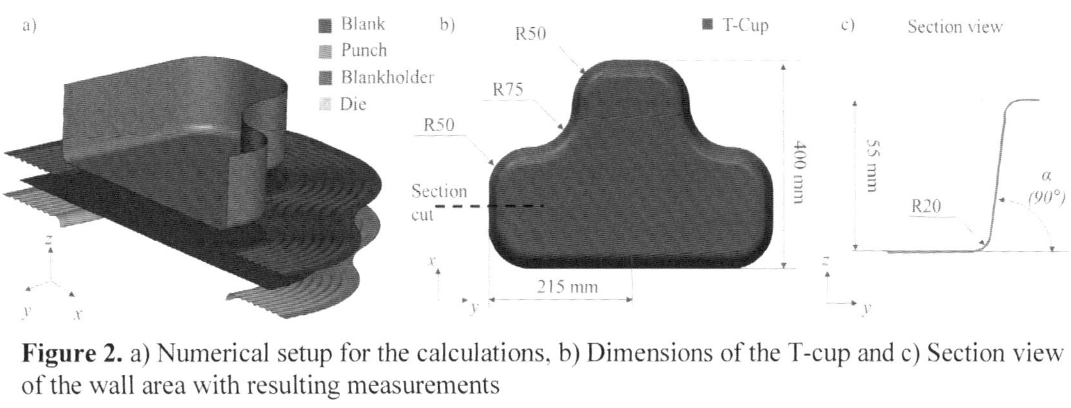

Figure 2. a) Numerical setup for the calculations, b) Dimensions of the T-cup and c) Section view of the wall area with resulting measurements

The analyses were performed by finite element analysis, whereas the relevant parameters, influencing the product quality were changed in a certain range (see table 1b). The drawing process was calculated explicit, the springback process is performed implicit afterwards. The tools were set as rigid bodies and for the blank fully integrated shell elements with an initial element size of 2 mm was used. During the calculations adaptive remeshing is used, with a minimum element length of 0.5 mm. The used T-cup geometry was reduced to half of the cup to reduce the calculation time on the cluster.

The material used was a deep drawing steel DX54 (1.0952) with an initial blank thickness of $t_0 = 1.18$ mm. The parameters for the used material model MAT133 or Yld2000-2D, developed by Barlat et al. [27], were first analysed and validated by Küsters [28], whereas the hardening parameters are listed in table 1a. DX54 is hot-dip coated, is characterized by a high elongation and suitable for higher forming requirements.

Table 1. a) Material parameters for DX54 (t_0=1.18 mm, [27]) after wift and b) Parameter for the evaluation

a)

Parameter	Value
Youngs modulus E	184 GPa
Poisson's ratio ν	0.3
Flow curve parameter C	553.30MPa
Flow curve parameter ε	0.01
Flow curve parameter n	0.267

b)

Parameter	Tested values
Friction coefficient μ	0.05, 0.15 0.25
Lateral offset x_0	± 2, ± 4 mm
Immersion depth δ	0.10, 0.20, 0.30mm
Blankholder force F_{bh}	200kN (full model)

The friction coefficient μ ranges between 0.05 for well lubricated surfaces, 0.15 for normal conditions and 0.25 for near dry conditions with high friction. Offsets for the blank were defined in the x-direction with $x_0 = \pm 2$ mm and ± 4 mm. Calculations with macro-structured tools were tested for three different immersion depths $\delta = 0.10$ mm, 0.20 mm and 0.30 mm. To set the result in relation, virtual experiments with standard tools were carried out with a blankholder force F_{BH} of 100 kN for the half model.

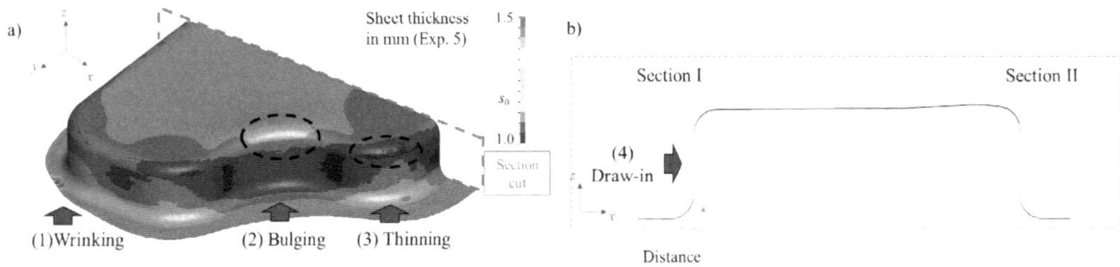

Figure 3. a) Resulting T-cup with the cut section (red) for the evaluation and the b) Different sections of the cut with focus on the resulting xz-displacement of the virtual experiments

To determine the influence of the changing parameters, three different areas of the cup were considered. The first is a section cut (see figure 3a) in the symmetry axis of the T-cup for the resulting geometry. The deviation from the reference contour is sketched (figure 3b). For the postprocessing of failures such as wrinkling, excessive thinning and bulging, known areas will be evaluated. In particular, the area prone to bulging in the process [4] will be analysed with respect to the bulge height measured from the bottom. The other area of interest is the backside of the cup (not visible on figure 3a). Bulges can occur within the wall area. The minimum sheet thickness is gathered and evaluated, whereas a range is given as it could result in unwanted bottom cracks. Usually, the bottom crack occurs at the marked position because of the induced stresses. Drawing forces for the punch will be examined as complementary.

4. Results and discussion

Table 2 provides an overview of the FE calculations performed and the parameters used in each case. Starting with the geometric deviations, the section cut of the wall area as shown in figure 4 with the resulting coordinates. The coordinates are equal to the node coordinates along the cup section and show the left and right cup walls. The section cuts plotted in figure 4 show the results for run 4 to 6 with constant immersion depth δ of 0.2 mm. Section I shows nearly matching plots, whereas in section II, the expected draw-in and small bulging is visible in the wall area. Here, less friction and low immersion depth lead to nearly free material flow to form the bulge. With the right immersion depth, the bulge can be reduced. For a removal, the material flow in the convex area has to be reduced further.

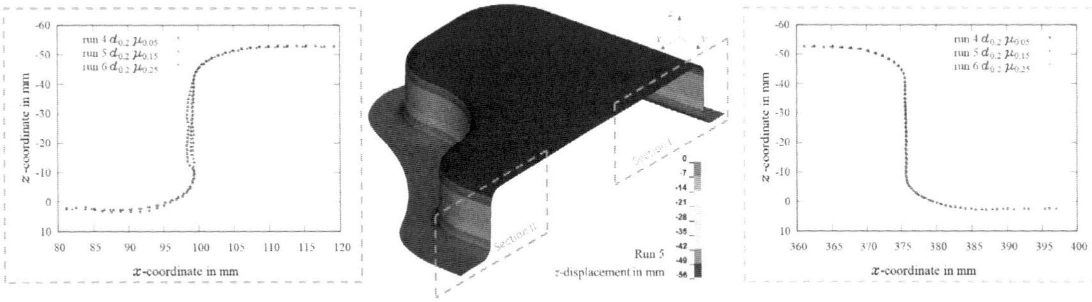

Figure 4. Resulting T-cup, showing the result for run 5 with minimal bulging and the cut sections (green and blue) for the evaluation of the different areas of the cut with focus on the deviation

The evaluation of the performed calculations with resulting data regarding drawing forces, bulge height and thinning values can be found in table 2. All calculations with macro-structured tools result in a drawable cup within a maximum thickness reduction of 18 % except run 9 (δ = 0.3 mm | μ = 0.25) was very likely to fail due to bottom cracking due to the high immersion depth and high friction. In this case, the process limit is reached and run 9 will be excluded for further evaluation.

Table 2. Evaluation of the experiments with macro-structured and standard tools

Number	Immersion depth δ [mm]	Friction coefficient μ	Offset x_0 [mm]	Bulge [%]	Drawing Force F_{BH} [kN]	Min. thickness t [mm]	[%]
run 1	0.1	0.05	0	14	77.9	1.056	11
run 2	0.1	0.15	0	10	97.8	1.063	11
run 3	0.1	0.25	0	11	123.7	1.045	13
run 4	0.2	0.05	0	12	87.8	1.049	12
run 5	0.2	0.15	0	5	127.4	1.019	15
run 6	0.2	0.25	0	2	161.1	0.998	18
run 7	0.3	0.05	0	7	108.6	1.012	14
run 8	0.3	0.15	0	4	132.9	1.022	13
run 9	0.3	0.25	0	-	-	crack	-
run 11	0.2	0.15	-2	7	120.9	1.014	16
run 12	0.2	0.15	-4	5	125.9	1.017	16
run 13	0.2	0.15	+2	12	89.8	1.057	11
run 14	0.2	0.15	+4	12	115.2	1.043	11
run 15	-	0.05	0	17	81.9	1.067	10
run 16	-	0.15	0	12	115.1	1.043	13
run 17	-	0.25	0	-	-	crack	-
run 18	-	0.15	-4	7	115.2	1.030	15
run 19	-	0.15	+4	6	114.4	1.042	13

Run 5 showed the best result in terms of minimal bulging and acceptable thinning with a realistic frictions coefficient. For this reason, the tests for the offset of the blank based on the parameters of this run. The results for the standard tools were within the expectations, whereas μ = 0.25 leads to a failure of the cup, likely due to a bottom crack. Furthermore, thinning is nearly equal with a small standard deviation. Due to the minimal differences in the offset calculations between x_0 = 2 mm and 4 mm, calculations for the standard tools were only performed for the highest offset. The drawing forces ranged from 78 kN up to 161 kN, whereas the mean was 99 kN. The drawing force is strongly dependent on the friction, whereas an offset of the blank has a higher impact on the forces for the macro-structured tools due to the changed draw-in behaviour. Figure 5 shows the different developments, divided in four diagrams for better clarity. Its noticeable that the curves for macro-structured tools decline faster than those for standard tools which points to a lower energy demand.

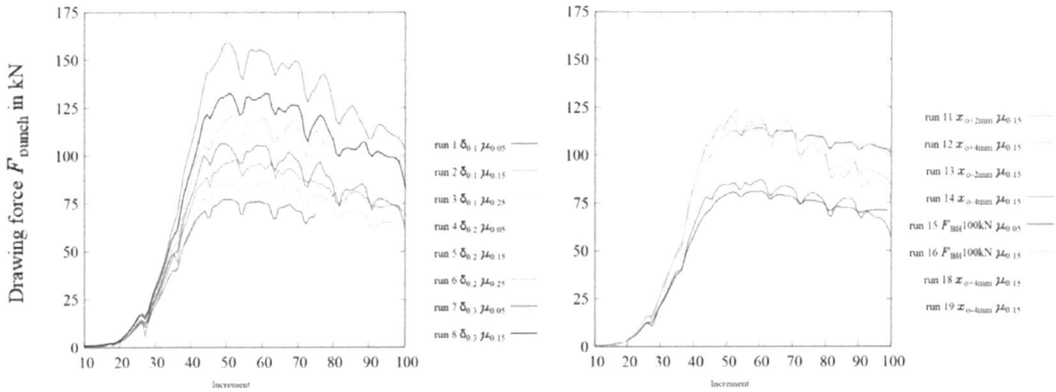

Figure 5. Drawing force for the different immersion depths and the experiments with offset

The displacement of the blank as cut section is also sketched along the unwound length of the T-Cup (refer to figure 4) and shows the distribution of the displacement in x and z for the different runs (see figure 6). run 5 as the best outcome of the experiments is used as a reference. For a differentiation, the tests with standard tool were dotted. In section I, next to the z-deviation, a higher x-deviation can be seen, which means higher draw-in. Macro-structured tools show an improved draw-in compared with standard tools, except for very low friction coefficients.

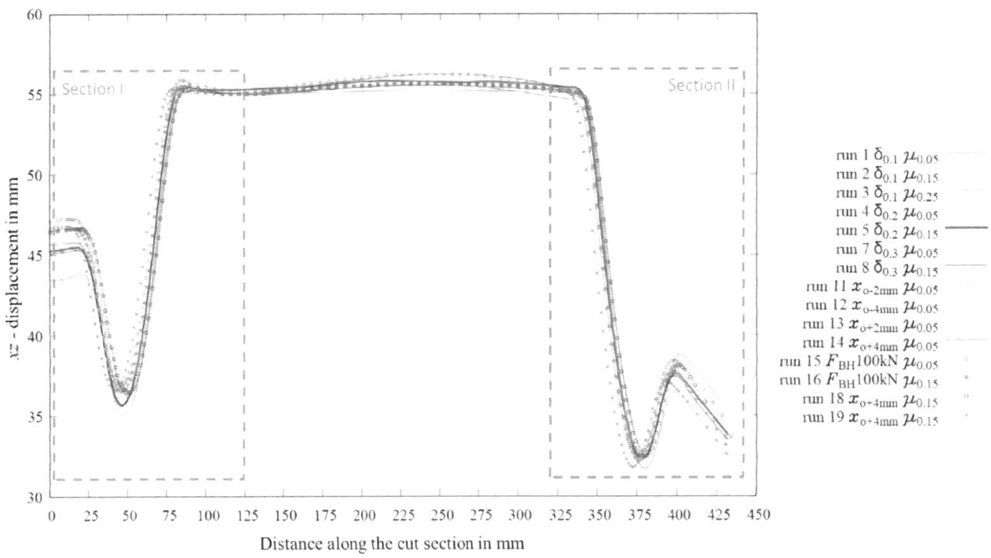

Figure 6. xz-displacement for the experiments with run 5 marked as reference (red line)

Regarding the range of deviation, run 1 to 14 show a lower deviation compared the four experiments with standard tools. To conclude the research, the springback of experiments with the same friction coefficient will be compared before (outlined marker) and after (filled marker) the springback calculation (see figure 7). Figure 7 shows section I where the draw-in and the resulting bulge within the wall is visible especially for the standard tool set. The experiments with macro-structured tools again show nearly similar results. Section II differs and run 8 shows the highest springback in comparison for this area.

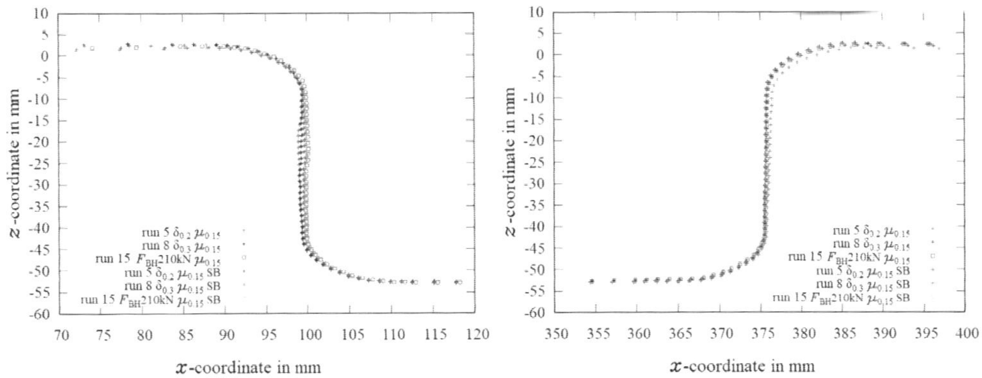

Figure 7. Wall area in both sections for three chosen experiments for comparison

5. Conclusion and outlook

By extracting and comparing the deviation in the geometry it can be concluded that deep drawing with macro-structured tools enable a robust process by tolerating changing friction parameters and offsets of the blank with an acceptable quality. Whereas wrinkling through the tests cannot be excluded, the influencing factor of friction is mitigated. Springback is another topic, in which the results were not consistent and a clear correlation could be detected by now. The draw-in is controllable by a variation of the immersion depth and can be fitted for the geometry. Here, the design process of the structure must focus on the different geometric sections. Further research will focus on a larger study altering the material and the design of the macro-structure in a larger scale as well as the validation of the findings on a servo-mechanical press.

Acknowledgments

The research project IFG 22945 BR / P 1528 is funded by the Research Association for steel Application (FOSTA) and by the Federal Ministry of Economic Affairs.

The authors are also grateful to the Center for Information Services and High Performance Computing at TU Dresden for providing its facilities for high throughput calculations.

References

[1] Harsch D, Renkci P, Heingärtner J and Hora P 2017 Influence of scattering material properties on the robustness of deep drawing processes *10th Forming Technology Forum. Model Based Control for Smart Forming Processes*

[2] Danckert J 2006 Analysis of Deep Drawing, Ironing and Backward Can Extrusion: With main emphasis on residual stresses and process robustness *PhD thesis* Aalborg University

[3] Mousavi A 2019 A novel approach towards a lubricant-free deep drawing process via macro-structured tools *PhD thesis* TU Dresden

[4] Wolf A, Tulke M and Brosius A 2022 Deep Drawing with Macro-structured Tools and Their Influence on Residual Stresses 10.1007/978-3-030-78424-9_11 pp 94–102

[5] Wiebenga J H 2014 *Robust design and optimization* (978-90-77172-96-4)

[6] Brosius A and Mousavi A 2016 Lubricant free deep drawing process by macro structured tools 10.1016/j.cirp.2016.04.060 *CIRP Annals* **65** pp 253–256

[7] Tiwari P R, Rathore A and Bodkhe M G 2022 Factors affecting the deep drawing process – A review 10.1016/j.matpr.2021.10.189 *Materials Today: Proceedings* **56** pp 2902–2908

[8] ul Hassan H, Maqbool F, Güner A, Hartmaier A, Ben Khalifa N and Tekkaya A E 2016

Springback prediction and reduction in deep drawing under influence of unloading modulus degradation 10.1007/s12289-015-1248-5 *Int J Mater Form* **9** pp 619–633

[9] Agrawal A, Reddy N V and Dixit P M 2007 Determination of optimum process parameters for wrinkle free products in deep drawing process 10.1016/j.jmatprotec.2007.03.050 *Journal of Materials Processing Technology* **191** pp 51–54

[10] Zein H, El-Sherbiny M, Abd-Rabou M and El Shazly M 2013 Effect of Die Design Parameters on Thinning of Sheet Metal in the Deep Drawing Process 10.12691/ajme-1 *AJME* **1** pp 20–29

[11] Figueiredo L, Ramalho A, Oliveira M C and Menezes L F 2011 Experimental study of friction in sheet metal forming 10.1016/j.wear.2011.02.020 *Wear* **271** pp 1651–1657

[12] Grüner M and Merklein M 2014 Determination of friction coefficients in deep drawing by modification of Siebel's formula for calculation of ideal drawing force 10.1007/s11740-014-0551-1 *Prod. Eng. Res. Devel.* **8** pp 577–584

[13] Vollertsen F and Schmidt F 2014 Dry metal forming: Definition, chances and challenges 10.1007/s40684-014-0009-0 *Int. J. of Precis. Eng. and Manuf.-Green Tech.* **1** pp 59–62

[14] Kathrin Grossenbacher 2008 *Virtuelle Planung der Prozessrobustheit in der Blechumformung*

[15] Patrick Larour Strain rate sensitivity of automotive sheet steels: influence of plastic strain, strain rate, temperature, microstructure, bake hardening and pre-strain

[16] Klocke F, Trauth D, Schongen F and Shirobokov A 2014 Analysis of friction between stainless steel sheets and machine hammer peened structured tool surfaces: Experimental and numerical investigation of the lubricated interaction gap 10.1007/s11740-013-0519-6 *Production Engineering* **8** pp 263–272

[17] Allen S J and Mahdavian S M 2008 The effect of lubrication on die expansion during the deep drawing of axisymmetrical steel cups 10.1016/j.jmatprotec.2007.08.005 *Journal of Materials Processing Technology* **199** pp 102–107

[18] Penter L 2017 *Qualifizierung von FE-Prozessmodellen zur Inbetriebnahme von Karosserieziehwerkzeugen*

[19] Agrawal A, Reddy N V and Dixit P M 2008 Optimal blank shape prediction considering sheet thickness variation: An upper bound approach 10.1016/j.jmatprotec.2007.05.046 *Journal of Materials Processing Technology* **196** pp 249–258

[20] Sekhara Reddy A C and Rajesham S Simulation Study on the Influence of Blank Offset in Deep Drawing of Circular Cups *Numerical Optimization in Engineering* pp 469–76

[21] Abdul Ghafar A, Abdullah A B and Mahmood J I 2021 Experimental and numerical prediction on square cup punch–die misalignment during the deep drawing process 10.1007/s00170-021-06595-5 *Int J Adv Manuf Technol* **113** pp 379–388

[22] Duchêne L and Habraken A M 2005 Analysis of the sensitivity of FEM predictions to numerical parameters in deep drawing simulations 10.1016/j.euromechsol.2005.04.007 *European Journal of Mechanics - A/Solids* **24** pp 614–629

[23] DIN ISO 22514-1:2019-07 Statistical methods in process management - Capability and performance - Part 1: General principles and concepts (Berlin: Beuth Verlag GmbH)

[24] Doumpos M, Zopounidis C and Grigoroudis E 2016 *Robustness analysis in decision aiding, optimization, and analytics* vol 241 (Switzerland: Springer. 978-3-319-33119-5)

[25] Allwood J M, Duncan S R, Cao J, Groche P, Hirt G, Kinsey B, Kuboki T, Liewald M, Sterzing A and Tekkaya A E 2016 Closed-loop control of product properties in metal forming 10.1016/j.cirp.2016.06.002 *CIRP Annals* **65** pp 573–596

[26] Wolf A, Tulke M and Brosius A Macro-structured Deep Drawing of Steel Alloys - Influenceability on Thinning and Drawing Forces *Production at the leading edge* pp 680–7

[27] Barlat F, Brem J C, Yoon J W, Chung K, Dick R E, Lege D J, Pourboghrat F, Choi S-H and Chu E 2003 Plane stress yield function for aluminum alloy sheets—part 1: theory 10.1016/s0749-6419(02)00019-0 *International Journal of Plasticity* **19** pp 1297–1319

[28] Küsters N 2020 *Halbanalytische Methode zur Charakterisierung der Fließortkurven von Blechwerkstoffen* (978-3-95908-209-9)

IOP Conf. Series: Materials Science and Engineering **1307** (2024) 012022 doi:10.1088/1757-899X/1307/1/012022

Multiscale design based on non-penalization smooth-edged material distribution for optimizing topology (SEMDOT)

Jiye Zhou[1], Yanan Wang[1], Louis N.S. Chiu[2] and Kazem Ghabraie[1,*]

[1] School of Engineering, Deakin University, Waurn Ponds, VIC 3217, Australia
[2] Department of Materials Science and Engineering, Monash University, Clayton, VIC 3800
*Correspondence: k.ghabraie@deakin.edu.au

Abstract. This paper presents a concurrent topology optimization method for macro and micro phases based on non-penalization smooth-edged material distribution for optimization topology (SEMDOT) method. Although there is existing research on the multiscale design method, grayscale elements are always emerged especially for penalization method for example the solid isotropic material penalization (SIMP) method, also high computational cost are required when large scale of elements are utilized for obtaining high resolution structures. The methodology proposed here aims to apply a new tech called non-penalization SEMDOT method to find the optimum layout on both scales of elements, it is assumed that the macro structure is composed of periodic materials and both element scales are optimized through their linearly interpolated grid points. The effective macroscopic properties are evaluated by the homogenization method. The approach could provide smooth and clear boundaries for multiscale system without grayscale elements or high computational cost. A series of numerical examples are presented to demonstrate the effectiveness and the efficiency of the proposed method.

1. Introduction

Topology optimization (TO) can reasonably design material distribution under the constraints of specific load, performance, and other conditions, to obtain the optimal structure configuration, which is an advanced structure optimization method [1]. Since Bendsøe and Kikuchi [2] proposed the seminal work, TO has been developed rapidly. Over the last 30 years, shortly after the initial homogenization method, other algorithms such as Solid Isotropic Material Penalization (SIMP) [3], Bi-directional Evolutionary Structural Optimization (BESO) [4], Level-set Method (LSM) [5], Moving Morphable Components (MMC) [6], Moving Morphable Voids (MMV) [7] have been created for variety of disciplines.

Topology optimization for lightweight design is a popular topic, most of them focused on macroscopic structures. In contrast to macroscopic design, the natural world provides numerous inherent microscopic cellular structures in biological entities like human bones and plant leaves. Each type of micro cellular structure not only possesses remarkable structural properties but also exhibits biocompatible characteristics. Moreover, cellular structures and materials have potential applications in the aerospace industry, civil engineering, biomedical sensors, optics, and semiconductors. In addition, 3D and 4D printed cellular structures are extensively used in the vibration isolation, buckling control and energy absorption [8-10].

To delve deeper into this field, many researchers, comes to the adoption of multiscale topology optimization. In contrast to single-scale methods where only the macroscopic structure is considered, multiscale design involves simultaneous topology optimization of both macroscopic and microscopic structures. Numerical homogenization method has been investigated by Sigmund [11], Rodrigues et al. [12] and Liu et al. [13] for the design of multiscale structures, where homogenized elastic tensor of macro cell will be derived from the inside micro cells. In addition, the energy-based homogenization method (EBHM) is of higher computational efficiency, more simplified programming, and more simple design sensitivity analysis procedure [14]. Zhang et al. [15], Xia and Breitkopf [16] investigate the design of material microstructures with extreme elastic properties using the energy-based homogenization method.

Grayscale elements remain an issue in multiscale design results based on the SIMP method. Additionally, achieving clear structural features often requires the use of large numbers of finite elements, which substantially increases computational cost. To solve this problem, Fu et al. [17-19] proposed a SEMDOT method based on SIMP, where a clear and smooth boundary could be obtained compared with traditional density-based methods. Moreover, the new type of non-penalization method further investigated by Huang [20] and Fu et al. [21] increases the efficiency of non-penalization method by discrete sensitivities and provide better performance. Following the unit cellular structure design investigated by Fu et al. [22]. This paper furtherly employ multiscale topology optimization based on the non-penalization SEMDOT, two discrete sensitivity values [23, 24] are used and eliminate the use of penalty factors commonly used in SIMP and BESO. Finally, smooth boundaries and comparable multiscale designs could be obtained for compliance minimization problem, thus proving the effectiveness of multiscale design based on non-penalization SEMDOT.

2. Methodology

2.1. Non-penalization SEMDOT method

Unlike the SIMP method, non-penalization SEMDOT no longer relies on grayscale elements. To be specific, it directly distributes the densities of grayscale elements linearly to each grid points, and then employs a smooth Heaviside function to project these gird points densities to 0 or 1 as shown in Figure *1*. Subsequently, the proportion (elemental volume fraction) of grid points with a density of 1 within a unit element is used as a new variable to replace the original elemental density.

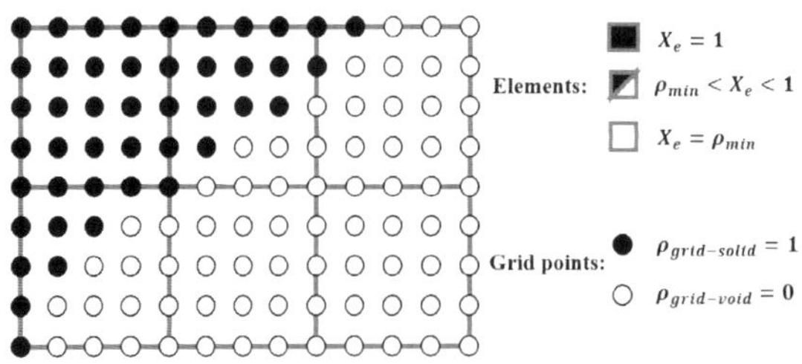

Figure 1. Illustration scheme of non-penalization SEMDOT

Due to the adoption of elemental volume fraction as new design variables, the sensitivity calculations for the corresponding element are discretised as solid and void phases, which is also called as discrete sensitivity calculations. This method has been investigated and validated by Liang et al.[23-26]. It could be expressed as a linear interpolation:

$$\begin{aligned} \frac{\partial C(\Psi_e)}{\partial \Psi_e} &= (1-\Psi_e)\frac{\partial C(\Psi_e)}{\partial \Psi_e}\bigg|_{\Psi_e=\rho_{min}} + \Psi_e \frac{\partial C(\Psi_e)}{\partial \Psi_e}\bigg|_{\Psi_e=1} \\ &= -[(1-\Psi_e)\rho_{min} + \Psi_e]\boldsymbol{u}_e^T\mathbf{K}^1\boldsymbol{u}_e \end{aligned} \quad (1)$$

where C is the objective compliance function, Ψ_e is the elemental volume fraction of eth element's solid region, $(1-\Psi_e)$ means the elemental volume fraction of eth element's void region, $\partial C(\Psi_e)/\partial \Psi_e|_{\Psi_e=\rho_{min}}$ (ρ_{min} is commonly defined as 0.001) represents the sensitivity of void elements and $\partial C(\Psi_e)/\partial\Psi_e|_{\Psi_e=1}$ is for solid elements, $\boldsymbol{u}_e$ represents the displacements of eth element and $\boldsymbol{K}^1$ represents the unit stiffness matrix.

2.2. Homogenization method and sensitivity analysis

Homogenization method is commonly used in multiscale design method with two requirements: 1) the microstructures should be much smaller than the macrostructures. 2) periodic microstructures are distributed among the structures. For traditional homogenization method, normally the first-order variation term is only considered for the displacement field, then the equivalent macroscopic stiffness tensor $\boldsymbol{D}^H$ could be calculated by the strain energy divided by the area (2D) or volume (3D):

$$\boldsymbol{D}^H = \frac{1}{|\Omega_m|}\int_{\Omega_M} \boldsymbol{D}(\varepsilon(\boldsymbol{u}_m^0) - \varepsilon(\boldsymbol{u}_m))(\varepsilon(\boldsymbol{u}_m^0) - \varepsilon(\boldsymbol{u}_m))d\Omega_m \tag{2}$$

where Ω_m is the area or volume of the microstructures, $\varepsilon(\boldsymbol{u}_m^0)$ is the predefined unit test strain field [11] and $\varepsilon(\boldsymbol{u}_m)$ is the unknown strain field happens on the microstructures.

Because of the discretization for FEA in SEMDOT, the homogenized elastic tensor in Equation (2) could be rewritten as:

$$\boldsymbol{D}_{ijkl}^H = \frac{1}{|\Omega_m|}\sum_{e=1}^{N} \left(\boldsymbol{u}_e^{(i,j)}\right)^T \boldsymbol{K}^1 \boldsymbol{u}_e^{(kl)} \tag{3}$$

where N is the total number of micro elements in one macro unit cell, $\boldsymbol{u}_e^{(kl)}$ is the elemental displacement that is derived from the unit test strain field.

Returning to the macroscopic mechanical compliance problem, the micro and macro design variables is expressed as:

$$\mathrm{C}(\Psi_M, \Psi_m) = \sum_{i=1}^{N} \boldsymbol{U}_i^T \boldsymbol{K}_i(\Psi_M, \Psi_m)\, \boldsymbol{U}_i \tag{4}$$

where Ψ_M, Ψ_m are the macroscopic and microscopic elemental volume fractions, respectively. $\boldsymbol{U}_i$ and $\boldsymbol{K}_i$ represent the nodal displacement and the stiffness matrix of the ith element with respect to the macrostructure of the total number of the element equal to N.

The general form of the elemental stiffness matrix is taken the form:

$$\mathbf{K} = \int_V \boldsymbol{B}^T \boldsymbol{D} \boldsymbol{B} dV \tag{5}$$

where $\boldsymbol{B}$ is the strain displacement matrix, and $\boldsymbol{D}$ is the elastic tensor of the element.

For microstructures, the elastic tensor $\boldsymbol{D}_{ijkl}^m$ is formulated to comply with the non-penalization SEMDOT method based on the constitutive elastic tensor $\boldsymbol{D}_{ijkl}^0$:

$$\boldsymbol{D}_{ijkl}^m = \Psi_m \boldsymbol{D}_{ijkl}^0 + (1 - \Psi_m)\rho_{min}\boldsymbol{D}_{ijkl}^0 \tag{6}$$

Then the macroscopic elastic tensor $\boldsymbol{D}_{ijkl}^M$ could be obtained through the $\boldsymbol{D}^H$ and expressed in a non-penalization SEMDOT mode:

$$\boldsymbol{D}_{ijkl}^M = [\Psi_M + (1 - \Psi_M)\rho_{min}]\boldsymbol{D}_{ijkl}^H \tag{7}$$

The sensitivity for the objective and constraint with respect to the design variable of the macroscopic cell could be illustrated as:

$$\begin{aligned} \frac{\partial C(\Psi_M)}{\partial \Psi_M} &= (1 - \Psi_M)\frac{\partial C(\Psi_M)}{\partial \Psi_M}\bigg|_{\Psi_M=\rho_{min}} + \Psi_M \frac{\partial C(\Psi_M)}{\partial \Psi_M}\bigg|_{\Psi_M=1} \\ &= -[(1 - \Psi_M)\rho_{min} + \Psi_M]\boldsymbol{U}_M^T \int_{\Omega_M} \boldsymbol{B}^T \boldsymbol{D}^H(\Psi_m)\boldsymbol{B} d\Omega_M\, \boldsymbol{U}_M \end{aligned} \tag{8}$$

The derivation of the first-order derivatives of the objective function and the micro volume constraint with respect to the micro design variables could be given as:

$$\frac{\partial C(\Psi_m)}{\partial \Psi_m} = -[(1-\Psi_m)\rho_{min} + \Psi_m]\boldsymbol{U}_M^T \int_{\Omega_m} \boldsymbol{B}^T \frac{\partial \boldsymbol{D}^H(\Psi_m)}{\partial \Psi_m} \boldsymbol{B} d\Omega_m \boldsymbol{U}_M \tag{9}$$

The first-order derivatives of the homogenized elastic tensor $\boldsymbol{D}^H$ with respect to the micro element densities are given as:

$$\frac{\partial \mathbf{D}^H(\Psi_m)}{\partial \Psi_m} = (1-\Psi_m)\frac{\partial \mathbf{D}^H(\Psi_m)}{\partial \Psi_m}\bigg|_{\Psi_m=\rho_{min}} + \Psi_m \frac{\partial \mathbf{D}^H(\Psi_m)}{\partial \Psi_m}\bigg|_{\Psi_m=1}$$
$$= -\frac{1}{2}\int_{\Omega_m} [(1-\Psi_m)\rho_{min} + \Psi_m]\mathbf{D}_{ijkl}^0(\varepsilon(\mathbf{u}_m^0) - \varepsilon(\mathbf{u_m}))\big(\varepsilon(\mathbf{u}_m^0) - \varepsilon(\mathbf{u_m})\big)d\Omega_m \tag{10}$$

Correspondingly, the sensitivity of volume constraint with respect to design variable could be simply illustrated as:

$$\begin{cases} \dfrac{\partial \boldsymbol{V_e}(\Psi_M)}{\partial \Psi_M} = \dfrac{\Psi_M \boldsymbol{V_0}}{\partial \Psi_M} = 1 \\ \dfrac{\partial \boldsymbol{V_e}(\Psi_m)}{\partial \Psi_m} = \dfrac{\Psi_m \boldsymbol{V_0}}{\partial \Psi_m} = 1 \end{cases} \tag{11}$$

where $\boldsymbol{V_e}(\Psi_M)$ and $\partial \boldsymbol{V_e}(\Psi_m)$ are the elemental volume fraction of macro and micro elements, respectively. $\boldsymbol{V}_0$ is the unit volume fraction which equals to 1.

The developed concurrent topology optimization formulation can be solved by well-established gradient-based MMA algorithms [27].

3. Experiment result

For the multiscale topology optimization design using the non-penalization SEMDOT, three compliance-based problems are selected as shown in Figure 2. It is worth noting that during the optimization process, the macroscopic design domain is uniformly distributed to achieve the desired volume fraction, while the microstructures need to be pre-defined with voids regions to avoid sensitivity uniformity issues arising from periodic boundary conditions [28] during optimization.

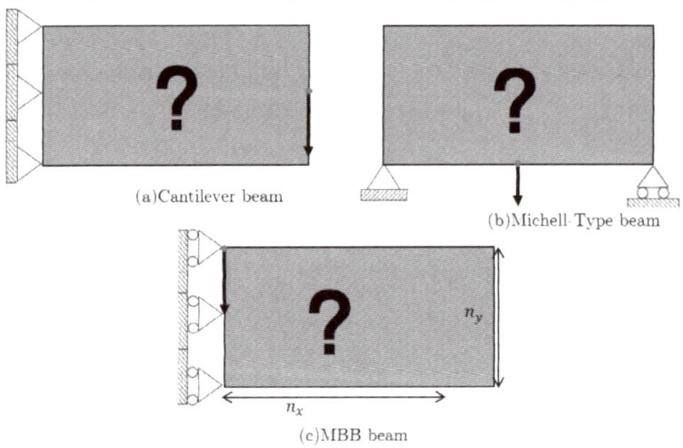

Figure 2. Three types of compliance-based design problems

3.1 Cantilever beam

In this illustrative case, the cantilever beam is examined, as depicted in Figure 2(a), wherein an external force (F = -1N) is applied at the midpoint of the right side. The dimensions of the beam are specified as follows: Length $L = 10mm$ and Height $H = 4mm$, the macrostructure is discretised into a grid

IOP Conf. Series: Materials Science and Engineering **1307** (2024) 012022 doi:10.1088/1757-899X/1307/1/012022

comprising 80 × 40 four-node plane stress finite elements, while the microstructure is represented by a grid of 50 × 50 plane stress finite elements.

Table 1. Multiscale design of cantilever beam under the same macroscopic design volume fractions and different microscopic design volume fractions

Design volume fractions	Result from non-penalization SEMDOT	Compliance $(N \cdot mm^2)$
$\begin{cases} \mathbf{V_M = 0.5} \\ \mathbf{V_m = 0.3} \end{cases}$	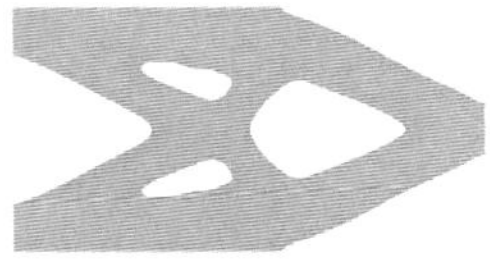 	347.1799
$\begin{cases} \mathbf{V_M = 0.5} \\ \mathbf{V_m = 0.6} \end{cases}$	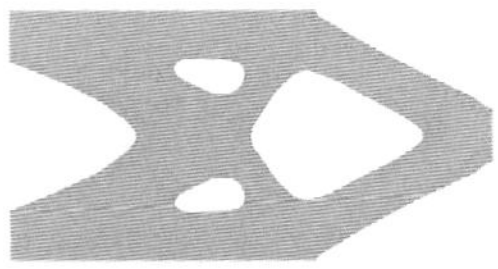 	143.6471
$\begin{cases} \mathbf{V_M = 0.5} \\ \mathbf{V_m = 0.8} \end{cases}$	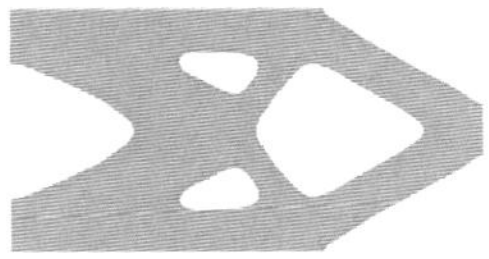 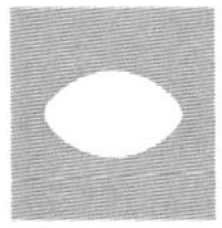	88.0369
$\begin{cases} \mathbf{V_M = 0.5} \\ \mathbf{V_m = 1.0} \end{cases}$	 	53.2086

Table 1 demonstrates the design results under different volume fractions of microstructures. It could be noticed that microstructures do not change significantly, while the microstructures will achieve the larger volume fraction until it converts to a pure solid. It could also be recognised a single design when the microstructures reach to 1.0 design volume fractions.

3.2 Michell type beam design

The classic Michell-type structure as displayed in Figure 2(b) is investigated. A force is applied at the midpoint of the lower edge, with the lower right corner being fixed and the lower left corner supported by a roller. The prescribed dimensions are as follows: $L = 10mm$ and $H = 4mm$. The numerical analysis employs four-node plane stress finite elements (100 × 40), while the microscale finite mesh remains unchanged with (50 × 50).

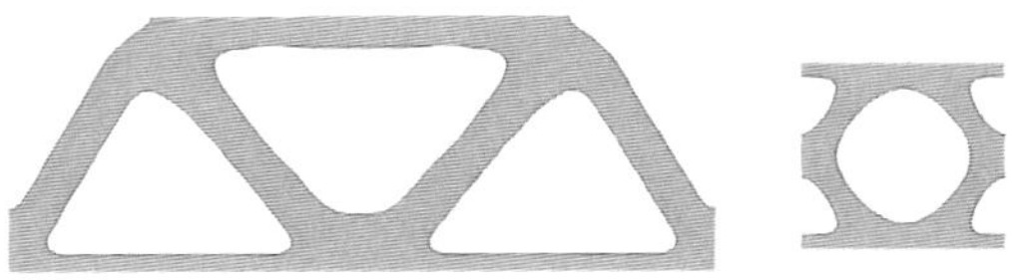

Figure 3. The optimal result for multiscale design of Michell-type beam based on non-penalization SEMDOT with Compliance equals to $93.89N \cdot mm^2$.

Figure 4. The optimal result for multiscale design of Michell-type beam based on SIMP with

$$Compliance = 94.26N \cdot mm^2 \text{ [29]}$$

Figure 3 and Figure 4 present a comparison between the results based on non-penalization SEMDOT and SIMP when both the aiming volume fractions are set to 0.4. It is evident that there are significant differences in both macroscopic and microscopic structures. However, the ultimate optimal compliance values are approximately equal ($93.89\, N \cdot mm^2 \approx 94.26N \cdot mm^2$). This demonstrates that non-penalization SEMDOT and SIMP are, in fact, two distinct types of optimization algorithms, and the application of non-penalization method in topology optimization design holds feasibility.

3.3 MBB beam design

The MBB design problem as shown in Figure 2(c) is also solved by non-penalization method, its length and width of the design domain are L =15mm and H = 3mm. Four-node plane stress finite elements (150 × 30) are utilized to discretise the macrostructure, and the microstructure is discretised into 50 × 50 plane stress finite elements. The maximums in the macro and micro volume constraints are defined as 0.4 and 0.5, respectively. A circle area with radius equals 1/3 of the total micro element size.

Table 2. Multiscale design of MBB beam under non-penalization SEMDOT and SIMP methods

Algorithm	Design optimal structure	Compliance ($N \cdot mm^2$)
Non-penalization SEMDOT		303.14
SIMP		315.43

Table 2 illustrates a comparison between the results of multiscale topology optimization using non-penalization SEMDOT and the p=3.0 SIMP method. It is evident from the results that the microstructure shapes obtained through both approaches exhibit considerable similarity. However, non-penalization SEMDOT yields better compliance value. Furthermore, the smooth boundary obtained with non-penalization SEMDOT is not influenced by the presence of grayscale elements.

4. The comparison of computational costs

Table 3. Comparison of computational cost of both methods under same design problems

Design problems	Parameters and specifications	SIMP	Non-penalization SEMDOT
Michell-Type beam	N_{macro}: (100×40); N_{micro}: (50×50) $\begin{cases} V_M = 0.4 \\ V_m = 0.4 \end{cases}$	Iterations:442 Time(s): 146	Iterations:50 Time(s): 73
MBB beam	N_{macro}: (150×30); N_{micro}: (50×50) $\begin{cases} V_M = 0.4 \\ V_m = 0.5 \end{cases}$	Iterations: 537 Time(s): 211	Iterations:37 Time(s): 80

For the comparison of computational costs, the computer configuration is Intel Core i7-12700K processor, NVIDIA GeForce RTX 3080 graphics card and 32GB RAM. Table 3 displays computational costs for both methods under different design problems, the grid points number used for the non-penalization SEMDOT is 10, N_{macro} and N_{micro} are the number of finite elements for each problem. It could be noticed that mutiscale design based on non-penalization SEMDOT requires less iterations and computational time towards convergence.

5. Conclusion

This work extended non-penalization SEMDOT to a homogenized version for multiscale design considering periodic boundary conditions. The compliance-based problem has been investigated for MBB, Michell-type and Cantilever beam design problems. Some simple comparison with the results obtained by SIMP was conducted. The numerical results show the capability of SEMDOT in forming smooth 2D multiscale structures. In the future, more design problems like heat transfer, natural frequency or 3D cases will be created for multiscale design based on non-penalization SEMDOT.

6. Declarations

The authors declare that they have no known competing financial interests or personal relationships that could have appeared to influence the work reported in this paper.

7. Acknowledgements

Mr JIYE Zhou appreciates Dr. Yun-Fei Fu (University of Alberta, Canada) and Prof. Bernard Rolfe (Deakin University, Australia) for providing non-penalization SEMDOT codes. Mr JIYE Zhou appreciates the support from the school funding: School of Engineering HDR Support.

8. Reference

[1] M. P. Bendsoe and O. Sigmund, *Topology optimization: theory, methods, and applications*. Springer Science & Business Media, 2013.

[2] M. P. Bendsøe and N. Kikuchi, "Generating optimal topologies in structural design using a homogenization method," *Computer methods in applied mechanics and engineering,* vol. 71, no. 2, pp. 197-224, 1988.

[3] O. Sigmund, "A 99 line topology optimization code written in Matlab," *Structural and multidisciplinary optimization,* vol. 21, no. 2, pp. 120-127, 2001.

[4] X. Y. Yang, Y. M. Xie, G. P. Steven, and O. Querin, "Bidirectional evolutionary method for stiffness optimization," *AIAA journal,* vol. 37, no. 11, pp. 1483-1488, 1999.

[5] M. Y. Wang, X. Wang, and D. Guo, "A level set method for structural topology optimization," *Computer methods in applied mechanics and engineering,* vol. 192, no. 1-2, pp. 227-246, 2003.

[6] X. Guo, W. Zhang, and W. Zhong, "Doing topology optimization explicitly and geometrically—a new moving morphable components based framework," *Journal of Applied Mechanics,* vol. 81, no. 8, 2014.

[7] W. Zhang *et al.*, "Explicit three dimensional topology optimization via Moving Morphable Void (MMV) approach," *Computer Methods in Applied Mechanics and Engineering,* vol. 322, pp. 590-614, 2017.

[8] A. Zolfagharian, M. Bodaghi, R. Hamzehei, L. Parr, M. Fard, and B. F. Rolfe, "3D-printed programmable mechanical metamaterials for vibration isolation and buckling control," *Sustainability,* vol. 14, no. 11, p. 6831, 2022.

[9] R. Hamzehei, A. Zolfagharian, S. Dariushi, and M. Bodaghi, "3D-printed bio-inspired zero Poisson's ratio graded metamaterials with high energy absorption performance," *Smart Materials and Structures,* vol. 31, no. 3, p. 035001, 2022.

[10] R. Hamzehei, A. Serjouei, N. Wu, A. Zolfagharian, and M. Bodaghi, "4D metamaterials with zero poisson's ratio, shape recovery, and energy absorption features," *Advanced Engineering Materials,* vol. 24, no. 9, p. 2200656, 2022.

[11] O. Sigmund, "Materials with prescribed constitutive parameters: an inverse homogenization problem," *International Journal of Solids and Structures,* vol. 31, no. 17, pp. 2313-2329, 1994.

[12] H. Rodrigues, J. M. Guedes, and M. Bendsoe, "Hierarchical optimization of material and structure," *Structural and Multidisciplinary Optimization,* vol. 24, pp. 1-10, 2002.
[13] P. Liu, Z. Kang, and Y. Luo, "Two-scale concurrent topology optimization of lattice structures with connectable microstructures," *Additive Manufacturing,* vol. 36, p. 101427, 2020.
[14] W. Zhang, F. Wang, G. Dai, and S. Sun, "Topology optimal design of material microstructures using strain energy-based method," *Chinese journal of aeronautics,* vol. 20, no. 4, pp. 320-326, 2007.
[15] W. Zhang, G. Dai, F. Wang, S. Sun, and H. Bassir, "Using strain energy-based prediction of effective elastic properties in topology optimization of material microstructures," *Acta Mechanica Sinica,* vol. 23, no. 1, pp. 77-89, 2007.
[16] L. Xia and P. Breitkopf, "Design of materials using topology optimization and energy-based homogenization approach in Matlab," *Structural and multidisciplinary optimization,* vol. 52, no. 6, pp. 1229-1241, 2015.
[17] Y.-F. Fu, K. Ghabraie, B. Rolfe, Y. Wang, and L. N. Chiu, "Smooth design of 3D Self-supporting topologies using additive manufacturing filter and SEMDOT," *Applied Sciences,* vol. 11, no. 1, p. 238, 2020.
[18] Y.-F. Fu, B. Rolfe, Y. Wang, X. Huang, and K. Ghabraie, "SEMDOT: Smooth-edged material distribution for optimizing topology algorithm," *Advances in Engineering Software,* vol. 150, p. 102921, 2020.
[19] Y.-F. Fu, K. Ghabraie, B. Rolfe, Y. Wang, and L. N. Chiu, "Smooth design of 3d self-supporting topologies using additive manufacturing filter and semdot," *Applied Sciences,* vol. 11, no. 1, p. 238, 2021.
[20] X. Huang and W. Li, "Three-field floating projection topology optimization of continuum structures," *Computer Methods in Applied Mechanics and Engineering,* vol. 399, p. 115444, 2022.
[21] Y.-F. Fu, K. Long, and B. Rolfe, "On Non-Penalization SEMDOT Using Discrete Variable Sensitivities," *Journal of Optimization Theory and Applications,* vol. 198, pp. 644-677, 2023.
[22] Y.-F. Fu, K. Long, A. Zolfagharian, M. Bodaghi, and B. Rolfe, "Topological design of cellular structures for maximum shear modulus using homogenization SEMDOT," *Materials Today: Proceedings,* 2023.
[23] Y. Liang, K. Sun, and G. Cheng, "Discrete variable topology optimization for compliant mechanism design via Sequential Approximate Integer Programming with Trust Region (SAIP-TR)," *Structural and Multidisciplinary Optimization,* vol. 62, pp. 2851-2879, 2020.
[24] Y. Liang, X. Yan, and G. Cheng, "Explicit control of 2D and 3D structural complexity by discrete variable topology optimization method," *Computer Methods in Applied Mechanics and Engineering,* vol. 389, p. 114302, 2022.
[25] Y. Liang and G. Cheng, "Topology optimization via sequential integer programming and canonical relaxation algorithm," *Computer Methods in Applied Mechanics and Engineering,* vol. 348, pp. 64-96, 2019.
[26] Y. Liang and G. Cheng, "Further elaborations on topology optimization via sequential integer programming and Canonical relaxation algorithm and 128-line MATLAB code," *Structural and Multidisciplinary Optimization,* vol. 61, pp. 411-431, 2020.
[27] K. Svanberg, "MMA and GCMMA-two methods for nonlinear optimization," *vol,* vol. 1, pp. 1-15, 2007.
[28] B. Hassani and E. Hinton, "A review of homogenization and topology optimization I—homogenization theory for media with periodic structure," *Computers & Structures,* vol. 69, no. 6, pp. 707-717, 1998.
[29] J. Gao, Z. Luo, L. Xia, and L. Gao, "Concurrent topology optimization of multiscale composite structures in Matlab," *Structural and Multidisciplinary Optimization,* vol. 60, pp. 2621-2651, 2019.

Integrating Simulation and Experimentation for Optimized Compaction Powder Forming: A Study on RVE Size, Friction, and Particle Size Distribution

Hossein Ghorbani-Menghari[1], Hwi-Jun Kim[2], Hyunjoo Choi[3], Pil-Ryung Cha[3] and Ji Hoon Kim[1]*
[1] School of Mechanical Engineering, Pusan National University, South Korea

[2] Korea Institute of Industrial Technology, Incheon, South Korea

[3] School of Advanced Materials Engineering, Kookmin University, South Korea

* Corresponding author's E-mail: kimjh@pusan.ac.kr

Abstract. This work comprehensively represents multi-particle finite element simulations for powder compaction, including examining representative volume element (RVE) size, friction effects, and particle distribution. The analysis provides valuable insights into the correlation between RVE size and relative density, facilitating a comprehensive understanding of the sensitivity of process parameters to relative density. Furthermore, the research investigates the impact of particle size distribution. Moreover, the study investigates the influence of friction between powder particles and the die wall. Results demonstrate that increased friction leads to a significant reduction in relative density.

Keywords: Compaction, Wall friction, Density distribution, Finite element method

1. Introduction

Powder compaction is fundamental in various industries, such as pharmaceuticals, ceramics, and metallurgy. The process involves transforming powder into a solid form using mechanical forces. The quality of the compacted material can be influenced by a myriad of factors, namely representative volume element (RVE) size, friction, and particle size distribution. Michrafy et al. [1] investigated the impact of friction between powder and die walls during the compaction of pharmaceutical binders across three lubrication techniques: lubricated die, non-lubricated die, and powder mixed with lubricant. Using the principles of Coulomb friction, the wall friction coefficient was assessed based on transmission, transfer, and aspect ratios. Frictional behavior was found to vary depending on the powder and lubrication method. While adding a lubricant to the powder reduced friction compared to a lubricated die, the friction coefficient changed during densification. A model combining the Janssen–Walker analysis with the Heckel equation was used to assess the effect of wall friction on the axial density distribution of tablets. However, this model was deemed unsuitable for powder mixed with lubricant due to its lack of relevant physical parameters. Korim et al. [2] developed a Cu–15Fe alloy using powder metallurgy, incorporating solid lubricants paraffin wax (PW), stearic acid (SA), and their blend. The compacts were shaped under varying pressures and sintered at 1050 ◦C for an hour. Assessments included densification, porosity, hardness, and morphological changes. Tribological properties were analyzed through pin-on-disk tests. Lubrication type and compaction pressure significantly affected the alloy attributes. PW achieved the best results at 50 MPa, decreasing with pressure, while SA performed best at 350 MPa, improving with increased pressure. PW samples demonstrated reduced friction and wear, but SA enhanced wear resistance, reducing the wear rate by roughly 50% compared to other samples. Guner et al. [3] examined metal powder compaction, focusing on 200 μm copper particles using the multi-particle finite element method (MPFEM). The von Mises model analyzed friction and contact dynamics in the compaction process. Three friction models, Amonton-Coulomb, Wanheim-Bay, and Levanov, were evaluated. Validation involved compaction experiments and microscopy comparisons. Friction coefficients for Wanheim-Bay and Levanov were between 0.04–0.07. The Wanheim-Bay model was suited for high-density compaction, while Levanov was preferable for low-density cases. Kahhal et al. [4] discuss how loading conditions, geometry, and wall friction impact punch force and stress distribution. They found that Increased wall friction (from 0.0 to 0.2) raises compaction force by 14.70%, and differences between upper and lower face forces increase. Changing the loading path from compression to hydrostatic compaction reduces compaction force by 57.7% (upper) and 65.75% (lower), decreases axial stress by 27.70%, and increases transverse stress by 40.41%. Compaction volume strain is 35.33% smaller than hydrostatic compression.

This study performs multi-particle finite element simulations for powder compaction, emphasizing the size of the RVE, friction, and particle distribution. The influence of RVE size is analyzed, aiming to discern its relationship with relative density. The impact of particle size distribution and the effects of friction between powder particles and the die wall are studied.

2. Material and Method

The modeling framework was devised by integrating both DEM and traditional FEM. In the initial phase, particle center coordinates from the DEM experiments, informed by empirical data, were utilized as the foundational configuration for the MPFEM. This led to the MPFEM-based execution of powder compaction simulations as the final step. The construction and execution of both DEM and MPFEM models were carried out using the commercial FEM software Abaqus (Version 2020). Due to the intensive computational nature of the model, the Abaqus explicit algorithm was employed. A detailed breakdown of this modeling approach is presented in the following sections.

2.1. Material Properties

The Fe-Si-Al-P alloy powder compaction was evaluated utilizing the elastic-plastic material framework. Table 1 illustrates the material characteristics of this model.

Table 1. Material parameters of Fe-Si-Al-P alloy powder.

Parameter	Value
Elastic modulus (GPa)	170
Poisson's ratio	0.3
ρ (kg/m^3)	7,130

Compression testing was executed using the Instron 8501 Servo Hydraulic Machine, which utilizes a servo-regulated hydraulic mechanism to exert compressive forces on specimens. The peak pressure sustained during the study was set to 1.2 GPa. Additionally, compressive assessments were conducted on samples sintered at 950 °C. By employing the power law swift hardening technique, extending the experimental compression results and identifying the stress-strain trajectory became feasible. The equation below represents the flow stress-plastic strain correlation for the given material:

$$\sigma = 1380 * (\varepsilon + 0.0354)^{0.254}$$

2.2. Details of MPFEM Models

The DEM simulation, employed to define particle properties and initialize configurations for MPFEM simulations, utilized the Edinburgh-Elasto-Plastic-Adhesion (EEPA) model with a model size of 1 mm^3. The EEPA model governed the interaction forces between particles, encompassing elastic, plastic deformation, and adhesive forces. After completing the DEM simulation, coordinate data of particle centers were extracted and transferred to the corresponding MPFEM model through Python scripts.
The MPFEM model incorporates both a punch and a die in addition to the meshed particles. Fig. 1 illustrates the generated particles and dies. Each particle is depicted as a solid component and is meshed utilizing C3D4 elements. These elements are determined by a seeding size of around 20 micrometers. The other parts are depicted as discrete 3D rigid structures. Initial relative density was considered 40 percent. Interparticle friction and particle-die friction coefficient were chosen at 0.2.

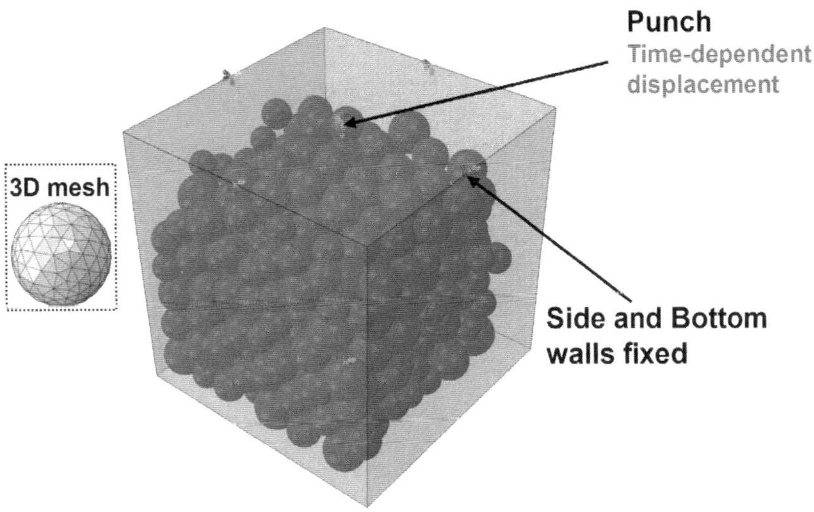

Fig. 1 MPFEM model

A customized MATLAB code, complemented by a Python script, was crafted specifically for Generating Representative Volume Elements (RVEs). This dual-software approach facilitated the import of particle data directly from the discrete element method results, particularly focusing on the positions and radii of each particle. The process for integrating particles into the RVE was multifaceted. If the designated position of a particle was within the die boundaries, it was included in the RVE. Conversely, any particle outside these confines was excluded, and a new particle was imported. This method ensured that all particles populated the RVE without overcrowding. Furthermore, a particle in the RVE had to meet three stringent criteria. Firstly, it had to be fully encapsulated by the die. Secondly, the initial relative density of the RVE should be maintained at 40 percent. Lastly, the RVE composition is needed to reflect the target distribution accurately. This rigorous approach to RVE generation ensures that the subsequent analyses are grounded in realistic and representative models of the material under investigation.

3. Results and Discussion

3.1. Influence of Friction Coefficient on Density

A thorough examination was conducted to understand how the friction coefficient between particles affects the compaction process. In our study, the friction between particles was varied, with values of 0.1 and 0.2 adopted. Fig. 2 presents the results demonstrating the significant impact of changing the friction coefficient between particles. A clear pattern can be observed: as the friction decreases, the density under load increases, and the spring back reduces, leading to a higher final density. This indicates that the friction coefficient of powder should be reduced to increase the density difference between various mixtures.

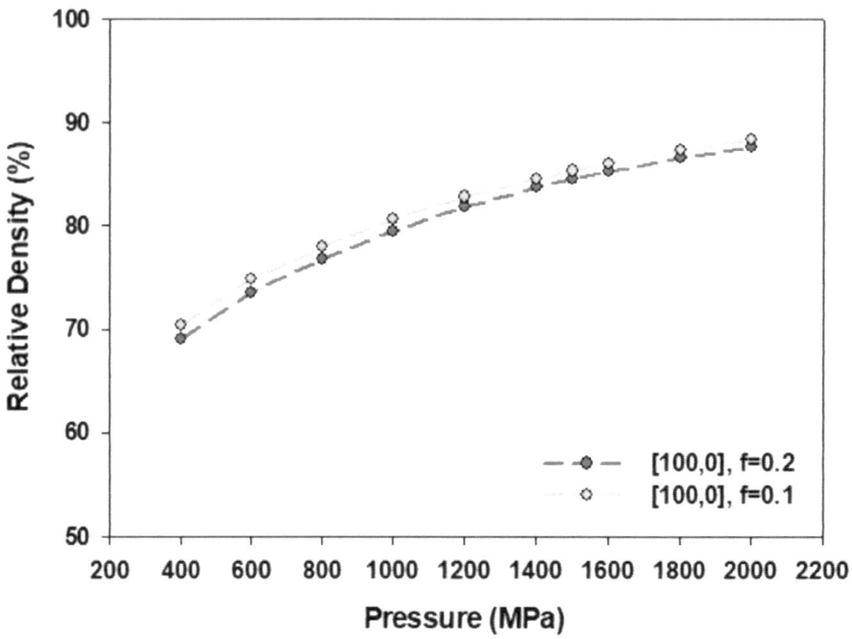

Fig. 2 Effect of friction between particles on relative density

In the subsequent analysis, the influence of inter-particle friction is investigated. For the kinetic friction evaluation, a coefficient of 0.2 was applied. In the case of static friction, two distinct coefficients, 0.2 and 1, were considered. Referring to Fig. 3, an evident trend was observed: with an increase in static friction, there was a decline in density under applied load.

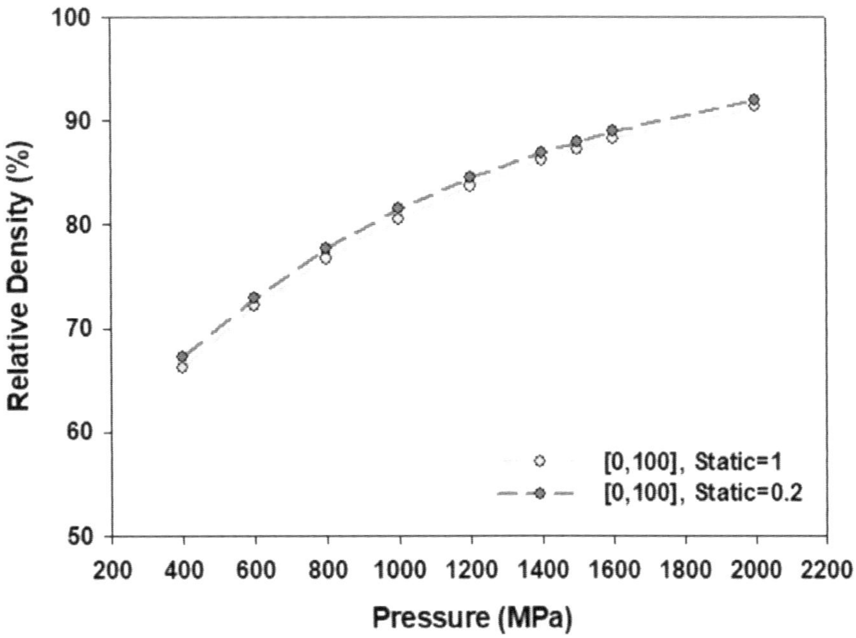

Fig. 3 Influence of static and kinetic friction on the density

3.2. Influence of RVE Size on Stress-Relative Density Curve

This section investigates the impact of the RVE size on relative density. For our study, RVE sizes ranging from 350 to 800 micrometers were chosen. Notably, smaller RVE sizes can introduce significant fraction errors, underscoring the importance of selecting an appropriate RVE size. On the other hand, with larger RVE sizes, there is an increase in the particle count, leading to a rise in computational demands.

Fig. 4 illustrates the impact of RVE size on relative density. Increasing the RVE size results in a corresponding increase in initial density. However, it is important to note that while this variation becomes more significant at lower pressures, the difference decreases and becomes relatively insignificant in areas where higher pressures are applied. This subtle interplay between RVE size and density, particularly at varying pressure levels, highlights the complex dynamics of powder compaction.

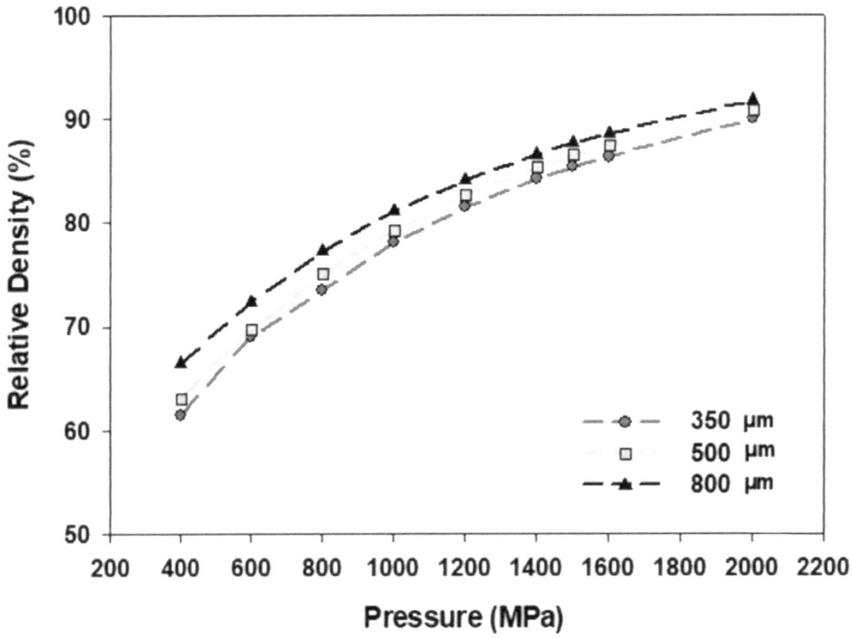

Fig. 4 Effect of RVE size on the relative density

3.3. Find the relative density for a given pressure for the three mixtures.

Fig. 5 depicts the pressure-relative density relationship at the compaction stage as predicted by the MPFEM across three cases: the coarse powder, the fine powder, and the equal mixture of the two powders, which are denoted as [0,100], [50,50], and [100,0]. In this updated analysis, distinct trends among the different mixtures were observed. Notably, the relative density of the [0,100] mixture is lower compared to both [50,50] and [100,0], while [100,0] exhibits a lower relative density than [50,50]. As the pressure increases, there is a consistent upward trajectory in relative density for all three cases. Understanding the relationship between pressure, deformation, and relative density is crucial when compacting powder mixtures. The comparison between the three cases provides valuable insights into how fine and coarse particles respond during compaction, with higher pressure consistently leading to increased relative density.

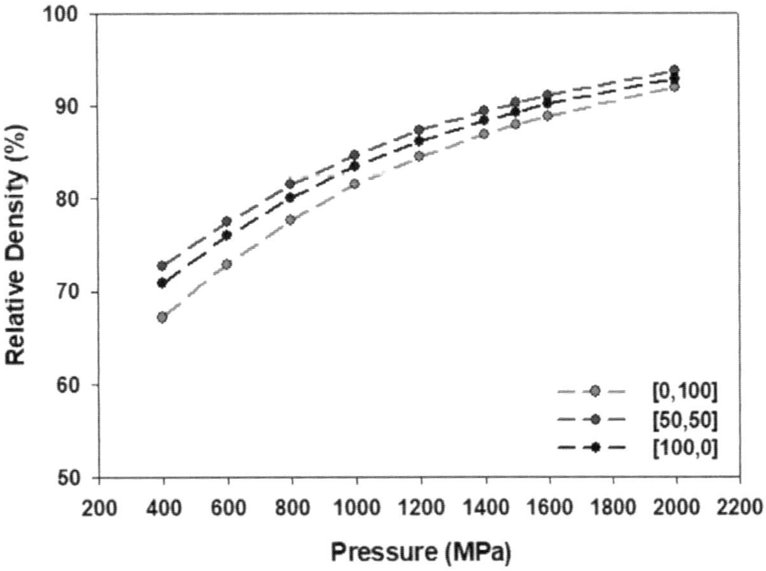

Fig. 5 The pressure-density relationship as predicted by MPFEM at the loading stage

Fig. 6 illustrates a relative density comparison between EXP and MPFEM at 1600 MPa during the unloading stage. The results obtained from MPFEM closely align with the experimental findings, demonstrating a similar trend. In both MPFEM and EXP results, the pure coarse powder mixture [0,100] exhibits a lower relative density than the pure fine powder mixture [100,0]. This consistency in trends between the simulation and experimental data reinforces the reliability of the MPFEM results in capturing the relative density behavior during unloading at 1600 MPa.

Fig. 6 Relative density comparison within EXP and MPFEM at 1600 MPa at the unloading stage

4. Conclusion

Throughout our extensive study of powder compaction, various factors that affect the final relative density of powder mixtures under various conditions were examined. The MPFEM results have revealed several critical insights into the behavior of these mixtures. Notably, as pressure increased, there was a consistent rise in relative density. Fine powders exhibited a higher density compared to their coarser counterparts. As the friction decreases, the density under load increases, and the spring back reduces, leading to a higher final density.

References

[1] A. Michrafy, M.S. Kadiri, J.A. Dodds, Wall Friction and its Effects on the Density Distribution in the Compaction of Pharmaceutical Excipients, Chemical Engineering Research and Design, Volume 81, Issue 8, 2003, Pages 946-952.

[2] N.S. Korim, A. Elsayed, L. Hu, Impacts of Lubricant Type on the Densification Behavior and Final Powder Compact Properties of Cu-Fe Alloy under Different Compaction Pressures. Materials, Volume 15, Issue 16, 2022, Page 5750.

[3] F. Güner, Ö.N. Cora, H. Sofuoğlu, Effects of friction models on the compaction behavior of copper powder, Tribology International, Volume 122, 2018, Pages 125-132.

[4] P. Kahhal, J. Jung, Y.C. Hur, Y.H. Moon and J.H. Kim, Analysis of Powder Compaction Process Using Multi-Particle Finite Element Method, Materials Transactions, Volume 63, Issue 11, 2022, Pages 1576-1582.

[5] P. Kahhal, H. Ghorbani-Menghari, H.J. Kim, H. Choi, P.R. Cha, J.H. Kim, Metaheuristic Optimization of Powder Size Distribution in Powder Forming Process Using Multi-Particle Finite Element Method Coupled with Artificial Neural Network and Genetic Algorithm, Materials Transactions, Volume 64, Issue 11, 2023, Pages 2648-2655.

IOP Conf. Series: Materials Science and Engineering **1307** (2024) 012024 doi:10.1088/1757-899X/1307/1/012024

Anisotropic-asymmetric mechanical behaviour for an Mg-Gd-Y alloy under various temperatures: Experiment and modelling

P Wu[1], Y Lou[1] and Q Chen[2]

1 School of Mechanical Engineering, Xi'an Jiaotong University 28 Xianning West Road, Xi'an, Shanxi 710049, China

2 Southwest Technology and Engineering Research Institute 33 Yuzhou Road, Chongqing 400039, China

ys.lou@xjtu.edu.cn

Abstract. The study focuses on the anisotropic-asymmetric behaviour of an Mg-Gd-Y alloy at the temperature range of 25~300 ℃. The mechanical experiments were carried out, including the uniaxial tension and compression specimens along rolling direction (RD), diagonal direction (DD), and transverse direction (TD). The deformation behaviour of all specimens was recorded through adopting the digital image correlation technology, obtaining the corresponding force-stroke curves. It is obviously seen that the variation of true stress-plastic strain curve is related with the coupling effect of temperature and strain. The Mg-Gd-Y alloy possesses the remarkable tension-compression asymmetry and anisotropy under various temperatures. A SVT model is proposed to characterize the temperature-related plastic flow behaviour under various loading condition. The evolution of yield loci at different equivalent plastic strain levels and temperatures of the Mg-Gd-Y alloy is accurately characterized by the A-Yoon2014.

1. Introduction

As a lightweight and high-strength metals, Magnesium (Mg) alloys have a huge application prospect in the fields of aerospace, military industry and vehicles due to its excellent property. In general, the strong anisotropy and remarkable strength differential (SD) effect are the common characteristic of Mg alloys due to the sensitivity of the texture orientation and stress state on the deformation mechanism [1]. In recent years, adding rare earth (RE) elements to the Mg crystal becomes one of the effective methods to improve the mechanical property and corrosion resistance of Mg alloy. For example, the RE element Gd, Y and Zn are added to achieve the enhanced strength owing to the additional reinforcement mechanism, such as fine grain strengthening and precipitation strengthening [2]. However, the anisotropy and SD effect become very complex because of the component difference of the Mg-RE alloys. Besides, the temperature has a remarkable effect on the plastic behaviour of Mg alloys [3]. The hybrid experimental-numerical is an effective way to uncover the plastic behaviour of metals. The material model is very important for achieving the numerical simulation with high accuracy [4].

In the study, the mechanical experiments of an Mg-Gd-Y alloy at 25~300 ℃ were firstly conducted, including the uniaxial tension and uniaxial compression along rolling direction (RD), diagonal direction (DD), and transverse direction (TD). A SVT model is proposed to characterize the temperature-related plastic flow behaviour under various loading condition. An analytical Yoon2014 (A-Yoon2014) yield function is established to capture the evolving yield behaviour of the Mg-Gd-Y alloy.

2. Mechanical experiment

The chemical composition of the Mg-Gd-Y alloy includes the 8.5% Gd, 4.5% Y, 0.8% Zn, 0.4% Zr and balanced Mg. The parent material of the Mg-Gd-Y alloy was prepared to the size of 190 mm × 105 mm × 10 mm by the hot rolling process and the T6 heating treatment. As shown in Figure 1, two specimens were manufactured along RD, DD and TD, which are named as UT and UC, respectively. The mechanical experiments were conducted on the universal testing machine by adopting the displacement control mode under the condition of 25~300 ℃ and strain rate of 0.001 s^{-1}. Each experiment was carried out at least three times. The loading process was recorded by adopting the digital image correlation (DIC) technology.

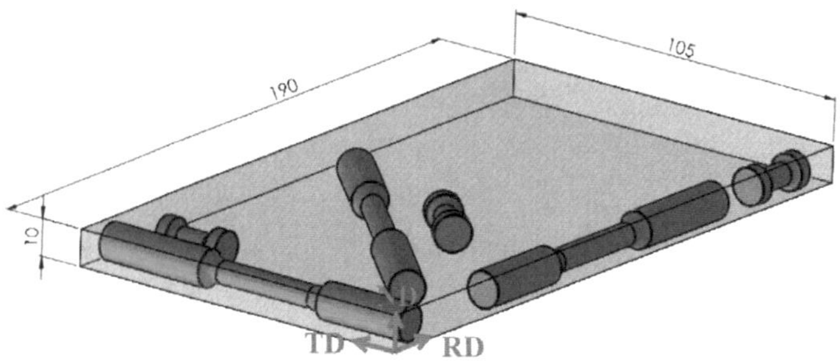

Figure 1 Sampling diagram of UT and UC specimens.

The force-stroke curve of each experiment was obtained based on the initial gauge of 20 mm (UT) and 6 mm (UC), as shown in Figure 2(a). The force-stroke curve before the maximum force is selected to calculate the uniform deformation behaviour of the Mg-Gd-Y alloy. The corresponding true stress-plastic strain relationship is presented in Figure 2(b). It is clearly seen that the Mg-Gd-Y alloy possesses the anisotropy and SD effect under different temperatures. The strength of the Mg-Gd-Y alloy presents a monotone downward trend with non-uniform variation as the temperature increases. The result indicates that the yield behaviour is related with the coupling effect of temperature and strain.

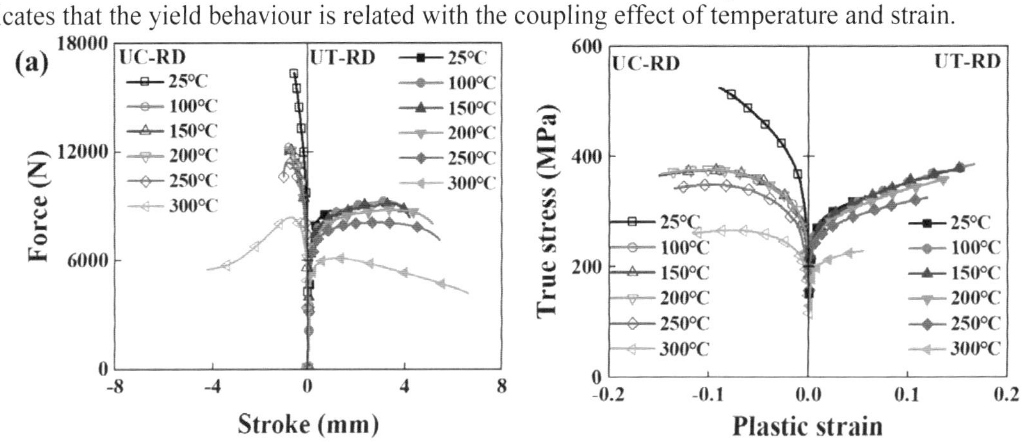

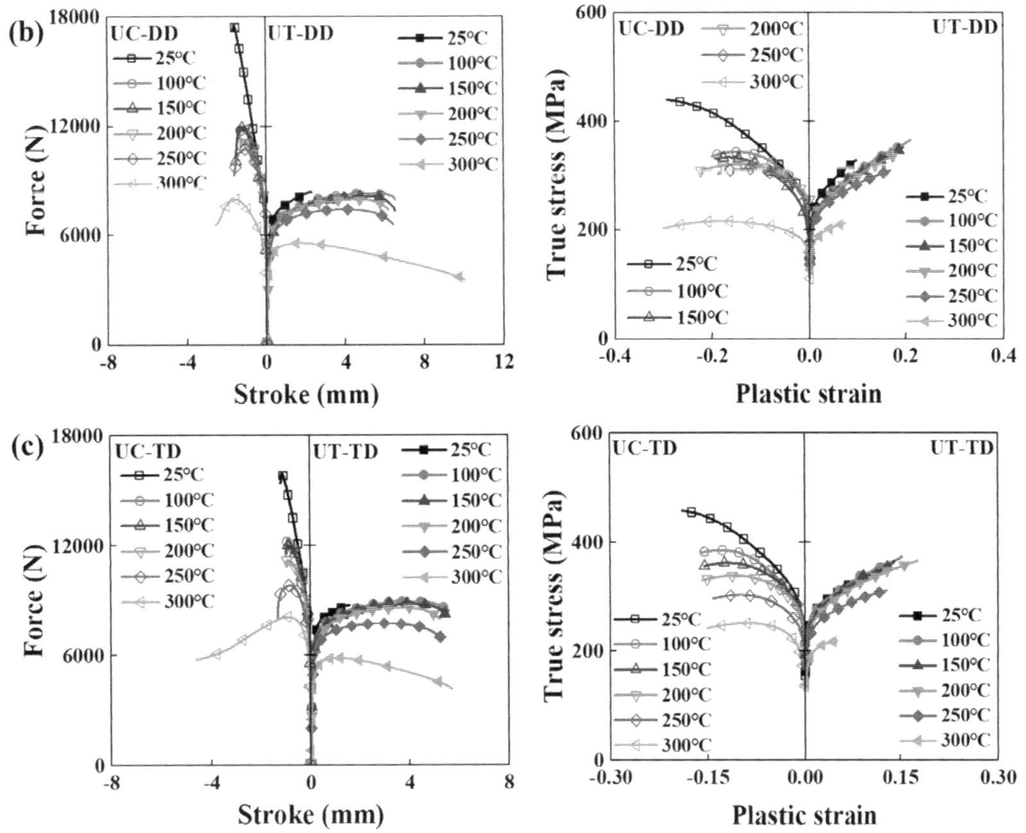

Figure 2 Force-stroke curves and true stress-plastic strain curves along (a) RD, (b) DD and (c) TD.

3. Material model

3.1. A-Yoon2014 yield function

A large number of anisotropic yield function were proposed to describe the yield behaviour of metals [5-13]. In this section, the A-Yoon2014 yield function [10] is adopted as the characterization model, which is expressed as below:

$$f(s_{ij}) = I_1 + (J_2^{'3/2} - J_3^{''})^{1/3} \tag{1}$$

with

$$I_1 = h\left(\sigma_{xx} + \sigma_{yy} + \sigma_{zz}\right) \tag{2}$$

$$J_2^{'} = \frac{1}{2} s^{'} : s^{'} = -s_{xx}^{'} s_{yy}^{'} - s_{yy}^{'} s_{zz}^{'} - s_{zz}^{'} s_{xx}^{'} + s_{xy}^{'2} + s_{yz}^{'2} + s_{zx}^{'2} \tag{3}$$

$$J_3^{''} = \det(s^{''}) = s_{xx}^{''} s_{yy}^{''} s_{zz}^{''} + 2 s_{xy}^{''} s_{yz}^{''} s_{zx}^{''} - s_{xx}^{''} s_{yz}^{''2} - s_{yy}^{''} s_{zx}^{''2} - s_{zz}^{''} s_{xy}^{''2} \tag{4}$$

where I_1, $J_2^{'}$ and $J_3^{''}$ denote the anisotropic first stress invariant, second stress invariant of stress tensor $s^{'}$ and third stress invariant of stress tensor $s^{''}$, respectively. Based on the linear transformation tensor, the $J_2^{'}$ and $J_3^{''}$ are expressed as

$$\begin{aligned} J_2^{'} &= -s_{xx}^{'} s_{yy}^{'} - s_{yy}^{'} s_{zz}^{'} - s_{zz}^{'} s_{xx}^{'} + s_{yz}^{'2} + s_{zx}^{'2} + s_{xy}^{'2} \\ &= l(\sigma_{xx} - \sigma_{zz})^2 + E(\sigma_{yy} - \sigma_{zz})^2 + (F - E - l)(\sigma_{xx} - \sigma_{zz})(\sigma_{yy} - \sigma_{zz}) + c_4^{'2}\sigma_{yz}^{2} + c_5^{'2}\sigma_{zx}^{2} + (4G - F)\sigma_{xy}^{2} \end{aligned} \tag{5}$$

$$J_3^{"} = \left[c_{66}\sigma_{xy}^2 - (2\sigma_{xx} - \sigma_{yy} - \sigma_{zz})(2\sigma_{yy} - \sigma_{xx} - \sigma_{zz})\right]\left[c_{11}\sigma_{xx} + c_{22}\sigma_{yy} - (c_{11} + c_{22})\sigma_{zz}\right]$$
$$+2\sqrt{|c_{44}c_{55}c_{66}|}\sigma_{xy}\sigma_{yz}\sigma_{xz} - c_{44}(2\sigma_{xx} - \sigma_{yy} - \sigma_{zz})\sigma_{yz}^2 - c_{55}(2\sigma_{yy} - \sigma_{xx} - \sigma_{zz})\sigma_{xz}^2 \quad (6)$$

with

$$l = \frac{1}{9}(c_2^{'2} + c_2^{'}c_3^{'} + c_3^{'2}), E = \frac{1}{9}(c_1^{'2} + c_1^{'}c_3^{'} + c_3^{'2}), F = \frac{1}{9}(c_1^{'2} + c_1^{'}c_2^{'} + c_2^{'2}), 4G - F = c_6^{'2},$$
$$c_{11} = \frac{c_{12}^{"}c_{23}^{"}(2c_{12}^{"} - c_{23}^{"})}{27}, c_{22} = \frac{c_{12}^{"}c_{23}^{"}(2c_{23}^{"} - c_{12}^{"})}{27}, c_{66} = \frac{9c_{66}^{"2}}{c_{12}^{"}c_{23}^{"}}, c_{44} = \frac{1}{3}c_{12}^{"}c_{44}^{"2}, c_{55} = \frac{1}{3}c_{23}^{"}c_{55}^{"2} \quad (7)$$

It is assumed that the anisotropic characteristic in thickness direction is similar with that in plane state, namely, $c_4^{'}=c_5^{'}=c_6^{'}$ and $c_{44}^{"}=c_{55}^{"}=c_{66}^{"}$. Therefore, the Eq. (9) is transformed as below:

$$J_2^{'} = \left[(\sigma_{xx} - \sigma_{zz}) - (\sigma_{yy} - \sigma_{zz})\right]\left[l(\sigma_{xx} - \sigma_{zz}) - E(\sigma_{yy} - \sigma_{zz})\right]$$
$$+F\left[(\sigma_{xx} - \sigma_{zz})(\sigma_{yy} - \sigma_{zz}) - (\sigma_{yz}^2 + \sigma_{zx}^2 + \sigma_{xy}^2)\right] + 4G(\sigma_{yz}^2 + \sigma_{zx}^2 + \sigma_{xy}^2) \quad (8)$$

The parameters c_{44} and c_{55} can be written in the form of

$$c_{44} = c_{55} = \frac{c_{11} + c_{22}}{2}c_{66} \quad (9)$$

Under the plane stress, the A-Yoon2014 yield function is as follows:

$$\sigma_{Yoon2014} = h(\sigma_{xx} + \sigma_{yy}) + (J_2^{'3/2} - J_3^{"})^{1/3} = \sigma_0^T(\varepsilon_{peeq}, T^*) \quad (10)$$

with

$$J_2^{'} = (\sigma_{xx} - \sigma_{yy})(l\sigma_{xx} - E\sigma_{yy}) + F\sigma_{xx}\sigma_{yy} + (4G - F)\sigma_{xy}^2 \quad (11)$$

$$J_3^{"} = (c_{66}\sigma_{xy}^2 + 2\sigma_{xx}^2 + 2\sigma_{yy}^2 - 5\sigma_{xx}\sigma_{yy})(c_{11}\sigma_{xx} + c_{22}\sigma_{yy}) \quad (12)$$

Based on eight hardening curves, material parameters l, E, F, G, c_{11}, c_{22}, c_{66} and h are expressed as below:

$$l = \left[\frac{(1-h)^3 + (\phi_{C0} + h)^3}{2}\right]^{2/3}, c_{11} = \frac{\left[(\phi_{C0} + h)^3 - (1-h)^3\right]}{4}, E = \left[\frac{(\phi_{T90} - h)^3 + (\phi_{C90} + h)^3}{2}\right]^{2/3},$$
$$c_{22} = \frac{\left[(\phi_{C90} + h)^3 - (\phi_{T90} - h)^3\right]}{4}, G = \left[\frac{(\phi_{T45} - h)^3 + (\phi_{C45} + h)^3}{2}\right]^{2/3}, c_{66} = \frac{4\left[(\phi_{C45} + h)^3 - (\phi_{T45} - h)^3\right]}{c_{11} + c_{22}} + 1, \quad (13)$$
$$F = \left[\frac{(\phi_{Tb} - 2h)^3 + (\phi_{Cb} + 2h)^3}{2}\right]^{2/3}, h = -\frac{B}{3A} + (-\frac{q}{2} + \sqrt{\Delta})^{1/3} + (-\frac{q}{2} - \sqrt{\Delta})^{1/3}$$

with

$$\phi_{C0} = \frac{\sigma_0^T}{\sigma_0^C}, \phi_{C45} = \frac{\sigma_0^T}{\sigma_{45}^C}, \phi_{C90} = \frac{\sigma_0^T}{\sigma_{90}^C}, \phi_{Cb} = \frac{\sigma_0^T}{\sigma_b^C}, \phi_{T45} = \frac{\sigma_0^T}{\sigma_{45}^T}, \phi_{T90} = \frac{\sigma_0^T}{\sigma_{90}^T}, \phi_{Tb} = \frac{\sigma_0^T}{\sigma_b^T},$$
$$A = -34, B = 24\phi_{Tb} - 24\phi_{Cb} - 3\phi_{C0} + 3 - 3\phi_{C90} + 3\phi_{T90}, C = -12\phi_{Cb}^2 - 12\phi_{Tb}^2 - 3\phi_{C0}^2 - 3 - 3\phi_{C90}^2 - 3\phi_{T90}^2, \quad (14)$$
$$D = 2\phi_{Tb}^3 - 2\phi_{Cb}^3 - \phi_{C0}^3 + 1 - \phi_{C90}^3 + \phi_{T90}^3, q = \frac{27A^2D - 9ABC + 2B^3}{27A^3}, \Delta = \frac{q^2}{4} + \frac{p^3}{27}, p = \frac{3AC - B^2}{3A^2}$$

where, σ_0^T, σ_{45}^T and σ_{90}^T denote the yield stresses along RD, diagonal direction (DD) and transverse direction (TD) under uniaxial tension stress state, σ_0^C, σ_{45}^C and σ_{90}^C denote the yield stresses along RD, DD and TD under uniaxial compression stress state, and σ_b^T and σ_b^C are the yield stresses under equi-biaxial tension and equi-biaxial compression, respectively.

3.2. SVT plastic flow equation

A SVT model is adopted to characterize the effect of temperature on the plastic flow behaviour of the Mg-Gd-Y alloy, which is expressed as below:

$$\sigma = \sigma_{Swift} \exp(-lT) + \sigma_{Voce}\left(1 + eT + fT^2 + gT^3\right) \tag{15}$$

with

$$\sigma_{Swift} = K\left(\varepsilon_0 + \varepsilon_p\right)^n \tag{16-1}$$

$$\sigma_{Voce} = \left[A - (A - B)\exp\left(-C\varepsilon_p\right)\right] \tag{16-2}$$

where, K, A and B are the hardening constant, ε_0 and C denote the strain coefficient, n represents the hardening exponent, and e, f, g and l describe the temperature effect.

4. Evolving yield behaviour

4.1. Characterization of plastic flow behaviour

On the basis of the experimental data presented in Figure 1, the material constants of the SVT model are calibrated through adopting the least square method with Levenberg-Marquardt optimization algorithm. The predicted true stress-plastic strain curves of the SVT model under different loading conditions are shown in Figure 3. By the comparison between the predicted values with experimental data, it is obviously seen that the predicted result is very close to the actual one. The result indicates that the SVT model is appropriate for depicting the temperature-dependent mechanical behaviour of the Mg-Gd-Y alloy under various conditions. The determination coefficient (DC or R^2) is selected as the predicted accuracy, as shown in Figure 3(b).

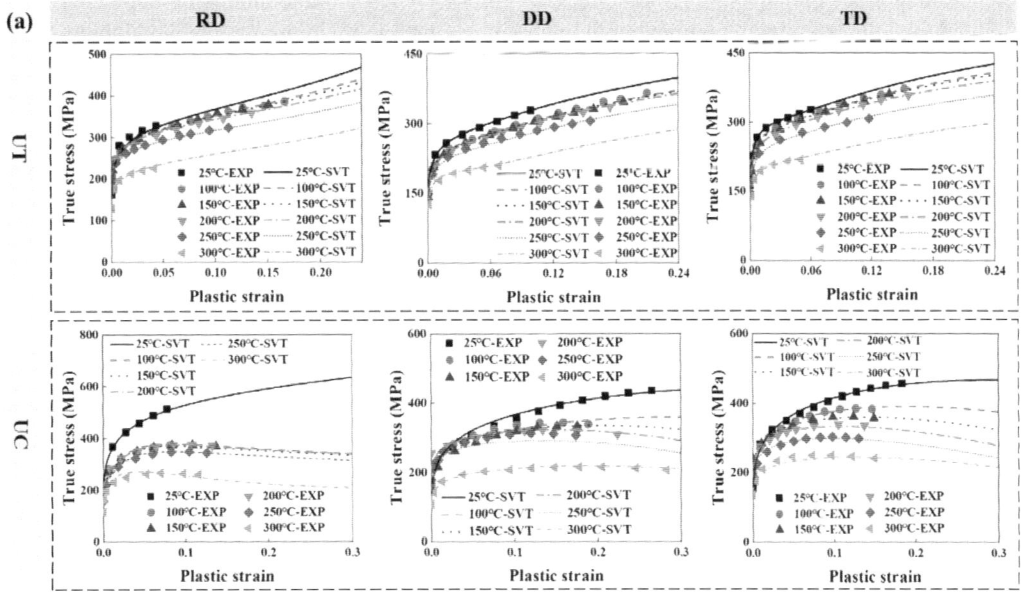

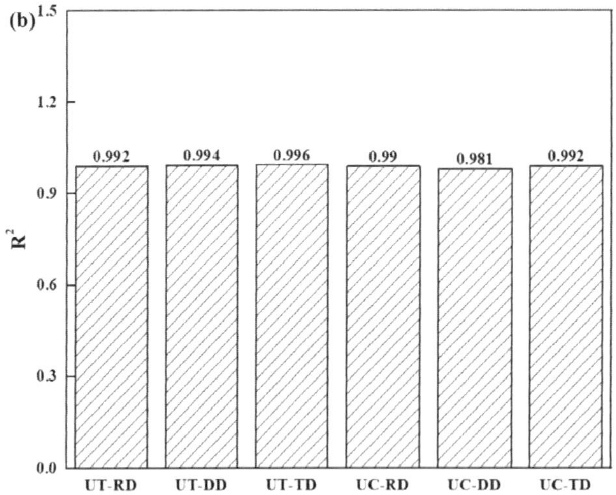

Figure 3 (a) Comparison between the predicted result of the proposed SVT model and experimental data, and (b) prediction accuracy.

4.2. *Prediction of yield evolving behaviour*

The yield stress under equi-biaxial loading condition is computed as [7]

$$\sigma_b^T = \frac{\sigma_0^T + \sigma_{90}^T + 2\sigma_{45}^T}{4}, \sigma_b^C = \frac{\sigma_0^C + \sigma_{90}^C + 2\sigma_{45}^C}{4} \tag{17}$$

On the basis of the analytical Yoon2014 yield function, the initial plane yield locus under different temperatures is presented in Figure 4. It is found that the temperature significantly influences the yield behavior of the Mg-Gd-Y alloy. The A-Yoon2014 yield function predicts the experimental yield stress. This indicates that the yield behavior of the Mg-Gd-Y alloy is related with the temperature, loading direction and stress state.

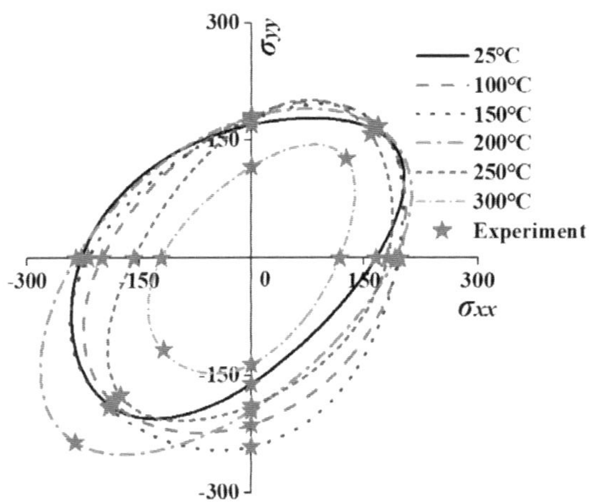

Figure 4 A-Yoon2014 plane yield locus under different temperatures.

In this section, the A-Yoon2014 yield function is adopted to characterize the evolving yield behavior of the Mg-Gd-Y alloy under different plastic strain levels. Based on the calibrated hardening curves of the SVT model, the evolution of the yield locus under different temperatures are presented in Figure 5. It is clearly observed that the A-Yoon2014 can efficiently predict the temperature- and strain-dependent yield behavior under axial loading. The result proves that the A-Yoon2014 is accurate to characterize the evolution of yield surface under various temperatures and plastic strain levels without any optimization means. This indicates that the analytical form of yield function avoids errors caused by interpolation methods and improves the characterization efficiency.

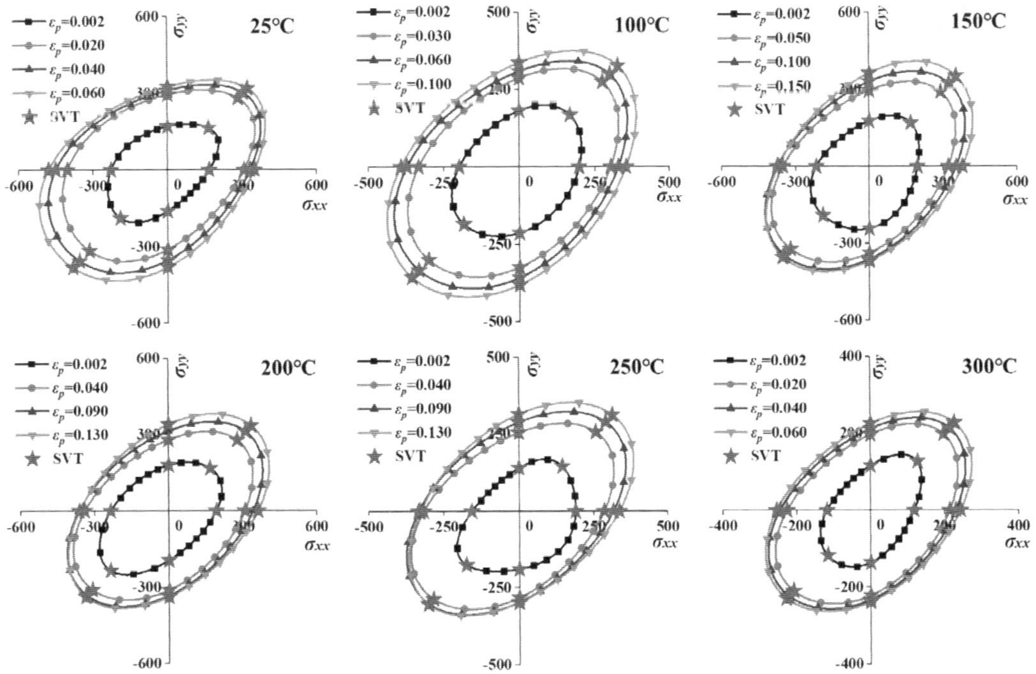

Figure 5 Evolving yield loci of the A-Yoon2014 yield function with plastic strain under various temperatures [unit: MPa].

5. Conclusions

In the study, the effect of temperature, anisotropy and stress state on the deformation behaviour of an Mg-Gd-Y alloy is investigated by the mechanical experiment. The result indicates that the anisotropy and SD effect of the Mg-Gd-Y alloy present a nonlinear change trend with temperature and strain. The plastic flow behaviour of the Mg-Gd-Y alloy is related with a coupling effect of temperature and strain. The SVT model is proposed to accurately characterize the temperature-dependent mechanical behaviour of the Mg-Gd-Y alloy. The evolution of yield behaviour under different plastic strain levels and temperatures is predicted with high accuracy by the A-Yoon2014 model, further proving the characterization ability and efficiency of the A-Yoon2014 yield function without optimization means.

References

[1] Hantzsche K, Bohlen J, Wendt J, Kainer KU, Yi SB and Letzig D 2010 *Scr. Mater.* **63(7)** 725-730
[2] Du YZ, Qiao XG, Zheng MY, Wu K and Xu SW 2015 *Mater. Sci. Eng. A* **620** 164-171
[3] Wu P, Lou Y, Chen Q and Ning H 2022 *Int. J. Mech. Sci.* **229** 107506
[4] Shang H, Wu P, Lou Y, Wang J and Chen Q 2022 J. Mater. Process Technol. ***302*** 117501
[5] Barlat F, Brem JC, Yoon JW, Chung K, Dick RE, Lege DJ, Pourboghrat F, Choi SH and Chu E 2003 *Int. J. Plast.* **19(9)** 1297-1319
[6] Cazacu O and Barlat F 2004 *Int. J. Plast.* **20(11)** 2027-2045
[7] Cazacu O, Plunkett B and Barlat F 2006 *Int. J. Plast.* **22(7)** 1171-1194
[8] Yoon JW, Lou Y, Yoon J and Glazoff MV 2014 *Int. J. Plast.* **56** 184-202
[9] Lou Y and Yoon JW 2018 *Int. J. Plast.* **101** 125-155
[10] Lou Y, Zhang S and Yoon JW 2020 *Int. J. Plast.* **134** 102813
[11] Hu Q and Yoon JW 2021 *Int. J. Plast.* **140** 102978

IOP Conf. Series: Materials Science and Engineering **1307** (2024) 012024 doi:10.1088/1757-899X/1307/1/012024

[12] Lou Y and Yoon JW 2023 *Int. J. Plast.* **166** 103647
[13] Lou Y, Zhang C, Wu P and Yoon JW 2024 *Int. J. Solids Struct.* **286-287** 112582

Acknowledgments

The authors acknowledge the financial support by the National Natural Science Foundation of China (Grant No. 52075423), the State Key Laboratory of Mechanical System and Vibration (Grant No. MSV202009), and the State Key Laboratory of High Performance Complex Manufacturing (Grant No. Kfkt2019-02).

IOP Conf. Series: Materials Science and Engineering **1307** (2024) 012025 doi:10.1088/1757-899X/1307/1/012025

Ductile fracture of tubular specimens under non-proportional loading condition

Madhav Baral[1*] and Yannis P. Korkolis[2]

[1]Dept. of Mechanical and Aerospace Engineering, University of Kentucky, Paducah, KY, USA
[2]Institut für Umformtechnik und Leichtbau (IUL), TU Dortmund, Dortmund, 44227, Germany

*Email: madhav.baral@uky.edu

Abstract. The ductile fracture behaviours of two tubular specimens, a AA6260-T4 macrotube (60 mm diameter and 2 mm thickness), and a SS-304L microtube (2.38 mm diameter and 0.15 mm thickness) under non-proportional loading condition are studied using a combined experimental-numerical approach. The experiments are conducted by loading the tubes under axial force and internal pressure along various non-proportional (i.e., corner) paths until failure by controlling the force/pressure ratio. The plastic behaviours of the tubes are characterized using the non-quadratic anisotropic yield criteria Yld2000-2D and Yld2004-3D. The material models are employed in the finite element (FE) simulation of the tube experiments using Abaqus/Standard (implicit). The FE models, which include thickness imperfection to capture the failure modes observed in the experiments, are then used to probe the fracture parameters inside the tube wall where fracture is likely to initiate. It is observed that the fracture strains from the non-proportional loading results are noticeably different when introduced to the proportional fracture strains, revealing the path-dependence fracture behaviour in tubular specimens.

1. Introduction

Fracture is the separation of a material into two or more pieces under the action of stress. Ductile fracture is one of the basic mechanisms of material failure, that terminates the useful life of a product or makes its manufacture unfeasible. Thus, the establishment of the fracture locus of a given material is critical in design and optimization of parts and products. From the micromechanical point of view, ductile fracture is associated with void nucleation, growth, and coalescence of voids, leading to a crack [1-3]. This has been extensively studied both numerically and experimentally [4-10]. The fracture characterization is often combined with experimental and numerical analyses because of non-homogenous deformation and fracture initiation from inside of the specimen [8-10]. The numerical method requires advanced plasticity models that can be implemented in finite element (FE) studies [11-13]. Typically, the fracture locus of a material is determined from experiments conducted under proportional loadings [8-10, 14]. Studies have shown that fracture behaviour can be path dependent [15]. Despite this understanding, the fracture locus under nonproportional loadings have received limited attention in the literature.

In this study, the ductile fracture of two tubular specimens, AA6260-T4 macrotube (60 mm diameter and 2 mm thickness), and SS-304L microtube (2.38 mm diameter and 0.15 mm thickness), under nonproportional loadings are investigated using a combined experimental-numerical approach. The experiments reported here are taken from the earlier work of the authors [11, 16]. Section 2 details the experimental procedure including the prescribed non-proportional (i.e., corner) loading paths. Section 3 includes the modelling of the plastic behaviours of the tubes using non-quadratic anisotropic yield criteria. Section 4 details the finite element (FE) models of the tubes including the comparison of the predictions with the experiments. Section 5 shows the fracture strains and loading paths from the nonproportional and proportional loadings to establish the path-dependence fracture behaviour of the tubes. Section 6 summarizes the main conclusions and findings of this study.

2. Experiments

The tubes were loaded under axial force and internal pressure using respective apparatuses that can apply axial-force and internal-pressure in a fully coupled way, so that the non-proportional loading paths can be followed. The corner paths (perpendicular stress-paths in the axial/meridional-hoop stress space) are designed by selecting the axial (meridional in microtube) or hoop stress at failure of a proportional path and using this as the predefined stress at the corner. The axial and hoop stress in the macrotube is calculated from the measured pressure and force considering the thin-walled cylinder theory. However, the microtubes deform into an axisymmetric bulge for some of the stress paths followed, having curvatures in both hoop and meridional directions. Due to this, the membrane stresses are referred as meridional and hoop stresses and are calculated based on the equilibrium requirements for a material element at the centre of the specimen.

The non-proportional stress paths in the nominal stress space from the AA6260-T4 macrotube and the SS-304L microtube are shown in Figures. 1a and 1b respectively.

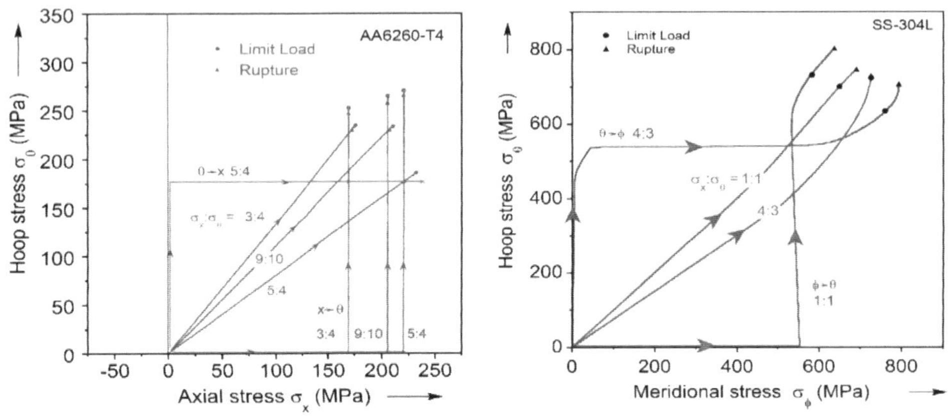

Figure 1 – Non-proportional loading paths in the engineering stress space, including the corresponding proportional ones for (a) AA6260-T4 tube and (b) SS-304L tube. Note the different scales of the axes in the two plots.

The two types of paths, i.e., x → θ and θ → x are included along with the corresponding proportional stress paths (note "x" is replaced with "φ" in the microtube). For the macrotube, four corner paths (three along x → θ and one along θ → x) are shown in Figure. 1a. For the microtubes, only the extreme cases of two corner paths, i.e., 1:1 x → θ and 4:3 θ → x from the experiments are shown. As seen in the Figure. 1a, the engineering stress at fracture can depend significantly on the loading path, e.g., $\sigma_x = 222$ MPa and $\sigma_\theta = 176$ MPa for σ_x: $\sigma_\theta = 5{:}4$ with 220 MPa and 264 MPa for the corresponding x → θ path in the

IOP Conf. Series: Materials Science and Engineering **1307** (2024) 012025 doi:10.1088/1757-899X/1307/1/012025

macrotube. Unlike the macrotube, the stresses at the corner of non-proportional paths for the microtubes deviate comparatively from the intended failure stress of the proportional paths, as seen in Figure. 1b.

3. Constitutive modelling

The constitutive model adopted plays an important role in the accurate prediction of the fracture strains during numerical simulations. The plastic behaviour of the macrotube is modelled using two nonquadratic, anisotropic yield criteria, i.e., Yld2000-2D and Yld2004-3D while that of the microtube is modelled with Yld2004-3D criterion [13]. Both 2D and 3D yield criteria are needed in the FE simulation of the macrotube, since that model consists of both shell and solid sections while the microtube consists of only solid section, as described in the next section. A rate-independent associated flow rule and the isotropic hardening assumption is implemented in the respective models. The anisotropy parameters are chosen for the best fit of the experimental data obtained from the proportional tests using least-squares method. The corresponding yield loci at the chosen plastic work contour level are shown in Figures. 2a and 2b respectively.

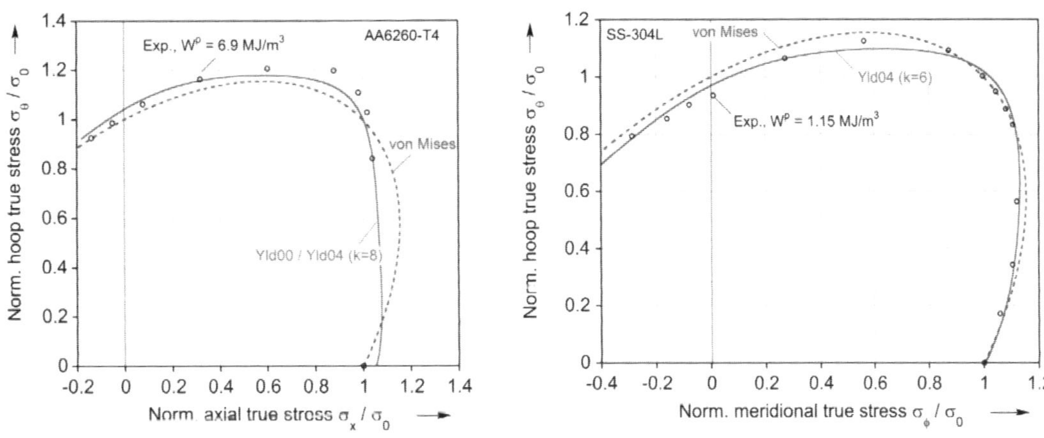

Figure 2 – Yield loci for (a) AA6260-T4 tube and (b) SS304-L tube, including the experiments and the von Mises and the Yld2000-2D/Yld2004-3D criteria.

4. Finite Element (FE) Simulations

The finite element (FE) models of the non-proportional loading experiments are developed in the nonlinear code Abaqus/Standard ver. 6.13-3 using the material modelling framework described above. To take advantage of the symmetry, only 1/4 of the tube is meshed. The macrotube is meshed with linear, reduced-integration solid elements (C3D8R) and shell elements (S4R) as shown in Figure. 3a. A surface-based shell-to-solid coupling is implemented in the FE model, to enable a fully-3D analysis at the local fracture initiation site, while the rest of the structure is represented as a shell, to increase computational efficiency. The microtube is meshed with the solid elements (C3D8R) as shown in Figure. 3b. A local imperfection in the form of an axial or circumferential groove of reduced thickness is added in the model to simulate the corresponding failure mode observed in most experiments (Note: only the axial imperfection model shown in Figs. 3a and b) [14, 17].

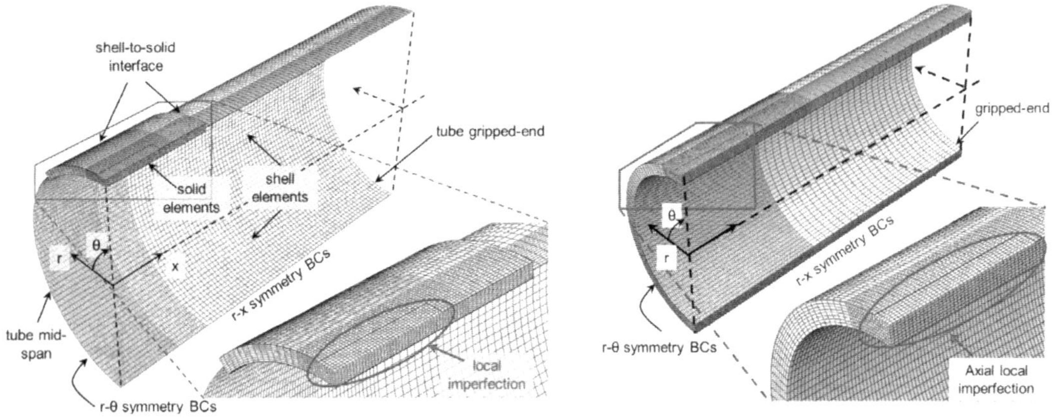

Figure 3 – FE models including the imperfection and meshes for (a) AA6260-T4 tube and (b) SS-304L tube [14, 17]

The results from the FE models are now compared to experiments to establish the fidelity of the numerical models. The predictions from one of the non-proportional stress paths, along with the experimental responses for the macrotube and the micro tube are shown in Figures. 4a and 4b respectively. As seen in the figures, the comparison of the axial (meridional in microtube) responses is more favourable than the hoop responses, which can possibly be attributed to the isotropic hardening assumption implemented in the FE model.

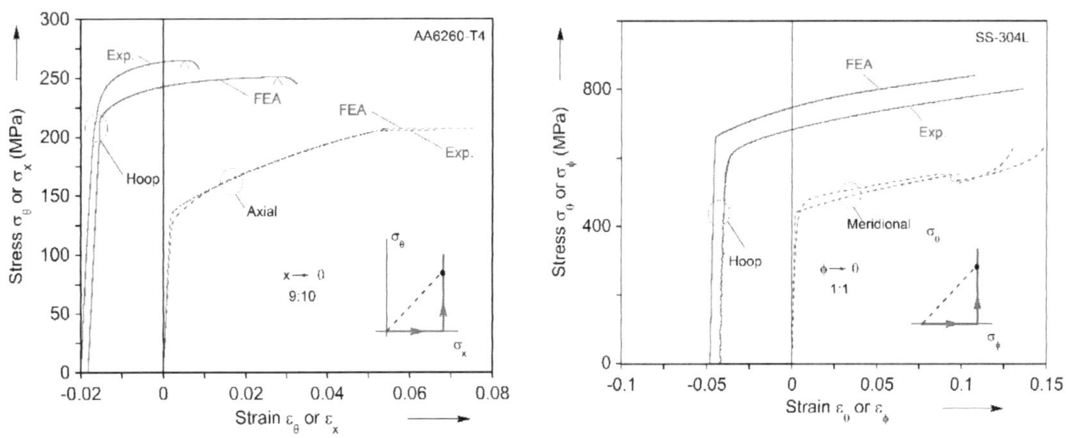

Figure 4 – Comparison of stress-strain curves from FE predictions and experiments (a) AA6260-T4 tube and (b) SS-304L tube.

Furthermore, the performance of these models is assessed by comparing the induced strain paths from FE simulations with the corresponding experiments. The results from the macrotube and microtube are shown in Figures. 5a and 5b respectively. The end of each path in the figures corresponds to the load maximum. The model reproduces the cusps observed at the path change, due to the elastic deformation occurring right after the change, and before re-yielding commences. As seen in the figures, the comparison between experiment and analysis is favourable for each path.

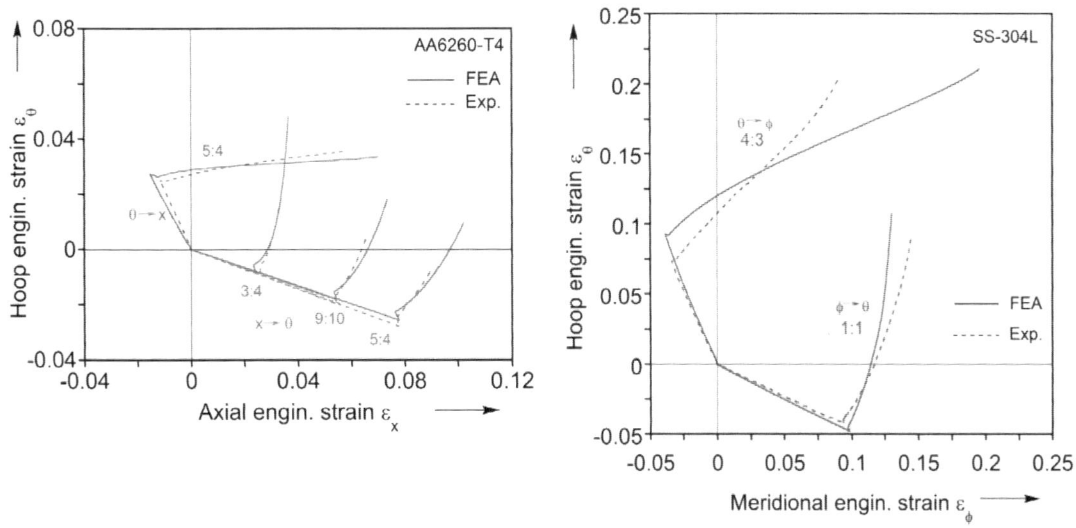

Figure 5 – Comparison of induced strain paths from FE predictions and experiments for (a) AA6260-T4 tube and (b) SS-304L tube.

5. Ductile fracture

The loading paths until the limit-load are probed based on the predictions of the FE models. The fracture parameters, i.e., stress triaxiality, Lode-angle-parameter and equivalent plastic strain are extracted from a material point adjacent, but outside of the imperfection region at mid-thickness. The evolution of equivalent plastic strain is shown with respect to stress triaxiality and Lode-angle-parameter for the macrotube and the microtube in Figures. 6 a and b respectively. These paths can be visualized in twodimensional space as triaxiality and Lode-angle-parameter vs. equivalent plastic strain, as shown in Figures. 7 a and b for the macrotube and Figs. 8 a and b the microtube respectively. The non-proportional results of Figures. 7 and 8 show clearly the abrupt path change that characterizes them. As seen in the figures, the loading paths from the non-proportional results (solid lines) are noticeably different from their proportional counterparts (dashed lines), indicating the path-dependent fracture behaviour of the tubes.

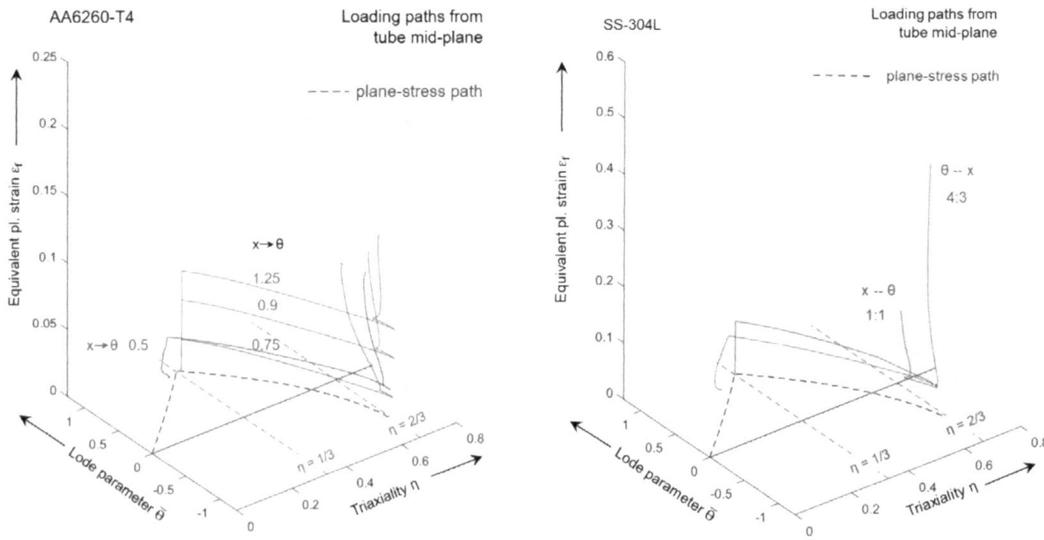

Figure 6 – Loading paths in the stress-triaxiality – Lode angle parameter – equivalent plastic strain space for (a) AA6260-T4 tube and (b) SS-304L tube.

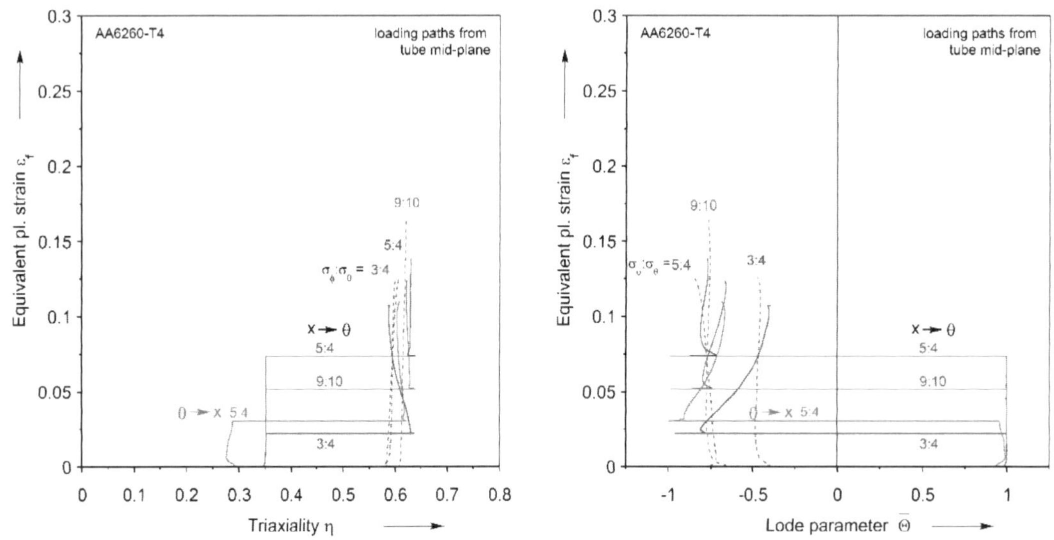

Figure 7 – Projection of Figure 6 (a) in the (a) stress triaxiality – equivalent plastic strain and, (b) Lode angle parameter – equivalent plastic strain spaces, along with the corresponding proportional paths for AA6260-T4 tube.

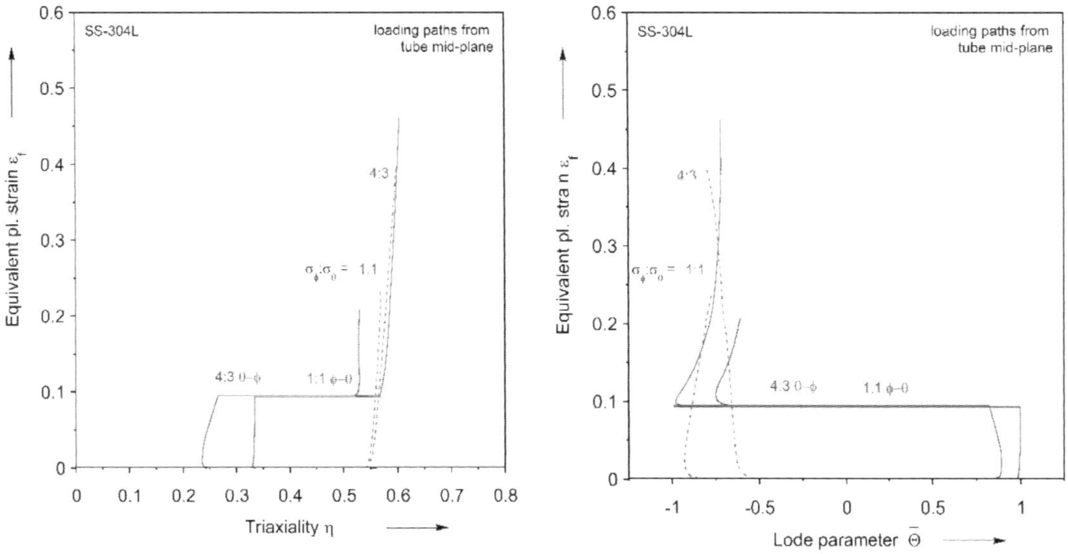

Figure 8 – Projection of Figure 6 (b) in the (a) stress triaxiality – equivalent plastic strain and, (b) Lode angle parameter – equivalent plastic strain spaces, along with the corresponding proportional paths for SS-304L tube.

6. Summary and Conclusions

The ductile fracture behaviour of two tubular specimens, AA6260-T4 macrotube and SS-304L microtube, were studied using combined experimental-numerical approach. The tubes were loaded until fracture under axial force and internal pressure along non-proportional loading paths. The FE models of the tubes were used to obtain the fracture parameters, i.e., triaxiality, Lode angle parameter and equivalent plastic strain to study the fracture behaviour. The results from the non-proportional loadings show the differences in the obtained loading paths when compared to the proportional results, indicating the possible path-dependent fracture behaviour of the tubular materials. In the future, the fracture strains from the non-proportional loadings will be introduced to the proportional fracture locus to quantify these deviations and capture the essential features of path-dependent ductile fracture behaviour.

References

[1] Rice, J.R., Tracey, D.M., 1969. On the ductile enlargement of voids in triaxial stress fields. J. Mech. Phys. Solids.
[2] Tvergaard, V., Needleman, A., 1984. Analysis of the cup-cone fracture in a round tensile bar. Acta Metall. 32, 157–169
[3] Benzerga, A.A., 2002. Micromechanics of coalescence in ductile fracture. J. Mech. Phys. Solids 50, 1331–1362.
[4] Benzerga, A.A., Surovik, D., Keralavarma, S.M., 2012. On the path-dependence of the fracture locus in ductile materials – Analysis. Int. J. Plast. 37, 157–170
[5] Bao, Y., Wierzbicki, T., 2004. On fracture locus in the equivalent strain and stress triaxiality space. Int. J. Mech. Sci. 46, 81–98.
[6] Coppola, T., Cortese, L., Folgarait, P., 2009. The effect of stress invariants on ductile fracture limit in steels. Eng. Fract. Mech. 76, 1288–1302.

[7] Mohr, D., Marcadet, S.J., 2015. Micromechanically-motivated phenomenological HosfordCoulomb model for predicting ductile fracture initiation at low stress triaxialities. Int. J. Solids Struct., 67–68, 40–55.
[8] Ha, J., Baral, M., Korkolis, Y.P., 2019. Ductile fracture of an aluminum sheet under proportional loading. J Mech Phys Solids 132, 103685.
[9] Baral, M., Ha, J., Korkolis, Y.P., 2019. Plasticity and ductile fracture modeling of an Al–Si–Mg diecast alloy. Int J Fract.
[10]Ha, J., Baral, M., Korkolis, Y.P., 2018. Plastic anisotropy and ductile fracture of bake-hardened AA6013 aluminum sheet. Int. J. Solids Struct.
[11]Korkolis, Y.P., Kyriakides, S., 2008. Inflation and burst of anisotropic aluminum tubes for hydroforming applications. Int. J. Plast.
[12]Korkolis, Y.P., Kyriakides, S., 2008b Inflation and burst of aluminum tubes. Part II: An advanced yield function including deformation-induced anisotropy Int. J. Plast.
[13]Barlat, F., Aretz, H., Yoon, J.W., Karabin, M.E., Brem, J.C., Dick, R.E., 2005. Linear transformation-based anisotropic yield functions. Int. J. Plast.
[14]Baral, M., Korkolis, Y.P., 2021. Ductile fracture under proportional and non-proportional multiaxial loading. Int J Solids Struct 210–211, 88–108.
[15]Thomas, N., Basu, S., & Benzerga, A. A. (2016). On fracture loci of ductile materials under nonproportional loading. International Journal of Mechanical Sciences, 117, 135-151.
[16]Ripley, P.W., Korkolis, Y.P., 2016. Multiaxial deformation apparatus for testing of microtubes under combined axial-force and internal-pressure. Exp Mech 56, 273–286.
[17]Baral, M., 2020. Experiments and Modelling of Plastic Anisotropy and Ductile Fracture Under Multiaxial Loading (Doctoral dissertation, University of New Hampshire).

Acknowledgments
This research was partially supported by the U.S. National Science Foundation through grants CMMI1563216 and 1929873. We also thank the Dept. of Mechanical Engineering at the University of New Hampshire for support during the course of this work. These supports are acknowledged with thanks.

IOP Conf. Series: Materials Science and Engineering **1307** (2024) 012026 doi:10.1088/1757-899X/1307/1/012026

Study on electric pulse-assisted plastic deformation behavior of 5182-O aluminum alloy

Hongchun Shang[1], Songchen Wang, Yanshan Lou[1]*

[1] School of Mechanical Engineering, Xi'an Jiaotong University, 28 Xianning West Road, Xi'an, Shaanxi 710049, China

ys.lou@xjtu.edu.cn

Abstract. Plastic model-based neural networks are emerging phenomenological models with excellent calibration accuracy and interpretability of parameters, and they do not significantly increase the finite element calculation time of neural network models with a simple structure optimized by the algorithm. Because aluminum and magnesium have low ductility at room temperature, complex components made from these alloys need to be forged at high temperatures. In comparison with traditional thermoforming, advanced current-assisted processing provides energy savings, efficiency and green production. Experimental and constitutive modeling are carried out in this study to investigate the coupling effects of electrical pulse, temperature, strain rate, and strain on flow behavior and ductile fracture. It has been shown that electric pulses induce Joule heating effects and electroplastic effects, which reduce deformation resistance and improve formability. Electrical pulses suppress negative strain rate effects caused by dynamic strain aging, which is the main reason for the non-monotonic relationship between temperature and strain rate. The behavior of plasticity and fracture initiation can be described by simulations based on a neural network-based evolving plasticity model that incorporates stress states, current density, temperature, and strain rates.

1. Introduction

Lightweighting is necessary to achieve energy saving and emission reduction in transportation, aerospace and other fields, and is also an inevitable requirement of China's carbon peak and carbon neutrality strategy [1-3]. Due to its low density, high specific strength, high specific stiffness, and strong corrosion resistance, 5182-O aluminum alloy is widely used in lightweight equipment. Although 5182-O aluminum alloy exhibits poor ductility and a large number of macroscopic defects, it is caused by the Portevin-Le Chatelier (PLC) effect [4]. Thus, how to enhance the plastic stability of 5182-O aluminum alloy by suppressing the PLC effect is currently a difficult and hot issue. Recent studies show that electric pulse-assisted forming reduces manufacturing difficulty, improves forming accuracy, improves microstructure, optimizes mechanical properties, and improves surface quality [5]. It is still difficult to identify how the electroplastic effect works and the impact of electric pulse-assist forming on plastic behavior, making it difficult to guide high-quality aluminum alloy forming [6].

Several factors contribute to the significant increase in manufacturing costs associated with thermoforming, including high energy consumption, long production cycles, complex working conditions, and high temperature resistance of mold materials. The most efficient and energy-saving way to achieve high quality and efficiency is by utilizing advanced electrically assisted processing. Free electrons move within metallic materials when electrical pulses are applied. Joule heating,

IOP Conf. Series: Materials Science and Engineering **1307** (2024) 012026 doi:10.1088/1757-899X/1307/1/012026

electromigration, skin effects, and other effects can occur as drift electrons interact with the metal's internal structure [7]. There are also electroplastic and polar effects that can be produced by coupling stress and electric fields [8]. As a result, metals develop internal structural characteristics that differ from those that are found in conventional processing and deformation techniques [9]. Electrically assisted processing involves many coupling factors, so it is critical for numerical simulation to take into account the effects of electric pulses and other factors. The modified Hollomon model and Johnson-Cook model are used to analyze and discuss the mechanical response of CP titanium during electrically assisted deformation [10]. The thermodynamic and mechanical behavior of AZ31 magnesium alloy at different densities can be predicted using an analytical model that considers Joule heat, solute dislocation interaction, and electron wind [11]. Lou et al. proposed fracture criteria of the DF series to simulate ductile fracture in the metal forming process [12-14]. In this study, the electric pulse-assisted plastic deformation behavior and electroplastic deformation mechanism of 5182-O aluminum alloy were studied by the fully connected neural network (FCNN) and traditional models.

2. Experiments

The electroplasticity, temperature and strain rate effects on the plastic behavior of 5182-O aluminum alloy were investigated by isothermal tensile test (ITT) and electrically-assisted isothermal tensile test (EAITT) at different stress states. Inverse engineering method is combined with neural network-based plasticity models to characterize coupling effects [15]. The FCNN model can predict the plastic behavior under large strains based on the stress-strain curves calibration obtained from the inverse engineering method [16]. In order to reveal how electric pulse improve ductility at different temperatures, the non-monotonic effect of fracture loading paths was analyzed experimentally and numerically. ABAQUS/Explicit is used to simulate ductile fracture and dynamic hardening behavior by using the FCNN model that is based on DF2014 [17].

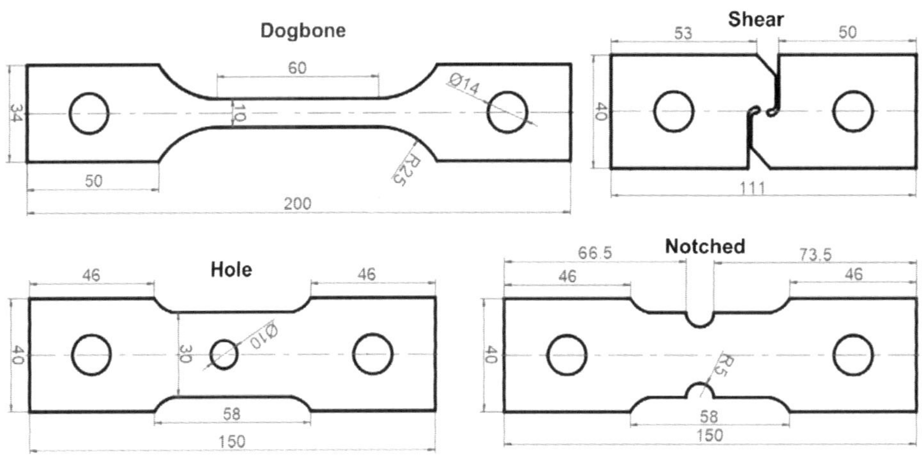

Figure 1. Specimen geometries [unit: mm].

Figure 1 shows the specimen's dimensions. A series of uniaxial tensile tests in the RD direction were conducted under a variety of stress states to determine the mechanical behavior of the material. To achieve strain rates of 0.001/s, 0.01/s and 0.1/s, the tensile speeds of dogbone specimens were set to 3.6, 36 and 360mm/min, and the tensile speeds of center hole, R5 notched and plane shear specimens are 0.5, 5 and 50 mm/min. The virtual extensometer of the dogbone specimen measures 30mm, while the extensometers of the other three specimens measure 20mm.

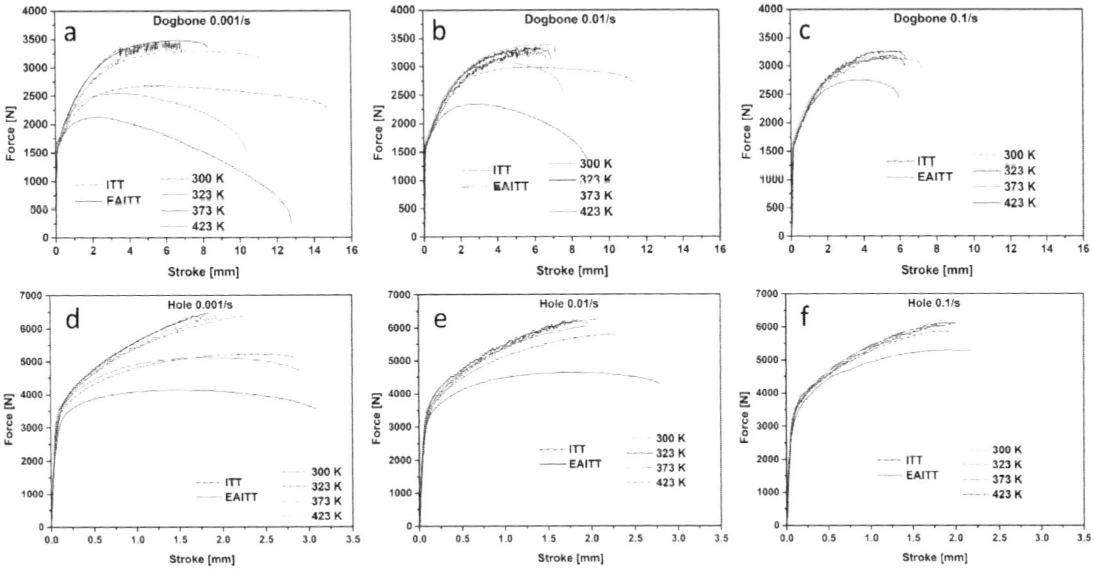

Figure 2. Experimental load-displacement curves obtained by ITT and EAITT: (a-c) Dogbone, (d-f) Hole.

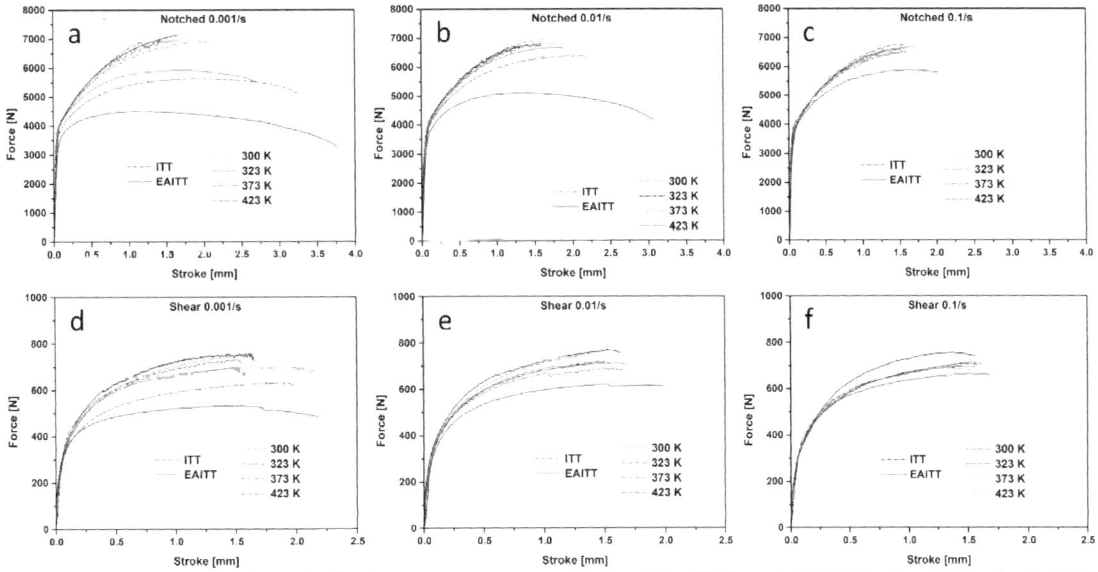

Figure 3. Experimental load-displacement curves obtained by ITT and EAITT: (a-c) Notched, (d-f) Shear.

In order to determine the effect of temperature and strain rate on load displacement, ITT and EAITT implemented a load-displacement curve for dogbone, center hole, notched, and shear specimens. The reliability of experimental data was verified by repeating all experiments at least three times. The onset of fracture is defined as a sharp decrease in force value of the force-displacement curve. It is evident from a comparison of different strain rates that non-monotonic dependence on the test temperature is demonstrated for all four specimens at low strain rates. However, fracture initiation at high strain rates is not significantly different from dynamic hardening behavior. Despite the negative strain rate sensitivity of the 5182-O aluminum alloy, the electrical pulse promotes the positive strain rate effect above 373 K, according to the results. It has been demonstrated that a high strain rate improves formability at 423 K by increasing necking strain, which indicates that high strain rates improve formability at 423 K. Some experiments found smaller serrations in force-displacement curves, particularly when shear specimens were subjected to large strains. Dynamic strain aging is widely accepted as the explanation for the PLC effect. Hardening behavior and fracture ductility are very complexly influenced by strain rate, temperature, and electrical pulses.

3. Calibration of fracture behavior

The fracture behavior is illustrated in Figure 4 under various loading conditions, including shear, uniaxial tension, plane strain tension, and equilibrium biaxial tension [18,19]. In comparison of formability prediction results at three strain rates, EAITT and ITT showed the greatest difference at 0.001/s. At a strain rate of 0.01/s, EAITT fracture strain in the 423 K plane stress tensile state is 1.7 times that of ITT, indicating that at the same strain rate, high temperature further enhances ductility. A high strain rate results in a short action time for temperature and current, so the difference between calibrated forming limit curves at 0.1/s is relatively small. There is a strong relationship between current density, strain rate, heat, and other factors in 5182-O aluminum alloy that complicates ductile fracture. However, the DF2014 criterion cannot cope with the influence of strong coupling effects, which can exacerbate fracture characteristics under a wide range of loading conditions. To achieve better prediction results under complex loading conditions, it is recommended to adopt a fracture model based on neural networks.

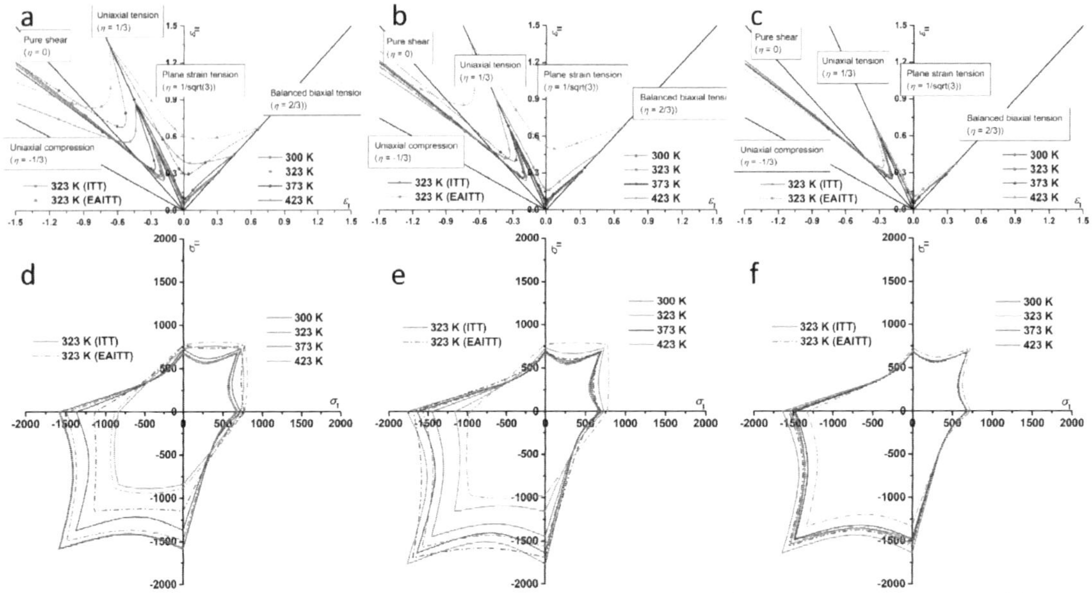

Figure 4. Fracture curves of ITT and EAITT in stress space and strain space constructed by the DF2014 criterion: (a), (d) 0.001/s; (b), (e) 0.01/s and (c), (f) 0.1/s.

FCNN is considered as a highly nonlinear model composed of different activation functions, which can be applied to describe the macroscopic response of materials under complex loading conditions. Moreover, the FCNN model can be further exploited when experimental data and traditional constitutive models are combined. Due to strong coupling effects in 5182-O aluminum alloy, the FCNN model based on DF2014 fracture criterion accurately describes its highly nonlinear properties. PSO algorithm and FCNN work together to achieve global optimization and high search efficiency simultaneously. A constant learning rate of 0.01 and a constant momentum of 0.9 are used to achieve fast convergence speed with training accuracy, while the number of iterations is set to 3000 to prevent overfitting. The weights and thresholds are then iteratively updated after the structure and parameters are set. Single-layer FCNN models optimized with PSO algorithm are highly accurate and efficient in simulation [15]. In this experiment, 90% of the experimental data are used for training and 10% for testing.

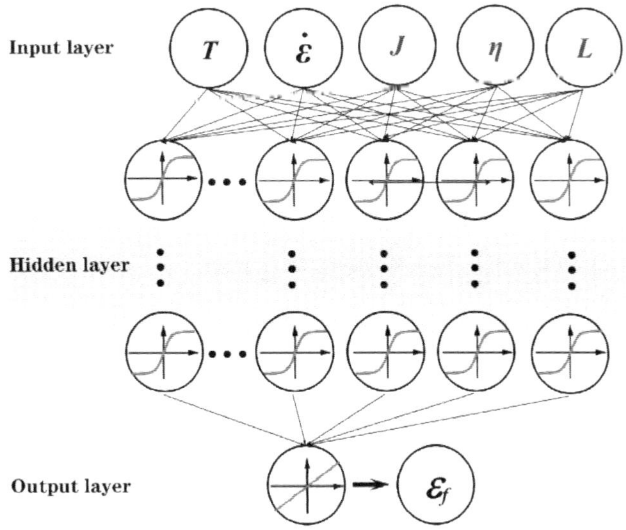

Figure 5. Structural diagram of FCNN model.

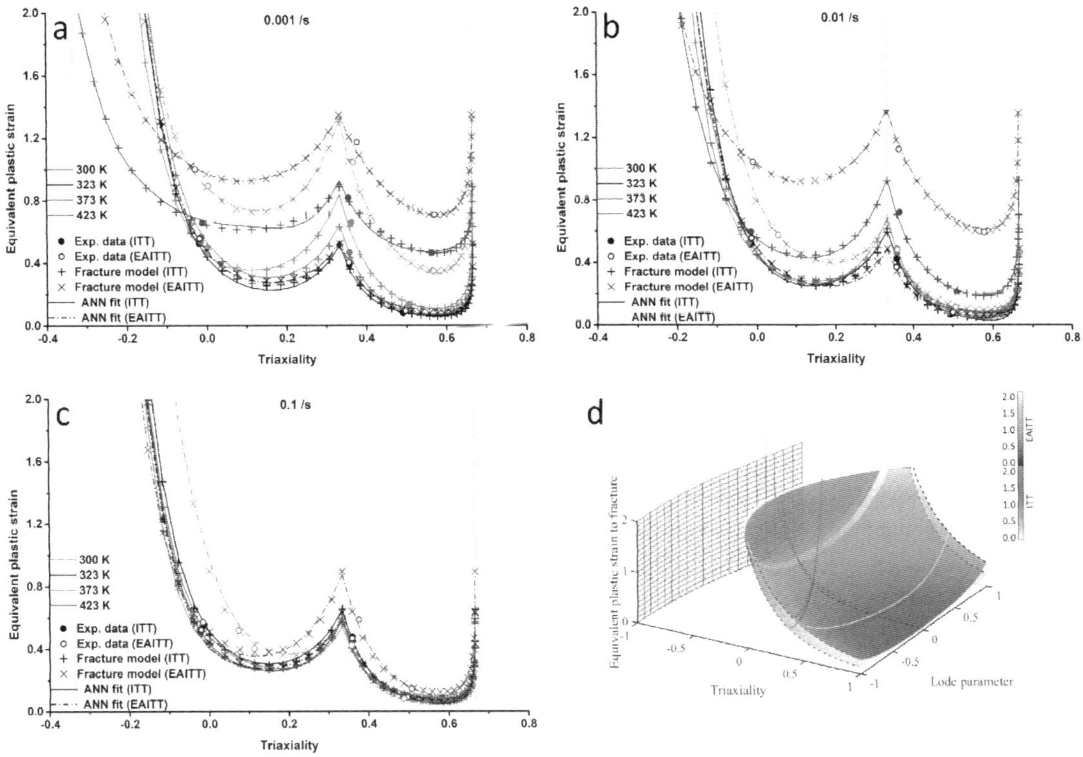

Figure 6. ANN predicted fracture loci for plane stress conditions at a strain rate of (a) 0.001/s, (b) 0.01/s and (c) 0.1/s. (d) Fracture surfaces of ITT and EAITT at 423 K-0.1/s.

The FCNN model for predicting fracture behavior is shown in figure 5. The fracture locations predicted by the FCNN model for the ITT and EAITT in the plane stress state are shown in figure 6 (a), (b), and (c). DF2014 criterion calibration results and FCNN model prediction results are represented by crossed lines and points, respectively. The fracture strain is well approximated by the model over the range of strain rates and temperatures tested. The FCNN model can accurately capture the negative strain rate effect of 5182-O aluminum alloy from Figure 6 (a). The comparative results of the three groups at different strain rates in Figure 6 (a), (b) and (c) show that the difference between the results obtained by ITT and EAITT is the largest at 423 K, suggesting that electroplastic effects on ductile fracture are more significant at higher currents. Figure 6 (d) illustrates that the 5182-O aluminum alloy exhibits good ductility under low stress triaxiality, and the difference in fracture strain is small at lower temperatures between ITT and EAITT.

4. Conclusion

The safety performance of the product depends on avoiding fractures as much as possible when achieving lightweight manufacturing. It is difficult to realize lightweight concepts such as plastically formed parts for 5182-O aluminum alloy due to its fracture failure. Finite element analysis of plastic forming processes can be improved by using advanced constitutive models to accurately predict plastic deformation and fracture behavior under complex loading conditions.

References

[1] Thackeray M. M., Wolverton C. and Isaacs E. D. 2012 Electrical energy storage for transportation- approaching the limits of, and going beyond, lithium-ion batteries *Energy Environ. Sci.* **5** 7854-7863.

[2] Zheng K., Politis D. J., Wang L. and Lin J. G. 2018 A review on forming techniques for manufacturing lightweight complex-shaped aluminium panel components *International Journal of Lightweight Materials and Manufacture* **1** 55-80.

[3] Henriksson F. and Johansen K. 2016 On Material Substitution in Automotive BIWs-From Steel to Aluminum *Body Sides Procedia CIRP* **50** 683-688.

[4] Min J. Y., Hector Jr L. G., Carsley J. E., Stoughton T. B., Carlson B. E. and Lin J. 2015 Spatio-temporal characteristics of plastic instability in AA5182-O during biaxial deformation *Mater. Des.* **83** 786-794.

[5] Nguyen-Tran H. D., Oh H. S., Hong S. T., Han H. N., Cao J., Ahn S. H. and Chun D. M. 2015 A review of electrically-assisted manufacturing *INT J PR ENG MAN-GT* **2** 365-76.

[6] Shang H. C., Wang S. C. and Lou Y. S. 2024 Electroplastic effect on plastic deformation of 5182-O aluminum alloy: experiments and ANN based modelling/analytical modeling *T NONFERR METAL SOC* (Preprint).

[7] Conrad H. 2002 Thermally activated plastic flow of metals and ceramics with an electric field or current *Mater. Sci. Eng. A* **322** 100-107.

[8] Antolovich S. D. and Conrad H. 2004 The Effects of Electric Currents and Fields on Deformation in Metals, Ceramics, and Ionic Materials: An Interpretive Survey *Mater. Manuf. Process.* **19** 587-610.

[9] Pan D., Zhao Y., Xu X., Wang W. and Chong X. 2019 A novel strengthening and toughening strategy for T250 maraging steel: Cluster-orientation governed higher strength-ductility combination induced by electropulsing *Mater. Des.* **169** 107686.

[10] Magargee J., Morestin F. and Cao J. 2013 Characterization of Flow Stress for Commercially Pure Titanium Subjected to Electrically Assisted Deformation *J Mater Sci Technol* **135** 041003.

[11] Wang X., Xu J., Shan D., Guo B., and Cao J. 2016 Modeling of thermal and mechanical behavior of a magnesium alloy AZ31 during electrically-assisted micro-tension *Int. J. Plast.* **85** 230-257.

[12] Lou Y., Huh H., Lim S. and Pack K. 2012 New ductile fracture criterion for prediction of fracture forming limit diagrams of sheet metals *Int J Solids Struct* **49** 3605-3615.

[13] Lou Y., Yoon J.W. and Huh H. 2014 Modeling shear ductile fracture of advanced metals considering a changeable cut-off value for the stress triaxiality *Int. J. Plast.* **54** 56-80.

[14] Lou Y., Chen L., Clausmeyer T., Tekkaya A.E. and Yoon J.W. 2017 Modeling of ductile fracture from shear to balanced biaxial tension for sheet metals *Int J Solids Struct* **112** 169-184.

[15] Shang H. C., Wu P. F., Lou Y. S.,Wang J. Z. and Chen Q. 2022 Machine learning-based modeling of the coupling effect of strain rate and temperature on strain hardening for 5182-O aluminum alloy *J Mater Process Technol* **302** 117501.

[16] Shang H. C., Zhang C., Wang S. C. and Lou Y. S. 2022 Large strain flow curve characterization considering strain rate and thermal effect for 5182-O aluminum alloy *Int. J. Mater. Form.* **16** 1-20.

[17] Shang H. C., Wang S. C., Zhou L. C. and Lou Y. S. 2023 Neural network-based ductile fracture model for 5182-O aluminum alloy considering electroplastic effect in electrically-assisted processing *Eng Fract Mech* **290** 109476.

[18] Lou Y. S. and Huh H. 2013 Prediction of ductile fracture for advanced high strength steel with a new criterion: Experiments and simulation *J Mater Process Technol* **213** 1284-1302.

[19] Lou Y. S. and Huh H. 2013 Extension of a shear-controlled ductile fracture model considering the stress triaxiality and the Lode parameter *Int J Solids Struct* **50** 447-455.

IOP Conf. Series: Materials Science and Engineering **1307** (2024) 012027 doi:10.1088/1757-899X/1307/1/012027

Study on GISSMO fracture model of hot-stamped Tailored-Tempering-Properties parts

B Y Qiu[1], Y L Wang[1]*, B Zhu[1], W Liu[1], Y S Zhang[1], X Q Ren[2] and X P Ren[2]

[1] State Key Laboratory of Materials Processing and Die & Mould Technology Huazhong University of Science and Technology, Wuhan 430074, China
[2] DongYing Goodthinking Robot Co., Ltd, Dongying 257000, China

*wangyilin@hust.edu.cn

Abstract. To accommodate the varying mechanical demands in different areas of vehicle components, the Tailored-Tempering-Properties technology of hot-stamped parts of high-strength steel came into being. This research crafted a segregated hot and cold partition die system to fabricate U-shaped parts from 22MnB5 with Tailored-Tempering-Properties. The sides and bottom surfaces of the hard and soft zones of the U-shaped parts were sampled, and uniaxial tensile experiments were carried out to obtain the stress-strain curve, tensile strength, and elongation. In addition, notch, shear, and tensile shear sampling were carried out in the soft area at the bottom flange of the U-shaped part, and the force-displacement curve and fracture strain under different stress conditions were determined by tensile test. The GISSMO fracture model was selected to simulate the failure fracture of the material, and the parameters were obtained by Hypermesh pretreatment, LS-DYNA solution calculation, and LS-OPT optimization. Finally, the GISSMO fracture model suit for the soft zone of 22MnB5 U-shaped parts was obtained to characterize and predict its fracture behavior, which is an essential basis for guiding the design and optimization of parts.

1. Introduction

Today, lightweight, energy efficiency, high efficiency, and safety have become the development trends in various industrial sectors. High-strength steel, characterized by its high tensile strength and good plasticity, can enhance components' load-bearing capacity and safety. It is widely used in automotive manufacturing, aerospace, building structures, and other fields. Within the manufacturing realm, the emergence of Tailored-Tempering-Properties technology fulfills the distinct mechanical property requisites across various component zones, thereby augmenting their safety performance. In automotive manufacturing, high-strength steel materials, which serve as critical structural components of the vehicle body, endure tremendous impact energy. This significantly increases the risk of fracture failure during collision events. Research into characterizing and predicting the fracture performance of high-strength steel materials is a crucial basis for guiding automotive component design, optimizing, and enhancing overall vehicle safety.

The concept of damage was initially proposed by Kachanov [1] while studying the fracture of metals caused by creep, referring to the gradual development of macroscopic defects from microscopic flaws under external influences [2]. Building upon this, Robotnov [3] introduced the concept of the damage factor, D. Johnson and Cook [4] utilized the stress triaxiality parameter η to characterize the complex stress states of materials and proposed the Johnson-Cook model, abbreviated as the J-C model. To further enhance the accuracy of collision impact fracture simulations in metal structures, Neukamm [5],

based on the J-C model, introduced instability criteria and fracture standards related to the loading path, addressing issues with nonlinear strain path loading, and proposed a stress-state-dependent generalized incremental damage model (referred to as the GISSMO model). Acquiring parameters for this model is relatively simple and has a clear physical meaning. It can be coupled with various material constitutive relationships [6]. Highly flexible and practical in finite element simulation applications, the GISSMO model demonstrates high simulation accuracy and finds widespread application in fields such as automotive collisions and metal forming.
This study designed and implemented hot stamping cold-hot zone molds to manufacture variable-strength U-shaped parts using 22MnB5 high-strength steel. The U-shaped parts were formed through hot stamping at 500°C. Uniaxial tensile tests were conducted on samples from the side and bottom surfaces to represent the hard and soft zones, respectively. Stress-strain curves and tensile strengths were obtained. Additional tensile tests were performed under four different stress states in the soft zone to determine force-displacement curves and fracture strains. The GISSMO fracture model was applied for material failure and fracture simulation. The experiment involved preprocessing with Hypermesh, solving and calculating using LS-DYNA, and optimizing fracture parameters via LS-OPT, ultimately yielding the GISSMO fracture model for the soft zone of 22MnB5.

2. Experimental methods

2.1. Experimental material

The steel used in this experiment is the non-coated New Uncoated Steel CR1000/1500HS plate, 22MnB5, composed of ferrite and pearlite structures. The dimensions of the plate are 820 × 426 × 1.6 mm. Its chemical element composition is presented in Table 1, and according to the technical datasheet, the yield strength and ultimate tensile strength are 360 MPa and 530 MPa, respectively.

Table 1. Chemical composition of 22MnB5 (wt. %).

C	Si	Mn	P	S	Cr	Ti	B
0.209	0.249	1.342	0.009	0.006	0.204	0.031	0.002

2.2. Experimental apparatus and hot stamping forming process

The experiment employs a U-shaped strength hot-forming die developed independently, as shown in Figure 1. Due to varying temperatures during the quenching process, the die is divided into a cold die part and two hot die parts. The middle part serves as the cold die, corresponding to the hard zone of the formed part, while the two sides function as the hot die, corresponding to the soft zone of the formed part. There are two transitional zones where the U-shaped part transitions from the hard zone to the soft zone. Their properties lie between those of the hard and soft zones. The left transitional zone measures 40 mm, and the correct transitional zone measures 60 mm.
As shown in Figure 2, the hot stamping forming process involves heating the sheet to 930°C and holding it for 330 seconds. After complete Austenitization, within 10 seconds, it is transferred from the roller hearth furnace to the die cavity. It is quenched and formed under a pressure of 300 tons for 15 seconds, then removed from the die cavity.

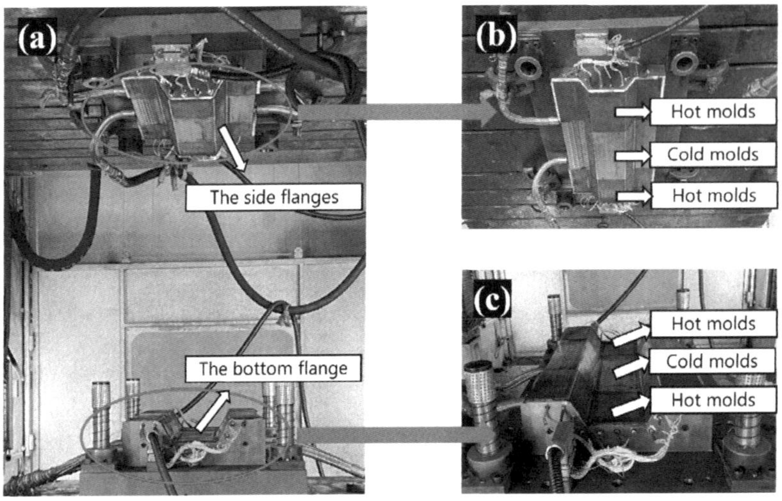

Figure 1. Hot-stamping equipment: (a) U-shaped strength hot stamping device, (b) upper die, (c) lower die.

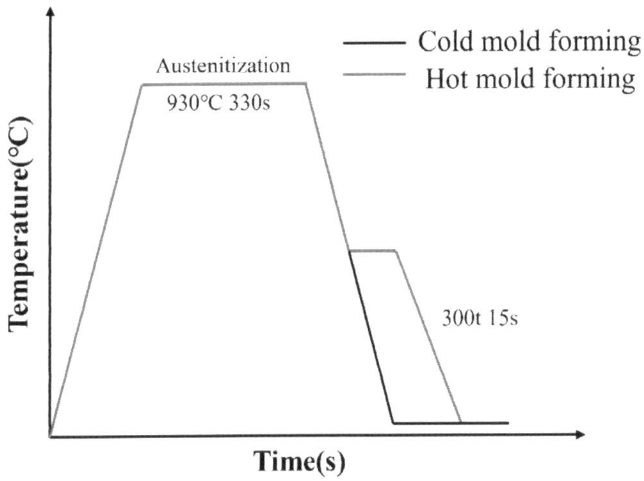

Figure 2. Schematic diagram of the hot stamping process.

3. Experimental results and analysis

3.1. Uniaxial tensile test

As shown in Figure 4(a), Samples were taken from the bottom flange and side flanges of the hard and soft zones of the U-shaped part, and quasi-static uniaxial tensile tests were conducted according to the GB/T 228 standard. The specimen dimensions are shown in Figure 3. Tensile tests were performed using a Shimadzu tensile testing machine (AG-IC) at a speed of 4.5 mm/min.

Figure 4(b) depicts the engineering stress-strain curves for different regions of the U-shaped part formed by stamping at a hot die temperature of 500°C. In the hard zone, the side surface exhibits a tensile strength of approximately 1500 MPa, and the bottom surface has around 1200 MPa. In the soft zone, the side surface has a tensile strength of about 800 MPa, while the bottom surface is approximately 780 MPa. During the hot stamping process, the side surface of the part, compared to the bottom flange, conforms more tightly to the molds, experiencing more significant pressure. Consequently, a higher proportion of austenite transforms into martensite during the forming process, leading to a higher tensile strength on the side surface of the part. Table 2 summarizes the material's tensile strength and elongation at break for different regions of the U-shaped strength part.

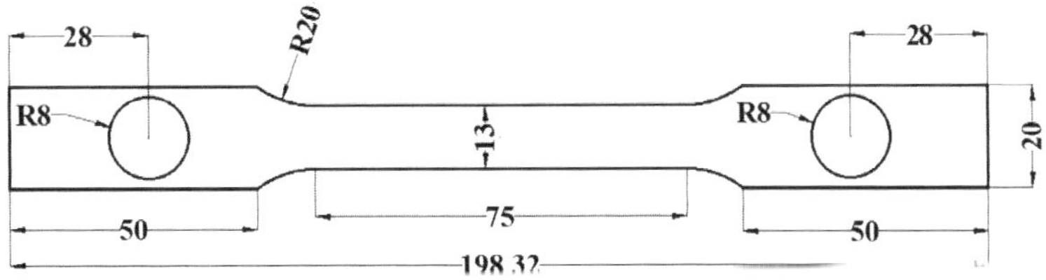

Figure 3. Schematic diagram of uniaxial tensile specimen.

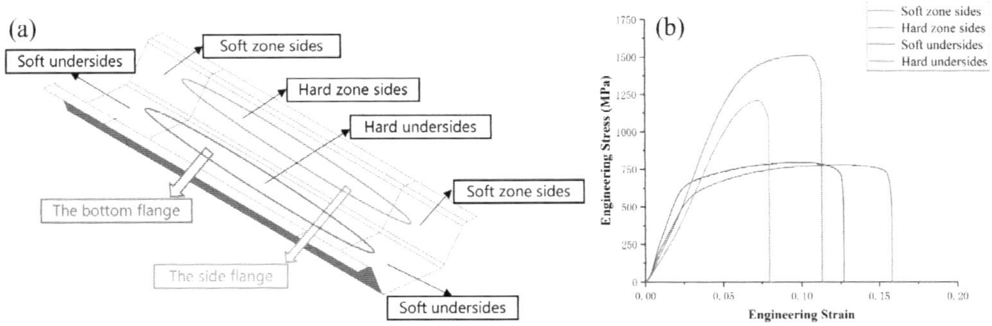

Figure 4. (a) Sampling diagram of U-shaped parts, (b) The engineering stress-strain curves for different sections of the U-shaped part.

Table 2. The results of uniaxial tensile tests.

Part location	Soft zone		Hard zone	
	Tensile strength (MPa)	Elongation (%)	Tensile strength (MPa)	Elongation (%)
The bottom flange	778.125	15.9	1207.969	4.90
The side flanges	794.688	12.9	1509.375	5.78

3.2. Fracture tensile tests under different stress conditions

Numerous studies have delved into fracture models for hard regions (1500 MPa), yet investigations into fracture models for soft regions are less frequent. Consequently, this research is dedicated to solving and refining fracture models specifically for the bottom of soft regions. Samples were taken from the bottom surface of the soft zone to investigate the fracture performance of the soft zone of the component further, and four tensile specimens were designed to conduct fracture tests under various stress conditions. While conducting tensile tests, there is some slippage between the specimen and the gripping ends of the testing machine. Therefore, we utilize Digital Image Correlation (DIC) technology to obtain more accurate tensile testing data (force-displacement curve). The specimen dimensions are shown in Figure 5, where (a) represents a notched specimen with a radius of 5 mm, (b) a notched specimen with a radius of 20 mm, (c) a 45° shearing specimen, and (d) a 45° tensile shearing specimen. The force-displacement curves obtained from the experiments are illustrated in the experiment curve of Figure 7. The subsequent section will involve parameter determination and optimization of the GISSMO fracture model based on the force-displacement curves obtained from the four fracture tensile tests under different stress conditions.

IOP Conf. Series: Materials Science and Engineering **1307** (2024) 012027 doi:10.1088/1757-899X/1307/1/012027

4. Solving and optimizing the GISSMO fracture model

4.1. Description of the GISSMO fracture model

The GISSMO fracture model introduces damage factors D and unstable deformation factor F to describe the failure behavior of materials. As shown in Figure 6, the damage process occurs in stages e0~e2, where the material is assumed to start accumulating damage as soon as it enters the plastic deformation stage until the material fractures when $D = 1$. The instability process occurs in the stages e0~e1, starting from the onset of plastic deformation, and accumulates until $F = 1$ when material instability and necking phenomena appear, at which point the damage factor is $D=D_{crit}$. Then, it enters the stress attenuation process, considering the Lemaitre coupling between cumulative damage and equivalent stress and correcting the equal stress of the material.

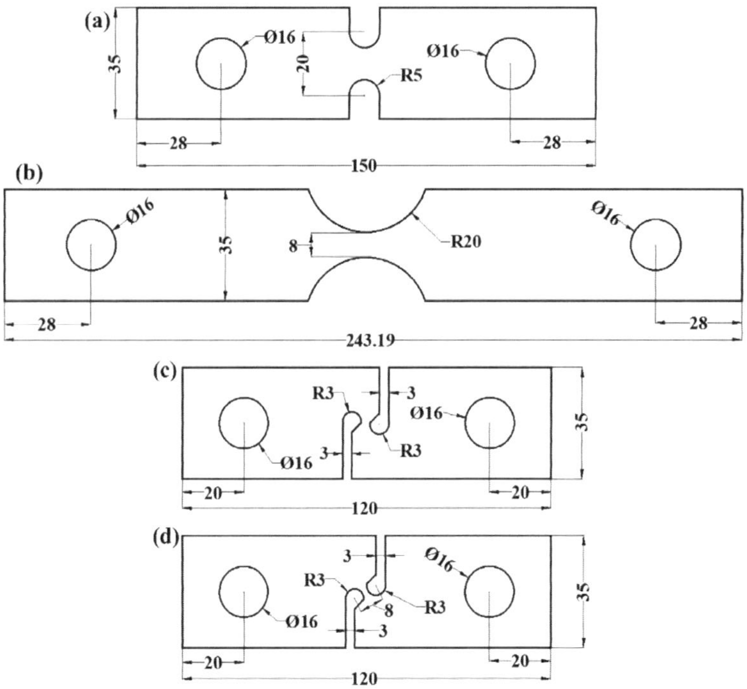

Figure 5. Schematic diagram of fracture tensile specimen dimensions under different stress conditions (a) R5 notched specimen, (b) R20 notched specimen, (c) shearing specimen, (d) tensile shearing specimen.

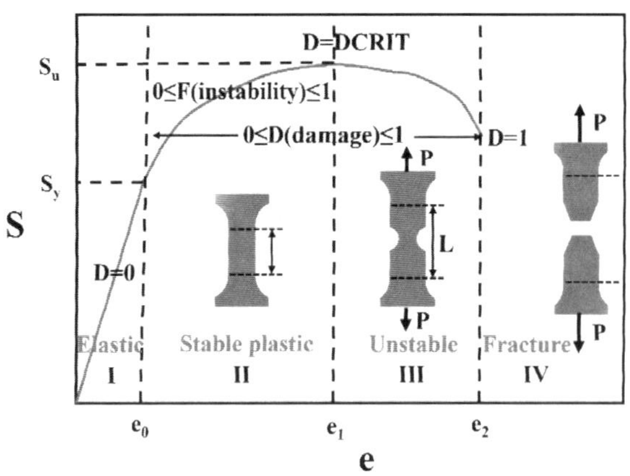

Figure 6. Depicts the damage process defined by the GISSMO fracture model.

(1) Accumulation process of damage

$$\Delta D = \frac{n}{\varepsilon_{f(\eta)}} D^{\left(1-\frac{1}{n}\right)} \Delta\varepsilon_p \tag{1}$$

Where ΔD is the increment of the damage factor, $\Delta\varepsilon_p$ is the increment of equivalent plastic strain, n is the cumulative damage exponent, and $\varepsilon_{f(\eta)}$ is the equivalent plastic strain at which the material fractures under different triaxial stresses.

(2) Instability process

$$\Delta F = \frac{n}{\varepsilon_{p,loc(\eta)}} F^{\left(1-\frac{1}{n}\right)} \Delta\varepsilon_p \tag{2}$$

where F represents the instability deformation factor, ΔF is the increment of the instability deformation factor, and $\varepsilon_{p,loc(\eta)}$ is the equivalent plastic strain at which the material becomes unstable under different triaxial stresses.

(3) Stress attenuation process

$$\sigma^* = \sigma\left[1 - \left(\frac{1-D_{crit}}{D-D_{crit}}\right)^m\right] \quad D \geq D_{crit} \tag{3}$$

where σ^* is the corrected equivalent stress, σ is the actual stress before softening, m represents the stress decay exponent, and D_{crit} is the damage factor or damage threshold at material instability.

In summary, the GISSMO fracture criterion model comprises a total of 5 material parameters: the triaxial stress factor 'η', cumulative damage exponent 'n', stress decay exponent 'm', as well as the equivalent plastic strain $\varepsilon_{f(\eta)}$ at which the material fractures under different stress states and the equivalent plastic strain $\varepsilon_{p,loc(\eta)}$ at material instability.

4.2. The solution and optimization of the GISSMO fracture model

Firstly, the finite element models of the four types of specimens were preprocessed in Hypermesh. The clamping end was ignored to simplify the simulation process, and only the notched section of the patterns was modeled, as shown in Figure 7. The mesh size was 0.5 mm, and the element type was C3D6 elements, representing six-node linear hexahedral elements. Figure 7(a)-(d) respectively present the force-displacement curves of fracture tests for R5 notch specimens, R20 notch specimens, shear specimens, and tensile-shear specimens. They also display the force-displacement curves obtained from the simulation without the fracture model alongside the corresponding stress triaxiality and equivalent plastic strain. The average stress triaxiality is calculated from the start of plastic deformation to the fracture stage of the specimen, and the equivalent plastic strain at the moment of specimen fracture is considered the fracture plastic strain. The obtained data for the equivalent fracture plastic strain under different stress states are summarized in Table 3.

Using LS-OPT software for fitting and optimization, the force-displacement curves of fracture tests under different stress conditions were set as target curves. The equivalent fracture plastic strains under different stress conditions were considered input parameters. In contrast, the cumulative damage exponent 'n', stress decay exponent 'm', and equivalent unstable plastic strain were set as parameters to be determined. The Sequential with Domain Reduction (SRSM) algorithm is applied for 12 optimization iterations. When assessing the error between the two curves, the Mean Square Error (MSE) method, which relies on vertical coordinates, is employed to calculate the Euclidean distance between the experimental and simulated curves. A tolerance of 10^{-5} is set, and upon reaching the specified accuracy criteria, the final GISSMO crack model parameters can be obtained. The optimized cumulative damage exponent 'n' was determined as 3.927, and the stress decay exponent 'm' was 3.71.

The comparison between experimental and simulation results, as depicted in Figure 8, reveals a strong alignment between the force-displacement curve, with minimal deviation observed at the point of specimen fracture. GISSMO can precisely forecast the failure behavior of metal materials under intricate stress conditions, as well as the accumulation of damage under non-proportional loading paths. The derived failure parameters from the study can be applied for subsequent material failure analysis, offering robust assistance to engineers in selecting suitable materials for lightweight structural design while ensuring safety. Nonetheless, this study has some things that could be improved. Future endeavors could entail calibrating fracture parameters through cup-cone experiments and delving deeper into the

IOP Conf. Series: Materials Science and Engineering **1307** (2024) 012027 doi:10.1088/1757-899X/1307/1/012027

impact of strain rate and mesh size on fracture strain. Such enhancements would bolster the model's predictive capacity for real-world car body fracture scenarios during crash tests.

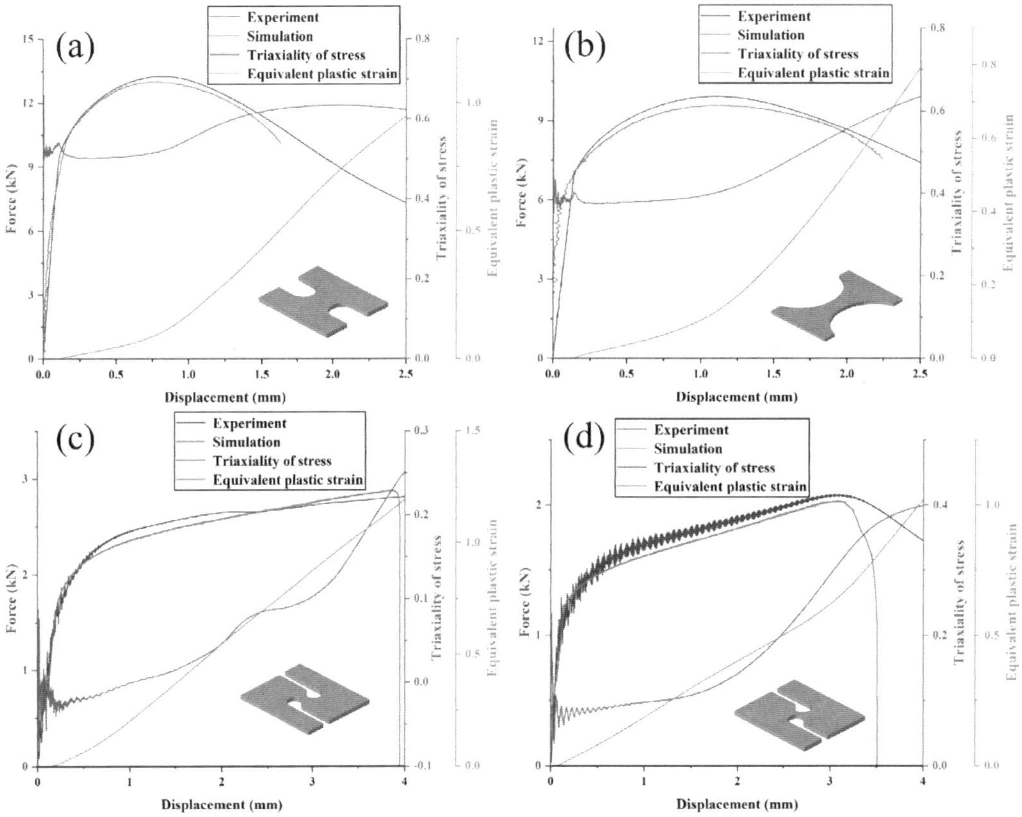

Figure 7. The force-displacement curves, stress triaxiality, and equivalent plastic strain: (a) R5 notched specimen, (b) R20 notched specimen, (c) shearing specimen, (d) tensile shearing specimen.

Table 3. Equivalent plastic strains at fracture under different stress states.

Tensile specimens	Average stress triaxiality	Equivalent plastic strain at break
R5 notch	0.557	0.4535
R20 notch	0.473	0.714
Shear	0.0201	0.7935
Tensile shear	0.155	0.771

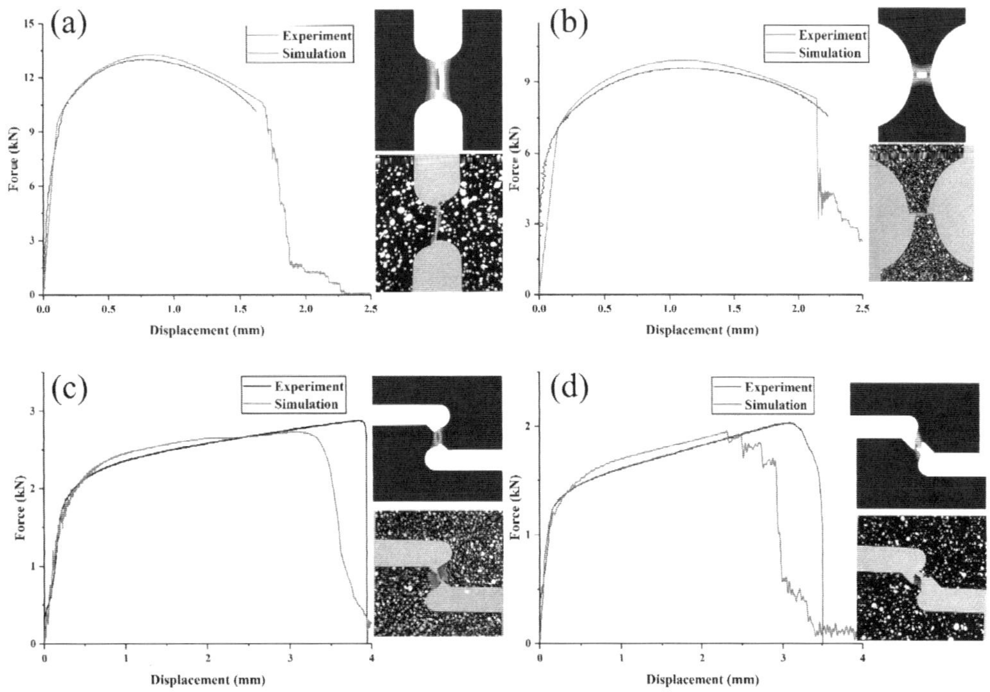

Figure 8. The optimization results from LS-OPT.

5. Conclusion

- Uniaxial tensile tests on samples from the U-shaped parts revealed a tensile strength of 1509.375 MPa and an elongation at break of 5.78% in the hard zone, while the soft zone exhibited a tensile strength of 794.688 MPa and an elongation at break of 12.9%.
- Four types of fracture specimens were designed under different stress states. Combined experimental and simulation data yielded stress triaxialities of 0.557, 0.473, 0.0201, and 0.155 for R5 notch specimens, R20 notch specimens, shear specimens, and tensile-shear specimens, respectively.
- The fracture parameters for the GISSMO model are: the damage exponent '*n*' is 3.927, and the stress decay exponent '*m*' is 3.71. The GISSMO-based fracture model accurately predicts the fracture behavior of 22MnB5 high-strength steel, offering valuable technical references for predicting its failure behavior.

References

[1] Johnson G R and Cook W H 1985 Fracture characteristics of three metals subjected to various strains, strain rates, temperatures and pressures *Eng. Fract. Mech.* **21** 31–48
[2] Ibijola E A 2002 On some fundamental concepts of Continuum Damage Mechanics Comput. Methods Appl. Mech. Eng. **191** 1505–20
[3] Rabotnov Y N, Leckie F A and Prager W 1970 Creep Problems in Structural Members J. Appl. Mech. **37** 249–249
[4] Johnson G R and Cook W H 1985 Fracture characteristics of three metals subjected to various strains, strain rates, temperatures and pressures *Eng. Fract. Mech.* **21** 31–48
[5] Neukamm F, Feucht M and Haufe A 2009 Considering damage history in crashworthiness simulations *Ls-Dyna Anwenderforum*
[6] Effelsberg J, Haufe A, Feucht M, Neukamm F and Du Bois P 2012 *On parameter identification for the GISSMO damage model* (Metal Forming)

Deep drawing simulation of dual phase steel using hardening curves and anisotropic parameters from uniaxial and biaxial tensile tests

Archit Shrivastava and D Ravi Kumar *

Department of Mechanical Engineering, Indian Institute of Technology Delhi, New Delhi, India

* E-mail: dravi@mech.iitd.ac.in

Abstract. Among the various factors, the accuracy of the predictions from the numerical simulation of sheet metal forming processes depends on the material model used to define the mechanical behavior of the blank material. The coefficients of the hardening model to define the flow curve and the plastic strain ratios are commonly determined using the uniaxial tensile tests. The advanced anisotropic yield criteria incorporate material flow behaviour and plastic strain ratio in equi-biaxial tension. In this work, deep drawing of a flat bottom cylindrical cup has been simulated using dual-phase steel (DP600) sheets. The biaxial material properties obtained by conducting hydraulic bulge test and cruciform specimen test are used in the anisotropic yield criteria in the simulations. The hardening curves are extrapolated using different hardening laws in which the coefficients are determined from the stress-strain curves obtained from both uniaxial tensile and hydraulic bulge tests. The predicted peak drawing load and thickness variation in the drawn cups are compared with the experimental results of deep drawing.

1. Introduction

In the automotive industry, the major focus has been towards reducing weight and minimizing exhaust gas emissions. A significant area of emphasis has been the careful selection of materials for automotive bodies and other sheet metal parts [1]. High strength steels, particularly advanced high strength steels like dual-phase (DP) steel, are being used in the automotive industry for manufacturing critical parts. DP steel is a low carbon steel containing both soft ferrite and hard martensite and hence is highly suitable for meeting the dual requirements of high strength and good formability [2]. Due to the complexity of the components in automotive industries, the state of stress is usually biaxial. The standard uniaxial tensile tests can not be used to determine the behaviour of materials in this condition. Material anisotropy leads to change in the deformation behaviour, subsequently influencing the predicted stress and strain distributions in sheet metal forming operations analysed through finite element analysis [3]. So, choosing an appropriate material model that consists of the hardening model to define the flow curve and the yield criterion that incorporates anisotropy is crucial for accurate predictions in simulation of processes such as deep drawing. The Hill48 quadratic yield criterion [4] was the first formulation that incorporated anisotropy in the yielding behaviour of the material. Hassannejadasl et al [5] indicated that changes in the material anisotropy coefficients with respect to strain rate alters the yield surface of DP600. Lenzen et al. [6] conducted elliptical hydraulic bulge tests to get the material properties in the plane strain condition in DP600 specimens. The modified yield surface led to better predictions in the drawing behaviour of the square cups. A non-quadratic anisotropic yield function referred to as Barlat89 was proposed in which the anisotropic coefficients are obtained by conducting the uniaxial tensile tests [7]. Some advanced yield criteria have been proposed by Barlat (2000) [8] and Banabic (2005) [9] to predict the yield locus more accurately. The advanced yield criteria incorporate the material properties obtained from the biaxial tensile tests also in which the specimens are subjected to equi-biaxial tension. F. Ozturk et al. [10] demonstrated the effect of different yield functions on the prediction of the limit strains of dual-phase (DP600) steel sheets in FE simulation using Marciniak-Kuczynski model.

In the present work, deep drawing of a flat bottom cylindrical cup has been simulated using dual-phase (DP600) steel sheets of 1.4 mm thickness. The combined effect of yield criterion and hardening model on the predicted load and thinning in deep drawing of DP600 steel sheets has been studied. The biaxial material properties obtained by conducting hydraulic bulge test and cruciform specimen test are used to determine the coefficients in the anisotropic yield criteria used in the simulations. The hardening curves are extrapolated using different hardening laws in which the coefficients are determined from the stress-strain curves obtained from both uniaxial tensile and hydraulic bulge tests. The predicted peak load and thickness variation in the drawn cups are compared with the experimental results of deep drawing.

2. Methodology

2.1. Uniaxial tensile test

Uniaxial tensile (UT) tests were conducted on dual phase (DP600) steel sheet specimens of 1.4 mm thickness in accordance with the ASTM E8M Standard [11] specification. The samples were prepared by blanking process on a mechanical press. The tests were conducted on a Instron make universal testing machine at a constant speed of 5 mm/min. The specimens were tested in three different orientations: parallel (0°), diagonal (45°), and perpendicular (90°) to the rolling direction of the sheet. For each orientation, three specimens were tested to determine the tensile properties. Anisotropy plays an important role in the formability of the sheet metal especially in the deep drawing process. Anisotropy of sheet metals is characterised by determination of plastic strain ratio (R) which is defined as the ratio of true strain in width to true strain in thickness before the onset of necking in a uniaxial tension test. The plastic strain ratio (R) of the DP600 sheet was determined as per the ASTM E517 [12] specification. For determining the R-value, specimens were elongated up to 15% longitudinal strain before the maximum load was reached. The R values were determined in all the three orientations as mentioned earlier in the tensile tests.

2.2. Hydraulic bulge test

The hydraulic bulge tests (HBT) were conducted using a hydraulic press with a capacity of 630 tons at LFT, FAU, Erlangen, Germany. The hydraulic bulge test setup is shown in Figure 1(a). In this test, a circular blank is deformed by fluid pressure until failure. High hydrostatic pressure acting on the sample during deformation delays the onset of necking and improves ductility. Thus, a much higher degree of uniform deformation can be achieved resulting in equivalent stress vs. equivalent strain curve up to a much larger strain than that obtained in uniaxial tensile test.

In this test, a circular blank with a diameter of 395 mm was clamped between an upper die and a lower die using a blank holding force of 3500 kN. This clamping force ensured the complete restriction of material flow from the flange region. The inner diameter of the die is 200 mm with an entry radius of 25 mm to reduce the bending effects that can occur with a smaller entry radius. The applied fluid pressure was monitored during the test using a pressure transducer. An online 3-D optical strain measurement system ARAMIS was used for measuring the major and minor strains at the pole throughout the test.

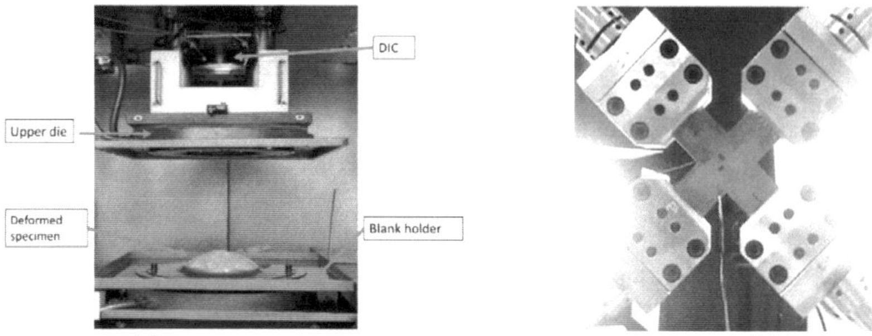

(a) (b)

Figure 1. Determination of biaxial flow curves by (a) hydraulic bulge test and (b) cruciform test

2.3. Cruciform test

The biaxial tensile tests (BCT) using cruciform specimens were performed on a biaxial tensile machine at Tata steel, R&D Jamshedpur as shown in Figure 1(b). The geometry of the biaxial test specimen used in this work is based on the design proposed as per ISO 16842:2021 standard [13]. Each arm of the specimen has seven slits, 60 mm long and 0.2 mm wide at 5.6 mm interval. The slits are provided to relax the geometric constraint against the uniform deformation in 45 × 45 mm^2 gauge area. The slits were prepared through laser cutting. The specimens were cut in such a way that the rolling and transverse directions of the sheet material coincide with the specimen arms. Two load cells attached in along the two in-plane mutually perpendicular directions were used for measuring the force values in rolling direction (RD) and transverse direction (TD) directions, respectively. The test was performed in equibiaxial condition that is, with a force ratio (Fx/Fy) of 1.0. The strains along the rolling and transverse directions were measured using the optical strain measurement system (GOM).

2.4. Numerical analysis of cylindrical deep drawing

Numerical simulations of flat bottom cylindrical cup deep drawing of 1.4 mm thick DP600 steel sheets were carried out using AutoForm software to predict the thinning and the drawing load. The FE model used for the simulation is shown in Figure 2. Circular blanks of 100 mm diameter were deep drawn with a 50 mm diameter flat bottom cylindrical punch. The elasto-plastic shell (EPS-5) elements of size 1.0 mm were used for meshing the blank with five integration points. The refining mesh option with a maximum refinement level of 7 was used for better accuracy in the predicted results. A blank holding force of 30 kN was applied to prevent wrinkling during deep drawing which corresponds to a blank holding pressure of 4 MPa. A punch speed of 1 mm/s was used in the simulations.

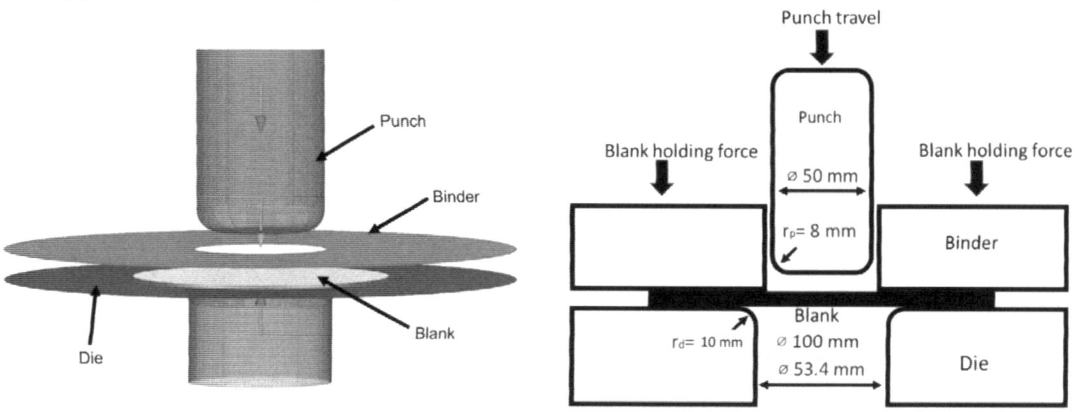

(a) (b)

Figure 2. (a) FE model used in simulation of Swift flat bottom cup test and (b) dimensions of the tools used in the deep drawing.

The coefficients of the Swift hardening law determined using the flow curve of UT and HBT-ISO were used to define the hardening behaviour in the simulation. The simulations were performed using different yield criteria (Mises, Barlat89, and BBC2005). An advanced yield criteria (BBC2005) using the equibiaxial properties determined using both cruciform tests (BBC2005 (CT)) and hydraulic bulge tests (BBC2005 (HBT)) were used in the simulation.

2.5. Deep drawing experiments

To validate the predicted results from the above simulations, Swift flat bottom cup deep drawing tests were conducted using circular blanks of 1.4 mm thick Dual Phase steel sheet (DP600). The setup consists of flat bottom cylindrical punch with a diameter of 50 mm, a blank holder to apply the blank holding force to prevent wrinkling, and the drawing die, designed to accommodate the increase in thickness of the sheet to avoid ironing. Blanks diameter of 100 mm (draw ratio of 2.0) were deep drawn in the experiments. Figure 2(b) illustrates the dimensions of the tools used in the deep drawing process. The tests were carried out on a 100 ton double action hydraulic press maintaining a constant punch speed of 1 mm/s. To prevent wrinkling, blank holding force was applied such that the initial blank holding pressure was 4 MPa. The cups were drawn to a depth of 35 mm, and the measurements of punch load and displacement were performed using a load cell and an LVDT, respectively.

3. Results and discussion

3.1. Prediction of flow stress model

The stress strain curves obtained from the uniaxial tensile tests and the hydraulic bulge tests are shown in Figure 3(a). It clearly shows that the stress-strain curve of DP600 steel can be plotted up to a much larger plastic strain (> 0.8) using hydraulic bulge tests when compared to the uniaxial tensile tests in which early onset of necking/instability takes place. However, the material undergoes large plastic strains during deep drawing. Therefore, if uniaxial flow curve is used to predict the hardening behavior by extrapolation to large plastic strains, the accuracy of such curves depends on the hardening law used. The difference between the flow stress approximated by different hardening laws could be significantly high at large plastic strain. In the present work, three models (Swift hardening law, Hockett-Sherby law, and Voce hardening law [14]) were used to plot the flow curves up to a large plastic strain as shown in Figure 3(a). The coefficients in different hardening laws were determined by regression analysis and minimizing the Sum of Squares with a minimum R^2 values of 99.9%. Using the equations of Swift hardening law, Hockett-Sherby law, and Voce hardening law, the flow curves were generated up to a true plastic strain of 1.0.

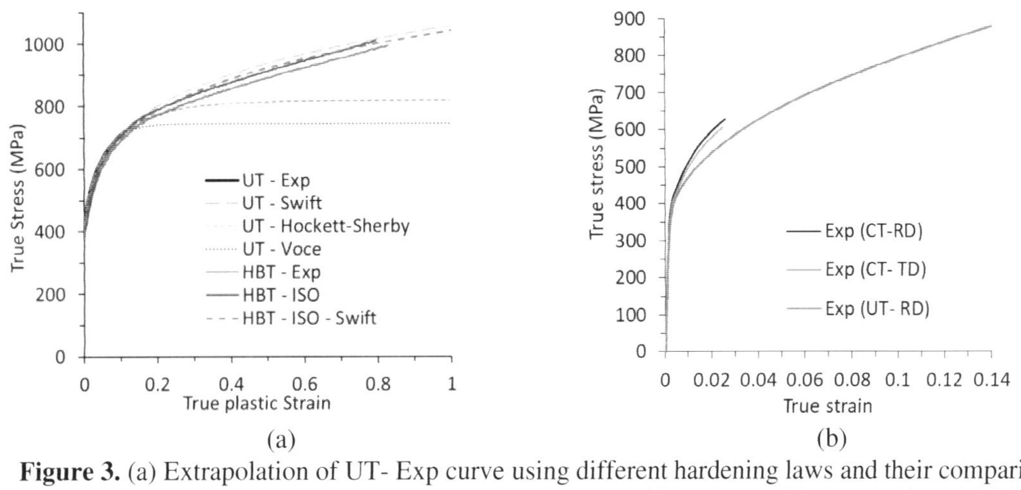

Figure 3. (a) Extrapolation of UT- Exp curve using different hardening laws and their comparison with HBT, (b) True stress-strain curve obtained from cruciform test and its comparison with UT.

The results of hydraulic bulge test were analysed, and the biaxial plastic strain ratio $(R_b)_{HBT}$ was determined in accordance with the circular bulge test DIN EN ISO 16808 standard [15]. The equivalent biaxial stress has been determined using the membrane theory. It is valid for a small ratio of sheet thickness to bulge diameter, where bending effects can be neglected. The equivalent strain is determined using Hill's 1948 anisotropic material behavior. To extrapolate the uniaxial stress-strain curve, the biaxial stress strain curve is scaled by a factor obtained from the plastic work principle [16]. The transformed equivalent stress - equivalent strain curve is referred to as HBT-ISO curve in this work. Biaxial yield stress $(\sigma_b)_{HBT}$ is determined using the plastic work equivalence corresponding at the plastic work dissipated in the uniaxial tensile test at the plastic strain of 0.004 in the rolling direction. The yield locus at this low plastic strain is necessary for generation of the initial yield surface. The HBT-ISO curve is already up to a large plastic strain, but it is further extrapolated upto a strain of 1.0 using the Swift hardening law (referred to as HBT-ISO-Swift). Therefore, the flow curve from uniaxial tensile tests extrapolated using Swift hardening law (UT-Swift) and the flow curve from HBT–ISO–Swift method are used in the simulation. The coefficients of the Swift hardening law for UT and HBT-ISO curves are shown in Table 1.

Table 1. Coefficients of swift hardening law predicted using uniaxial tensile and hydraulic bulge tests.

Material	UT- Swift			HBT- ISO-Swift		
	ε_0	n	K (MPa)	ε_0	n	K (MPa)
DP600 steel	0.0046	0.174	1056	0.0041	0.176	1041

The stress strain curves obtained from the cruciform tests along rolling and transverse direction are shown in Figure 3(b). The limit strain reached in the cruciform tests is very small compared with the tensile tests due to strain localization at cross corners and necking in arms. The true stress-true strain curve in RD and TD directions is converted into true stress-true plastic strain curves using ISO 16842:2021 standard. The biaxial plastic strain ratio $(R_b)_{CT}$ is calculated from the ratio of plastic strain in the transverse direction to the rolling direction. Plastic work per unit volume was calculated from the true stress - true plastic strain curves obtained from the cruciform tests. The summation of plastic work in the RD and TD directions of the cruciform specimen at the yield corresponds to the total plastic work dissipated in the uniaxial tensile test at a plastic strain of 0.004 in the rolling direction. The equivalent biaxial yield stress $(\sigma_b)_{CT}$ is calculated from the stresses in the RD and TD directions using the Hill48 anisotropic yield criterion.

Table 2. Material properties determined to plot the BBC2005 yield surface in AutoForm.

Uniaxial						Biaxial (HBT)		Biaxial (CT)	
$\frac{\sigma_0}{\sigma_0}$	$\frac{\sigma_{45}}{\sigma_0}$	$\frac{\sigma_{90}}{\sigma_0}$	R_0	R_{45}	R_{90}	$(R_b)_{HBT}$	$\left(\frac{\sigma_b}{\sigma_0}\right)_{HBT}$	$(R_b)_{CT}$	$\left(\frac{\sigma_b}{\sigma_0}\right)_{CT}$
1.0	1.0739	1.056	0.96	0.71	0.93	1.033	0.961	1.061	0.975

σ_0 = 405 MPa

3.2. Prediction of different yield surfaces

The ratios of yield stresses and plastic strain ratios obtained from uniaxial and biaxial tests are summarized in Table 2. Figure 4 shows the predicted yield surfaces obtained from Mises, Barlat89, and BBC2005 yield criteria. Stress ratios derived from both uniaxial and biaxial tensile tests are represented as dotted points. The figure reveals disparities between Mises and Barlat89 under both uniaxial and equibiaxial stress conditions, particularly in equibiaxial and uniaxial transverse scenarios. This discrepancy arises because Mises assumes the material to be isotropic in the material, while Barlat89 considers anisotropy but with only uniaxial properties (R_0, R_{45}, R_{90}, and σ_0) for defining the yield function. In equibiaxial stress conditions, both Mises and Yld89 tend to overestimate yielding compared to the experimental values. On the other hand, BBC2005 utilizes both uniaxial and biaxial material properties to define its yield function as given in Table 2. The BBC2005 yield function plotted using the biaxial properties (biaxial plastic strain ratio (R_b) and the equibiaxial flow stress ratio (σ_b)) obtained from the cruciform tests and the hydraulic bulge tests are represented as BBC2005 (CT) and BBC2005 (HBT).

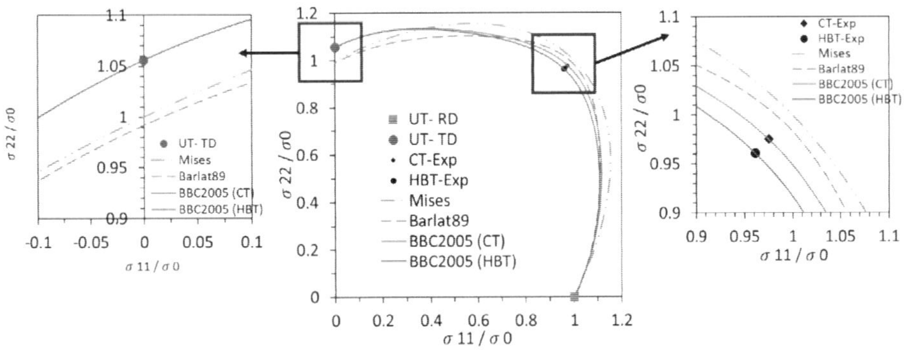

Figure 4. Predicted yield surfaces using different yield criteria

3.3. *Deep drawing simulation using different yield criteria*

The predicted load–displacement curves from the numerical simulations of deep drawing are compared with the experimental results at a draw depth of 35 mm. Figure 5(a) shows the effect of different yield models on the load-displacement curves. The resistance against deformation in the flange region leads to an increase in the drawing force. The error in the prediction of maximum load is 5% on using the isotropic yield model (Mises), whereas it reduced to less than 3% when anisotropic Barlat89 yield model is used. The error in prediction further reduced to less than 1% when the advanced non quadratic BBC2005 yield function is used. Figure 5(b) shows the thickness distribution in the drawn cup from the centre of the cup to the flange. The maximum thinning is observed in the punch corner radius near to the cup bottom. The maximum thinning is predicted with an accuracy of 5% using the BBC2005 yield model. Combining the advanced yield criterion (BBC2005) with a reliable hardening model of the material up to large strains leads to a more accurate prediction of peak load and thinning behaviour.

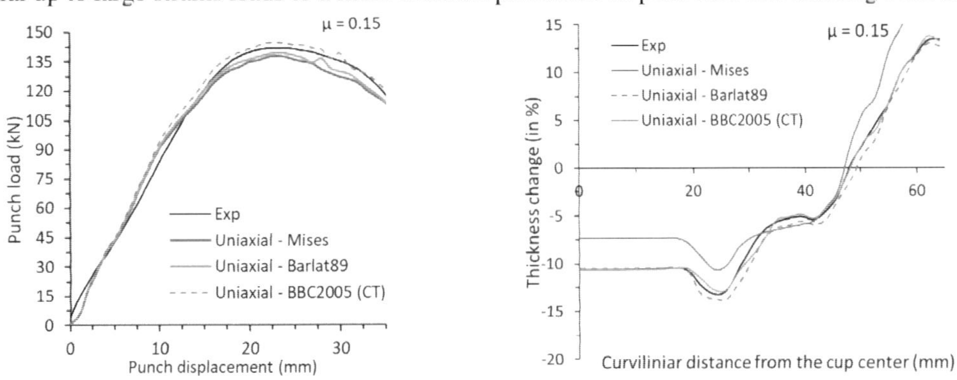

(a) (b)

Figure 5. Comparison of (a) load-displacement curves and (b) variation of change in thickness in the deep drawn cup predicted using different yield models.

The cup is drawn to a depth of 35mm successfully as shown in Figure 6(a). An experimentally determined forming limit curve, shown in Figure 6 (b), is used in the FE simulation to predict the failure. The simulation also shows successful forming of the cup. The planar anisotropy present in the material leads to earing in the flange region during deep drawing. Figure 6 (c) shows comparison of the earing

IOP Conf. Series: Materials Science and Engineering **1307** (2024) 012028 doi:10.1088/1757-899X/1307/1/012028

profile predicted using the Mises and BBC2005 yield models. The effect of planar anisotropy is visible in the flange region clearly and it leads to earing at the top of the cup wall when drawn completely. The profile obtained from the von-mises yield model is circular due to assumption that the material is isotropic. BBC2005 model incorporates the anisotropic material behaviour and hence resulted in better prediction of the earing profile.

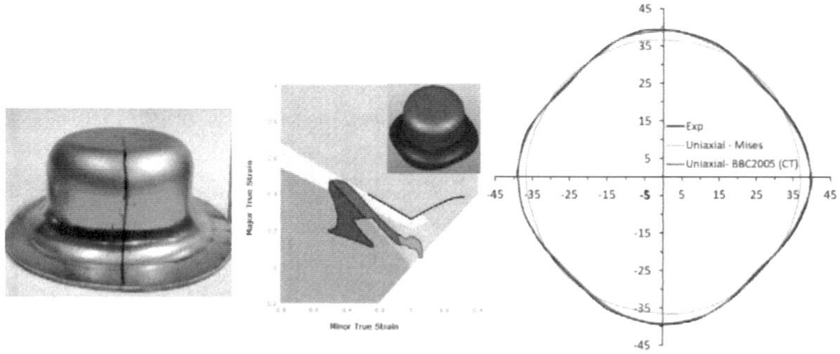

(a) (b) (c)

Figure 6. (a) Experimentally drawn cup, (b) Simulation result of deep drawing, and (a) Comparison of the earing profile using different yield models at a draw depth of 35 mm.

3.4. Effect of biaxial properties in deep drawing simulation

The load-displacement curves and variation of % thickness change obtained from simulations using uniaxial (UT-Swift) and biaxial (HBT-ISO-Swift) flow curves are shown in Figure 7(a) and 7(b) respectively. These curves are obtained with BBC2005 yield surface using the equibiaxial properties determined using both cruciform tests (BBC2005 (CT)) and hydraulic bulge tests (BBC2005 (HBT)). The load-displacement curves show a good agreement with the experimental results. The difference in the predicted results is negligible when the hardening laws used are UT-Swift and HBT-ISO-Swift. The error in predictions would have been higher if the uniaxial flow stress-strain was extrapolated using other flow stress models. Also, the flow curve obtained from the hydraulic bulge tests is up to a large plastic strain and hence the uncertainty related to the choice of hardening law is very low when hydraulic bulge test flow curve is used.

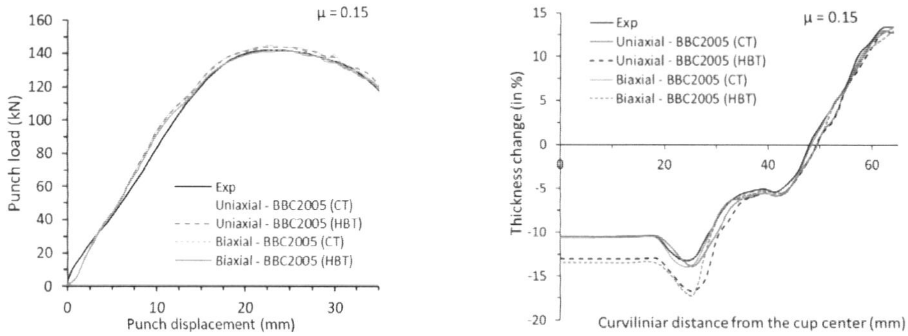

(a) (b)

Figure 7. Comparison of predicted (a) load-displacement curves, (b) % change in thickness in the drawn cup using BBC2005 yield criterion with experimental curves.

A significant difference between the thinning behaviour predicted using the BBC2005 (CT) and BBC2005 (HBT) yield criteria is observed as shown in Figure 7(b). This could be attributed to the fact that thinning in the cup wall is due to biaxial stretching. The BBC2005 yield surface obtained using the biaxial properties from the cruciform tests shows better predictions with the experiments than the one obtained from the hydraulic bulge test. The in-plane deformation in cruciform tests leads to a more accurate determination of biaxial properties than the out-of-plane deformation in hydraulic bulge tests [17]. At the location where thinning is maximum, the deviation of the predicted value using the BBC2005 (CT) from the experimental value is nearly 5%, whereas the deviation is more than 24% when BBC2005 (HBT) is used.

4. Conclusions

It clearly shows that the stress-strain curve of DP600 steel can be plotted up to a much larger plastic strain using hydraulic bulge tests when compared to the uniaxial tensile tests in which early onset of necking/instability takes place. The uncertainty related to the choice of hardening law decreased when hydraulic bulge test flow curve is used. The extrapolation of the stress strain curve obtained from the uniaxial tensile tests using Swift hardening law predicts similar hardening behaviour as obtained by the hydraulic bulge tests even at large plastic strains. BBC2005 yield criterion predicted the yield surface more accurately while Mises and Barlat89 overestimated yield stress in equi-biaxial tension compared to the experimental values. In deep drawing simulations, prediction of peak load, thickness variation and earing profile is closer to the experimental observations with BBC2005 yield function than with Mises, and Barlat89. BBC2005 yield surfaces were compared using the equibiaxial properties determined using both cruciform tests (CT) and hydraulic bulge tests (HBT). A significant difference between the thinning behaviour predicted using the BBC2005 (CT) and BBC2005 (HBT) yield criteria is observed. Simulation with BBC2005 yield surface obtained using the biaxial properties from the cruciform tests showed better prediction than the one obtained from the hydraulic bulge test.

Acknowledgments

The authors would like to thank Prof. Marion Merklein and her research group at LFT, FAU, Erlangen, Germany for the hydraulic bulge tests and Dr. Rahul Kumar Verma and his research group at Tata Steel R&D, Jamshedpur for providing the material and the biaxial tensile tests.

References

[1] Nasser A, Yadav A, Pathak P, and Altan T 2010 *J. Mater. Process. Technol.* **210** pp 429-436.
[2] Sarraf IS and Green DE 2018 *IOP Conf. Ser.: Mater. Sci. Eng.* **418** 012089.
[3] Suttner S and Merklein M 2016 *J. Mater. Process. Technol.* **235** pp 121–133.
[4] Hill R 1948 *Proc. R. Soc. Lond. Ser. A,* **193** pp 281-297.
[5] Hassannejadasl A, Rahmaan T, Green DE, Golovashchenko SF, and Worswick MJ 2014 *Procedia Eng.* **81** pp 1378 – 1383.
[6] Lenzen M and Merklein M 2018 *IOP Conf. Series: J. Phys. Conf. Ser.* **1063** 012161.
[7] Barlat F and Lian JI 1989 *Int. J. Plast.* **5** (1) pp 51-56.
[8] Barlat F, Brem JC, Yoon JW, Chung K, Dick RE, Lege DJ, Pourboghrat F, Choi SH, and Chu E 2003 *Int. J. Plast.* **19** pp 1297–1319.
[9] Banabic D, Kuwabara T, Balan T, and Comsa DS 2004 *J. Mater. Process. Technol.* **157–158** pp 462–5.
[10] Ozturk F, Torosb S, and Kilic S 2014 *Procedia Eng.* **81** pp 760 – 765.
[11] ASTM E8/E8M – 13a: Standard Test Methods for Tension Testing of Metallic Materials.
[12] ASTM E517 – 19: Standard Test Method for Plastic Strain Ratio r for Sheet Metal.
[13] ISO 16842:2021: Biaxial tensile testing method using a cruciform test piece.
[14] Djouabi M, Ati A, and Manach PY 2019 *Int. J. Damage Mech.* **28** (3) pp 427-454.
[15] ISO 16808:2022 : Determination of biaxial stress-strain curve by means of bulge test with optical measuring systems.
[16] Prakash V, Kumar DR, Horn A, Hagenah H, and Merklein M, 2020 *Int. J. Adv. Manuf. Technol.* **106** pp 1133–1148.
[17] Mulder J, Vegter H, Aretz H, Keller S, van den Boogaard AH 2015 *J. Mater. Process. Technol.* **226** pp 169-187.

Comparison of homogeneous anisotropic hardening models in the case of the direct redrawing of a DP600 steel

N Gautam[1], S Yoon[2,3], F Barlat[1,2] and S Thuillier[1]

[1] Univ. Bretagne Sud, UMR CNRS 6027, IRDL, F-56100, Lorient, France
[2] Pohang University of Science and Technology, Pohang, Republic of Korea

E-mail: nivesh.gautam@univ-ubs.fr, sandrine.thuillier@univ-ubs.fr

Abstract. The use of DP600, an advanced high strength steel, has gained significant attention in automotive industry, especially for complex structures that require multi-step forming operations, leading to non-linear strain path changes. From a numerical modelling perspective, the use of advanced constitutive equations has enabled a precise representation of a large range of behaviors, encompassing reverse and orthogonal strain path changes. Within this context, this study is dedicated to the numerical simulation of a two-step deep drawing process based on distortional plasticity. Two models developed within the Homogeneous Anisotropic Hardening (HAH) framework are considered. This study presents a comparison of the model predictions, calibrated over the same experimental database, in terms of their ability to predict the strain path changes and mechanical behavior of the material during the forming process. Several outputs like the punch load evolution and the strain field are compared with experimental data.

1. Introduction

The evolution of dislocation structures during sheet metal forming has been the driving motivation for numerous studies related to complex strain paths since the 1980s. As a consequence, the mechanical behavior of thin sheet metals has been investigated under multiaxial loadings and strain path changes [1]. The first studies were dedicated to the microstructure evolution at the grain scale and later, to the modelling of the specific features at the macroscopic level in reverse and cross loadings [2]. In a simplified way, though representative of the forming process, sequences composed of combinations of tension, simple shear, equibiaxial tension, plane strain tension have been used. A unified approach has led to a quantification of the strain path change magnitude, originally via the strain-based Schmitt's indicator [3].

Several phenomenological mechanical models have been developed to reproduce the features associated with strain path changes, among which distortional hardening based models referred to as Homogeneous Anisotropic Hardening (HAH) models [4, 5, 6]. HAH-2011 is dedicated to reverse loading features, such as the Bauschinger effect and was further enhanced (HAH-2014) to consider cross loading, as well as permanent softening. Further modifications extend the model (HAH-2020) to pressure-sensitive yield criterion and improve the evolution laws of several internal variables, to provide a better description of the permanent softening. They were

[3] Present address: Max-Planck-Institute für Eisenforschung GmbH, Düsseldorf, Germany

implemented in a user subroutine for Abaqus with a non-iterative stress projection method [7, 8]. Though HAH-2020 was successfully used for the virtual forming of automotive parts, there is still a need to compare the performance of both models when dealing with multi-step processes.

Indeed, although strain path changes may occur in single stage process such as the cross-die test [9], multiple stage forming is closer to real industrial process and may exhibit a large range of strain path change magnitudes [10]. A dedicated forming tool in two stages, inspired from the reverse redrawing test and proposed as one of Numisheet 1999 benchmark [11], was designed and used to produce cylindrical cups of DP600 [12]. Convergence issues were encountered in the numerical simulation of the two stage process, using HAH-2014 model. To overcome this issue, a modification of the tool geometry in stage 2 was performed but has an influence on the load evolution.

In this paper, HAH-2014 and HAH-2020 models, both implemented with a non-iterative stress projection method, are used for the virtual forming of the 2 stage drawing process of cylindrical cups. The prediction of the mechanical behavior of a DP600 steel under reverse and cross loadings, using these 2 models, are compared. Then, the numerical predictions of the two stage forming process are compared with experimental data.

2. Homogeneous Anisotropic Hardening models

HAH-2014 model, previously referred to as e-HAH, is an enhanced version of the original HAH model [4] developed as an alternative to the kinematic hardening model to capture the Bauschinger effect and account for cross loading effects. HAH_{20}, called HAH-2020 in this paper, includes several enhancements related to permanent softening after reverse loading and cross loading singularity highlighted in [13]; moreover, HAH-2020 has shown its superiority in the cross-loading predictions [14]. A pressure-independent formulation is considered in this study. HAH-2014 and HAH-2020 depend on respectively 13 and 23 material parameters. The equations for both models are briefly described below.

A tensorial internal variable $\mathbf{h}$, also called microstructure deviator, is introduced in HAH models. The deviatoric stress tensor $\mathbf{s}$ is projected on the microstructure deviator to quantify the magnitude of the strain path change. Normalized quantities $\hat{\mathbf{h}}$ and $\hat{\mathbf{s}}$ are also defined. The deviatoric stress tensor is additively split into two components, namely $\mathbf{s_c}$, which is collinear to $\hat{\mathbf{h}}$ and $\mathbf{s_o}$ which is orthogonal to it, such that:

$$\mathbf{s_c} = H(\mathbf{s} : \hat{\mathbf{h}})\hat{\mathbf{h}} \tag{1}$$

The scalar H is equal to respectively 8/3 and 1 for HAH-2014 and HAH-2020. For latent hardening and cross loading contraction effects, two additional tensors, defined by linear transformations of the stress deviator, are defined:

$$\mathbf{s}_L = \frac{1}{\xi_L(g_L - 1) + 1}\mathbf{s}_c + \frac{1}{g_L}\mathbf{s}_o \tag{2}$$

$$\text{(HAH-2014)} \quad \mathbf{s}_X = \xi_C\,(1 - g_C)\,\mathbf{s}_o \qquad \text{(HAH-2020)} \quad \mathbf{s}_X = \xi_C \left(\frac{1 - g_C}{g_L}\right)\mathbf{s}_o \tag{3}$$

For HAH-2014, the values $\xi_L = 0$ and $\xi_C = 4$ are fixed. g_L and g_C are internal variables.

2.1. *Yield function*

The yield function contains a stable component $\bar{\xi}(\mathbf{s})$, which is based on the anisotropic yield function Yld2000-2d noted $\overline{\phi}$ in this study, and a fluctuating component ϕ_h, which distorts the yield surface:

$$\overline{\sigma}\left(\mathbf{s}, f_-, f_+, \hat{\mathbf{h}}\right) = \left[\bar{\xi}(\mathbf{s})^q + \phi_h\left(\mathbf{s}, f_-, f_+, \hat{\mathbf{h}}\right)\right]^{\frac{1}{q}} = g_P\sigma_r(\bar{\varepsilon}) \tag{4}$$

where $\overline{\sigma}$ is the effective stress, $\overline{\varepsilon}$ the plastic work-based equivalent strain, σ_r a reference flow stress and q a scalar. g_P is a state variable that defines the amount of permanent softening for HAH-2020; it is fixed constant and equal to unity for HAH-2014. ϕ_h is defined as follows:

$$\phi_h\left(\mathbf{s}, f_-, f_+, \hat{\mathbf{h}}\right) = f_-^q \left|\hat{\mathbf{h}} : \mathbf{s} - \left|\hat{\mathbf{h}} : \mathbf{s}\right|\right|^q + f_+^q \left|\hat{\mathbf{h}} : \mathbf{s} + \left|\hat{\mathbf{h}} : \mathbf{s}\right|\right|^q \tag{5}$$

The stable component $\overline{\xi}$ is given by:

$$\overline{\xi}(\mathbf{s}) = \left[\overline{\phi}(\mathbf{s}_L)^p + \overline{\phi}(\mathbf{s}_X)^p\right]^{\frac{1}{p}} = \sigma_r(\overline{\varepsilon}) \tag{6}$$

where p is a scalar, fixed equal to 2 for HAH-2014. The two functions f_ω, with $\omega \equiv -$ *or* $+$ depend on internal variables g_ω and have different expressions for the two models:

$$(\text{HAH-2014}) \quad f_\omega = \left[\frac{\sqrt{6H}}{4}\left(\frac{1}{g_\omega} - 1\right)\right]^{\frac{1}{q}} \qquad (\text{HAH-2020}) \quad f_\omega = \sqrt{\frac{3}{8}}\left[\frac{1}{g_\omega^q} - 1\right]^{\frac{1}{q}} \tag{7}$$

2.2. *Strain path change indicator*

To quantify the strain path change magnitude, a parameter $\cos\chi$ is defined in a similar way to the strain-based one proposed in [3] but with a stress-based definition:

$$\cos\chi = H\hat{\mathbf{s}} : \hat{\mathbf{h}} \tag{8}$$

This parameter is equal to 1 for monotonic loading, -1 for reverse loading and 0 for the most abrupt case of cross loading.

For HAH-2014, the evolution of the microstructure deviator is given by

$$d\hat{\mathbf{h}} = ksgn(\cos\chi)\left[\left|\frac{\cos\chi}{H}\right|^{\frac{1}{z}} + g_R\right]\left(\hat{\mathbf{s}} - \cos\chi\hat{\mathbf{h}}\right)d\overline{\varepsilon} \tag{9}$$

where

$$\frac{dg_R}{d\overline{\varepsilon}} = k_R[k_R'(1 - \cos^2\chi) - g_R] \tag{10}$$

In HAH-2020, a second microstructure deviator $\mathbf{h}'$ is defined, as well as $\cos\chi'$ defined by Equation 8. The evolution equations for these two microstructure deviators $\hat{\mathbf{h}}$ and $\hat{\mathbf{h}}'$ are given by:

$$\frac{d\hat{\mathbf{h}}'}{d\overline{\varepsilon}} = \lambda' k' \left(\hat{\mathbf{s}} - \cos\chi'\hat{\mathbf{h}}'\right) \tag{11}$$

$$\frac{d\hat{\mathbf{h}}}{d\overline{\varepsilon}} = \lambda k \left(\cos\chi'^{2\xi_R} + \cos\chi^{2\xi_R}\right)\left(\hat{\mathbf{s}} - \cos\chi\hat{\mathbf{h}}\right) \tag{12}$$

where $\lambda = sgn(\cos\chi)$ and $\lambda' = sgn(\cos\chi')$. $\hat{\mathbf{h}}$ evolves only when $\cos\chi$ or $\cos\chi'$ is close to one. The coefficient ξ_R controls the evolution.

2.3. *Internal variable evolution laws for HAH-2014*

State variables g_- and g_+ corresponds to reverse loading and g_3 and g_4 are associated with permanent softening. The evolution laws are detailed below:

$$\begin{array}{rcll} & & \cos\chi \geq 0 & \cos\chi < 0 \\ \frac{dg_-}{d\overline{\varepsilon}} & = & k_2\left(k_3\frac{\sigma(0)}{\sigma(\overline{\varepsilon})} - g_-\right) & k_1\frac{g_4 - g_-}{g_-} \\ \frac{dg_+}{d\overline{\varepsilon}} & = & k_1\frac{g_4 - g_+}{g_+} & k_2\left(k_3\frac{\sigma(0)}{\sigma(\overline{\varepsilon})} - g_+\right) \\ \frac{dg_3}{d\overline{\varepsilon}} & = & 0 & k_5(k_4 - g_3) \\ \frac{dg_4}{d\overline{\varepsilon}} & = & k_5(k_4 - g_4) & 0 \end{array}$$

The two state variables g_L and g_C correspond respectively to latent hardening and cross loading contraction. In this paper, the latent hardening effect is neglected, based on experimental evidence for DP steels. As a consequence, g_L remains constant and equal to its initial value of 1. The evolution law of g_C is given by the following equation:

$$\frac{dg_C}{d\bar{\varepsilon}} = k_C[1 + (C-1)\cos^2\chi - g_C] \tag{13}$$

2.4. *Internal variable evolution laws for HAH-2020*

State variables related to the Bauschinger effect g_- and g_+ evolve as:

$$\frac{dg_-}{d\bar{\varepsilon}} = \frac{1+\lambda}{2}\left[k_1\frac{1-g_-}{g_-^{\xi_B'}}(1-\cos\chi^{2\xi_B}) + k_2(k_3\frac{\sigma_y}{\sigma(\bar{\varepsilon})} - g_-)\cos\chi^{2\xi_B}\right] + \frac{1-\lambda}{2}\left[k_1\frac{1-g_-}{g_-^{\xi_B'}}\right] \tag{14}$$

$$\frac{dg_+}{d\bar{\varepsilon}} = \frac{1-\lambda}{2}\left[k_1\frac{1-g_+}{g_+^{\xi_B'}}(1-\cos\chi^{2\xi_B}) + k_2(k_3\frac{\sigma_y}{\sigma(\bar{\varepsilon})} - g_+)\cos\chi^{2\xi_B}\right] + \frac{1+\lambda}{2}\left[k_1\frac{1-g_+}{g_+^{\xi_B'}}\right] \tag{15}$$

The evolution equation of g_C is defined by

$$\frac{dg_C}{d\bar{\varepsilon}} = k_C(C - g_C) \text{ if } |\cos\chi| \geq 0.996 \quad \text{or} \quad \frac{dg_C}{d\bar{\varepsilon}} = k_C'\frac{1-g_C}{g_C^{\xi_C'}} \tag{16}$$

The evolution of permanent softening in HAH-2020 model is based on the assumption that no permanent softening develops as long as $\cos\chi \geq 0$, mostly occurring in pure reverse loading. State variable g_P is introduced to account for permanent softening, as well as $g_3 \leq g_P$. When the sign of $\cos\chi$ changes, the variables g_P, g_3 and $\hat{\mathbf{h}}$ are denoted as g_P^*,g_3^* and $\hat{\mathbf{h}}^*$. One more intermediate variable g_P' is defined by

$$g_P' = g_P^* - \left(g_P^* - g_3^*\left|\mathbf{s} : \hat{\mathbf{h}}^*\right|^{2\xi_S}\right) \tag{17}$$

which becomes the new value of g_P with the condition $g_P' - g_P < 0$ which ensures that g_P always decreases. The evolution of g_3 and another intermediate variable g_3' is calculated as follows

$$\frac{dg_3}{d\bar{\varepsilon}} = k_5 g_S(k_4 - g_3) \quad \text{and} \quad g_3' = g_3 + dg_3 \tag{18}$$

g_3' becomes the new g_3 with the condition $g_3' - g_3 \leq 0$. In other cases, g_3 is set equal to g_P. The variable g_S in introduced to encounter the decrease of permanent softening when material is not subject to pure reverse loading and its evolution is given by

$$\frac{dg_S}{d\bar{\varepsilon}} = -k_S\left[1 - \left|\mathbf{s} : \hat{\mathbf{h}}^*\right|\right] g_S \tag{19}$$

2.5. *Reference flow stress*

In this study, hybrid Swift-Voce law is used to describe the reference flow stress $\sigma_r(\bar{\varepsilon})$, defined as

$$\sigma_r(\bar{\varepsilon}) = R\left[K(\varepsilon_0 + \bar{\varepsilon})^n\right] + (1-R)\left[Q_1(1 - Q_2\exp(-Q_3\bar{\varepsilon}))\right] \tag{20}$$

where $R, K, \varepsilon_0, n, Q_1, Q_2, Q_3$ are material parameters and take the values provided in [15].

3. Material and methodology

3.1. Material and mechanical behavior

A dual phase steel DP600, provided in sheets of thickness 1.2 mm by ArcelorMittal, is used in this study. The ferrite-martensite microstructure of the material was observed by scanning electron microscopy and a martensite volume fraction of $14\% \pm 4\%$ was measured, which is consistent with results from the literature. The mechanical behavior at room temperature and under quasi-static conditions was thoroughly characterized in uniaxial tension, biaxial tension (with hydraulic bulging), simple shear, both in monotonic loading but also with sequences involving strain path changes, like simple shear loading followed by unloading and then a reverse shear in the opposite direction. This sequence corresponds to $\cos\chi = -1$, while pre-strain in tension on large specimen followed by simple shear in the tensile direction, also included in the database for two pre-strain levels, corresponds to an abrupt change of strain path, associated with $\cos\chi = 0$.

The detailed description of the specimen geometry and boundary conditions is given in [15]. The coefficients for the reference flow stress (cf. Eq. 20) were optimized manually to describe both uniaxial and biaxial data, while the Yld2000-2d parameters were calculated using the wizard tool from the ESI Group. An inverse optimisation was then performed to determine coefficients for strain path changes, using forward-reverse shear tests with two values of forward shear strain, i.e., 0.084 and 0.212. Tests composed of a pre-strain in tension followed by a shear test in the same direction, with two values of tensile pre-strain, i.e., 0.049 and 0.193, were also considered. Concerning the calibration of HAH-2020, a manual identification of the parameters related to the strain path changes was performed.

The mechanical properties and plastic anisotropy coefficients are calculated from tensile tests performed for 7 directions between rolling and transverse ones. It was found that Young's modulus $E = 199\,\text{GPa}$, the ultimate tensile strength $R_m = 650\,\text{MPa}$, the normal and planar plastic anisotropy coefficients are respectively $\bar{r} = 0.99$ and $\Delta r = 0.09$.

3.2. Finite element model of direct re-drawing

A two step deep drawing process of a cylindrical cup in the same direction is considered in this study. The tool dimensions are shown in Figures 1 and 2. Starting from a circular blank of diameter 80 mm, a cup of inner diameter 48 mm is first formed and then redrawn to a final cup of inner diameter 36 mm. The ABAQUS explicit solver is used for numerical simulations and

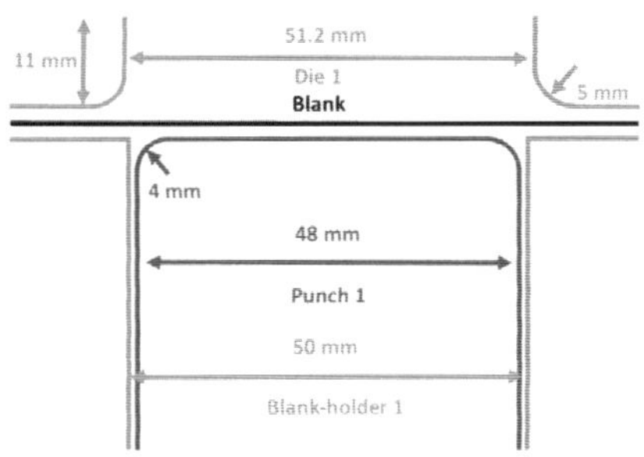

Figure 1. Tool dimensions for stage 1

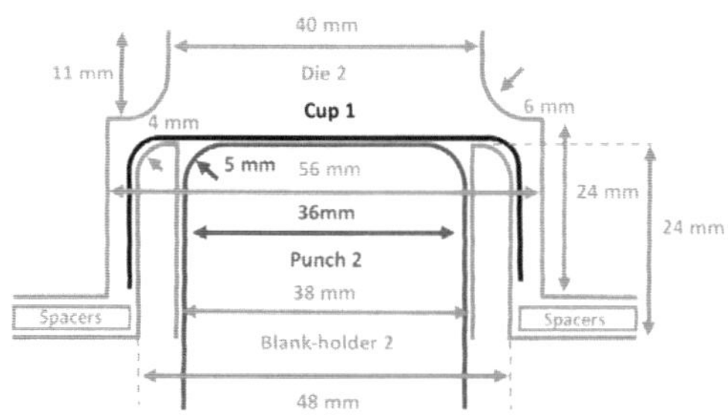

Figure 2. Tool dimensions for stage 2

both HAH models have been implemented in a user subroutines (VUMAT). In order to reduce the simulation time and taking into account of the low initial anisotropy of DP600, only a 5° slice of the blank is utilized, as well as von Mises yield criterion. This slice was meshed with shell element (S4R) with 9 thickness integration points. The element size for the radial end of the blank was set to 1 mm while a smaller size of 0.4 mm was used in the middle of the blank. The element size at the centre of the blank was increased up to 1 mm. The friction coefficient was set to 0.18 for all the contact areas. Symmetry boundary conditions in the global coordinate system were applied to one side of the blank, as well as to the rotated edge, in a local coordinate system.

4. Results

4.1. Model calibration

Calibrated values of material parameters for both models are shown in Table 1. Values highlighted in bold font were optimized for DP600 whereas the other ones correspond to suggested values from [5, 6]. A comparison of both HAH predictions with experimental results for the reverse simple shear loading case are shown in Figures 3 and 4. The experimental repeatability was investigated only in the forward direction, as the exact value of shear strain when inverting the shear direction is difficult to control. It can be seen that both models give a rather accurate description of the stress level, with some slight differences during reverse loading. In particular, HAH-2020 predicts a closer curvature in the re-yielding point than HAH-2014. A similar quality of the predictions is obtained for the other tests in the database.

Table 1. HAH-2014 and HAH-2020 parameters for DP600. Although some parameters have the same name for both models, they are related to different equations.

HAH-2014	HAH-2020	Comment
$q = 2$	$p = 2$, $q = 2$ $\xi_C = 4$, $\xi_L = 0.5$	Yield condition
$\mathbf{k} = \mathbf{130}$, $z = 5$, $k_R = 15$, $k_R^{'} = 0.2$	$k = k_1$, $k^{'} = k_1/2$ $\xi_R = 0.25$	Microstructure deviator
$\mathbf{k_1} = \mathbf{110}$, $\mathbf{k_2} = \mathbf{250}$ $\mathbf{k_3} = \mathbf{0.1}$	$\mathbf{k_1} = \mathbf{100}$, $\mathbf{k_2} = \mathbf{40}$, $\mathbf{k_3} = \mathbf{0.1}$ $\xi_B = 4$, $\xi_B^{'} = 1.5$	Reverse Loading
$\mathbf{k_4} = \mathbf{0.65}$ $\mathbf{k_5} = \mathbf{6.72}$	$\mathbf{k_4} = \mathbf{0.8}$ $\mathbf{k_5} = \mathbf{4.3}$	Permanent softening
$\mathbf{C} = \mathbf{0.77}$ $\mathbf{k_C} = \mathbf{135}$	$\mathbf{C} = \mathbf{0.74}$, $\mathbf{k_C} = \mathbf{60}$ $\xi_C^{'} = 6$, $k_C^{'} = k_C$	Cross loading contraction

4.2. Numerical simulation of direct re-drawing

Numerical simulations of the forming process were performed with the help of the non-iterative stress projection method [8], which provides a reliable convergence of the calculations. Previously [12], a modification was made in the tool geometry to overcome the convergence issues when using HAH-2014. However, the current study provides more consistent results with the original tool geometries. The influence of the material mechanical model on the punch force evolution with respect to the displacement throughout the forming process is illustrated in Figures 5 and 6 respectively for stage 1 and 2. Concerning the experiments, a good repeatability was observed with small differences in the load evolution, as highlighted in Figures 5 and 6. The gap is around 2 kN on the maximum level of 80 kN for the first stage and around 1 kN for the average plateau level of 60 kN for the second stage. In addition to both HAH models predictions, and out of comparison's sake, an isotropic hardening model based on a hybrid Swift-Voce law, referred to as HSV, is also used. It can be observed that there is a fair agreement of the load levels between the experimental and numerical results in the case of stage 1 cup. An over-prediction of the load

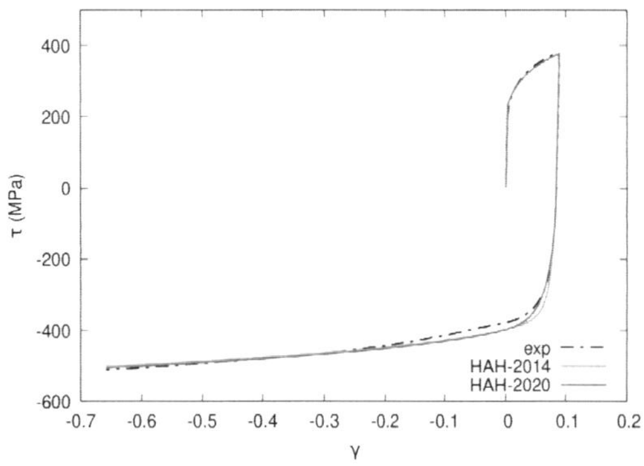

Figure 3. Forward-reverse shear test with $\gamma = 0.084$

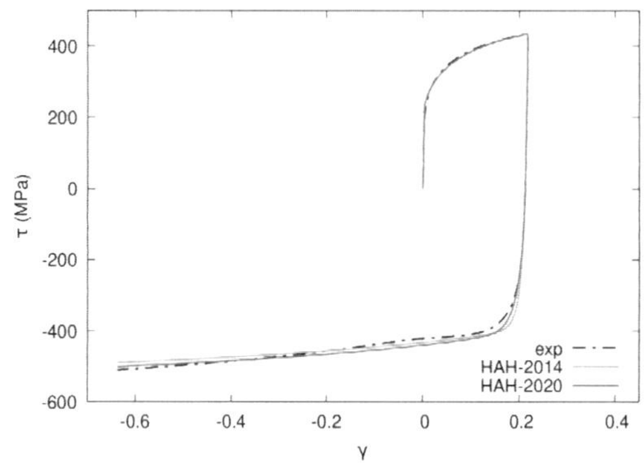

Figure 4. Forward-reverse shear test with $\gamma = 0.212$

level can be seen in the case of stage 2 for the 3 models, though HSV model prediction is slightly higher than the two other curves. The influence of the friction coefficient can be a reason behind it as it was set to a constant value of 0.18 for all the contact areas. Similarly to [12], a sensitivity analysis of the friction coefficient would be very interesting to carry out. Concerning the strain field, both HAH models provide very similar distribution of the plastic strain levels. Maximum value of the equivalent plastic strain was noted very close to respectively 0.52 for stage 1 and 1.12 for stage 2.

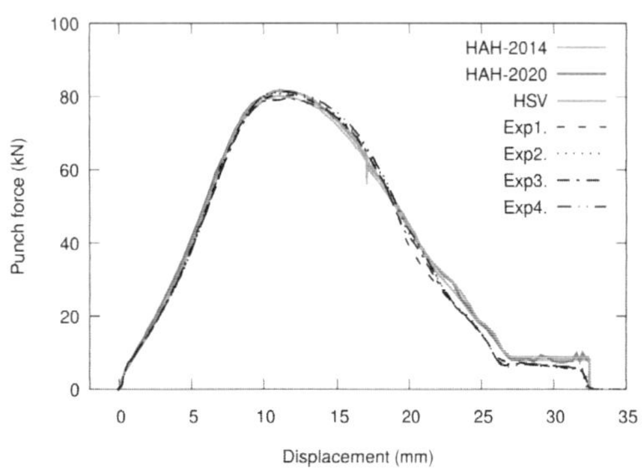

Figure 5. Punch force evolution for stage 1

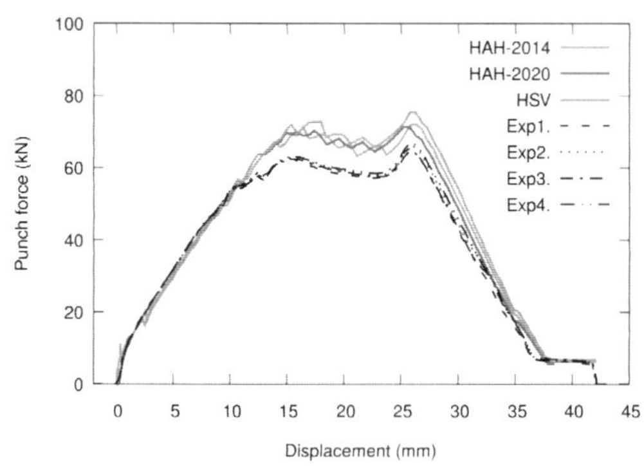

Figure 6. Punch force evolution for stage 2

4.3. Strain path change evolution

The stress-based expression for $\cos\chi$ used in HAH models has proven to be a good indicator of the series of changes in the strain path [15]. For stage 1, it seems that HAH-2014 over predicts the occurrence of cross loading, with the value of $\cos\chi$ approaching 0 in 2 areas in the cup wall, while the HAH-2020 model predicts mainly monotonic or reverse strain paths, which seems reasonable for a single stage cup drawing process. Figures 7 and 8 show the distribution of $\cos\chi$ at the end of stage 2. A similar over-prediction of cross loading scenario can be observed for HAH-2014 model. Moreover, the reverse strain path change observed with HAH-2020 is at a higher distance from the edge of the cup as compared with the HAH-2014 case. This comes from the principles of surface rotation by the loading path changes that are differently defined in the two models [16]. A comprehensive analysis is required to better understand these differences.

IOP Conf. Series: Materials Science and Engineering **1307** (2024) 012029 doi:10.1088/1757-899X/1307/1/012029

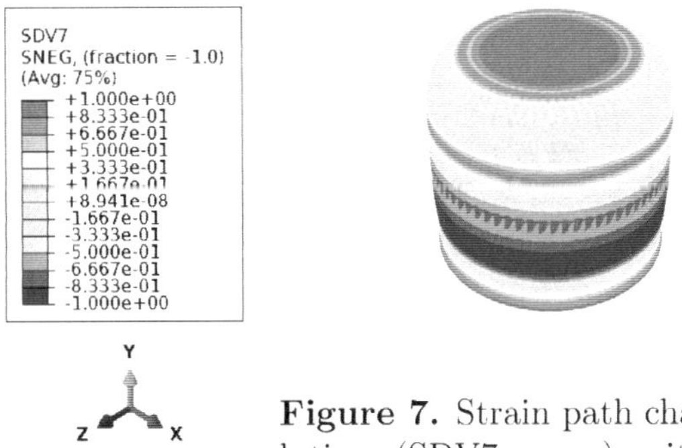

Figure 7. Strain path change evolution (SDV7=$\cos\chi$) with HAH-2014 for stage 2

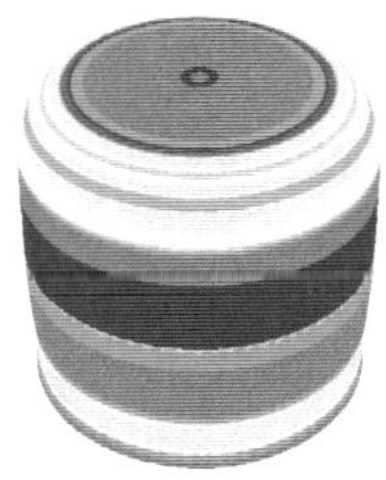

Figure 8. Strain path change evolution (SDV7=$\cos\chi$) with HAH-2020 for stage 2

5. Conclusions

Numerical simulations of the direct redrawing process of a DP600 steel were performed using HAH-2014 and HAH-2020 models implemented with a non-iterative stress projection method and numerical results are compared with experiments. Both HAH models estimate the behaviour of the material very accurately, however HAH-2020 seems to give a better prediction of the strain path change with the help of the improved formulation as compared to HAH-2014 model. The non-iterative stress projection method is proven to be stable and efficient, despite the high number of contact areas in this forming process.

Acknowledgments

This project has received funding from the Research Fund for Coal and Steel under grant agreement No 888153 (VForm xSteels) and from Institut Carnot ARTS (DAMAS).

Disclaimer

The results reflect only the authors' view, and the European Commission is not responsible for any use that may be made of the information it contains.

References

[1] Raphanel J L, Schmitt J H and Baudelet B 1986 *Int. J. Plast.* **2** 371–378
[2] Thuillier S and Rauch E F 1994 *Acta Metall. Mater.* **42** 1973–1983
[3] Schmitt J H, Aernoudt E and Baudelet B 1985 *Mater. Sci. Engineer.* **75** 13–20
[4] Barlat F, Grácio J J, Lee M G, Rauch E F and Vincze G 2011 *Int. J. Plast.* **27** 1309–1327
[5] Barlat F, Vincze G, Grácio J J, Lee M G, Rauch E F and Tomé C N 2014 *Int. J. Plast.* **58** 201–218
[6] Barlat F, Yoon S Y, Lee S Y, Wi M S and Kim J H 2020 *Int. J. Solids Struct.* **202** 947–962
[7] Yoon S Y, Barlat F, Lee S Y, Kim J H, Wi M S and J K D 2022 *J. Mater. Process. Technol.* **32** 117494
[8] Yoon S Y and Barlat F 2023 *Mech. Mater.* **183** 104683
[9] Carvalho-Resende T, Balan T, Bouvier S, Abed-Meraim F and Sablin S 2013 *Model. Simul. Mat. Sci. Eng.* **21** 015008
[10] Thuillier S, Manach P Y and Menezes L F 2010 *J. Mater. Process. Technol.* **210** 226–232
[11] Numisheet 1999 *Proceedings of the 4th International Conference and Workshop on Numerical Simulation of 3D Sheet Forming processes* (Besançon, France: J. C. Gelin, P. Picart)
[12] Hérault D 2021 *Virtual forming of metallic sheet materials : influence of complex strain paths on service life prediction* phdthesis Université Bretagne Sud
[13] Holmedal B 2019 *Int. J. Plast.* **123** 86–100
[14] Barlat F and Yoon S Y 2023 *Advanced Theories for Deformation, Damage and Failure in Materials* ed Altenbach H and Ganczarski A (Springer International Publishing) pp 63–119
[15] Hérault D, Thuillier S, Lee S Y, Manach P Y and Barlat F 2021 *Int. J. Mech. Sciences* **194** 106217
[16] Reyne B and Barlat F 2022 *Int. J. Plast.* **155** 103303

IOP Conf. Series: Materials Science and Engineering **1307** (2024) 012030 doi:10.1088/1757-899X/1307/1/012030

Forming of large scale bipolar plates for high power fuel cell stacks

Xiaolong Ma[1], Xianglu Zhang[1], Nan Guo[1], Li Qin[2], Yao Xiao[1, 3], Daijun Yang[4], Junying Min[1*], Pingwen Ming[4], Cunman Zhang[4]

[1] School of Mechanical Engineering, Tongji University, Shanghai 201804, China
[2] 3D Machinery (Shanghai) Co., Ltd, Shanghai 201615, China
[3] Institute for Advanced Study, Tongji University, Shanghai, 200092, China
[4] Clean Energy Automotive Engineering Center & School of Automotive Studies, Tongji University, Shanghai 201804, China

* Corresponding author: junying.min@tongji.edu.cn (J. Min)

Abstract. Developing high-power (e.g. megawatt-scale) single fuel cell stacks is of significance to extending the application of hydrogen fuel cells in high-energy-consumption fields such as aerospace, maritime, and rail transportation. Bipolar plate is one of the core components of hydrogen fuel cell stacks. Currently, the mainstream hydrogen fuel cell stacks achieve a maximum power of about 200 kW with a bipolar plate area of approximately 600 cm^2. While the megawatt-scale hydrogen fuel cell stacks requires large scale bipolar plates with an area of e.g. >2000 cm^2 and higher geometric complexity of flow channel. However, the structural design and manufacturing process for such large scale bipolar plates remain unexplored. Based on the concept of "partitioned modular manufacturing", the large scale bipolar plate is divided into multiple smaller scale bipolar plate modules in this work, and then integrated into a single component, which is then formed by applying multi-step stamping process to each module. Therefore, a so-called "partitioned multi-step stamping process" is proposed to form large scale bipolar plates with fine flow channels. Experimental validation was conducted using 0.1 mm thick titanium sheets and austenitic stainless steel sheets, demonstrating a prospective solution to manufacture large scale bipolar plates for high power hydrogen fuel cell stacks.

1. Introduction

Hydrogen possesses advantages such as zero emissions, high energy density, renewability, and abundant reserves. It has found extensive applications in various sectors, including transportation, energy storage, industrial production, and aerospace. Fuel cells can convert the chemical energy in hydrogen into electrical energy through electrochemical reactions. Among numerous types of fuel cells, Proton Exchange Membrane Fuel Cell (PEMFC) offers high power density and energy conversion efficiency, which enables it a crucial terminal application for hydrogen energy [1, 2].

Bipolar plate is one of the core components of PEMFC, serving several essential functions including: (1) providing channels for reactants/products; (2) collecting reaction currents; (3) offering support and enhancing overall structural strength [3]. Currently, materials used in the manufacturing of bipolar plates include graphite, metals and their alloys, composite materials, etc. Metal materials such as stainless steel, titanium, and their alloys are among the primary directions for the development of bipolar plate materials due to their characteristics of being lightweight, cost-effective, and excellent electrical and thermal conductivity [4]. The design and manufacturing processes of bipolar plate components are closely linked to the performance, lifespan, and cost of the fuel cell stack. The structure of the bipolar plate is illustrated in Fig. 1.

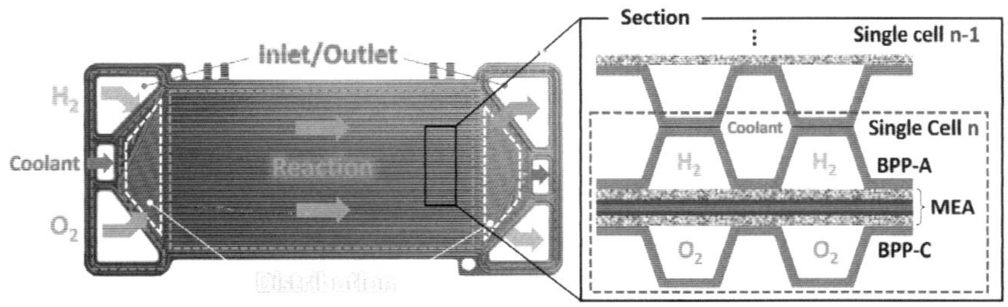

Fig. 1. A schematic diagram of metal bipolar plate structure

With the continuous promotion of the "green transportation" concept, the application of hydrogen fuel cells in high-energy consumption sectors is becoming increasingly prominent. However, the current mainstream PEMFC operates in a power range of 100 to 200 kW, which is challenging to meet the high-power demands under specific conditions. The output power of PEMFC is closely related to the area of the bipolar plate, increasing the effective area of the reaction zone can enhance the contact area between fuel and oxidant, thereby increasing the power output of a single fuel cell stack. For instance, a PEMFC with an output power of 200 kW usually requires a bipolar plate area of 600 cm^2, while megawatt-level PEMFC demand a bipolar plate area exceeding 2000 cm^2. Therefore, one direct and effective method to enhance the output power of a fuel cell stack is to increase the area of the bipolar plate.

To meet the power density requirements of the next generation of high-performance fuel cells, bipolar plates must possess more complex flow channel structures and higher forming precision. According to the forming characteristics of ultra-thin metal sheets, traditional cold stamping processes for bipolar plates are prone to serious defects such as springback, local thinning, and cracking [5]. To overcome these limitations, many scholars have proposed advanced forming processes, such as multi-step stamping [6], electromagnetic bulging [7], rubber forming [8], warm stamping [9], etc. Compared the above process, parts formed using the multi-step stamping process exhibit improved aspect ratio, draft angles, and smaller corner radii. Moreover, this process has advantages such as high forming precision, low cost, and high efficiency.

As the area of bipolar plates increases to megawatt level, several challenging bottlenecks arise: (1) The larger forming area results in a linear increase in the pressure required for stamping process, and the commonly used stamping machine are unable to meet the forming requirements. (2) As the area of bipolar plate increases, the difference in channel forming depth between the border and central regions gradually arises. This leads to poor consistency in the overall bipolar plate dimensions, causing differences in local contact pressure between the membrane electrode assembly and the bipolar plate. Consequently, this variation can adversely impact the efficiency of power generation in the fuel cell. [10-13]. (3) Tools suitable for manufacturing large-scale bipolar plates are expensive, and their processing is more challenging. Therefore, innovative approaches to the design methods and manufacturing processes for large-scale bipolar plates are still urgently needed to address these issues.

In this study, an innovative design method for large-format bipolar plates is introduced. Based on modular design concept, a large-scale bipolar plate is divided into a holistic structure composed of multiple small-scale bipolar plate modules. Correspondingly, a partitioned multi-step stamping process is proposed. This method involves multiple iterations of modular stamping on metal sheets to achieve high-precision manufacturing of large-scale bipolar plates. Finally, stamping tools for modular large-scale bipolar plates were developed, and experimental validation was conducted using austenitic stainless steel (SUS316L) and commercial pure titanium (CP-Ti) ultra-thin sheets, which proved the advancement of the design and manufacturing method proposed in this study.

2. Large-scale bipolar plate design and manufacturing method

2.1. Materials

Stamping experiments were conducted using ultra-thin sheets (0.1mm) of SUS316L and CP-Ti. The mechanical properties of the two materials in the rolling direction (RD), diagonal direction (DD), and transverse direction (TD) were characterized through uniaxial tensile tests. Detailed experimental methods and specimens size refer to the ASTM-E8M standard. The tested mechanical properties of the two materials are summarized in Tables 1 and 2.

Table 1. Mechanical Properties of the ultra-thin SUS316L sheet along various orientations.

Orientation	Young's modulus (GPa)	Yield Strength (MPa)	Tensile Strength (MPa)	R-value	Elongation (%)
RD	177.0	270.7	625.4	0.73	52.0
DD	165.2	276.0	610.5	1.09	68.1
TD	188.2	282.4	625.2	1.06	69.3

Table 2. Mechanical Properties of the ultra-thin CP-Ti sheet along various orientations.

Orientation	Young's modulus (GPa)	Yield Strength (MPa)	Tensile Strength (MPa)	R-value	Elongation (%)
RD	86.8	205.0	297.5	1.89	39.2
DD	98.0	197.5	283.2	3.67	40.2
TD	81.2	189.9	297.5	6.18	36.2

2.2. Large-scale bipolar plate design

To address the issue related to large-scale bipolar plates at the end of Section 1, this study proposes a novel design approach for large-scale bipolar plates. The concept of "modularity" is integrated into the design of large-scale bipolar plates for the first time, by dividing large-scale bipolar plates into multiple small-scale modules with identical characteristics. Through adjusting the number of modules, the size (length and width) of the large-scale bipolar plate can be customized. However, to minimize the size consistency fluctuations between modules during forming process, the number of modules composing the large-scale bipolar plate should be reduced. Since each module is independent, it must possess complete bipolar plate functionality, including gas inlets and outlets, distribution regions, flow channel regions, and other functional features. Based on the bipolar plate design method of "single plate with multiple chambers", the small-scale modules designed in this study consist of two flow channel regions. Each flow channel region is equipped with corresponding gas inlets and outlets, as well as distribution regions. Specifically, a common gas inlet and outlet are placed in the central region of the module, supplying gas simultaneously to the two side flow channel regions. This design enhances the effective utilization of the bipolar plate's area.

Through the modular design method for large-scale bipolar plates described above, we evenly distribute the large characteristic dimensions (length and width) of an integrated large-scale bipolar plate onto each small-format module. This approach helps reduce the difference in channel depths between the border and central regions (along and perpendicular to the flow channel direction) during the forming process, optimizing the uniform distribution of reactants and products within the flow channel regions. A test bipolar plate is designed in this study as shown in Fig. 2a, where the blue dashed box represents a single small-scale module, and the entire plate is composed of 2 modules. Taking Module-1 as an example, the yellow dashed box represents the gas inlets and outlets, the white dashed box represents the distribution regions, and the green dashed box represents the flow channel regions. Theoretically, by increasing the quantity of modules, bipolar plate with an unlimited area can be manufactured. The red dashed box represents the stamping overlap region between the two modules, which will be discussed in subsequent sections.

The designed area of the test bipolar plate is 530 cm^2. The general dimensions and cross-sectional dimensions of the flow channels are illustrated in Fig. 2a and Fig.2b, respectively. The characteristic dimensions (length and width) of this test plate are half of the target large-format bipolar plate. After validating the feasibility of the large-scale bipolar plate design method proposed in this study through experiments, it can be further applied to large-scale bipolar plate design projects.

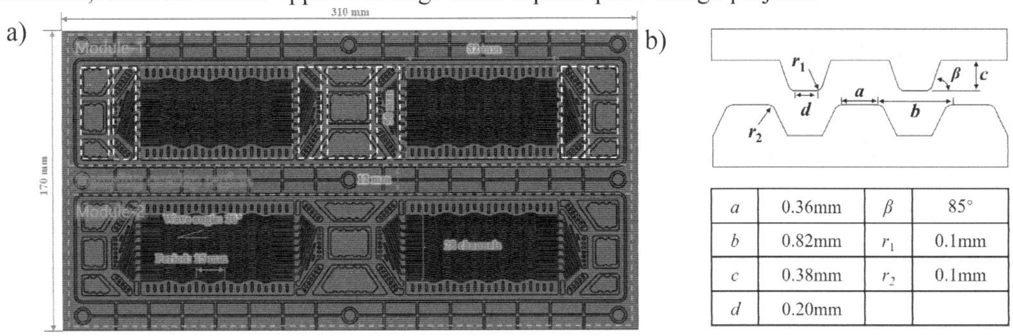

a	0.36mm	β	85°
b	0.82mm	r_1	0.1mm
c	0.38mm	r_2	0.1mm
d	0.20mm		

Figure. 2. a) Structural design of partitioned bipolar plate, b) cross-sectional dimensions of flow channels

2.3. *Partitioned multi-step stamping process*

Based on the modular design scheme for large-scale bipolar plates introduced in Section 2.2, this section outlines the corresponding partitioned multi-step stamping process, as illustrated in Fig. 3. Traditional stamping processes involve the integral forming of an entire large-scale bipolar plate in a single step. However, the large-scale bipolar plate is partitioned into multiple independent small-scale modules within the modular design. Hence, this study proposes a novel forming process for large-scale bipolar plates. During this process, each module undergoes independent stamping. After forming one module at a station, the metal sheet is then moved to the next module position, aligning it with the stamping station, and the stamping process is repeated. This iterative procedure enables the partitioned stamping of large-scale bipolar plates.

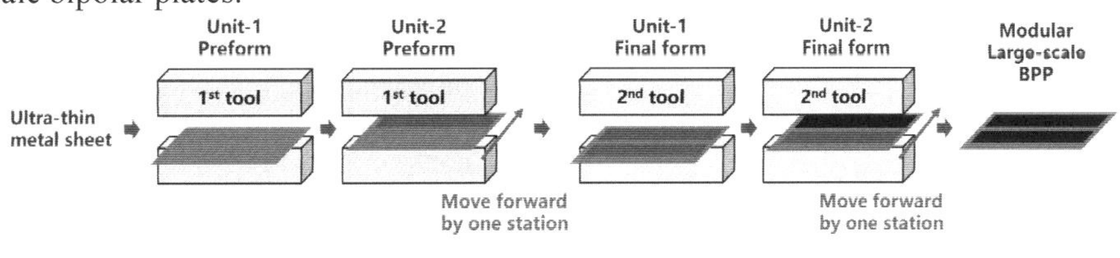

Figure. 3. A schematic diagram of partitioned multi-step stamping process

Due to the challenging forming requirements of flow channels in the test bipolar plate, single-step cold stamping is insufficient to meet the forming demands. Therefore, a multi-step stamping process is employed for each module's forming. Taking into account the improvement of the number of forming steps on the formability of ultra-thin metal sheets and considering manufacturing costs, this study determines that the multi-step forming process involves two steps (pre-forming and final forming), as shown in Fig. 3. It should be noted that during the pre-forming step, the formed flow channel structure does not merely about decreasing depth below the design value; instead, it involves changing the cross-sectional shape of the channel. This modification induces plastic deformation of the ultra-thin metal sheets in the entire region, optimizing the plastic flow of materials during the final forming process. This approach aims to reduce local thinning and cracking of the bipolar plate, enhancing the formability of the ultra-thin metal sheets under complex geometric shapes [14].

The tools for the partitioned multi-step stamping process of large-scale bipolar plates were developed, including preforming and final forming (Fig. 4). In this study, the flow channel corner radii of the preform tool was enlarged from 0.1 mm from 0.25 mm, compared to the final form tool. Since

IOP Conf. Series: Materials Science and Engineering **1307** (2024) 012030 doi:10.1088/1757-899X/1307/1/012030

this process targets only a single small-scale module for stamping within a single step, it can reduce the tonnage requirements on the stamping machine compared to traditional integral cold forming processes. Additionally, it can decrease the difficulty in manufacturing the tools and the associated machining costs.

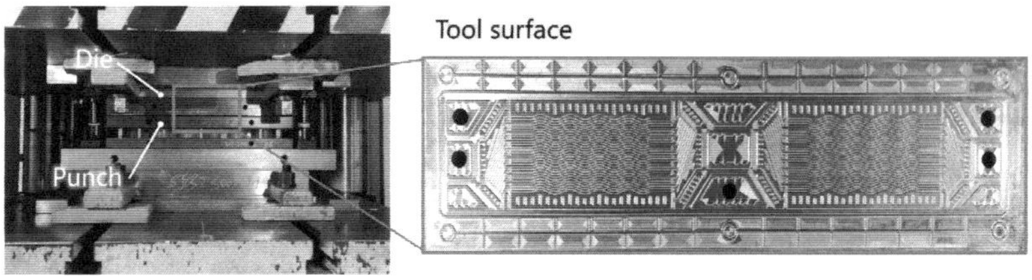

Figure. 4. Partitioned multi-step stamping tools (final machined tool surface)

Furthermore, a "stamping overlap region" (refer to the feature inside the red dashed box in Fig. 2a) is positioned between two adjacent modules. This region is shared by the two modules and is repeatedly stamped in partitioned forming steps of the two modules. Setting up a "stamping overlap region" offers several advantages: (1) It enhances the effective utilization of the bipolar plate's area. (2) During the partitioned stamping process, manual movement of the blank is required for forming adjacent modules. This can lead to positioning errors in the metal sheets, resulting in lower positional accuracy between the two modules after forming. By setting special geometric features in the stamping overlap region, it serves as a precise positioning reference for adjacent modules after the initial stamping, which can effectively improve the positional accuracy between modules. (3) Reinforcement features can be introduced in the stamping overlap region to enhance the overall rigidity of the large-scale bipolar plate.

2.4. Experimental methodology

Stamping experiments were conducted on ultra-thin SUS316L and CP-Ti sheets to validate the feasibility of the large-scale bipolar plate design and manufacturing method proposed in this study. A pressure test is conducted before experiments. Gradually increase the pressure and measure the forming depth of flow channel after each stamping. When there is no significant change in depth, it indicates that the pressure is sufficient, and this pressure is used for subsequent experiments. Firstly, single and multiple stamping process were both conducted using only the final forming tool as a reference. Then the ultra-thin metal sheets were subjected to multi-step stamping using both the pre-forming and final forming tools for a single module to confirm the formability improvement.

In response to the uneven channel forming depth issue caused by large characteristic dimensions, comparative experiments on multi-step stamping of ultra-thin CP-Ti sheets were conducted using both the partitioned multi-step stamping tools (Fig. 4) and the testing tools (Fig. 5), which also include preforming and final forming. Measurements of channel forming depth were then conducted on the bipolar plate specimens after forming. Ten channels were selected from the border to the central region, sequentially numbered from 1 to 10, and their forming depths were measured. For each waved channel, data collection was performed at intervals of one period along the direction of channel extension (Fig. 5a). The displayed data is the mean value of three repeated stamping specimens.

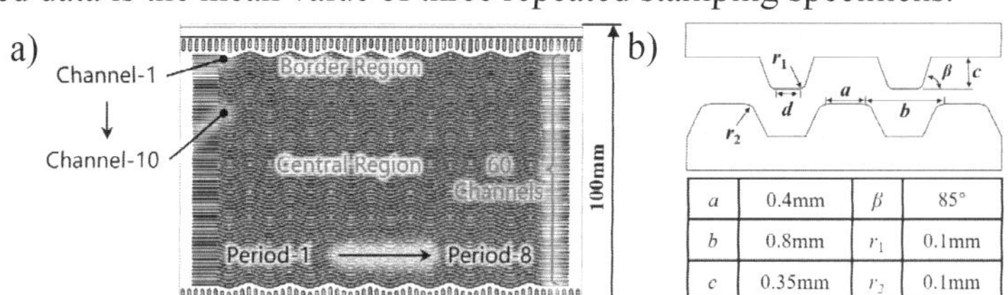

a	0.4mm	β	85°
b	0.8mm	r_1	0.1mm
c	0.35mm	r_2	0.1mm
d	0.23mm		

Fig. 5. a) Schematic diagram of testing tool, b) cross-sectional dimensions of flow channels

3. Result and discussion

3.1. Mechanical properties of ultra-thin metal sheets

From the mechanical properties data in Section 2.1, it can be seen that the yield strength, tensile strength, and elongation parameters of ultra-thin SUS316L sheet are significantly higher than those of ultra-thin CP-Ti sheet. Additionally, the ultra-thin SUS316L sheet exhibits lower anisotropy (as indicated by the R value) compared to the pronounced anisotropy observed in the ultra-thin CP-Ti sheet. This suggests that the formability of ultra-thin SUS316L sheet is much greater than that of ultra-thin CP-Ti sheet. Consequently, during the forming process of intricate flow channel structures, ultra-thin CP-Ti sheet tends to experience premature occurrences of local thinning and cracking. The use of two materials with significantly different mechanical properties aims to validate the wide applicability of the proposed bipolar plate design and manufacturing method across different metal materials.

3.2. Formability improvement of multi-step stamping process

For ultra-thin CP-Ti sheets, the results revealed cracking in flow channel region (Fig. 6a) for the single step stamping bipolar plate, while the multi-step stamping bipolar plate did not exhibit any cracking defects across the entire range (Fig. 6b). Therefore, it can be inferred that the multi-step stamping process, by incorporating a pre-forming step with different flow channel cross-sectional shapes, engages as much material as possible within the flow channel structure to participate in plastic deformation, thereby enhancing the formability of ultra-thin metal sheets. Changing only the number of forming steps without altering the flow channel cross-sectional shape of the tool, has a relatively small impact on the formability of ultra-thin metal sheets. Repeating the above experiments with ultra-thin SUS316L sheets yielded results similar to those obtained with ultra-thin CP-Ti sheets (Fig. 6c, 6d).

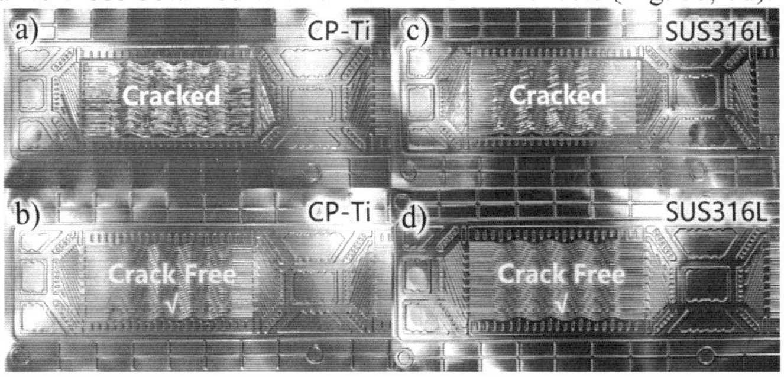

Fig. 6. Single module BPP specimen for CP-Ti: a) and b); and SUS316L: c) and d) on single/multi-step stamping

3.3. Channel depth consistency improvement of partitioned multi-step stamping process

The partitioned multi-step stamping process was applied to form large-scale bipolar plates using ultra-thin SUS316L and CP-Ti sheets. The formed large-scale bipolar plate specimens are depicted in Fig. 7. The results indicate that the partitioned multi-step stamping process allows multiple modular stamping on a single metal sheet, facilitating the formation of large-scale bipolar plates.

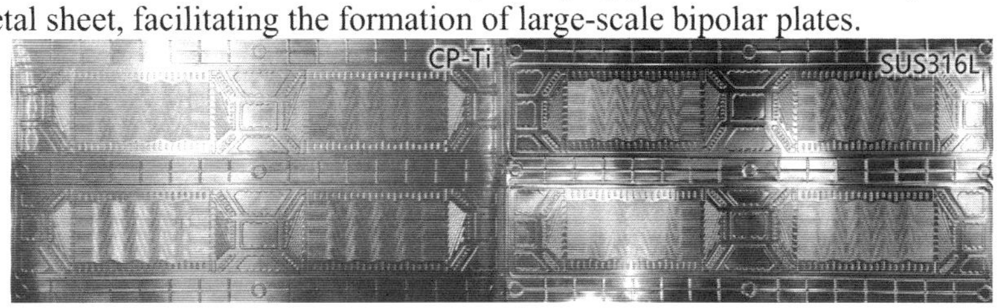

Fig.7. Partitioned multi-step stamping specimen for CP-Ti and SUS316L

For bipolar plates formed by testing tool, the measurements results (Fig. 8a) indicate a gradual decrease in channel forming depth from border region to central area. For a single channel, the forming depth in the middle is greater than at the two ends, consistent with trends observed in previous research[10, 11]. Across the entire measurement region of the specimen (distance in the channel width direction from channel 1 to 10 is 12mm, channel length is 105mm, with a total of 8 periods), the maximum observed difference in channel forming depth reaches 60μm.

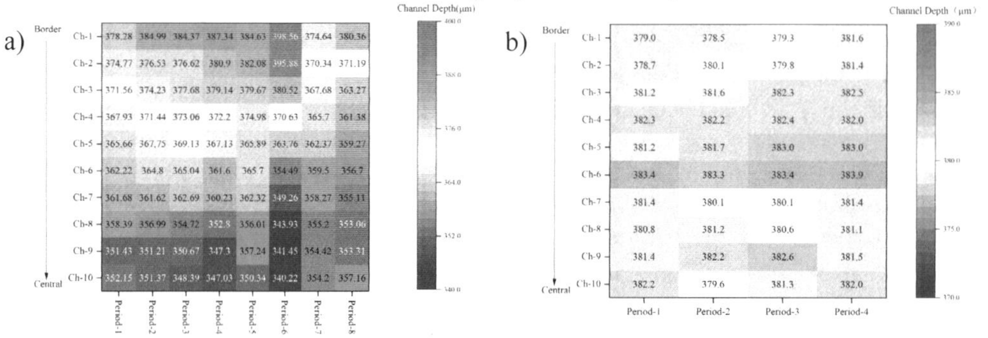

Fig. 8. Distribution diagram of channel forming depth for a) testing specimen and b) partitioned multi-step stamping specimen

For the partitioned bipolar plate, the characteristic dimensions (length and width) of a single module are relatively smaller. The measurements result of a single active area (Fig. 8b) indicate that, across the entire measurement region of the specimen (distance in the channel width direction from 1 to 10 is 12mm, channel length is 45mm, with a total of 4 periods), the maximum observed difference in channel forming depth is less than 10μm.

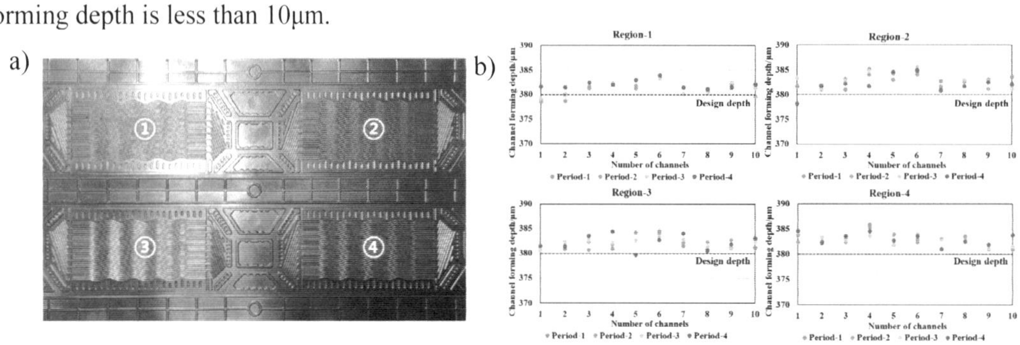

Fig. 9. Consistency test of partitioned multi-step stamping specimen: a) Sampling region, b) channel forming depth data

Simultaneously, a comparison was made between the forming depths of channels in the adjacent 4 modules of the large-scale bipolar plate specimen (Fig. 9a), sampling method is the same as before. The results (Fig. 9b) demonstrate a consistent trend in the distribution of channel forming depth data among multiple modules of the large-format plate specimen. Across the entire measurement region of the specimen (690mm in length and 380mm in width), the maximum difference in channel forming depth is less than 10μm. It is noteworthy that the partitioned multi-step stamping process of the large-scale bipolar plate underwent four times stamping processes on the same ultra-thin metal sheet, indicating the excellent consistency of the proposed partitioned multi-step stamping process.

4. Conclusions

In this study, a modular design approach for large-scale bipolar plates used in high-power PEMFC is proposed for the first time. Correspondingly, a partitioned multi-step stamping process is designed. Process test tools were developed, and stamping experiments were conducted using ultra-thin SUS316L and CP-Ti sheets to manufacture a test bipolar plate with an area of 530 cm^2. The test bipolar plate were then subjected to dimensional measurements to validate the feasibility of the design and manufacturing method. The main conclusions are summarized as follows:

(1) The partitioned multi-step stamping process has demonstrated success in manufacturing large-scale bipolar plates and is applicable to various metal materials such as SUS316L and CP-Ti. The consistency of channel depths in the formed parts is satisfactory (deviation less than 10μm for the 530 cm^2 test bipolar plate), and the channel dimensions remain stable even after four times stamping processes.

(2) The modular design approach for large-scale bipolar plate can reduce the requirements on stamping machine and tools during stamping process. Besides, it addresses the consistency issues in channel depth arising from larger characteristic dimensions. Theoretically, by adjusting the size (length and width) as well as the quantity of the modules, it is possible to design and manufacture large-scale bipolar plate suitable for high-power PEMFC applications.

Acknowledgments

This study is supported by the National Key Research and Development Program of China (No. 2022YFE0207500).

References

[1] Lebrouhi B E, Djoupo J J, Lamrani B, et al. Global hydrogen development-A technological and geopolitical overview. International Journal of Hydrogen Energy, 2022, 47(11): 7016-7048.

[2] Wang Y, Pang Y, Xu H, et al. PEM Fuel cell and electrolysis cell technologies and hydrogen infrastructure development–a review. Energy & Environmental Science, 2022, 15(6): 2288-2328.

[3] Wilberforce T, El Hassan Z, Ogungbemi E, et al. A comprehensive study of the effect of bipolar plate (BP) geometry design on the performance of proton exchange membrane (PEM) fuel cells. Renewable and sustainable energy reviews, 2019, 111: 236-260.

[4] Asri N F, Husaini T, Sulong A B, et al. Coating of stainless steel and titanium bipolar plates for anticorrosion in PEMFC: A review. International Journal of Hydrogen Energy, 2017, 42(14): 9135-9148.

[5] Karacan K, Celik S, Toros S, et al. Investigation of formability of metallic bipolar plates via stamping for light-weight PEM fuel cells. International Journal of Hydrogen Energy, 2020, 45(60): 35149-35161.

[6] Nakagaki N. The newly developed components for the fuel cell vehicle, Mirai. SAE Technical Paper, 2015.

[7] Dong P, Li Z, Feng S, et al. Fabrication of titanium bipolar plates for proton exchange membrane fuel cells by uniform pressure electromagnetic forming. International Journal of Hydrogen Energy, 2021, 46(78): 38768-38781.

[8] Lee K H, Jin C K, Kang C G, et al. Fabrication of titanium bipolar plates by rubber forming process and evaluation characteristics of TiN coated titanium bipolar plates. Fuel Cells, 2015, 15(1): 170-177.

[9] Modanloo V, Alimirzaloo V, Elyasi M. Manufacturing of titanium bipolar plates using warm stamping process. Arabian Journal for Science and Engineering, 2020, 45(11): 9661-9667.

[10] Barzegari M M, Khatir F A. Study of thickness distribution and dimensional accuracy of stamped metallic bipolar plates. International Journal of Hydrogen Energy, 2019, 44(59): 31360-31371.

[11] Zhou Z. Sringback prediction in multichannel forming of ultrathin stainless steel sheet. Shanghai Jiao Tong University, 2017.

[12] Peng L, Wan Y, Qiu D, et al. Dimensional tolerance analysis of proton exchange membrane fuel cells with metallic bipolar plates. Journal of Power Sources, 2021, 481: 228927.

[13] Qiu D, Yi P, Peng L, et al. Channel dimensional error effect of stamped bipolar plates on the characteristics of gas diffusion layer contact pressure for proton exchange membrane fuel cell stacks. Journal of Fuel Cell Science and Technology, 2015, 12(4): 041002.

[14] Xu Z, Li Z, Zhang R, et al. Fabrication of micro channels for titanium PEMFC bipolar plates by multistage forming process. International Journal of Hydrogen Energy, 2021, 46(19): 11092-11103.

IOP Conf. Series: Materials Science and Engineering **1307** (2024) 012031 doi:10.1088/1757-899X/1307/1/012031

Investigation on the limp properties of formed foils for hydrogen applications

S Wituschek*, D Römisch, M Lechner and M Merklein

Friedrich-Alexander-Universität Erlangen-Nürnberg, Institute of Manufacturing Technology, Egerlandstr. 13, 91058 Erlangen, Germany

E-mail: *simon.wituschek@fau.de

Abstract. The mass production of bipolar plates for electrolyzers and fuel cells is a central step towards the realization of efficient and cost-effective energy systems of the future. However, current production processes are reaching their limits and can hardly realize the quantities that will soon be demanded, nor can they scale up to the required volumes. Particularly for the handling of half-plates and bipolar plates, major challenges are to be expected, especially with regard to production rates. Existing handling systems have restricted scalability and precision. Therefore, new stacking technologies are necessary, which have to be adaptable to the mechanical properties of the components and maintain tight tolerances during stacking to ensure hydrogen sealing for safety and efficiency. An important property in the handling of the plates is their limpness, which is distinguished by instability of the components as well as plastic deformation at low forces and moments. Therefore, the limp behavior of the components must be analyzed. To investigate the limpness of foil components, a flowfield is first formed using a 1.4404 stainless steel foil with a sheet thickness of 0.075 mm. Subsequently, the workpieces are analyzed in terms of their limp properties by means of a 3-point bending test.

1. Introduction

The role of hydrogen in the energy supply of the future is of central importance. As a adaptable energy carrier, it offers solutions to the challenges of renewable energy, mobility and heat supply [1]. Its potential to decarbonize and store renewable energy makes it a key element in a sustainable energy future [2]. Manufacturing technology together with electrochemistry plays a crucial role on the way to mass production and deployment of hydrogen applications [3]. In order to develop this environmentally friendly energy resource, efficient production processes are essential. The proton exchange membrane (PEM) fuel cell consists of several components. In the center layer is a proton exchange membrane, which is coated with a catalyst and forms the catalyst coated membrane (CCM) from the composite. The following layer is the gas diffusion layer, which ensures an even distribution of the media on both sides of the cell and is coated with a micro-porous layer to improve the lifetime of the system. An important characteristic is the tightness of the entire system, and this must also be taken into account during assembly [4]. The single cell is completed with the bipolar plate. From a production perspective, this forms a large part of the PEM fuel cell. The bipolar plate consists of two joined half plates and is subject to a variety of requirements. These include the distribution of fuel and oxidant, the separation of the individual cells, the transport of charge and water and the cooling of the overall system. Basic requirements can be derived from this for the material utilized. These include chemical resistance in the form of corrosion resistance, the impermeability of the formed components, electronic and thermal

IOP Conf. Series: Materials Science and Engineering **1307** (2024) 012031 doi:10.1088/1757-899X/1307/1/012031

conductivity, adequate mechanical properties and economic manufacturability. In today's PEM fuel cell applications, stainless steel is predominantly used, as it is characterized by high corrosion resistance, comparatively low costs, low electrical resistance and good coating properties. The plates are coated in a subsequent process step [5]. Based on current production capacities and possibilities, the metallic bipolar plate offers a high potential for industrialization and cost reduction and can therefore promote the progress of hydrogen applications to the mass market.

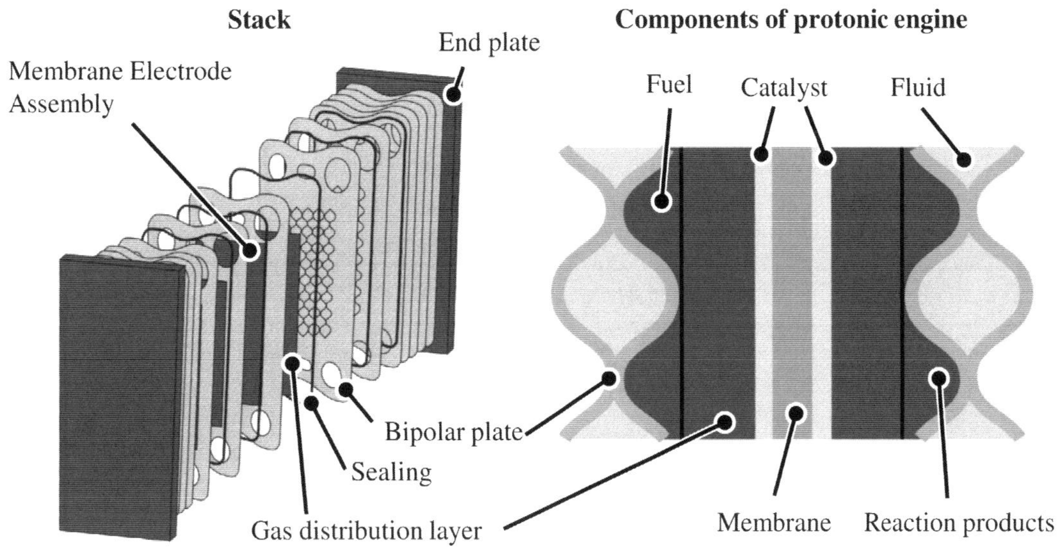

Figure 1: Stack configuration and components of a protonic engine

The flow field design is the central element for the design of the bipolar or half plates. This significantly influences the throughput of the media flowing through and therefore the performance potential of the entire system [6]. Homogeneous distribution of the media and minimization of pressure losses during operation are important. When designing the flowfield, the focus is therefore on the operating phase of the component, but the manufacturability of the half plates must also be taken into account for bipolar plates produced using forming technology. The main factors influencing this are the course of the channels in the plane of the plate and the channel cross-section aspect ratio [7]. In addition to the flowfield properties, where the manufacturing technology is a limiting factor, the foil thickness is another significant characteristic for the production technology. Since the bipolar plates account approx. 80 % of the total mass of fuel cells, the reduction of the foil thickness leads to significant mass losses [8]. In current applications, the foils are approx. 75 µm thick, but there is further research potential for reducing the foil thickness to reach a lower overall weight of the stack. This is particularly relevant for mobile solutions such as heavy-duty applications and the costs can be reduced as less material is required. A major challenge in the production of electrolyzers and fuel cells is stacking the various components [9]. A widely used method in this context are pick and place systems with robotic solutions [10]. The foil thickness also plays a role here. Due to the low thickness of the foil, the bending stiffness is very low and the components can be described as limb [11]. This means that deformation of the foil before the forming process and the resulting strain hardening is a problem for handling in the production process. As the foil thickness is to be further reduced in the future in order to achieve a lower total mass of the stack, new challenges arise for forming technology and handling. In particular, the impermeability of the systems requires low tolerances, which are also reflected in the tolerance chain, due to the large number of stacked components [12]. As a result, the handling of half and bipolar plates requires precision technology to avoid deforming the parts at high handling speeds. Another challenge that affects both the handling and the forming technology is the surface quality of the components. As the foil thickness is low, there is little scope for surface-protecting measures, as the forming capability of the foils must be utilized to a large extent. Handling technologies that cause changes to the surfaces cannot be used either. Despite increasing automation in the industry, the handling of flexible parts is often still carried out manually and automated processes reach their limits with flexible gripping systems capable of gripping foils, rubbers or cylinders in a process-robust manner. The geometric and mechanical properties make the components difficult to handle and have so far largely prevented the automation potential from being utilized. In view of the expected increase in applications with hydrogen, significantly higher production capacities are expected and therefore a better fundamental understanding of the production processes is necessary. However, there is enormous potential here for reducing costs through automated solutions [10].

2. Methodology

Manufacturing the half plates from bipolar plates, a 3-dimensional polymer-printed punch and die combination was produced as part of this study. For fundamental investigations, the rapid production and the great research potential for geometric properties offer a significant advantage. Figure 2 shows the die of the tool. The use of half plates produced by forming technology enables the characterization of factors influencing the individual components from industrial processes and the rapid production of samples. In previous tests, plates machined from solid material were often used, which require significantly more effort to produce. As already described, an important feature is the distribution of the media in the cell through the bipolar plate. This distribution is influenced in particular by the design of the flow field, in which the media are channeled to the subgasket. The geometry of the flow field enables not only embossing processes to be used during forming, but also other basic forming processes such as stretch forming and deep drawing portions. The flowfield is rectangular and has an active area of 2500 mm^2. The number of channels is 21 and is distributed linearly over the entire active area. The channel depth is 0.5 mm and the channel width is 2.2 mm. For these tests, the distribution between the channel and the bar is also distributed equally to reduce stress peaks. The angle between the sheet level of the output foil and the flank angle of the channel is 125°. This flowfield design prevents embossing during the forming process and reduces localized stress peaks. Figure 2 shows a schematic depiction of the channel geometry. As part of the investigations, different forming forces are analyzed during the forming of the half plates and the influence on the bending stiffness is then determined.

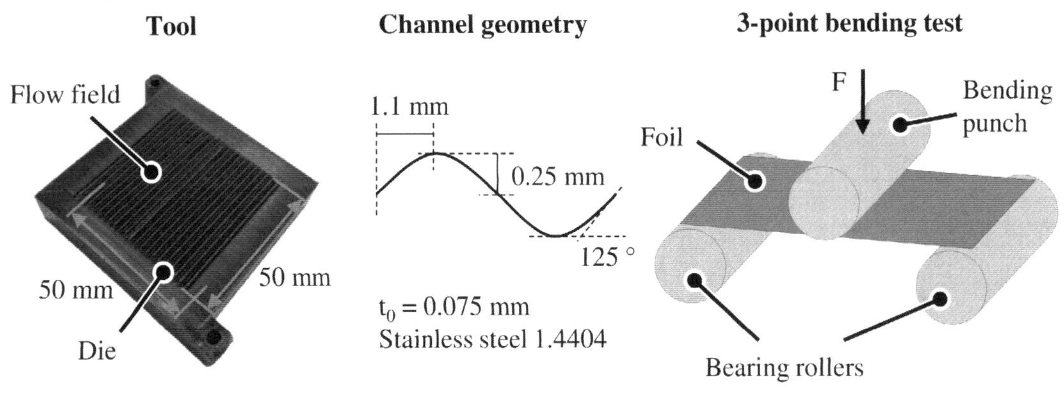

Figure 2: Tool design and geometry, schematic channel geometry and 3-Point-Bending test according to ISO 7438:2020 [13]

The bending stiffness of the components is an important characteristic parameter for the handling of the half and bipolar plates. In order to be able to describe this and to obtain reference points for these properties, 3-point bending tests are carried out as part of these investigations. Based on ISO 7438 [13], the foil is placed on two bearing points and formed with the bending punch to determine the plastic deformability of metallic materials in foil size under bending stresses. Influences on the bending stiffness of the foil can be determined from the force measured on the bending punch during the test.

In addition to analyzing the force-displacement curves during 3-point bending, geometric analyses of the samples are also carried out before the mechanical characterization. The three-dimensional profilometer Alicona InfiniteFocus G5 is used for this purpose. The investigations into the bending stiffness of formed half-sheets will be used to determine the extent to which formed and flat foils pose challenges for handling in the production of stacks of bipolar plates.

3. Results

The investigation of the limp properties of semi-finished plates produced using forming technology can be divided into two process steps. In the first step, the panels are formed with a forming tool. Subsequently, the formed specimens are tested in a 3-point bending test.

3.1. Forming process of the half plates

In the first stage, the half plates are manufactured with different maximum forming forces. For this purpose, the two additive manufactured tool components are placed between two upsetting trajectories and the two tools are aligned. For the tests, half plates are formed in ten steps between 5 kN and 70 kN. Based on this, differently shaped channel geometries are achieved and this results in an influence on the bending stiffness in the 3-point bending test investigated subsequently. Figure 3 shows an overview of the geometric properties of the overall half plate for a forming force of 70 kN. The result shows the half plate with channels at the upper side. The illustration indicates a homogeneous shaping of the channels in the channel direction as well as alongside the channels.

The intakes, which are placed to indicate the introduction of gas into the flow field, also show homogeneous height distributions. However, the overview of the entire half plate also shows clearly inhomogeneously distributed heights at the edges of the plate. These are up to 0.25 mm more than the channel depth. It can also be seen that the height reduction decreases from the edge towards the forming zone. This indicates that the edges are subject to bending and have been introduced either by the forming process or the subsequent manual handling of the specimens. As the process is carried out without a blank holder, a smoothing of the sample can only be observed at high force levels.

At lower forming forces, a clear wrinkling can be identified, which is comparatively pronounced in the corners of the foil. However, the sample in Figure 3 was formed at 70 kN and the foil is in full contact with the die and punch surfaces. This means that the deviations from the target geometry can be ascribed to a significant extent to factors not related to the forming process. It can also be seen that the deviations from the target geometry are only in the non-formed area of the sample. This shows that foils with a thickness of 0.075 mm are already plastically formed when very low forces and moments are applied. Figure 3 also shows a comparison between a sample formed with 40 kN and 70 kN.

The contour is recorded from above with a three-dimensional profilometer and thus shows the profile of the foil on the upper side. The geometric limits of the channel geometry from +0.25 mm to -0.25 mm are shown next to the curves of the foil surfaces so that a comparison can be made with the target geometry of the channels. The results indicate that the desired depth of the channels is not achieved at both force levels. Disregarding the foil thicknesses that cannot be mapped due to the measurement method, this is also evident. It can be assumed that the tools are deformed due to the surface pressures and that the full channel geometry cannot be achieved. However, it can also be seen that the channel depths are evenly distributed over the individual channels and that the channels only drop off at the edges.

However, in addition to the influence of the forming process, the influence of the edge zone on the final geometry can also play a role here. Thus, the bending over the surface section may have been introduced by the subsequent manual handling of the sample. Figure 3 also shows a comparison of the surface geometries of different maximum force levels during forming. It can be seen that the higher forming force of 70 kN achieves a deeper forming of the channels in the flow field. The profile curve shows a greater geometrical difference between the force levels in the inner area of the foil, except for the first channel. The better forming in the outer area is assumed to be caused by more material flowing out of the flange area of the foil, since there are no channels. In addition, it can be stated that die failure was observed in forming tests with higher forces.

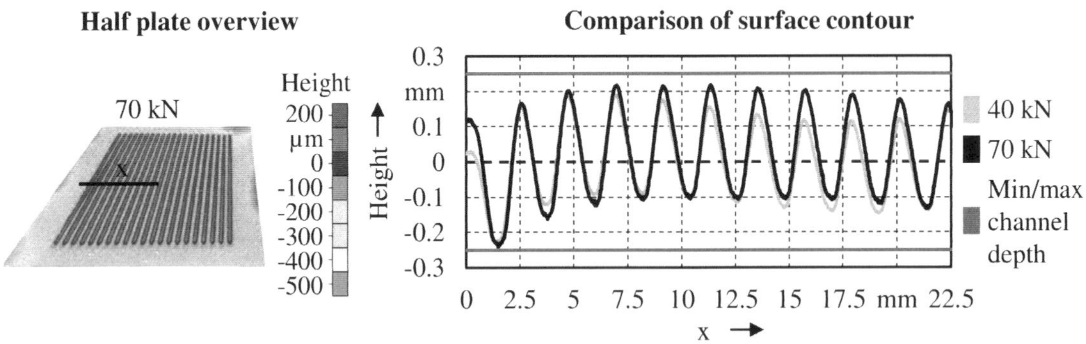

Figure 3: Geometric properties of formed half plates as overview and in section in comparison of 40 kN and 70 kN

3.2. *Testing the bending stiffness with a 3-point bending test on limp characteristics*

Following the manufacturing and geometric testing of the formed half plates, the 3-point bending tests are carried out. The 3-point bending tool is installed in a universal testing machine ZwickRoell Z10 for this purpose, as shown in Figure 4. The tool is clamped and aligned with the clamping jaws. For the tests, the formed half-plates with channels at the bottom are placed on the bearing points of the lower tool. This means that the flange areas, which are outside the active surface of the channels, are in contact with the tool.

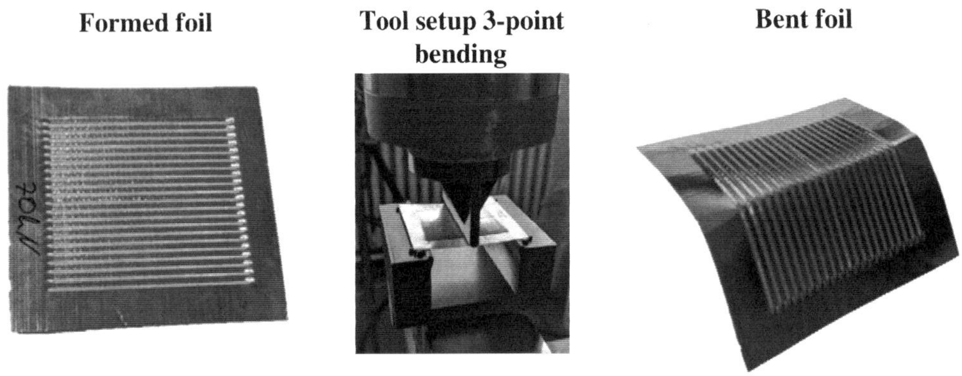

Figure 4: Experimental test procedure of forming and bending the foil

The bending punch comes into contact with the specimen on the bottom side, which is the side where the channels do not protrude. During the process, the upper tool moves from a defined position above the overlying half plates at a constant speed of 10 mm/min. The force-displacement curves in Figure 5 have differently distributed heights due to the different geometric formations, in particular the flange areas. As already described, these can be seen in the false color image in Figure 3. At the start of the process, the inhomogeneous deflections of the individual foils, which were formed with different force levels, are aligned. The contact between the punch and the foil, as well as the bearing points and the foil, is also not completely levelled out due to the bends. The unevenness between the individual samples is only evened out from a force level of approx. 1 N. After complete contact is achieved over the entire tool length, very different gradients of the force displacement curves can be identified. However, it can be determined that the gradient is dependent on the maximum force levels during the forming process of the half plates and with increasing force there is also a steeper increase in the 3-point bending test at the start of the test. It can also be concluded from the diagram that an increase in the forming force at lower force levels causes a greater increase in the force during the 3-point bending test. At the same time, the force during 3-point bending increases less in half plates that have been formed with higher forming forces. This is possibly because of the shape of the half plates, which become saturated as the force increases and the geometric differences between the half plates are smaller. The maximum force in the 3-point bending test is achieved with the half plate, which was formed with the highest forming force. At approx. 12 N, this force is significantly higher than the force achieved with non-formed foils, which is approx. 1.8 N maximum punch force. The maximum force also tends to be achieved at a lower stroke with increasing forming force, which means that an increase in the stiffness of the half plate is achieved. The force-displacement curves show that the half plates manufactured with a higher forming force also exhibit a greater drop in force after reaching the maximum bending force. However, a higher bending force can also be seen at the end of the process for half plates with a higher forming force. The increase in force in the 3-point bending test can be explained by the larger formed channels of the active flow field of the half plates. Due to the higher forming force, the channels become deeper and more material flows from the flange area into the channel structure. This increases the moment of area and the bending stiffness. Furthermore, a higher strain hardening of the material can be expected with deeper channels, as higher levels of plastic strain are achieved. This also results in higher forces during further forming, such as bending. The combination of these two factors therefore causes an increase in the force required to form the half panels. It is important for the half plates that the stiffness can be increased, which is reflected in the gradient of the individual force-displacement curves as the forming force increases. This means that at higher forming forces, the limp properties of the components are reduced and a less fragile component is manufactured.

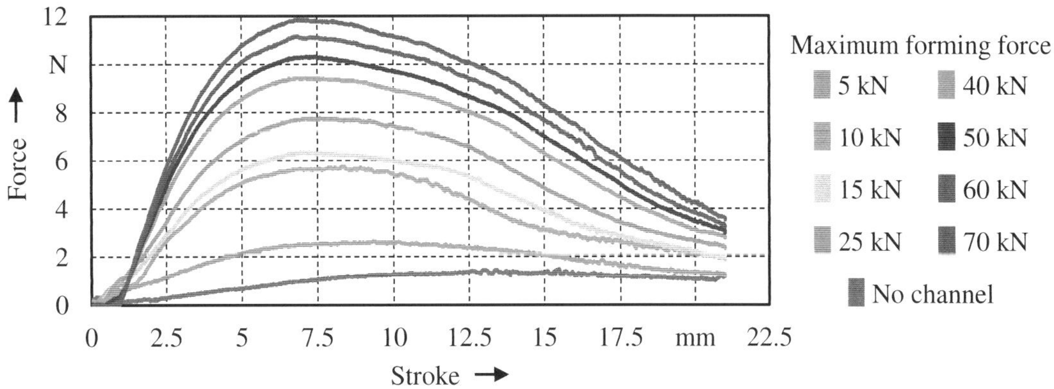

Figure 5: Force-displacement curves of 3-point-bending tests with half plates formed with varying maximum forming forces

The investigations show that higher bending stiffness can be achieved through higher forming forces that result in deeper channels. These findings have a positive effect on subsequent handling operations. Due to the increased bending stiffness, the half panels can be handled with greater ease and the risk of unintentional plastic deformation of the half panels is reduced.

Furthermore, the findings show that handling the foil before the forming operation leads to an increase in process difficulty, as lower forces and torques result in plastic forming of the foil. It can therefore be deduced from the present results that trimming operations using shear cutting or laser cutting processes should, if possible, only be carried out after the forming and molding of the flow field and other functional geometries of the half plates.

4. Summary and Outlook

Future energy supply will require new technologies both in terms of generation and consumption. One possibility is the use of hydrogen as an energy carrier, as it can be used flexibly and can achieve sufficient energy densities with the help of current technologies. However, there are a number of challenges for the production of systems for the utilization of hydrogen applications, particularly from a production technology perspective. These include the handling of the limp components. In order to meet this challenge, the bending stiffness of half plates as parts of bipolar plates with different forming forces was analyzed in this study.

In a first step, a geometry similar to a flow-filed of a bipolar plate was formed from a foil made of 1.4404 stainless steel with a foil thickness of 0.075 mm. An additively manufactured plastic die and punch are utilized for this forming process. With this manufacturing process, the geometric properties of the active flow field can be adapted quickly and half and bipolar plates produced using forming technology can be used in the chemical and electronic characterization of electrolyzers and fuel cells. For the investigation of bending stiffness and limp properties, half plates with varying maximum forming forces are produced. It is shown that the channel depths are dependent on the forming force and that deeper channels can be formed at higher forces. Optical measurements also show that the samples exhibit deformation at the edges of the flanges. This may be due to manual handling operations of the samples.

In the subsequent step, the half plates produced with different maximum forming forces are characterized using a 3-point bending test. It is found that a higher forming force of the half plates has a greater gradient in the force-displacement curve and shows higher maximum forces at the bending punch. The forces increase from approx. 1.8 N for non-formed foils to approx. 12 N at a forming force of 70 kN. This shows that the combination of a higher moment of area due to the deeper channels and the increasing work hardening of the stainless steel at higher channel depths have a positive influence on the bending stiffness and limp properties. The result is that process stages before the forming process of the foil must pay particular attention to the limp properties of the foil, whereas this challenge is lower in subsequent processes.

For future investigations, additional geometries in the form of beads, which are already present in industrially used half-sheets due to media feeds and sealing systems, can also be integrated. Furthermore, an investigation of the work hardening of the foil through the forming process in combination with the thinning of the foil is a possible field of research that could provide more information about the processes during the production process. Furthermore, automated gripping and stacking of the half-sheets is necessary in order to be able to evaluate the effect of the bending stiffness of the half-sheets on the handling process. This involves the use of different gripping systems based on different mechanisms of action.

Acknowledgement
This work was supported by the Friedrich-Alexander-Universität Erlangen-Nürnberg within the scope of the FAU profile center "Hydrogen".

References

[1] European Commission, 2020. A hydrogen strategy for a climate-neutral Europe. Communication from the commission to the european parliament, the council, the european economic and social committee and the committee of the regions, Brussels, 24 pp.
[2] Carrette, L., Friedrich, K. A., Stimming, U., 2001. Fuel Cells - Fundamentals and Applications. Fuel Cells 1 (1), 5–39.
[3] European Parliament. EU hydrogen policy. 2021
[4] 2018. Toyota plans mass production of fuel cell stacks and hydrogen tanks, 10-fold increase from 2020. Fuel Cells Bulletin 2018 (5), 1.
[5] Li, X., Sabir, I., 2005. Review of bipolar plates in PEM fuel cells: Flow-field designs. International Journal of Hydrogen Energy 30 (4), 359–371.
[6] Liu, H., Li, P., Juarez-Robles, D., Wang, K., Hernandez-Guerrero, A., 2014. Experimental Study and Comparison of Various Designs of Gas Flow Fields to PEM Fuel Cells and Cell Stack Performance. Front. Energy Res. 2.

[7] Manso, A.P., Marzo, F.F., Mujika, M.G., Barranco, J., Lorenzo, A., 2011. Numerical analysis of the influence of the channel cross-section aspect ratio on the performance of a PEM fuel cell with serpentine flow field design. International Journal of Hydrogen Energy 36 (11), 6795–6808.
[8] Mehta, V., Cooper, J.S., 2003. Review and analysis of PEM fuel cell design and manufacturing. Journal of Power Sources 114 (1), 32–53.
[9] Bobka, P., Gabriel, F., Dröder, K., 2020. Fast and precise pick and place stacking of limp fuel cell components supported by artificial neural networks. CIRP Annals 69 (1), 1–4.
[10] Williams, M., Tignor, K., Sigler, L., Rajagopal, C., Gurau, V., 2014. Robotic Arm for Automated Assembly of Proton Exchange Membrane Fuel Cell Stacks. Journal of Fuel Cell Science and Technology 11 (5).
[11] Wilberforce, T., El Hassan, Z., Ogungbemi, E., Ijaodola, O., Khatib, F.N., Durrant, A., Thompson, J., Baroutaji, A., Olabi, A.G., 2019. A comprehensive study of the effect of bipolar plate (BP) geometry design on the performance of proton exchange membrane (PEM) fuel cells. Renewable and Sustainable Energy Reviews 111, 236–260.
[12] Barzegari, M.M., Khatir, F.A., 2019. Study of thickness distribution and dimensional accuracy of stamped metallic bipolar plates. International Journal of Hydrogen Energy 44 (59), 31360–31371.
[13] International Organization for Standardization. Metallic materials - Bend test. Beuth Verlag GmbH, Berlin. doi:10.31030/3167638.

Feature based bipolar plate forming

D. Meelkop[1], H. Janssen[1] and C. Brecher[1,2]

[1] *Fraunhofer Institute for Production Technology IPT, Aachen, Germany*
[2] *Laboratory for Machine Tools and Production Engineering (WZL) of RWTH Aachen University*

Email: dennis.meelkop@ipt.fraunhofer.de

Abstract. By the end of the 1990s, the global peak of energy supply from oil and gas deposits had already been reached. Since then, there has been a clear trend towards the development of new, more sustainable ways of supplying energy. To realize this reduction, a holistic and rapid restructuring of the energy supply towards CO2-free and renewable energy production, storage and processing across all areas of life is necessary. The development and implementation of a functioning hydrogen economy can be crucial with respect to the goal of climate neutrality. The scale-up and commercialization of technologies for the cost-effective production and use of hydrogen carry significant importance. Accordingly, innovative production technologies and process chains must be developed to be able to manufacture fuel cells reliably in high volumes and at optimized costs. In this context, BPPs in the PEM-FC are responsible for about 80 % of the total weight, just over 50 % of the system volume, and up to 40 % of the total cost. Current research approaches in the field of BPP forming focus mainly on the optimization of a pre-selected production technology. Thereby, the possibility of combining different forming technologies to represent a holistic process chain is mostly disregarded. This research gap is to be closed by the developments within the scope of this work. The bipolar plate is not to be regarded as a coherent component, but as a composition of different geometric features.

1. Introduction

Not only due to the growing world population, but also due to developments in the context of digitalization and electrification, increasing amounts of energy will be needed globally in the future. Despite the initial efforts towards more sustainable generation and use of energy, production is still heavily influenced by fossil fuels and thus contributes massively to climate change [1]. To counteract this fact, the installation of a functioning hydrogen economy is considered a key element. Hydrogen can be used as an element itself for decarbonization, for example in the steel industry, or by conversion in a fuel cell as an energy carrier for electricity [2]. Compared to other fuels, the ratio of energy density to mass in kilograms is about three times higher for hydrogen (140 J/kg) than for conventional gasoline (48 J/kg) [3].

In this context, the low volumetric energy density of hydrogen evidently cannot be neglected, so that it leads to greater expenses in the entrainment and transport of hydrogen in gaseous or liquid form [3]. In order to reduce emissions in the field of mobility, the use of green hydrogen is nevertheless promising and considered one of the best complements to purely electric drive trains [2]. In this field of application, the proton exchange electrolyte membrane fuel cell in particular has become the focus of development due to its compact design with high energy efficiency and dynamic control behavior [4]. The core of the PEMFC consists of three components, the external current collectors and the repeating unit consisting of the membrane electrode assembly and the bipolar plate. In this context, the bipolar plate itself is responsible for up to half of the system costs and over two-thirds of the system weight [1]. The tasks of the bipolar plate in the stack are firstly to direct and distribute the electrons produced, secondly to guide the operating gases oxygen and hydrogen on the outer sides, thirdly to dissipate process heat by means of the internal conduction of a cooling medium, and fourthly to provide the necessary mechanical stability to ensure the integrity of the overall system [4]. To meet these requirements, the bipolar plate consists of two precisely formed half plates made of corrosion-resistant base material that must be cut and joined under high accuracy requirements. In addition, the manufacturing of this component is subject to enormous cost pressures due to the large contribution of bipolar plate costs to the overall system. The simplicity of scaling up and parallelizing existing production technologies and routes makes it unlikely that this complexity will be resolved and that the fuel cell system will achieve economic competitiveness [1, 5]. To overcome these challenges, there is an urgent need to revise the conventional production process for manufacturing bipolar plates and to apply new approaches and technologies, starting with the forming process [6]. In this context, the approach in this publication is intended to contribute towards deriving a geometric characteristic-compliant production chain for forming, depending on the specific bipolar plate design, which reflects optimum manufacturing conditions for different areas of the bipolar plate.

2. State of the art in fuel cell production

2.1. Functionality of a PEM Fuel Cell

The PEMFC electrochemically converts the introduced hydrogen with the addition of oxygen to produce electric power [6]. To represent a power of about 100 kW, the fuel cell stacks usually consist of up to 400 repeating units of the MEA as well as the BPP [6]. The anode side of the single cell is supplied with hydrogen. At the junction of the proton exchange membrane, the hydrogen releases individual electrons and the positive hydrogen ions react with the oxygen on the cathode side to form water. In addition to the electricity generated, this exothermic reaction also releases heat, which is dissipated via the conduction of a cooling medium within the bipolar plates. The bipolar plates guide the process media in the stack and are in direct contact with the gas diffusion layer (GDL). This interface has a significant influence on the performance of the overall system due to the contact resistance created at the transition [7]. Furthermore, the performance of the bipolar plate can be related to the aspect ratio, i.e., the ratio of channel depth to channel width. Increasing this ratio and thus increasing the energy density of the individual cell is a production challenge, especially for forming [8]. The complete process chain for manufacturing of bipolar plate is described in the following section.

To fulfill the tasks of a BPP as mentioned in the previous description, a basic distinction can be made between three different types of base material for the fabrication of bipolar plates. Non-metallic materials such as graphite, metallic materials such as stainless steel or titanium, and composites that can be either metal- or carbon-based are utilized [1]. Due to the corrosion resistance, composite materials have generated a lot of attention in research and industry at the beginning of the industrialization of the fuel cell. However, the traditional manufacturing methods for these materials in injection molding and compressing are unsuitable for mass production of fuel cells together with bipolar plates. Thus, the most promising base material for realizing the cost targets and geometric requirements is metallic [6].

2.2. Bipolar plate production

The production chain for manufacturing a metallic bipolar plate basically consists of three necessary processing steps and at least two optional ones. The forming of the half-shells is usually at the beginning of the chain, followed by a trimming of the half-shells and a joining operation, which can also be interchanged in the sequence. In addition, coating of the joined bipolar plate is conceivable, or the use of already pre-coated material. Depending on the composition of the individual cell, a gasket application can follow on both halves of the bipolar plate, for example by screen printing or dispensing. Accordingly, the options for describing the process chain for the manufacturing of metallic bipolar plates have not been clearly standardized to date. In addition, there is a wide range of technology options in each of the individual process steps, showing, for example, an implementation of the forming of the half-plates.

Consequently, the literature contains well-established and more recent technologies for forming thin metal sheets. These include stamping, hydroforming, hydrostatic forming and rotary embossing [6, 8]. Among the technologies mentioned, stamping is one of the fastest and most cost-efficient variants in forming. Nevertheless, with this technology, deformation is often accompanied by introduced stress in the material, which in turn can lead to unwanted thinning, wrinkling or even cracking [6]. In comparison, hydroforming or hydrostatic forming offers a more uniform pressure distribution due to the technology, which reduces thinning behavior, but usually involves a higher cycle time [7]. The way in which the material is formed also influences the springback behavior of the half-plate and ultimately of the individual channel geometries. If this springback is not controlled by the process technology, this can lead to losses in the performance of the overall system [6]. The selection of the technologically optimal process is therefore a decisive factor for component quality, especially in forming but also in regard to the overall process chain.

2.3. Geometrical description of bipolar plates

Bipolar plates have intricate geometries. As shown in Figure 1, a typical bipolar plate can be divided into regions according to their function. At the outermost area, a loop serves as the contact point for joining the bipolar plate halves. The through holes at the ends form the manifolds for the anode gas, coolant, and cathode gas. Following the manifolds, a distribution area homogenizes the flowrate. Finally, the center portion contains the flow field, which is responsible for distributing the reactants across the MEA.

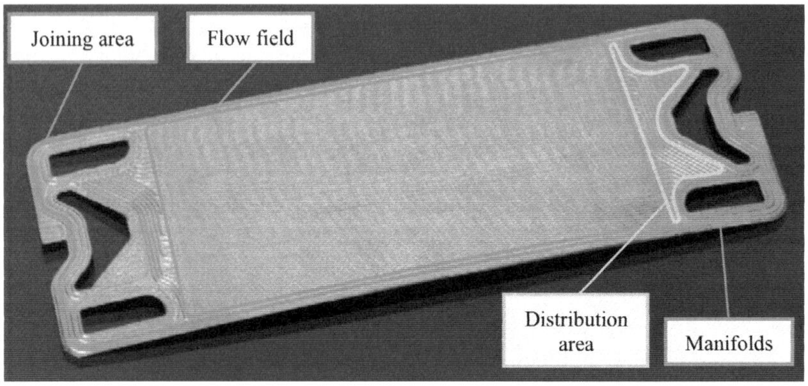

Figure 1: Typical bipolar plate and its main areas. Design from research project FlyGO.

Such flow field commonly feature the cross-sectional geometry displayed in Figure 2. In the present work, the side with reactant flow has been taken as the upper side.

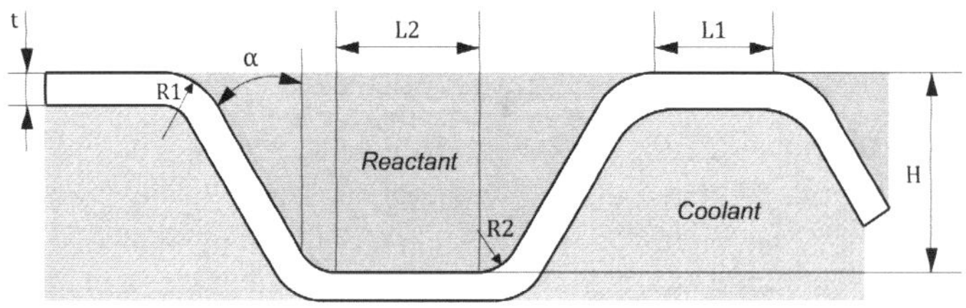

Figure 2: Typical bipolar plate dimensions.

IOP Conf. Series: Materials Science and Engineering **1307** (2024) 012032 doi:10.1088/1757-899X/1307/1/012032

Currently, the names of a cross section typical dimensions vary between publications. By examining the terms used in a literature sample, the following observations have been made: L1 is commonly named *rib width* [9]; L2 usually appears as *channel width* [10]; α is often named *draft angle* [9]; for H, *channel height* and *channel depth* are used interchangeably [10]; R1 and R2 as channel radii have no standard naming. Some authors describe them in relation to the tool parts (*die radius* and *punch radius* [11], respectively), but they do not form a majority within the analyzed sample. In fact, in some cases a name might be used for R1 by one author, and R2 by another (example: *internal fillet radius* – R1 [12], R2 [13]). For that reason, in the present work they'll be referred to as top and bottom radius.

2.4. *Feature based manufacturing*

Decomposition is the breakdown of the problem into smaller pieces, which are functional elements that together form the entire system [14]. In this work, such smaller pieces are features, which can be defined as a single face or set of connected faces with characteristic combinations of geometry and topology [15]. A bipolar plate frequently has more than 100 features. Therefore, in order to have a time and cost-effective process, it is necessary to automate feature recognition. In essence, feature recognition systems classify the geometric features from low-level entities using common high-level information [16]. There are two major categories of feature recognition methods: based on boundary matching or volume decomposition. Methods based on volume decomposition are less efficient and cannot recognize features with complex and unfixed topological relations. Thus, the present work will focus on boundary matching. One useful tool is the concept of seed face, defined as a combination of the key face and its edges which best represent the feature. Together with a hybrid method, which combines the Holistic Attribute Adjacency Graph (HAAG) with hint and rule-based approaches, a robust procedure for feature recognition can be achieved [17]. Finite Element Analysis is used to save time and resources during the design phase [18]. However, due to the small radii, the number of elements required to analyze a bipolar plate is large [9]. Hence, it must be computationally advantageous to minimize the evaluated area. Nevertheless, the distance of the channel to the edge of the blank has an impact on the local process conditions. As an example, thinning is lower near the edges and higher in the center region. In fact, the boundary conditions have a negligible effect on the thinning near the center of the bipolar plate. Therefore, the analysis of a single channel is sufficient for the evaluation of the deformation on the straight channel section of the BPP [9]. Furthermore, results from simulations conducted by [10] demonstrate that the same effect can be observed on curved channels, suggesting that single feature simulation might be capable of assessing the forming conditions for other features as well.

3. Summary and aim of research

As explained in the previous sections, a need for innovation in the forming of metallic bipolar plates can be deduced. As explained in section 2.1, the bipolar plate is a core component of a fuel cell for the distribution of media and the transfer of thermal and electrical energy. In addition, a fuel cell stack typically requires 300-400 single cells connected in series to provide sufficient system output. This means that 300-400 bipolar plates per system, i.e., 600-800 bipolar plate halves, have to be manufactured with the highest precision. Accordingly, an efficient and volume-capable production strategy is necessary for the ramp-up of the hydrogen economy. In section 2.2, the differences between the individual bipolar plate segments and features were highlighted, as well as the manufacturing methods in the state of the art. With the information from Section 2.3 on the inconsistent designation of the features of a bipolar plate, it can be concluded that none of the systems presently used in the state of the art take into account the different feature requirements during production. In order to setup the production of bipolar plates resilient and cost-effective, all optimization options must be considered. Accordingly, this research presents a method for forming bipolar plates according to the features.

4. Methodology of feature based bipolar plate forming

The methodology to be presented can be fundamentally divided into three building blocks and follows the problem-solving approach presented in section 2.4. The overall methodology can be seen in Figure 3.

IOP Conf. Series: Materials Science and Engineering **1307** (2024) 012032 doi:10.1088/1757-899X/1307/1/012032

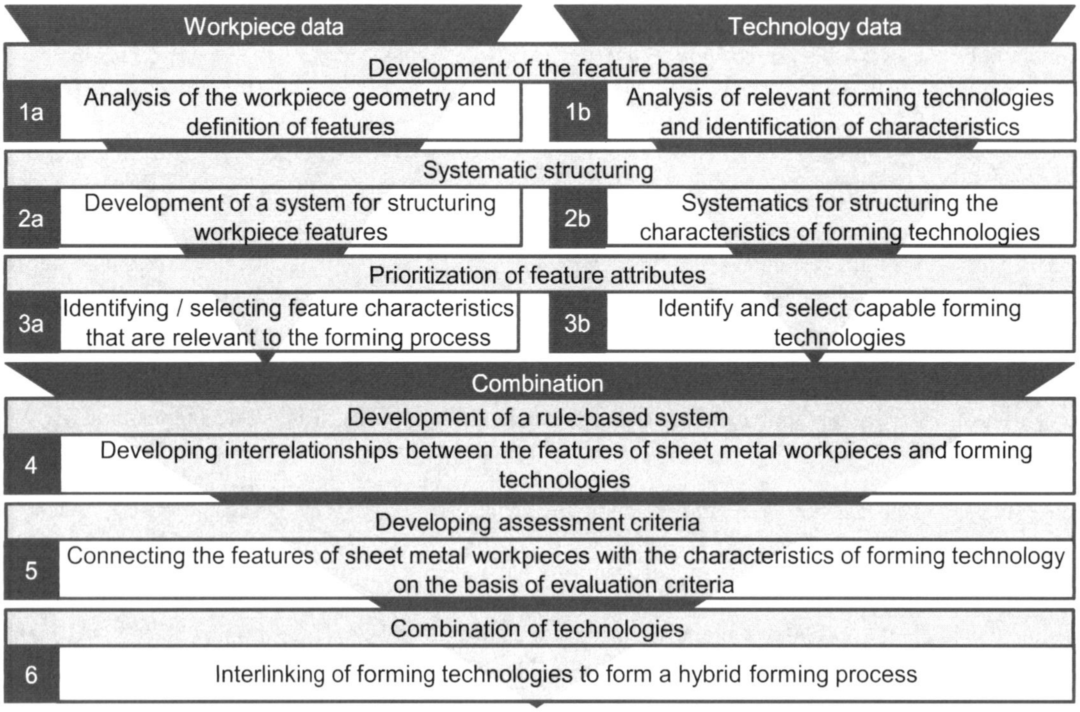

Figure 3: Methodology for feature-based manufacturing.

The first two components describe the workpiece data to be analyzed as a manufacturing target and the associated technology data with the description and evaluation of the technology capabilities. In step 1a, the part to be produced is analyzed and geometric features which are essential for forming are determined as described in section 2.4 and applied in the following section 5.1. Complementary, in step 1b, the technological solutions are described in terms of their characteristic capabilities. As discussed in Section 2.2, forming technologies have different operating principles and therefore place different stresses on the materials depending on the process, from which in turn different influences on the component can be derived. Following the basic analysis of the workpiece requirements and the technological capabilities, those are systematically structured in a second step in order to be able to prioritize within the meaning of an evaluability in the third step. On this basis, the combination of the data to derive a feature-based process chain can be started in step 4. Individual features, which are considerably reduced in their complexity, are examined in all technology alternatives and evaluated on the basis of a KPI, e.g. the maximum allowable thinning. The determination of the most suitable characteristic values for the application case is carried out in step 5. In this step, further system-related information, e.g. regarding the overall lifespan of the component, can be taken into account in order to define a set of KPIs. Finally, the collected information is used to define a feature-based and technology-optimized manufacturing process sequence. By analyzing and separating the individual geometric features, it is possible to assign the specific component and tolerance requirements to these features. This catalog of requirements is then matched with the technology that offers the best fit in terms of the predefined KPIs. In this way, a complex forming task can be broken down into the fabrication of the individual features to identify technological solutions at this simplified level. Accordingly, the solution to the complex overall forming challenges results from the sum of the individual technological solutions appropriate to the features.

5. Validation in feature recognition and hybrid forming

5.1. Feature analysis and separation for workpiece data

Based on the bipolar plate layouts summarized by [19], eleven features have been identified. As discussed in section 2.4, seed faces are a suitable way to define patterns that hint at the presence of a feature. Hence, in the present work these have been defined as shown in Figure 4a. For demonstration purposes, the process of identifying a straight channel will be described. Its main rules can be found in Figure 4b.

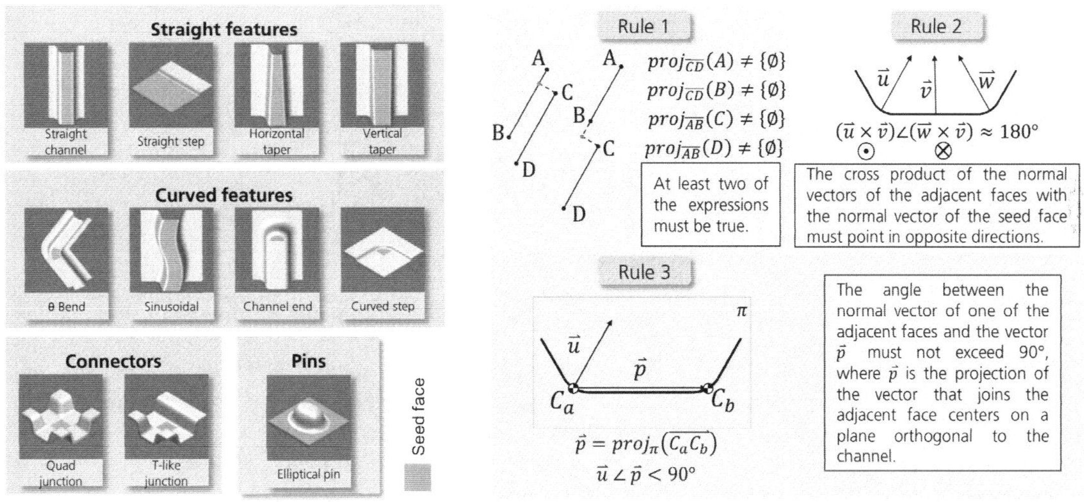

(a) (b)

Figure 4: (a) Bipolar plate features and their respective seed faces and (b) Typical straight channel graph.

In order to solve the issue of interacting features, the procedure proposed by [17] has been adapted and implemented using the software tools employed by [20]. As depicted in Figure 4a, two characteristics can be observed in all seed faces: they are located at the bottom of the feature, and are facing upwards (positive normal vector Y component). With these rules, the seed faces could be separated from the rest. The presence of two or more parallel, non colinear edges hints at the presence of straight channels in a seed face. To confirm it, a set of three rules Figure 4b and one logic procedure have been proposed. The first rule checks whether the line segments could be orthogonally projected onto each other, which eliminates misaligned parallel edges. The second rule compares the normal vectors of the faces adjacent to the real edges with the normal vector of the seed face. This filters edge pairs which could not form opposing walls. The third rule compares the normal vector of one adjacent face to the vector that joins their centers. Such step removes edge pairs whose walls do not face each other.

However, the proposed rules do not completely isolate the valid straight channels. In our tests, the straight channels had to be logically deducted from the remaining edge pairs. Therefore, the fourth filter has been designed to cover that gap. First, a relations matrix must be constructed. Each line and column of the relations matrix represents a real edge of interest. The element a_{ij} stores a 1 or a 0, which represents whether the geometrical relation of interest (parallelism, in the case of the straight channel) is present between the edges i and j. In the current work, lines have been defined as parallel to themselves, which implies that the main diagonal of the relations matrix is composed of ones. The next step involves calculating the sum of all elements in each row. Whenever a row sum equals 2, the indicated edges form a pair. Hence, all elements whose row contains a relation with only one edge from the pair should equal zero ($a_{ki} = 0 \; \forall \; k \neq j \neq i$ and $a_{kj} = 0 \; \forall \; k \neq j \neq i$). This step is repeated until no new pairs can be found.

5.2. Hybrid forming tool for technology data derivation

As already described in section 4, the technology-specific effects in the context of the production of individual geometric features are initially elaborated by simulation. However, a physical, hybrid forming tool was engineered considering the physical validation. A cross-sectional view of the hybrid tool with a side-by-side representation configuration can be seen in Figure 5.

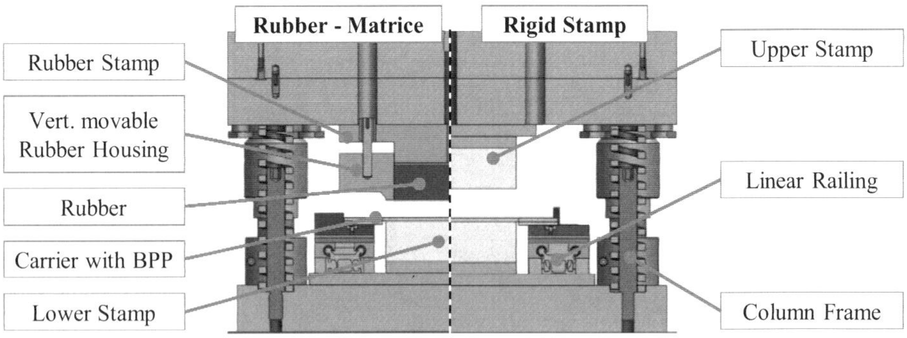

Figure 5: Hybrid forming tool. Cross section with rubber pad forming and stamping configuration.

Overall, the hybrid tool is illustrated in the setup of a column frame, which allows various configurations to test different forming technologies and their feature suitability. Starting with the left-hand version in hydrostatic forming, a rubber pad can be seen in the upper half of the tool, which is guided in a movable housing. However, in order to avoid a single stage forming process in a single technology for the technological validation, a universal carrier system was developed which is guided by a linear axis and thus positions the preformed half-shell exactly above the lower die. In doing so, the influence of the forming sequence in hydrostatic forming or stamping, for example, can also be investigated. With the use of the hybrid tool, the features marked in Figure 4 can be formed in different versions and their forming quality optimized by combining the different technologies. On the one hand, these tests validate the previous simulation results, and, on the other hand, they form a data basis for the derivation of feature-specific forming technologies through multiple test execution and evaluation.

6. Summary and Outlook

This research effort establishes the foundation for a new approach to manufacturing bipolar plates. The previous technological approaches, especially in forming, are an attempt to be able to use established technologies without component-specific process adaptations. However, as described in the present work, this approach cannot achieve the necessary scaling mechanisms, nor can it meet the cost targets worked out for marketability. Accordingly, the approach presented in this paper for the feature-based production of bipolar plates offers an alternative for technologically adapting the process chain to specific features and tolerance conditions. By solving a large number of smaller forming tasks at the feature level, both the numerical and the experimental optimization effort can be reduced. In addition, by dividing the overall forming task, it is feasible to use forming equipment with lower forming forces in order to reduce investment costs or to make existing equipment usable. In such a production scenario, individual companies could strengthen the resilience of their process chain through flexible integration or might choose to map the forming process in a supplier network that takes on feature-specific forming tasks. The validation approaches presented for feature recognition, separation and evaluation of the technological capability of individual forming processes in the hybrid tool constitute the first fundamental elements for the implementation of the overall methodology. In the further course of the research work, these individual method modules have to be integrated into the overall context, in order to create a design-independent but feature-specific procedure for the development of an optimized process chain for the forming of bipolar plates.

Acknowledgements
The authors would like to acknowledge the financial support by the "H2GO – National fuel-cell production action plan" project which is funded by the Federal Ministry for Digital and Transport Affairs (BMVD) as part of the National Innovation Program Hydrogen and Fuel Cell Technology Phase 2 (NIP II), funding code 03B11027.

References

[1] T. Wilberforce, O. Ijaodola, E. Ogungbemi, Z.E. Hassan, J. Thompson, A.G. Olabi, in *Reference Module in Materials Science and Materials Engineering,* Elsevier **2018**.
[2] I. Staffell, D. Scamman, A. Velazquez Abad, P. Balcombe, P.E. Dodds, P. Ekins, N. Shah, K.R. Ward, *Energy Environ. Sci.*, **2019**, *12*, 463.
[3] R.-A. Felseghi, E. Carcadea, M.S. Raboaca, C.N. TRUFIN, C. Filote, *Energies*, **2019**, *12*, 4593.
[4] Y. Awin, N. Dukhan, *MSF*, **2018**, *933*, 342.
[5] M.M. Whiston, I.L. Azevedo, S. Litster, K.S. Whitefoot, C. Samaras, J.F. Whitacre, *Proceedings of the National Academy of Sciences of the United States of America*, **2019**, *116*, 4899.
[6] A. Tang, L. Crisci, L. Bonville, J. Jankovic, *Journal of Renewable and Sustainable Energy*, **2021**, *13*, 22701.
[7] C.-J. Hung, W.-J. Chen, C.-A. Lin, H.-R. Shiu, B.-H. Chen, *Energies*, **2021**, *14*, 7109.
[8] P. Zhang, M. Pereira, B. Rolfe, W. Daniel, M. Weiss, *J. Phys.: Conf. Ser.*, **2017**, *896*, 12115.
[9] D.M. Neto, M.C. Oliveira, J.L. Alves, L.F. Menezes, *Metals*, **2019**, *9*, 810.
[10] H.T. Ghadikolaee, M.M. Barzegari, F.A. Khatir, S. Seddighi, *Iranian Journal of Hydrogen & Fuel Cell*, **2022**, *1*, 1.
[11] D. Qiu, L. Peng, P. Yi, X. Lai, W. Lehnert, *Energy Conversion and Management*, **2018**, *174*, 814.
[12] M.M. Barzegari, F.A. Khatir, *International Journal of Hydrogen Energy*, **2019**, *44*, 31360.
[13] N. Mohammadtabar, M. Bakhshi-Jooybari, S.J. Hosseinipour, A.H. Gorji, *Int J Adv Manuf Technol*, **2016**, *85*, 765.
[14] V.J. Shute, C. Sun, J. Asbell-Clarke, *Educational Research Review*, **2017**, *22*, 142.
[15] H. Sakurai, D.C. Gossard, *IEEE Comput. Grap. Appl.*, **1990**, *10*, 22.
[16] Y. Shi, Y. Zhang, K. Xia, R. Harik, *CADandA*, **2020**, *17*, 861.
[17] Y.G. Li, Y.F. Ding, W.P. Mou, H. Guo, *Proceedings of the Institution of Mechanical Engineers, Part B: Journal of Engineering Manufacture*, **2010**, *224*, 271.
[18] W. Thomas, T. Altan, *Steel research*, **1998**, *69*, 181.
[19] Kangning Xiong, Wei Wu, Shuangfeng Wang, Lin Zhang, *Applied Energy*, **2021**, *301*.
[20] C. Gonnermann, D. Gebauer, R. Daub, *Applied Sciences*, **2023**, *13*, 990.

IOP Conf. Series: Materials Science and Engineering **1307** (2024) 012033 doi:10.1088/1757-899X/1307/1/012033

Data-Driven Approach for Minimizing Springback in Roll Forming: An Adaptive Process Parameter Control Strategy

J. He[1], Y. Ren[1], Y. Sun[1], K. Wu[1], Z. Xiong[1], and J. Xiao[1,2]
[1] Shenzhen Institute for Advanced Study, UESTC, Shenzhen, 518000, Guangdong, China
[2] ShenSi Shape (Shenzhen) Technology Co., LTD, Shenzhen, 518000, Guangdong, China
E-mail: yong_sun@uestc.edu.cn

Abstract. Springback is the most common quality defect in roll forming. It is difficult to control springback efficiently because the forming process is very complex and difficult to analyze. This paper proposes a data-driven springback control method that can adaptively adjust process parameters to minimize springback. Firstly, a digital twin model was established for efficiently collecting equipment data, formed part data, and remote equipment control. Then, based on gathered springback data considering different sheet widths, materials, uphill volumes, and roll gaps, a high-precision springback prediction model was constructed using the SAPSO-SVR algorithm. Lastly, building upon the prediction model, optimization models were developed using bat algorithm, enabling the system to automatically adjust process parameters based on optimization results when forming different sheets.The experimental results show that after optimization, the springback at the two corners of the hat-shaped part is reduced by 36.8% and 14.3% respectively. Experimental findings initially demonstrate the efficacy of this approach in controlling springback across sheets of diverse materials and varying widths. The proposed adaptive springback control method significantly enhances production efficiency and forming quality, driving the intelligent advancement of sheet metal roll forming technology.

1. Introduction

Roll forming is a process that uses multi-pass forming rolls to gradually bend the sheet to make a metal profile with a specific cross-section (Figure 1). Roll forming has the characteristics of high efficiency, low cost and flexibility, and is widely used in new energy vehicles, aerospace and other industries. Springback is the most common and difficultly solved quality defect in roll forming. How to efficiently control springback has become a difficult problem.

The parameters affecting springback have been extensively studied. Based on previous theoretical and experimental analysis [2–5], the process parameters related to springback selected in this article include the uphill volume and the roll gap. The uphill volume is the vertical height difference between the centres of the upper (lower) rolls of two adjacent passes (Figure 2).

The common control methods for the springback in roll forming include theoretical analysis [7] and finite element simulation [8–12]. However, due to the complexity of the roll forming process, there are limitations in the accuracy and reliability of theoretical analysis results. Simulating complex cross-sections also incurs significant time costs. Both of these methods have constraints.

In other fields, some studies have explored the use of optimization algorithms to seek an optimal process. For instance, Gauntt [13] employed genetic algorithms in lightweighting

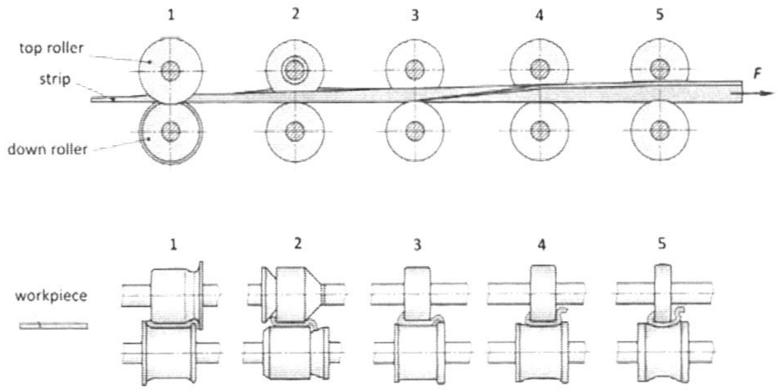

Figure 1. Roll forming process [1].

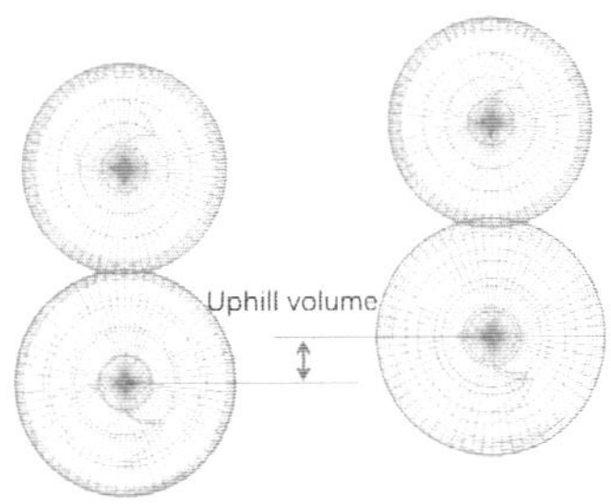

Figure 2. Roll forming process [6].

research on gears to search for the optimal gear structure. However, the optimization function relies on simulations for calculation. Each search iteration necessitates running a large number of simulations to obtain a single iteration result, leading to a final result that is relatively reliable but with low time efficiency.

Aiming at solving the limitations of previous methods, this paper proposes an adaptive springback intelligent control method for roll forming. First, a digital twin model is built for real-time monitoring and regulation. Then, a SAPSO (simulated annealing particle swarm optimization)-SVR (support vector regression) springback prediction model was constructed based on 32 sets of springback data obtained from actual experiments. In previous study [6], the SAPSO-SVR predictive model exhibited high accuracy in predicting springback. However, the focus of this study is on a part with a more intricate cross-sectional geometry, namely a hat-shaped component, introducing increased complexity to the forming process. Moreover, the model is constructed based on actual orthogonal experimental data, with a relatively limited quantity, potentially leading to some deviation in the results compared to previous studies. Based on the springback prediction model, the method takes the prediction function as the optimization function and the process parameters as the optimization parameters. And it obtains the optimal springback scheme in real time when different sheets are input. This method achieves higher accuracy and reliability than traditional theoretical analysis methods and demonstrates greater timeliness than FEA method. It promotes the intelligent upgrading of roll forming technology.

2. The adaptive springback optimization method

2.1. Data acquisition, control and visualization

To achieve real-time optimization, it is necessary to provide an optimization plan in advance before the sheet metal forming process. This requires a closed-loop method consisting of data acquisition, springback prediction, springback optimization, and remote equipment control. The overall framework is shown in Figure 3. Firstly, a prediction model is constructed based on previous experimental data. This prediction model serves as the optimization function for the optimization module. The optimization algorithm utilizes this function to search for the optimal process parameters under different material conditions. The results are then fed back to the digital twin system for real-time control of the equipment and recording of the final outcomes. The optimal solution and results can be re-inputted into the prediction model to expand its dataset, continuously improving the prediction accuracy.

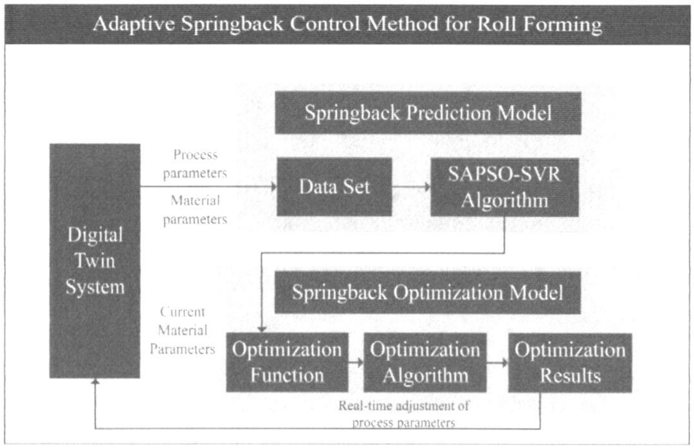

Figure 3. The framework of the intelligent optimization method.

Real-time reading and adjustment of key parameters, including roll gap and uphill volume, is crucial for achieving springback prediction and optimization. The roll gap can be calculated based on the positions of the upper and lower rolls for each pass, as well as the lateral positions of the passes. Therefore, the elemental parameters include the positions of the upper and lower rolls for each pass, the lateral positions of the passes, and the rotational speed of the motors.

In the digital twin system, the principle of real-time reading and adjustment of key parameters involves the following steps, as shown in Figure 4: Analyzing the code of the PLC control unit to determine the memory addresses and data types of the key parameters. Establishing a communication connection and using a suitable communication protocol such as Modbus TCP for data exchange with the PLC. By reading the real-time parameter data stored in the PLC memory, we can obtain the current values of the parameters. These parameter values are typically stored in memory as binary representations of floating-point numbers. To achieve precise control, appropriate conversions and processing may be required to obtain the actual values of roll gap, roll crown, and rotational speed.

Once the parameter values are obtained, adjustments can be made as needed, for example, setting target positions or adjusting the rotational speed. The adjusted parameter values are transmitted to the servo motors to achieve precise motion control. Servo motors provide higher accuracy and stability for controlling the motion of the upper and lower rolls and passes in the roll-forming equipment. The currently implemented digital twin system is shown in figure 5.

Real-time monitoring and adjustment of key parameters enable the optimization of roll-forming equipment operation, facilitating springback prediction and optimization. This provides crucial technical support for improving process accuracy and production efficiency.

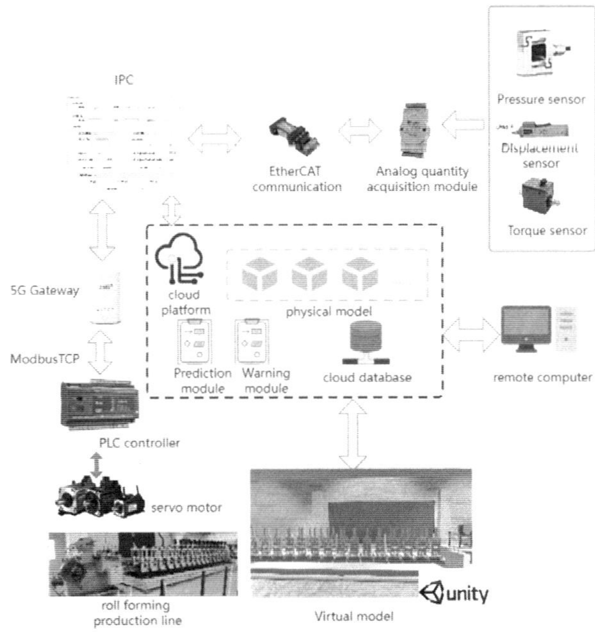

Figure 4. Roll forming digital twin construction diagram.

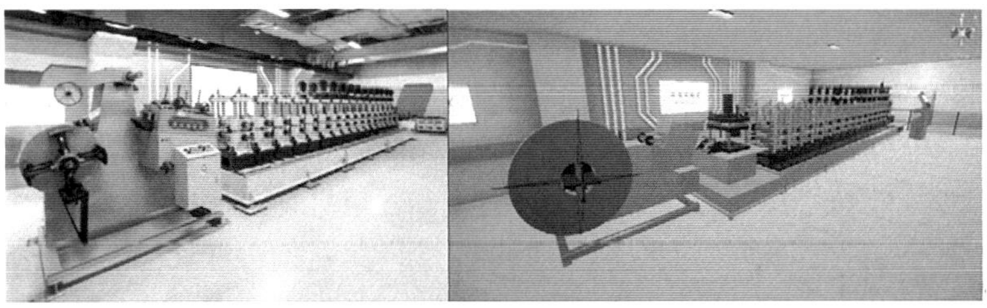

Figure 5. Virtual scene in the digital twin system of roll forming equipment.

2.2. Springback prediction model
This article uses the SAPSO-SVR algorithm to predict springback. SVR is a common machine learning model used to solve regression problems. SVR is very suitable for solving small sample, high-dimensional, non-linear problems. These characteristics make it feasible to build high-precision regression models at low cost and in a short time. In order to ensure the scalability of the model and avoid the unreliability of results and time costs caused by manual parameter adjustment, SAPSO is used to optimize the model. The improved SAPSO-SVR can perform adaptive parameter adjustment after the model database is expanded, so that the versatility and accuracy of the new prediction model are automatically improved. This facilitates the continuous addition of new data to the model during the roll forming process and taking into account the forming of new sections.

In previous study [6], the SVR kernel function used adopts RBF (radial basis function) kernel. The three key parameters to be determined are the insensitive loss coefficient ε, the penalty coefficient C, and the width coefficient γ. Optimization of the SVR model must optimize all three parameters simultaneously. Inappropriate parameters can make the accuracy of the trained model degrade and make it difficult to fit the real situation. And the algorithm parameters of the optimal prediction model corresponding to data sets with different data volumes are also different. Therefore, it is necessary to use SAPSO to optimize SVR. The SAPSO-SVR algorithm flow chart is shown in Figure 6. The temperature in the algorithm diagram is associated with the annealing process in the simulated annealing algorithm.

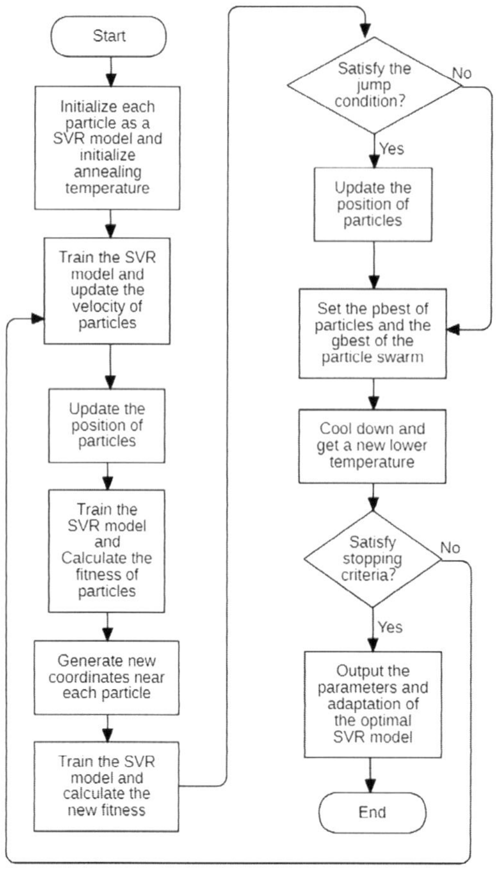

Figure 6. SAPSO-SVR algorithm flow diagram.

The optimized algorithm no longer needs manual parameter adjustment. Whenever the data set is expanded with a part of the data, the algorithm is automatically adjusted to improve the prediction accuracy and applicability.

2.3. Springback optimization model
On the basis of realizing high-precision springback prediction model, bat algorithm is used to seek optimal process parameters. The prediction model with the highest accuracy is saved as the optimization function. Before the optimization algorithm runs, the digital twin system needs to be given the current sheet data, and the optimization algorithm searches for the corresponding optimal process parameters. This optimization method realizes real-time optimization, and its significance is that when a part with different sheet widths and different materials needs to be

formed, this method can automatically obtain the optimal process parameters and transfer them to the digital twin model for adjustment without human intervention.

At the same time, it should be noted that the reliability of the optimization results depends on the accuracy of the prediction model. If the accuracy of the prediction model is low, the obtained optimization results may not achieve the expected results. It is therefore necessary to continuously record molding data and augment predictive model datasets.

3. Validation

3.1. Target part

The hat-shaped parts are used for verification. Its section is shown in the Figure 7. The flange part changes with the width of the original sheet.

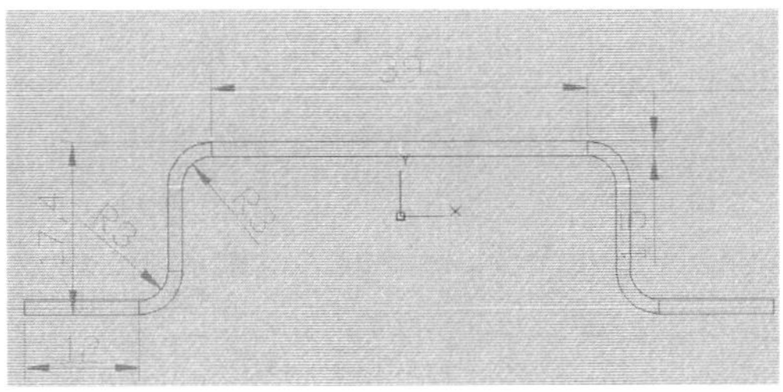

Figure 7. Hat-shaped parts.

3.2. Experimental design

The process parameters studied include uphill volume and roll gap. The values of the uphill volume are 0 mm, 2.5 mm, 5 mm, 7.5 mm, and 10 mm. The values of the roll gap are 1.48 mm, 1.49 mm, 1.5 mm, 1.51 mm, and 1.52 mm. Three materials, Al3003, Al5052, and Al6061, were used for research. There are 4 different sheet widths: 85 mm, 90 mm, 95 mm, and 100 mm.

In order to train the springback prediction model and optimization model in the shortest time and at the lowest cost, we designed an orthogonal experiment to accumulate the relatively most comprehensive hat-shaped part forming data. The data set contains 32 sets of experiments. In each experiment, due to the symmetrical characteristics of the hat-shaped part, the springback of the two corners on the left side of the hat-shaped part was taken as the forming evaluation index. The measurement method is to measure the two bending angles of the cross-section at 1/4, 1/2, and 3/4 of the molded part. These three cross-sections can be seen in Figure 8. The advantage of this is that the forming situation of the entire sheet is comprehensively considered. The deformation at both ends of the sheet is not stable, so the two sections at the starting point and end point of the length direction of the sheet are not taken.

3.3. Results and analysis

The prediction accuracy test was carried out based on the forming data of hat-shaped parts. All experiments were conducted using the Shensi Lab roll forming equipment. The dataset is available upon request from the authors. In each experiment, the data was shuffled and 70% of it was selected as the training set, while 30% was used as the testing set. And the optimal SVR parameter pairs obtained by SAPSO-SVR and the corresponding resultant metrics are shown in Table 1.

The results show that the optimized parameters are not stable. Common reasons for this phenomenon are too little data volume and some noisy data. The difference in R2 between

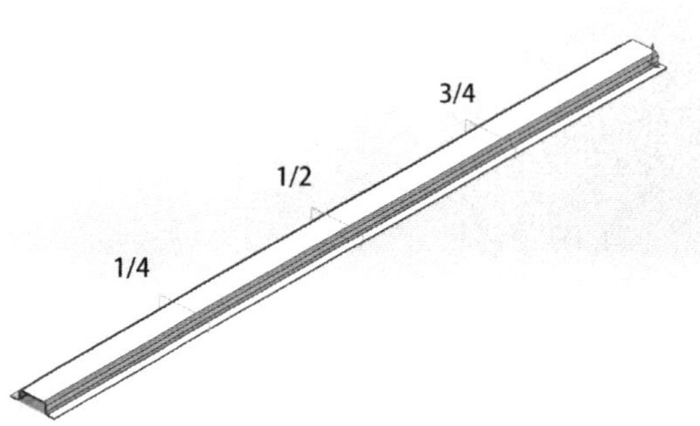

Figure 8. The cross-section at 1/4, 1/2, and 3/4 of the molded part.

Table 1. Optimal parameters and results found by SAPSO-SVR in 5 experiments.

No.	ϵ	C	γ	R2	MAE	MDAE
1	0.1985	328.79	0.0194	0.706	0.445	0.437
2	0.5320	4996.30	0.0032	0.940	0.223	0.172
3	0.1231	4696.15	0.0024	0.788	0.233	0.190
4	0.3886	4998.60	0.0054	0.899	0.235	0.207
5	0.4021	2270.87	0.1202	0.718	0.429	0.308

several experimental results are caused by the random filling of the test set and the training set. This also shows that some experimental results may be distorted and require further processing. The optimal model obtained in 5 experiments has the R2 of 0.94, the mean absolute error of 0.223, and the median absolute error of 0.172. The accuracy of this model initially meets the needs of the optimization model. At the same time, further expansion of the data set can be considered to improve the accuracy of the prediction model. It takes about 0.04s to use the trained prediction model to predict 500 pieces of data. The prediction speed of SAPSO-SVR prediction model is very fast and can meet the need of real-time prediction.

Based on the highest-precision springback prediction model obtained in Experiment No. 3 as the optimization function, the springback optimization model was obtained using the BA algorithm. An experiment was conducted on a sheet with a material model of Al6061 and a sheet width of 100. The initial unadjusted process and the optimized process were compared. The experimental results are shown in the Table 2.

Table 2. The comparison of the springback result of unadjusted process (the first row) and the optimized process (the second row).

Material	Sheet width	Roll gap	Uphill volum	springback1	springback2
Al6061	85	1.5	0	1.9°	2.1°
Al6061	85	1.52	9.83	1.2°	1.8°

The results show that the average springback of the two bending corners of the hat-shaped part obtained by the optimized process is reduced by 0.7° and 0.3° respectively. With the current small sample set, this result is considered to initially demonstrate the effectiveness of the model. The reason why the optimization results are not closer to 90° may be that the insufficient amount of data results in insufficient accuracy of the prediction model. It may also be caused by some noisy data. In the future, it is necessary to further expand the experimental collection data to obtain a better molding result.

4. Conclusion

This research is dedicated to solving the complex cross-section and springback defects in the sheet metal roll forming process through an adaptive springback intelligent control method. Through the collaboration of the digital twin module, springback prediction module and springback optimization module, this method realizes the fully automatic adjustment of process parameters, thereby significantly improving production efficiency and forming quality. Experimental results show that the cross-section after the optimized process has a springback reduction of 36.8% and 14.3% respectively at the two bend corners compared to the default process. The effectiveness of the springback optimization model proposed in this paper has been initially verified.

In the future, it is necessary to further expand the data set, enhance the reliability of the model, and consider more process parameters and more complex cross-sections to enhance the versatility of the model.

Acknowledgments

The authors wishes to thank Shenzhen Institute for Advanced Study, UESTC for the technical and financial support.

References

[1] Jurkovic M, Jurkovic Z, Buljan S and Obad M 2018 *Advances in Production Engineering & Management* **13** 57–68

[2] Ruan F, Feng Y and Liu W 2008 *2008 second international symposium on intelligent information technology application* vol 1 (IEEE) pp 157–161

[3] Fu Z and Mo J 2011 *The International Journal of Advanced Manufacturing Technology* **53** 473–483

[4] Paralikas J, Salonitis K and Chryssolouris G 2011 *The International Journal of Advanced Manufacturing Technology* **56** 475–491

[5] Paralikas J, Salonitis K and Chryssolouris G 2009 *The International Journal of Advanced Manufacturing Technology* **44** 223–237

[6] He J, Cu S, Xia H, Sun Y, Xiao W and Ren Y 2023 *Journal of Intelligent Manufacturing* 1–17

[7] Vrh M, Halilovič M, Starman B and Štok B 2009 Modelling of springback in sheet metal forming

[8] Ragai I and Nemes J A 2010 *International Manufacturing Science and Engineering Conference* vol 49477 pp 15–22

[9] Wang W 2001 *Proc. NUMIFORM 2001, Simulation of materials processing* 741–744

[10] Narasimhan N and Lovell M 1999 *Finite elements in analysis and design* **33** 29–42

[11] Fu Z, Chen W, Tian X and Hu B 2014 *The International Journal of Advanced Manufacturing Technology* **72** 561–570

[12] Yan Y, Wang H, Li Q, Qian B and Mpofu K 2014 *The International Journal of Advanced Manufacturing Technology* **72** 209–220

[13] Gauntt S M 2021 *Dynamic Analysis of Aerospace Gearboxes Featuring Hybrid Steel-Composite Gears* (The Pennsylvania State University)

Holistic springback compensation procedure

P Zimmermann[1]*, A Birkert[1], P Saup[2], F Marner[2] and M Häussermann[2]

[1] Center for Metal Forming and Car Body Manufacturing, Faculty of Mechanics and Electronics, Heilbronn University of Applied Sciences, Heilbronn, Germany
[2] inigence gmbh, Bretzfeld, Germany

*philipp.zimmermann@hs-heilbronn.de

Abstract. The elastic springback during the manufacturing process of stamped car body components causes dimensional deviations. To compensate these deviations, the common approach is to modify the tool surfaces in the opposite direction of the deviations – this is called springback compensation. In the procedure of springback compensation various issues must be solved. To achieve the main goal of a dimensionally accurate part, it must be ensured that, on the one hand the distance normal to the sheet surface between the springback part and the target geometry is within a specified dimensional tolerance. On the other hand, it must be ensured that the surface areas and characteristic lengths of the springback part and that of the target geometry match as closely as possible. In the past, approaches/methods, such as the physical compensation method and the physical scaling approach, have been presented which can successfully counteract these problems. Furthermore, in a multi-stage process in subsequent operations a stable part position must be achieved and unwanted plastic deformations must be avoided during blankholder closing. Therefore, different compensation strategies have been presented, which can fulfil these requirements. However, in the publication of these methods, the problems were always considered individually. This paper shows how all the named requirements can be achieved by combining the individual methods in springback compensation.

1. Introduction

During the forming process of car body components, the part geometry changes during the opening of the forming tools due to the elastic-plastic behavior of the part material. This effect is called springback. In general, forming tools are initially designed based on the target part geometry and because of that the part after springback deviates from the target geometry. If the deviations exceed a specified tolerance, the tool surfaces can be modified in the opposite direction of the springback, so that the part after springback is in tolerance. This procedure is called springback compensation.

The deviations resulting from the springback can be classified into two different categories. Due to elastic membrane stresses in the drawshell at the end of the drawing process, the drawshell contracts during springback. This leads to dimensional deviations in part surface direction on the one hand and to dimensional deviations in vertical direction to the part surface on the other. In addition, the part distorts and/or bends due to elastic bending stresses, which results in further dimensional deviations in vertical direction to the part surface. The part contraction after drawing is compensated by scaling the tool surfaces of the drawing tool, whereas the tool surfaces are overbent in an appropriate way to compensate the deviations in the vertical direction to the part surface.

However, car body components are manufactured in several consecutive operations, whereby the part geometry and thus its stiffness is changing in each operation. In addition to the main goal of producing a dimensionally accurate part at the end of the entire operation sequence, it must also be ensured that the arriving part in each operation is in contact with the tool surfaces as fully as possible to guarantee a stable part position.

To achieve both a part within dimensional tolerances and a stable part position in all operations, all compensation measures must be distributed in an appropriate way on the individual operations. The definition of which operation is compensated with which springback result is called compensation strategy.

2. State-of-the-art springback compensation methods and approaches

As already mentioned, there are different methods for solving the individual problems. Firstly, methods for compensating the dimensional deviations vertical to the part surface, then for compensating the part contraction and finally different compensation strategies are considered.

2.1. Methods to compensate deviations normal to the sheet surface (compensation methods)

Many methods to compensate the vertical dimensional deviations have been developed in the past. The most popular of these is the displacement adjustment method [1] or variants of it [2-11]. These methods are based on a mathematical inversion of the springback vector field to generate the compensation vector field. A method that differs from the methods mentioned is the so-called Physical Compensation Method (PC-Method) [12]. With this method, the vertical deviations between the part after springback and the target part geometry is used to generate the compensation vector field by overbending the target part geometry in a separate FE simulation. In this way, the lengths and surface area of the overbent part – in contrast to methods based on the mathematical inversion of the springback vector field [2-11] – can be almost completely preserved.

2.2. Methods to compensate surface area deviations (scaling approaches)

The change in the lengths and the surface area of the part resulting from springback should be compensated as best as possible by a suitable tool scaling. Because the stresses which cause this effect are mainly membrane tensile stresses, which are induced into the drawshell during the forming process in drawing, the part shrinks during springback. To compensate this part shrinkage, the global scaling approach is generally used, whereby the tool geometry is increased by a constant factor. This factor can be defined using the ratio between the biaxial yield stress and the Young's modulus of the part material, for example. Optionally, different factors can be defined in the two main surface directions. However, the global scaling approach is only suitable for compensating the part shrinkage in a more or less good approximation. Because the part shrinks locally different as a result of its strain distribution and locally different stiffness [13], another approach is required for more accurate scaling. Such a local scaling approach was presented in [13] and developed further in [14]. The results presented there showed significant advantages compared to using the global scaling approach.

2.3. Springback compensation strategies

In the usually multi-stage production process of car body parts, each operation affects the springback due to the introduced shape and geometry changes, which leads to different springback results after each operation. The compensation strategy defines which operation is compensated with which springback result. Roll et. al [15] have recommended that springback should be compensated in the operation in which it occurs. However, there was also mentioned that for processes consisting of more than a drawing and a trimming operation, the problem becomes significantly more complex. A unique and consistent approach was not presented. Such a consistent approach was presented in the so-called Improved Springback Compensation Strategy (Improved Strategy) [16]. In addition to a dimensionally accurate part at the end of the process, this strategy also makes it possible to achieve a stable part position when inserting the parts into the subsequent operation. Another modified compensation strategy that emerged from the first mentioned is the Advanced Springback Compensation Strategy (Advanced Strategy) [17].

IOP Conf. Series: Materials Science and Engineering **1307** (2024) 012034 doi:10.1088/1757-899X/1307/1/012034

3. Combination of selected springback compensation methods to generate a holistic procedure

The compensation process usually starts with the so-called basic simulation with initially uncompensated tools, also known as zero tools. Before starting springback compensation, it must be ensured that the feasibility of the part with regard to wrinkles and splits is given and that the springback characteristics are sufficiently stable with scattering process parameters [18].

A combination of the Advanced Strategy [17] and the Improved Strategy [16] is used here as compensation strategy. The optimized Physical Scaling Approach [14] is used for scaling the drawing operation and the physical compensation method [12] is used for compensating deviations normal to the sheet surface.

In the first step, the drawing operation is scaled. For this purpose, a simulation setup is generated in which the drawing operation is simulated by using zero tools. In addition, the following trimming operation is simulated as a "laser cut". After the trimming operation, the springback of the part is calculated (Figure 1). This springback result is now used to calculate the local scaling of the drawing operation. In order to calculate the required local scaling dimensions, the stress state before springback is being inverted and then adjusted in 20 iteration loops.

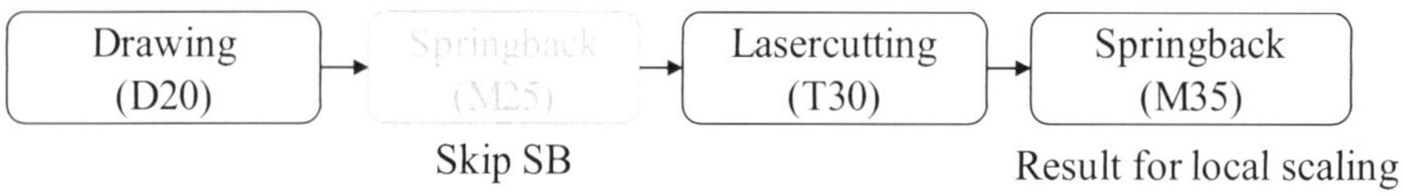

Figure 1. Simulation Setup for local scaling.

In the next step, the forming operations and the respective springback are calculated (Figure 2). The springback after drawing (D20) and trimming (T30) is calculated with scaled tools. For the simulation of the forming operations (F40 & F50), the respective part zero geometry is used for the arriving part. To generate these part zero geometries, the stresses in the respective part are eliminated before springback, with the consequence that the part does not spring back. The use of the part zero geometries prevents any (unwanted) influence of the closing of the blank holder on the springback of the part.

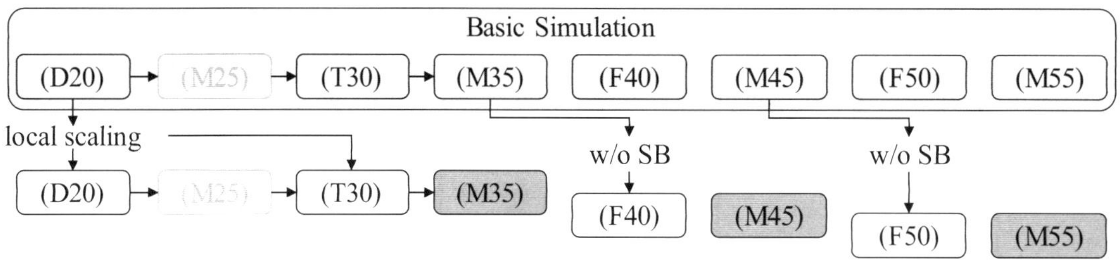

Figure 2. Calculating springback without unwanted influence of blankholder closing.

The single springback results are now used to sum up all dimensional deviations from the single operations (Figure 3, Total SB) in order to generate a compensation vector field by using the PC-Method [12]. The simulation for the application of the PC-Method has been done with the FE-Software Ansys Mechanical. With the generated compensation vector field all operations are compensated by the same amount (Figure 3, Compensation value), in order to create the precondition for the later adjustment of the subsequent tools after drawing to the in each operation arriving part.

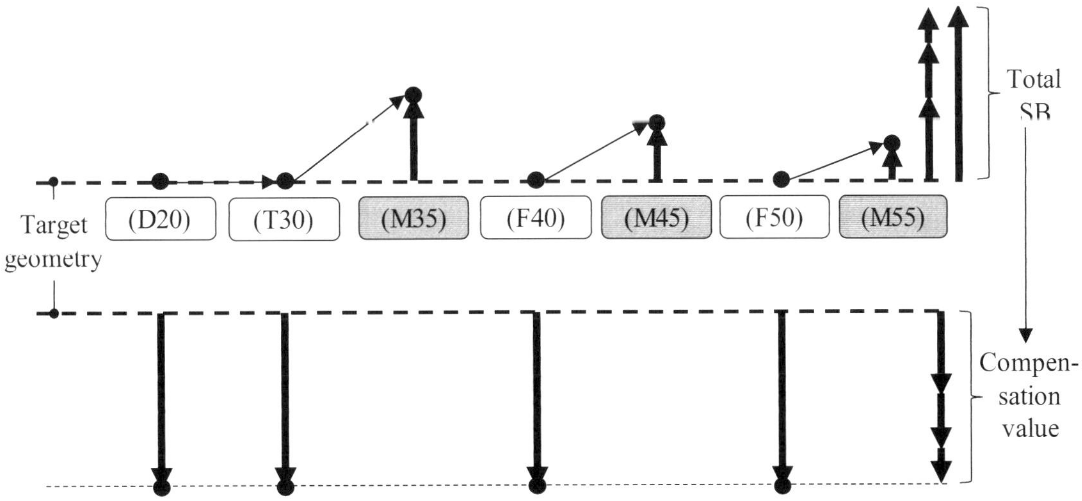

Figure 3. Generate a compensation vector field for all operations (Iteration 1).

After all operations have been compensated, a new forming simulation is performed with the compensated tools (Figure 4). Therefor initially only the drawing operation, the trimming operation and the subsequent springback are calculated. This springback result is now used to adjust the forming operations F40 and F50 to the part after springback (Adjustment M35). The first forming operation F40 is now simulated with the adjusted tools and the resulting springback M45 is calculated. This springback (M45) is now used to adjust the second forming operation F50 to the arriving part (Adjustment M45). Finally, the forming operation F50 is simulated and the resulting springback (M55) is calculated. As the springback usually changes due to the compensation of the tools, there are deviations between the part after springback and the target part geometry at the end of the process (Deviation IT1). If these deviations exceed the specified dimensional tolerances, the compensation process must be repeated.

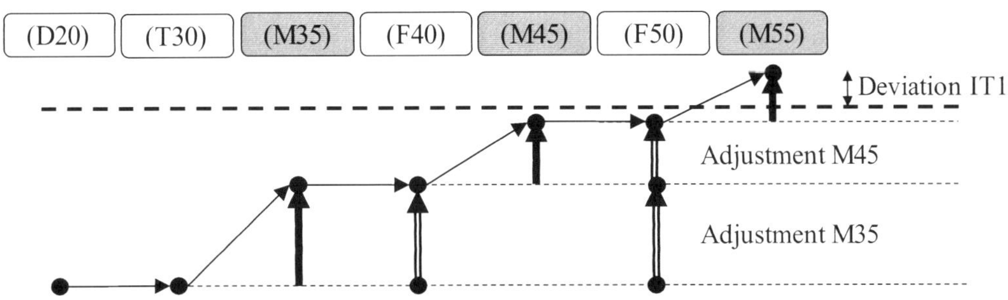

Figure 4. Tool modification based on springback results to fit them to the arriving part (Iteration 1).

To adjust the tool geometries again, the summed up dimensional deviations of the basic simulation (Figure 3, Total SB) are added to the dimensional deviations after IT1 (Figure 4, Deviation IT1) in order to compensate the tool surfaces in all operations (Figure 5). The tool surfaces are then adjusted to the respective arriving part as before. This procedure is repeated until the dimensional deviations are within the specified dimensional tolerances.

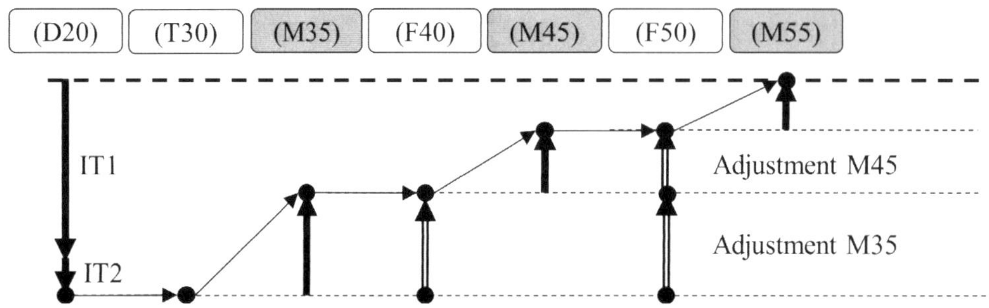

Figure 5. Compensation values in Iteration 2.

4. Results of the holistic springback compensation procedure

The described procedure is validated using the example of an aluminum headboard with a sheet thickness of 2.5 mm (Figure 6). The production process was simulated with AutoForm Forming R10. The process consists of a drawing operation (D20), a laser trimming operation (T30) and two forming operations (F40 & F50).

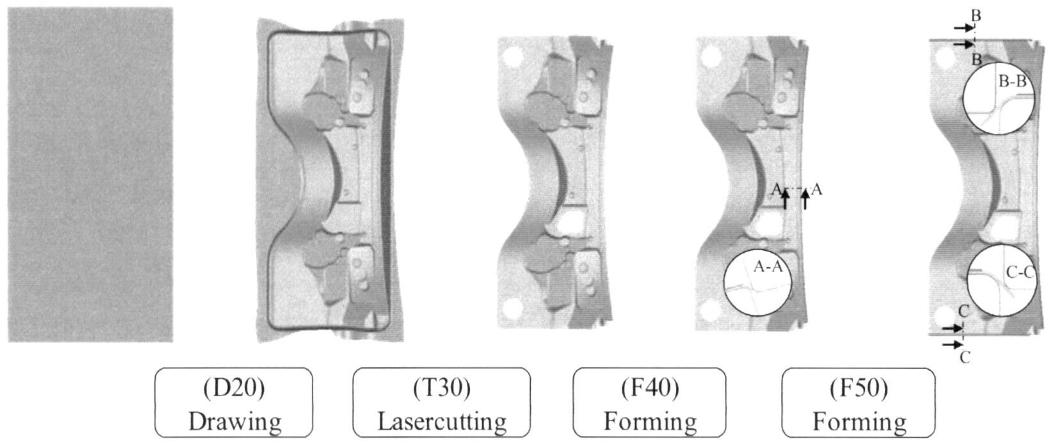

Figure 6. Manufacturing process of the headboard.

The production process was compensated in four compensation loops. At first, the dimensional deviations are analyzed (Figure 7). In addition to the graphical representation of the dimensional deviations, the maximum dimensional deviation (Δmax), the ratio of the part surface with deviations > 0.5 mm from the target part geometry (A > 0.5 mm) and the mean dimensional deviation (ΔØ) are also evaluated. After the first iteration (IT1), no improvement in dimensional accuracy can be seen. The sign inversion of the dimensional deviations from positive to negative and vice versa shows that with a lower compensation factor, a more dimensionally accurate part could probably have been achieved in the first iteration (IT1). This effect can also be seen at the transition from IT2 to IT3, although compensation was already made with a compensation factor of KF = 0.7. For this reason, the compensation factor in IT4 was reduced again to KF = 0.6.

The maximum dimensional deviation even increases slightly from IT0 (basic simulation) to IT1, but can then be gradually reduced from initial 3.8 mm (IT0) to final 0.8 mm. The surface area of the part with deviations > 0.5 mm from the target geometry can be reduced from the basic simulation 72.3% (IT0) to 2.4% (IT4). According to IT3, this was even slightly lower at just 1.5%. A similar convergence behavior can be observed when looking at the mean dimensional deviation. This evaluation value can be reduced from initially 1.03 mm to 0.12 mm, with a slightly lower value of 0.10 mm after IT3. The slightly worse values in IT4 compared to IT3 (except the maximum deviation) again indicate that an even smaller compensation factor should be used.

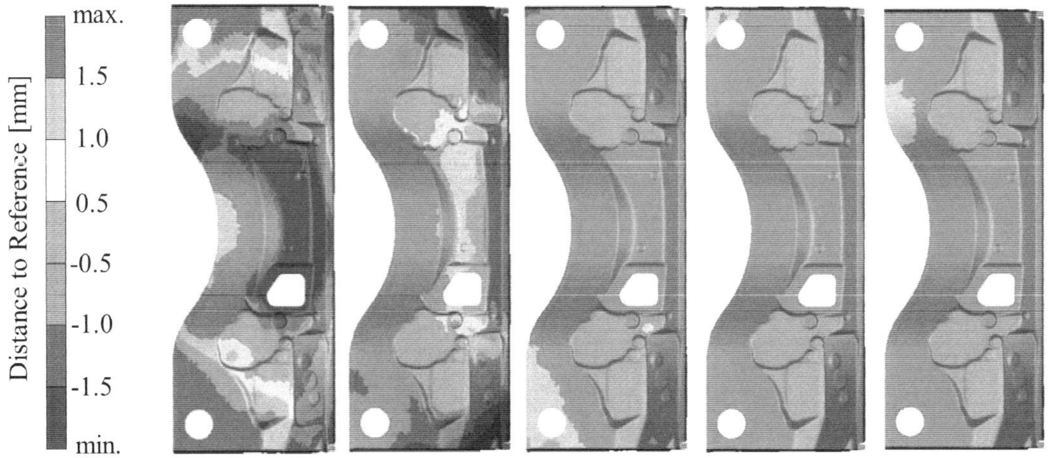

	IT0	IT1	IT2	IT3	IT4
Δmax [mm]	3.8	4.1	1.4	1.1	0.8
A > 0,5 mm	72.3%	53.9%	7.5%	1.5%	2.4%
ΔØ [mm]	1.03	0.74	0.21	0.10	0.12

Figure 7. Evaluation of the dimensional deviations caused by springback.

After evaluating the dimensional deviations, the evaluation of the part position in the forming operations is now carried out (Figure 8). For this purpose, the distance between the incoming part and the receiving tool geometry is evaluated. In the basic simulation, the distance between the part and the tool in both forming operations is over 2 mm across large areas. In the final simulation setup, the distance in both forming operations is less than 1 mm over the entire tool surface.

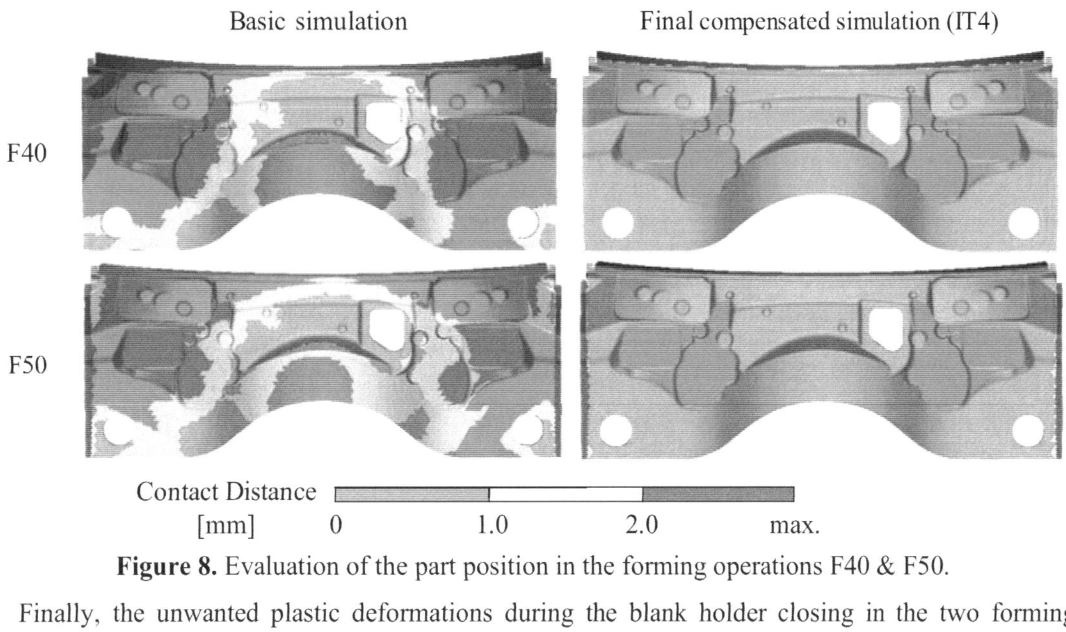

Figure 8. Evaluation of the part position in the forming operations F40 & F50.

Finally, the unwanted plastic deformations during the blank holder closing in the two forming operations are evaluated (Figure 9). This is done by using the evaluation variable Plastic Strain Rate (PSR) which is implemented in AutoForm Forming R10. Compared to the basic simulation, the unwanted plastic deformations in the final compensated simulation setup (IT4) were significantly reduced in both forming operations. In the demonstrated experiment full-surface tools were used. In

IOP Conf. Series: Materials Science and Engineering **1307** (2024) 012034 doi:10.1088/1757-899X/1307/1/012034

practice, strongly curved areas are often set back (radius clearances). The remaining unwanted plastic deformations could be further reduced by such radius clearances.

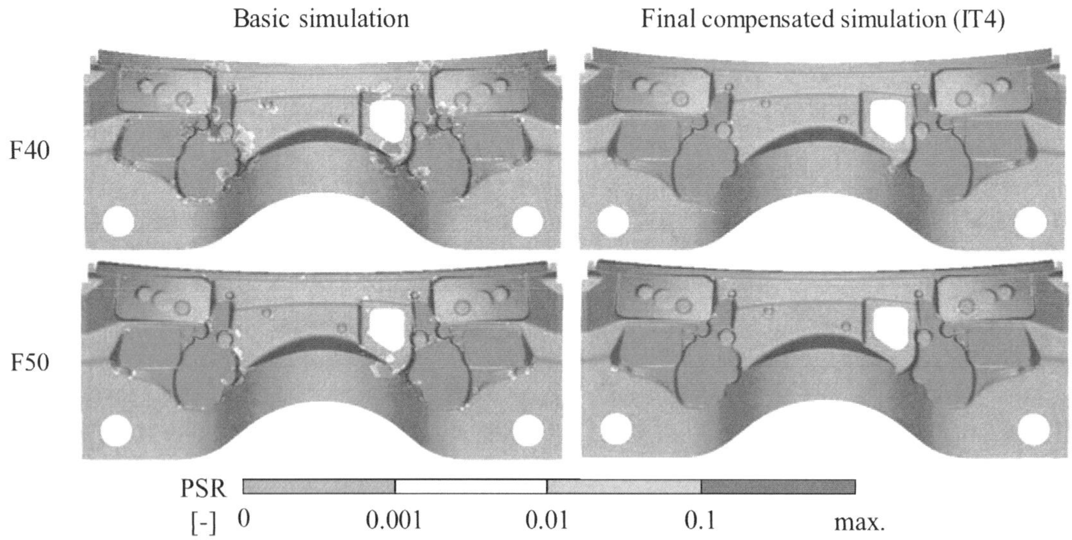

Figure 9. Evaluation of unwanted plastic deformations caused by blankholder closing in F40 & F50.

5. Conclusion

This paper describes how several methods for springback compensation are combined to generate a holistic approach. In a first step, the drawing operation is scaled locally. This compensates the part contraction ("shrinkage") resulting from springback and the related surface and length differences in the sheet surface between the part after springback and the target part geometry. In a next step, the single forming operations are simulated separately and without the influence of unwanted deformations during blank holder closing and the respective springback is calculated. The single springback results are then summed up using the Advanced Strategy [17]. After that the forming operations are adapted to the arriving part, as specified in the Improved Strategy [16]. The combination of both strategies was chosen because the springback result needed for compensation with the Advanced Strategy is less work-intensive and time-consuming than with the Improved Strategy. However, the Improved Strategy achieves a better part position than the Advanced Strategy. The PC-Method [12] was used to generate a compensation vector field for compensating the tool geometries using the (summed up) dimensional deviations, as the surface areas and lengths of the tools are retained here.

In order to validate the effectiveness of the procedure, it was tested on a structural car body component. The evaluation of the dimensional deviations shows that these can be continuously reduced. As the chosen part reacts to the compensation of the tools with a relatively strong change in springback, the compensation factor in IT3 and IT4 was reduced. Probably a lower compensation factor starting from IT1 would have reduced the dimensional deviations even further. With regard to both, the part position and unwanted plastic deformation during blank holder closing, a significant improvement in the quality of results was achieved compared to the basic configuration.

In future work, the procedure should be tested on the production process of other car body components. In addition, the sub-processes of the procedure should be further automated and interfaces should be optimized, so that a holistic software solution can be developed.

References

[1] Gan W and Wagoner RH 2004 Die design method for sheet springback *Int. J. Mech. Sci.* **46**(7) 1097-113

[2] Karafillis A and Boyce M 1992 Tooling design in sheet metal forming using springback calculations *Int. J. Mech. Sci.* **34** 113-31
[3] Yang X and Ruan F 2011 A die design method for springback compensation based on Displacement adjustment *Int. J. Mech. Sci.* **53** 399-406
[4] Weiher J, Rietman B, Kose K, Ohnimus S and Petzoldt M 2004 Controlling springback with compensation strategies *AIP Conf. Proc.* **712** 1011-15
[5] Lingbeek R, Huétink J, Ohnimus S, Petzoldt M and Weiher J 2005 The development of a finite elements based springback compensation tool for sheet metal products *J. Mater. Proc. Tech.* **169** 115-25
[6] Cheng HS, Cao J and Xia ZC 2007 An accelerated springback compensation method *Int. J. Mech. Sci.* **49** 267–79
[7] Cafuta G, Mole N and Štock B 2012 An enhanced displacement adjustment method: springback and thinning compensation *Mater.& Design* **40** 476-87
[8] Liao J, Xue X, Zhou C, Barlat F and Gracio JJ 2013 A springback compensation strategy and applications to bending cases *steel research international* **84**(5) 463-72
[9] Li G, Liu Y, Du T and Tong H 2014 Algorithm research and system development on geometrical springback compensation system for advanced high-strength steel parts *Int. J. Adv. Manuf. Tech.* **70** 413–27
[10] Siswanto WA, Anggono AD, Omar B and Jusoff K 2014 An alternate method to springback compensation for sheet metal forming *Sci. Wor. J.* **2014** 301271
[11] Zhang Z, Wu J, Zhang S, Wang M, Guo R and Guo S 2016 A new iterative method for springback control based on theory analysis and displacement adjustment *Int. J. Mech. Sci.* **105** 330-9
[12] Birkert A, Hartmann B and Straub M 2017 New method for springback compensation for the stamping of sheet metal components *J. Phys.: Conf. Ser.* **896** 012067
[13] Birkert A, Zimmermann P, Hartmann B, Scholle M and Straub M 2020 New physical scaling approach to compensate the part contraction in drawing operations *IOP Conf. Ser.: Mater. Sci. Eng.* **967** 012041
[14] Zimmermann P, Birkert A, Nowack M and Kraft T 2023 Optimization of the physical scaling approach through the adaption of the inverted membrane stresses *IOP Conf. Ser.: Mater. Sci. Eng.* **1284** 012081
[15] Roll K, Lemke T and Wiegand K 2004 *Simulationsgestützte Kompensation der Rückfederung* LS-Dyna Anwenderforum, Bamberg
[16] Birkert A, Hartmann B, Lepple F, Straub M and Zimmermann P 2019 Improved Spring-Back Compensation Strategy through Location Optimized Part Position in the Dies *IOP Conf. Ser.: Mater. Sci. Eng.* **651** 012036
[17] Birkert A, Zimmermann P, Hartmann B and Nowack M 2022 Advanced Springback Compensation Strategy through elimination of avoidable elastic strain energy *IOP Conf. Ser.: Mater. Sci. Eng.* **1238** 012083
[18] Birkert A, Hartmann B, Scholle M and Straub M 2018 Optimization of the process robustness of the stamping of complex body parts with regard to dimensional accuracy *IOP Conf. Ser.: Mater. Sci. Eng.* **418** 012107

IOP Conf. Series: Materials Science and Engineering **1307** (2024) 012035 doi:10.1088/1757-899X/1307/1/012035

Anomaly Detection in the Automotive Stamping Process: An Unsupervised Machine Learning Approach

J Zhang[1], D Creighton[1*], C P Lim[1], B Rolfe[2], M Weiss[3], A Neiat[4], A Zaslavsky[4], T Nguyen[5], J Navaei[6], R Gamasaee[6], B Barresi[6], and M Novak[6]

[1] Institute for Intelligent Systems Research and Innovation, Deakin University, Geelong, Australia

[2] Faculty of Science, Engineering and Built Environment, Deakin University, Geelong, Australia

[3] Institute for Frontier Materials, Deakin University, Geelong, Australia

[4] School of Information Technology, Deakin University, Geelong, Australia

[5] Department of Data Science and Artificial Intelligence, Monash University, Clayton, Australia

[6] Ford Motor Company, Dearborn, Michigan, USA

*douglas.creighton@deakin.edu.au

Abstract. In metal forming, such as stamping of automotive parts, unsupervised machine learning models offer a transformative approach to real-time quality control, especially when labelled data are scarce. Leveraging clustering algorithms and autoencoders, we develop a machine learning system capable of autonomously monitoring sensor data and identifying deviations suggestive of potential defects. The system offers multiple benefits including rapid intervention, reduced part defects and lower stoppages required to rectify defects. The use of unsupervised machine learning models also adds a layer of adaptability, allowing the system to continually refine its understanding of what constitutes a 'normal' operation. Empirical evaluation demonstrates the potential of the developed system in detecting anomalies in production data collected from dynamic automotive manufacturing environments.

1. Introduction

In the automotive industry, the stamping process plays a crucial role in transforming metal sheets into specific parts, such as body panels, interior connectors, and Body-In-White structural components [1-3]. On the production floor, large tonnage stamping presses are used to mold as-received blanked metal sheets into desired three-dimensional shapes based on predefined tooling. The stamping process often involves a series of operations such as blanking, drawing, piercing, and trimming, executed in a highly coordinated manner to ensure precision and consistency. Given the complexity and high-speed nature of these operations, quality control is of paramount importance. Even minor deviations in pressure, alignment, or material quality can result in defects, leading to waste of materials and the potential costly stoppages to rectify issues. Therefore, ensuring that each stamped part meets rigorous quality standards is a significant concern, demanding advanced technological solutions for effective monitoring and control.

Quality control in automotive stamping process presents unique challenges owing to the intricate nature of the operations and the high part throughput. Ensuring consistency and precision in stamping parts is not just about controlling the mechanical aspects like pressure and alignment; it may also involve monitoring material characteristics such as grain microstructure and yield strength/hardening, or process characteristics such as temperature and lubrication. Each of these factors can significantly influence the quality of the final product. Moreover, given the volume of parts produced, even a small percentage of defects can lead to substantial material waste, necessitating line stoppages that are both time-consuming and expensive. Traditional quality control methods, often based on post-production inspection, are inadequate for detecting defects early enough to avoid costly interruptions and may only be discovered at the point of assembly. As a result, there is a rising demand for real-time, data-driven approaches to quality control that can swiftly identify and rectify issues. The incorporation of machine learning models, central to the Industry 4.0 era, offers a powerful approach to achieving the demand [4-8]. These models can analyse real-time data to predict potential defects and enable immediate corrective actions, thereby elevating the efficiency and accuracy of quality control processes [9].

The cost implications of quality control in stamping processes are significant and multi-faceted. Indeed, frequent quality issues can lead to longer downtimes, increased labour costs for rework, and potential reputational damage, which could impact future business. On the other hand, investing in advanced quality control systems, such as machine learning solutions for real-time monitoring, requires initial capital expenditure and ongoing maintenance costs. However, these investments are often justified by the subsequent reduction in defects and associated savings in material and operational costs.

2. Related work and proposed framework

Process data harvested from real operational environments provide a goldmine for machine learning models to derive insights, offering advantages such as enhancing production efficiency, minimising downtime, and reducing waste [10]. However, the absence of labelled data poses a significant challenge for developing supervised learning-based solutions. Specifically, supervised learning models, which are commonly used for classification tasks like defect detection, require substantial number of labelled samples for training purposes [10-13]. Without these labelled samples indicating what constitutes a 'defect' or 'non-defect', a supervised learning model lacks the guidance needed to generalize and make accurate predictions on new, unseen data. Therefore, the lack of labelled data necessitates the exploration of alternative methods, like unsupervised learning, to effectively identify anomalies or defects in both historical and real-time data; therefore, the motivation of this research.

Autoencoders have been effectively used to evaluate defect and machine degradation conditions by analysing process signals like acoustic emission and force data in stamping [14, 15]. However, these investigations mainly considered stroke forces, and their findings were not cross validated using alternative unsupervised learning methods like clustering algorithms. In our study, we focus on the effects of lubrication depth in stamping processes, conceptualising the varying depths as an image-like data set for anomaly detection. Additionally, we cross-reference the outcomes from autoencoders with those obtained from clustering algorithms, aiming to mitigate the issue of false positives. The proposed detection framework is shown in Figure 1.

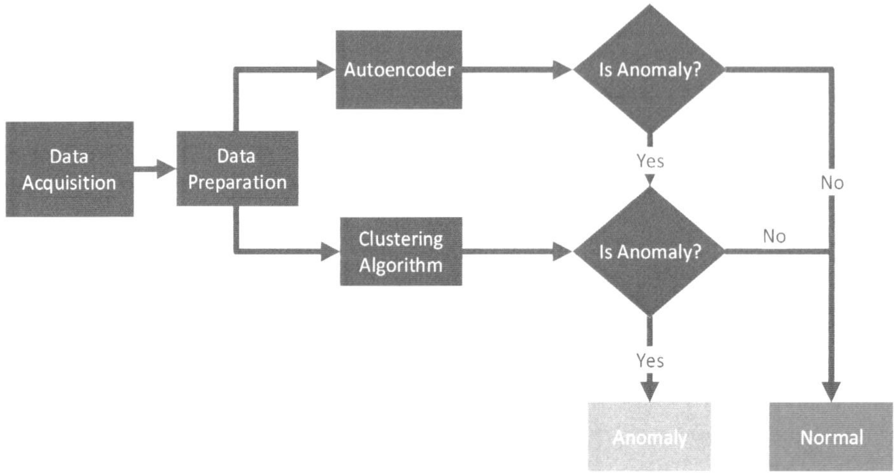

Figure 1. Framework for anomaly detection in stamping process.

3. Methodology

This study employs both clustering algorithms and autoencoders for anomaly detection in stamping processes in an automotive production plant. Data samples are gathered from three key areas: the parts, the machines, and the lubricants involved in the stamping process. We focus on the lubrication data for anomaly detection in this study. Anomalies or issues in automotive stamping processes can arise from various factors including material defects, tooling issues, process variations, and equipment malfunctions. Among these, lubrication and oiling play crucial roles in the stamping process, significantly influencing the quality of stamped parts. Proper distribution of lubrication and oiling reduces friction between the metal sheet and the die, promoting smooth material flow and preventing surface defects, ultimately enhancing the dimensional accuracy and surface finish of stamped parts while extending the tool life. Proper lubrication also minimizes the occurrence of common stamping defects such as wrinkles, tears, resulting in higher production yields and reduced scrap rates.

Lubrication distribution depends on factors such as part geometry and material properties. In some cases, uniform distribution across the part surface ensures consistent material flow, while selective application in specific areas may optimise lubrication effectiveness for complex geometries or high-friction zones. To identify the right distribution of lubrication, autoencoders are utilised to reconstruct lubrication sensor inputs and flag significant deviations as anomalies, while clustering algorithms are used to corroborate and visualise the results. The performance of our method is assessed through two primary means: comparative analysis and visual inspection. For comparative analysis, anomalies identified by both the clustering and autoencoder methods are compared for consistency.

Significant overlap in the anomalies detected by both methods would lend confidence to the outcome. For visual inspection, graphical representations of the normal and abnormal data samples are examined to qualitatively assess the outcome. These visual checks serve to corroborate the algorithmic findings and provide an additional layer of validation. Through this dual-prone evaluation approach, our study aims to provide a rigorous assessment of the effectiveness of the developed solution in real-time for anomaly detection, leading to reduction in defective parts and operational costs.

4. Overview of unsupervised learning

4.1. Clustering algorithm

Density-Based Spatial Clustering of Applications with Noise (DBSCAN) is an unsupervised learning model that identifies clusters in a data set based on the density of data samples. Unlike partitioning methods such as K-means, DBSCAN does not require the number of clusters to be specified in advance. The algorithm starts with an arbitrary sample and explores its neighbourhood based on a distance metric (usually Euclidean) and a density parameter, which is defined by the number of samples within a given radius. If the neighbourhood has sufficient density, a cluster is formed, and the algorithm continues to explore neighbouring samples, expanding the cluster. Samples that are insufficiently dense are labelled as outliers.

4.2. Autoencoder

An autoencoder is a type of neural network used for undertaking unsupervised learning tasks, primarily for data compression and reconstruction [16]. It consists of two main parts: an encoder, which compresses the input data into a lower-dimensional latent space, and a decoder, which reconstructs the data from the latent representation back to its original form. The network is trained to minimise the difference between the input and the reconstructed output, typically measured by a loss function like the mean squared error (MSE). Autoencoder is useful for feature learning, anomaly detection, and data denoising, among other applications. They are particularly useful in scenarios where labelled data samples are scarce, as they can learn to identify patterns and anomalies solely based on the input data structure.

5. Evaluation criteria

Comparative analysis between DBSCAN and autoencoder forms a critical part of the performance evaluation in this study. By comparing the anomalies identified by both methods, we aim to assess the robustness and reliability of each method in real-world environments. A high degree of overlap in the anomalies flagged by both methods would indicate a consistent and reliable identification of abnormal conditions, thus reflecting the effectiveness of our approach. On the other hand, discrepancies between the two methods could point to areas where one model may be more sensitive or specific than the other, especially in detecting anomalies related to lubrication. This comparative lens offers a more nuanced understanding of the strengths and weaknesses of each method, thereby providing valuable insights for refinements and implementations.

Visual inspection serves as a qualitative performance metric in our methodology, complementing the quantitative metrics obtained through comparative analysis. By graphically representing both the original and reconstructed data samples, as well as the clustering outputs, we can identify the abnormal data samples, facilitating the next step of validation and verification. Visual inspection provides an intuitive understanding of how well a method segregates normal operational data from anomalies, particularly in the context of lubrication metrics. Plots and charts are used to highlight possible misclassification, overlap, or conspicuous gap that may not be immediately evident through numerical evaluation alone. The visual representation can corroborate findings from the computational methods, therefore offering an additional layer of assessment, making it an indispensable component of our developed evaluation strategy.

6. Experimental Studies

We use a Python-based environment for implementing the DBSCAN and autoencoder models, relying on libraries such as scikit-learn for clustering and TensorFlow for building and training the autoencoder. A set of lubrication data was collected in May and June 2022. Based on an 18-by-20 grid, each cell records a value for the lubrication depth applied to the blank. A histogram of the collected data samples is depicted in Figure 2.

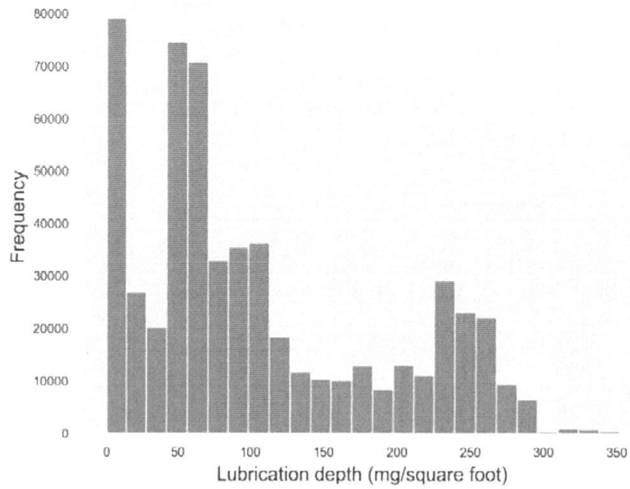

Figure 2. Distribution of lubrication depths.

The autoencoder model is trained and evaluated using a data set with a split of 80% for training and the remaining 20% for test. During training, the autoencoder minimises the reconstruction error, which is commonly measured using a loss metric like MSE. Several training epochs are used to iteratively adjust the model parameters for yielding an optimal performance. After training, the model is evaluated on the test set to assess its generalisation capability on unseen data samples. The test performance indicates the ability in accurately detecting anomalies, providing insights into real-world stamping processes in an automotive plant.

7. Results

7.1. Anomaly detection on test data

During training, the autoencoder learns to reconstruct the input from a latent space with a reduced dimension. Similarly in the test phase, the autoencoder reconstructs the test data from that latent space. The reconstruction error or loss on the test data set is assessed. Three instances with the highest reconstruction losses are flagged as anomalies for further inspection. These high-reconstruction-loss instances are deemed as deviations from the normal operating conditions, suggesting potential issues in the stamping process, particularly related to lubrication. The problem with lubrication could lead to defects such as wrinkles and tears. The identification of these anomalies indicates the model's ability to promptly flag deviations that could lead to defective parts or necessitate a production stoppage, thereby facilitating real-time monitoring of stamping processes in the production line. There is a clear separation between larger clusters of parts with small reconstruction loss and isolated clusters of parts with relatively large reconstruction loss. Those samples that lie above a nominal line are flagged as anomalies.

7.2. Clustering results

When applying the DBSCAN algorithm to the lubrication data, two distinct clusters appear when the Principal Component Analysis (PCA) is used to project the data into a two-dimension space, as shown in Figure 3. The first cluster, denoted as "normal cluster", consists of 706 parts that depict expected patterns pertaining to the standard operating conditions. The second (smaller) cluster, deemed as "anomaly cluster," contains only four parts. These four parts deviate significantly from the norm, based on the density-based metrics of DBSCAN, suggesting potential issues in the lubrication process that could lead to defects or

IOP Conf. Series: Materials Science and Engineering **1307** (2024) 012035 doi:10.1088/1757-899X/1307/1/012035

even require a halt in production. The stark difference in the number of parts between the two clusters serves to underscore the efficacy in the DBSCAN algorithm.

7.3. Comparison of autoencoder and DBSCAN
A comparison between the autoencoder and DBSCAN methods reveals highly correlated results, reinforcing the usefulness of each model in anomaly detection. Specifically, the bottom right three samples in the DBSCAN "anomaly cluster" correspond exactly to the top three samples with the highest reconstruction loss scores yielded by the autoencoder. Additionally, the data sample with the fourth-highest reconstruction loss score from the autoencoder matches the remaining top-left sample in the DBSCAN anomaly cluster. This strong alignment not only indicates the effectiveness of each method but also adds a layer of confidence in the finding. The correlation confirms that both methods are sensitive in identifying lubrication-related anomalies in the stamping production line.

7.4. Visualisation of results
To enhance our assessment of anomaly detection, we present the lubrication depth data using heatmaps. Figure 4 and Figure 5 depict the distribution of lubrication data pertaining to normal and anomaly parts, respectively. Both heatmaps share similar patterns of thick lubrication in the left half. However, for normal part, rows 4 to 6 of the right half part exhibit consistent deep lubrication and transition out to rows 3, 2, and 1 gradually. In contrast, for anomaly part, only rows 5 to 6 displays deep lubrication. Not only there is a significant drop in row 4, but row 1 also shows a sudden jump in lubrication depth. This visual inspection reinforces the distinctions made by both autoencoder and DBSCAN models.

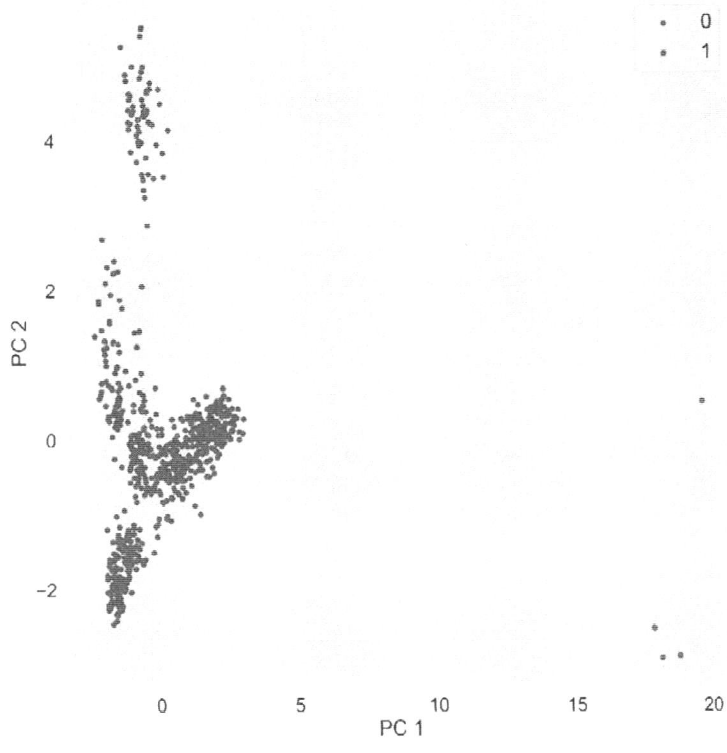

Figure 3. PCA projection of DBSCAN results.

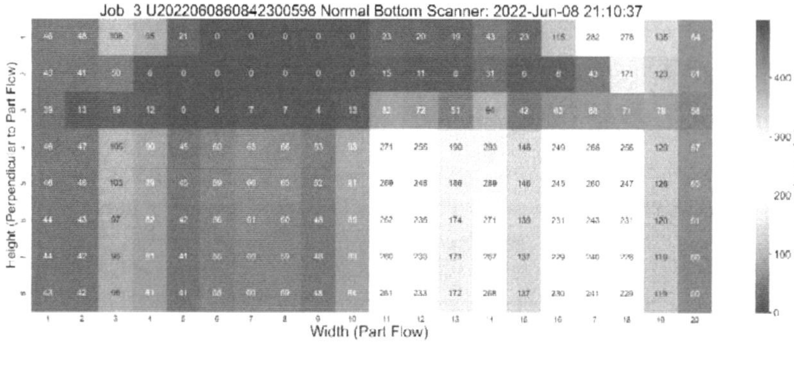

Figure 4. Lubrication depth data for normal part.

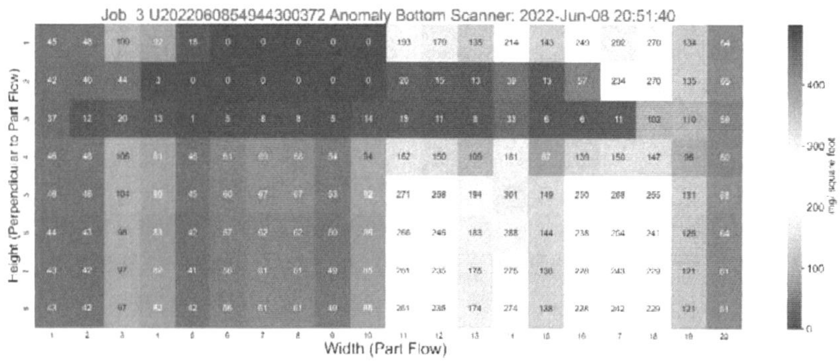

Figure 5: Lubrication depth data for anomaly part.

7.5. Limitations and possible improvements

While the results from both autoencoder and DBSCAN models demonstrate strong potential for anomaly detection, it is necessary to validate the detected samples by subject matter experts. The absence of such expert validation means that the model outcome is yet to fully corroborate with human-level quality inspection. Nevertheless, the consistency observed between the two models represents a significant step towards the development of a usable and useful recommender system for anomaly detection, facilitating screening by experts and leading to possible cost and time savings in stamping operations.

8. Conclusions and Future Direction

In summary, our study has demonstrated the applicability of unsupervised learning models, specifically DBSCAN and autoencoder, for anomaly detection in stamping processes in an automotive production plant. Focusing on the lubrication data, the results have indicated a high degree of correlation between the anomalies detected by both DBSCAN and autoencoder, reinforcing their reliability in anomaly detection. The detection outcomes present the clue for subject matter experts to pinpoint their attention on these anomaly parts. The next step of our work is to link these anomaly parts with process data such as material and machine status for further investigation.

For further research, we plan to integrate our developed methods into real-time context-aware monitoring systems, prompting the identified anomalies for validation by subject matter experts as well as ontology-based semantic knowledge base in the production line. A real-time context management platform, i.e., Context-as-a-Service (CoaaS) [17] developed at Deakin University, enables sensor data enrichment with real-time context to help validate the acquired data. CoaaS could be integrated with DBSCAN and autoencoder. It is also useful to explore the use of different machine learning models and combine the strengths of various algorithms in an ensemble form to improve the robustness and accuracy in anomaly detection.

9. References:

[1] Pereira M P, Weiss M, Rolfe B F, and Hilditch T B 2013 *Int. J. Mach. Tools Manuf.* **66** 44
[2] Chezan A, Atzema E, Pilthammar J, and Sigvant M, Year, *42nd Int. Conf. Deep Drawing Res. Group (IDDRG 2023)* (Luleå, Sweden) vol 1284 (IOP Publishing) p 012037
[3] Birkert A, Hartmann B, Nowack M, Petker A, Scholle M, Zimmermann P, and Kraft T, Year, *41st Int. Conf. Deep Drawing Res. Group (IDDRG 2022)* (Lorient, France) vol 1238 (IOP Publishing) p 012082
[4] Zhang J, Johnstone M, Le V, Khan B, Hosen A, Creighton D, Carney J, Wilson A, and Lynch M 2021 *Expert Syst. Appl.* **186** 115684
[5] Johnstone M, Le V T, Zhang J, Gunn B, Nahavandi S, and Creighton D 2015 *Expert Syst. Appl.* **42** 8078
[6] Seera M, Lim C P, Nahavandi S, and Loo C K 2014 *Expert Syst. Appl.* **41** 4891
[7] Lourenço R, Andrade-Campos A, and Georgieva P J M 2022 *Metals* **12** 427
[8] Mirandola I, Berti G A, Caracciolo R, Lee S, Kim N, and Quagliato L J M 2021 *Metals* **11** 833
[9] Penumuru D P, Muthuswamy S, and Karumbu P J J o I M 2020 *J. Intell. Manuf.* **31** 1229
[10] Dib M, Oliveira N, Marques A, Oliveira M C, Fernandes J, Ribeiro B, Prates P A J N C, and Applications 2020 *Neural Comput. Appl.* **32** 12335
[11] Cruz D, Barbosa M, Santos A, Miranda S, and Amaral R 2021 *Metals*
[12] Martinitz L and Hartmann C, Year, *42nd Int. Conf. Deep Drawing Res. Group (IDDRG 2023)* (Luleå, Sweden) vol 1284 (IOP Publishing) p 012052
[13] Finamor F, Wolff M, and Lage V, Year, *40th Int. Conf. Deep Drawing Res. Group (IDDRG 2021)* (Stuttgart, Germany) vol 1157 (IOP Publishing) p 012080
[14] Niemietz P, Unterberg M, Trauth D, and Bergs T, Year, *40th Int. Conf. Deep Drawing Res. Group (IDDRG 2021)* (Stuttgart, Germany) vol 1157 (IOP Publishing) p 012082
[15] Niemietz P, Kornely M J, Trauth D, and Bergs T 2022 *J. Intell. Manuf.* **33** 2143
[16] Bank D, Koenigstein N, and Giryes R, *Autoencoders* (Machine Learning for Data Science Handbook). Springer, 2023, pp. 353.
[17] Hassani A, Medvedev A, Delir Haghighi P, Ling S, Zaslavsky A, and Jayaraman P 2019 *Sensors* **19** 1478

IOP Conf. Series: Materials Science and Engineering **1307** (2024) 012036 doi:10.1088/1757-899X/1307/1/012036

Inline Monitoring of the Geometry of Free-form Bent Components

Lorenzo Scandola, Daniel Maier, Johannes Meier and Wolfram Volk
Chair of Metal Forming and Casting, TUM School of Engineering and Design, Technical University of Munich, Walther-Meissner-Str. 4, 85748 Garching, Germany
E-mail: lorenzo.scandola@tum.de

Abstract. The free-form bending process shows a high potential for the manufacturing of bent structural components. Especially in the automotive sector, the process can be employed for the production of arbitrary geometries evolving in the 3D-space, reducing tooling costs and production times. In addition, it is characterised by an outstanding flexibility, and allows the manufacturing of variants of a single component just with the adaption of the tool kinematics. Nevertheless, the reproducibility of the process is a challenging task, and a process monitoring approach is required. In this contribution, an inline-measurement strategy for evaluating the resulting geometry of free-form bent parts is developed. First, a review of the actual measurement system is given. Successively, the inline measurement of the bending radius using a laser displacement sensor, as well as of the bending angle by means of an IR-camera are presented. The obtained inline signals are then processed to retrieve the geometry of the obtained part, and the strategy is validated by comparison with the offline measurement results. The developed inline-system represents the first step towards the development and the monitoring of the free-form bending process.

1. Introduction

Free-form bending can represent a key technology in the automotive industry. The increasing demand for electric vehicles requires a rethinking of the design of the structural frame, which benefits in a significant way from a flexible manufacturing process. Due to its kinematics-based working principle [1] the process can produce variable bending radii and angles on different bending planes without the need of changing the tool, offering clear advantages over traditional bending processes such as rotary-draw bending [2] or three-roll-push bending. In addition, also spline geometries evolving freely in the 3D-space can be achieved. Nevertheless, the increase in flexibility is linked to a more complex design of the kinematics of the tool, and results in a lengthy trial-and-error compensation procedure [3]. To obtain a part within the desired tolerance range, a component is bent using calibrated machine instructions. The part is then measured with a laser scanner, and the deviations with the target are calculated through a surface comparison. At the moment, the compensation procedure takes place entirely offline. This represents the bottleneck in the design of free-form bending, and result in an open-loop process. In order to further develop free-form bending for series production and standard component application, an inline measurement strategy is required. In this paper, the inline measurement of the bending radius and of the bending angle for in-plane 2D bending are addressed. After a review of the state of the art and the description of the actual offline measurement system, the developed strategy is described. First, the use of a laser displacement sensor to measure the bending radius is proposed. In addition, the acquisition of the bending angle through an IR-camera is introduced. The two inline measurements are then validated through the comparison with the offline method. This shows the potential of the inline monitoring and control for the enhancement of the quality of free-form bent parts.

2. State of the art

The application of sensors and measurement strategies in metal forming processes can enhance the quality of the produced parts, in a significant way, as stated by Allwood et al. in [4]. In the framework of bending processes, a variety of measurement techniques has been proposed. Allwood et al. [4] discuss digital cameras, laser sensors and light section sensors as optical measurement systems for the bending angle for the case of V-bending. Simonetto [5] investigated the springback in rotary draw bending and proposed an inline inertial measuring unit system (IMU) enabling a closed-loop control for the process. In another contribution the author proves the strategy also for the three-roll-push-bending [6]. Ha addresses in [7] the same task, by embedding a laser sensor at the free end of the bent product. By comparing the travel of the laser beam on a board from the before and after springback, the springback angle can be retrieved. Regarding three-roll-push-bending Chatti [8] introduced a concept based on three laser triangulation sensor mounted on a movable frame. The use of CCD cameras to assess the forming angle after springback for the case of progressive die bending is investigated by Löbbe [9]. This strategy also allows to measure the geometry after warm forming. Regarding the acquisition of the bending radius, Pan [10] developed a device featuring three inclinometers to measure the curvature ovaliation in cyclic bending. The acquisition and evaluation of a complete curvature profile is shown by Groth [11]. By including regularisation techniques the curvature of the whole bending line can be retrieved [12]. Nevertheless, no inline measurement strategy has been up to now applied to free form bending. The complexity of the process and its flexibility make the development of a robust and universal inline measurement system for the acquisition of the bending radius and angle in real time difficult. In the following, the radius measuring concept based on a laser displacement sensor and the angle acquisition using an IR-camera are described.

3. Offline measurement strategy

In order to acquire the geometry of bent components after the bending operation has been performed, the part is fully digitised. This is achieved with the hand-held Zeiss T-Scan laser scanner, available at the department. The system allows for a measuring volume of 10 mm^3 and guarantees an accuracy of 0.033 mm+0.033 mm/m [13]. The measurement results in an discrete mesh which can be used for further evaluations. The determination of the bent radius is done by analysing the curvature of the obtained bending line. The measured `*.stl` mesh is examined with the freeware software GOM Inspect. A discrete number of sections are applied to mesh and the points belonging to the bending line are identified as centres of the fitting circles. These are interpolated by a spline curve and the corresponding curvature is extracted. Considering a single bending in the free-form bending process, two main process stages can be identified. The transient part of the process occurs during the time needed for the bending head to reach the target position. Here the curvature of the part increases continuously. After that, the bending head maintains the fixed position, and the process undergoes stationary conditions, showing a constant value in the curvature. The obtained bending radius is determined as the reciprocal of the mean curvature in the stationary part of the signal, as shown in Figure 1 (left). This method is taken as a convention, although it is not always trivial to distinguish the transient part from the stationary part of the geometry. On the other hand, the bending angle is determined by performing a simple cylinder fit on the linear segment of the beginning of the tube and on its end, and calculating the resulting angle between the two cylinders, as shown in Figure 1 (right).

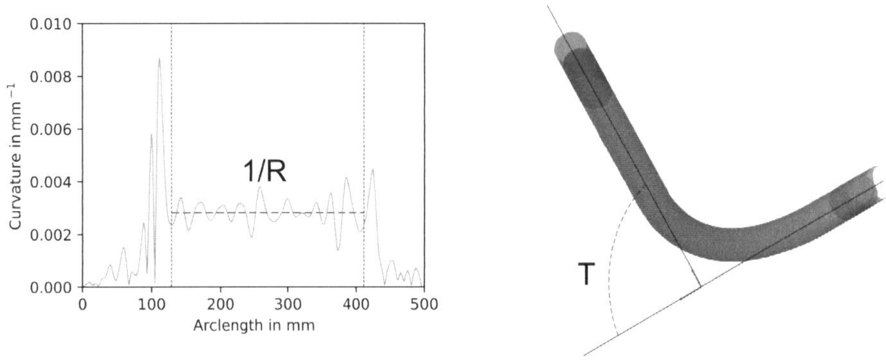

Figure 1. Offline evaluation of the bending radius (left) and bending angle (right)

4. Inline measurement strategy

The developed inline measuring strategy is realised by coupling a laser sensor for the determination of the radius and an infrared (IR) camera for the acquisition of the angle, which are described in the following.

4.1. Acquisition of bending radius signal

As described above, the offline determination of the bending radius is based on the averaging of the curvature signal of the obtained bending line. Due to the complexity of the extraction of the bending line and the calculation of the curvature, which correlates with the second derivative of the curve's points, a direct inline determination of the radius is extremely challenging. For this reason, a phenomenological approach is chosen. A laser triangulation displacement sensor optoNCDT 1420 by Micro-Epsilon [14] ensuring an accuracy of 2 μm at 2 kHz is positioned above the bending head right at the outlet of the tool. The sensor is kept in this fixed position and records the distance between itself and the tube while the bending occurs. Since the obtained signal is relative, the exact position of the sensor is not relevant, provided that it remains constant for all the conducted experiments. The only important information about the positioning of the sensor is the offset from the bending head δ and the centering of the measuring point η, which

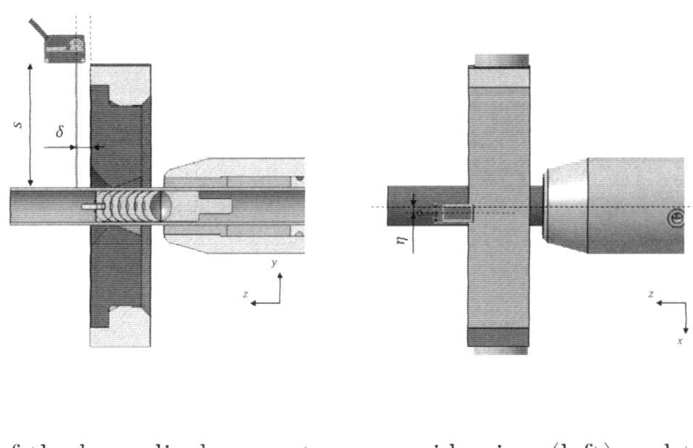

Figure 2. Setup of the laser displacement sensor, side-view (left) and top-view (right)

are kept smaller than respectively 5 mm and 1 mm. The recorded signal s is characteristic for a single bending and can be correlated to the achieved radius R:

$$R(t) \propto s(t) \tag{1}$$

More detail about the correlation will be presented in Section 6, after the analysis of the experimental results.

4.2. Camera tracking of bending angle

For the determination of the bending angle, the IR-camera Polaris Vicra by NDI [15] is used. The system enables the tracking of tools of a defined geometry, which consists of a distribution of markers in the 3D-space. The markers are passive spheres featuring a retro-reflective coating that reflects the IR light back to the source. The system can span a measuring volume of $491 \times 392 \times 887$ mm^3 and has an accuracy of 0.25 mm. In order to track the bending angle, the tool featuring four markers in Figure 3 is 3D-printed and positioned on the tube. The angle is retrieved by considering the spatial evolution in time of the vector linking the first and second point as follows:

$$T(t) = arccos\left(\frac{\boldsymbol{AB}(0) \cdot \boldsymbol{AB}(t)}{||\boldsymbol{AB}(0)|| \cdot ||\boldsymbol{AB}(t)||}\right) \tag{2}$$

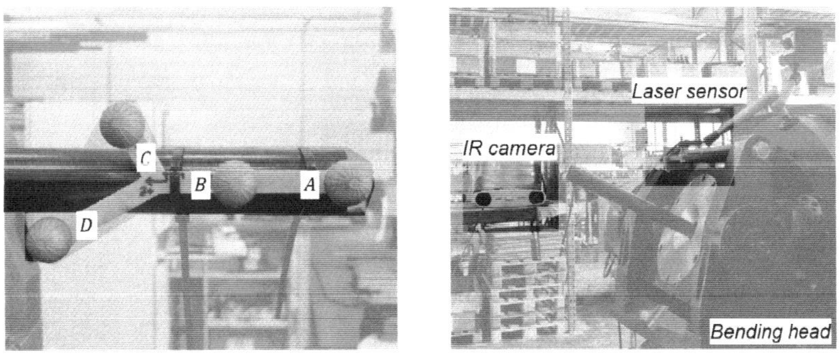

Figure 3. Tracking tool with four markers (left) and complete inline measurement setup, with highlight on laser sensor and IR-camera (right)

The complete measurement setup including the IR-camera and the laser displacement sensor is shown in Figure 3.

5. Experimental tests and reproducibility of the measurements

The performance of the inline measuring system are evaluated by conducting the experimental plan in Table 1. In order to maintain the analysis as general as possible, a severe, moderate and mild bending radius are investigated, each featuring respectively a small, medium and high bending angle. The dispersion of the results is evaluated by performing every test three times. Additionally, three other bending radii for every range are bent for a single angle configuration and performing one repetition, with the aim of enlarging the investigated radius span.

5.1. Sensitivity of the acquired inline radius signal

The reproducibility of the laser sensor signal acquisition is shown in Figure 4 for respectively the severe, mild and moderate radius range. For every radius range and target angle it can be

Table 1. Experimental plan covering wide span of bending radii and angles

	Severe				Moderate				Mild			
R_{tgt} in mm	100	115	130	145	160	175	190	200	250	300	400	500
T_{tgt} in deg	15, 30, 50	39.8	35.2	31.6	28.6	26.2	24.1	15, 30, 50	36.7	30.6	15, 30, 50	18.3

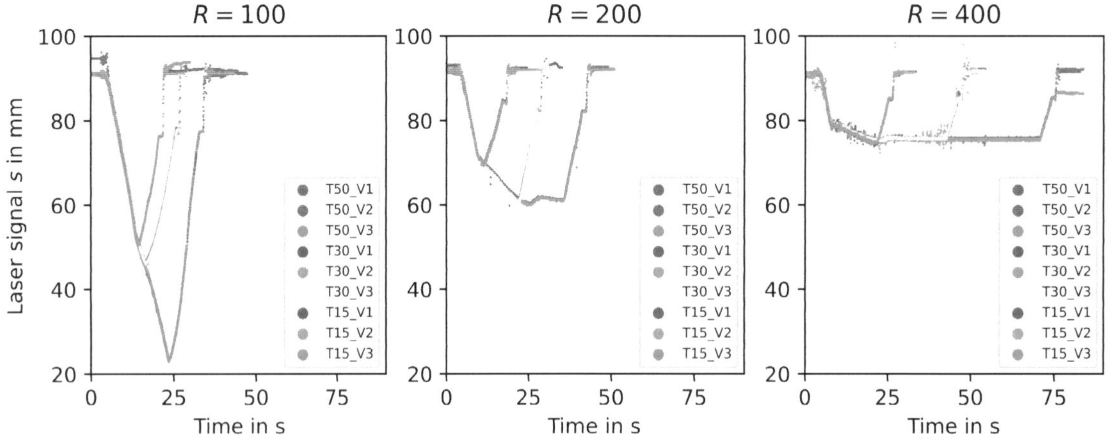

Figure 4. Acquired distance signals over 3 repetitions for small, 15 deg, medium, 30 deg, and big, 50 deg, bending angle in the case of a severe (left), moderate (centre) and mild (right) radius.

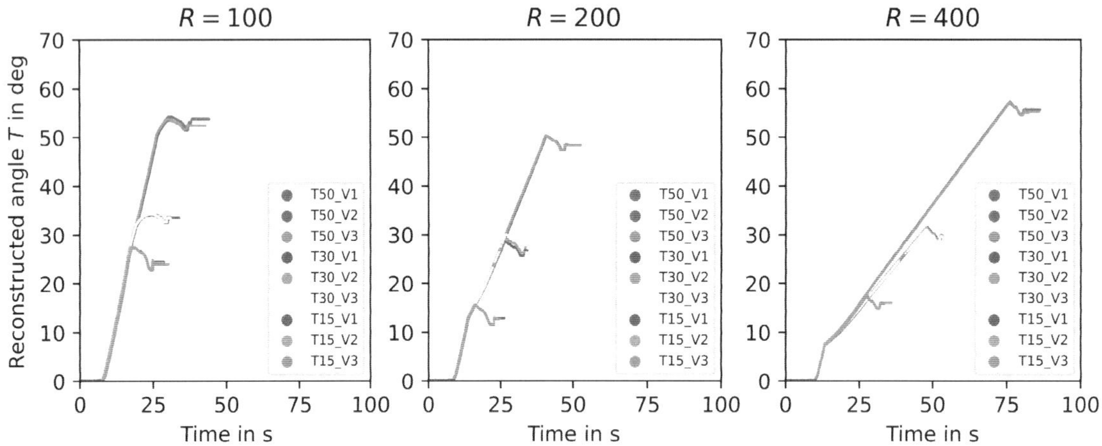

Figure 5. Reconstructed angles over 3 repetitions for small, 15 deg, medium, 30 deg, and big, 50 deg, bending angle in the case of a severe (left), moderate (centre) and mild (right) radius.

appreciated that the recorded signals for the three repetitions are indistinguishable one from each another. For this reason, the additional tests for enlarging the radius span can be conducted just once. This demonstrates the robustness of the laser displacement sensor setup and proves its suitability for the indirect assessment of the bent radius.

5.2. Sensitivity of the reconstructed bending angle

The reconstructed bending angle from the camera signals are shown in Figure 5. Also for this measurements it can be concluded that a high repeatability is achieved with the implemented setup. This is particularly evident considering the signal up to the maximum achieved bending angle, which remains very similar for the three iterations. Especially for the case of severe bending with a bending angle of 50 deg, the final bending angle differs of around 2 deg between the three iterations. This fluctuations is nevertheless due to the process and not to the measurement system, since the same results is achieved by employing the offline measurement technique.

6. Discussion and interpretation of the results

In this section the results of the inline acquired signals and their application to the description of the arising bending radius and angle are interpreted and commented.

6.1. Interpretation and exploitation of the laser distance signal

The signals acquired for all of the bending radii configurations show a recurring pattern. It can be observed, that the signal starts with a almost linear trend with constant steepness Figure 6 (left). This is a characteristic of the process and is only due to the relative velocity of the bending head and not to the target radius. After that, the signals remains linear but a reduction in the steepness is observed. The bigger the radius and hence the lower the bending head displacement, the lower the steepness. After this, a minimum is achieved, followed by a constant value. It is important to notice, that this stationary condition is only achieved if a sufficient bending angle, namely fed length, is provided. For all the severe and some moderate radii, namely from 100 mm to 190 mm, the bent angle is not enough to induce the stationary bending condition. This suggests, that a *minimum bending angle* can be defined, below which the target radius can never be achieved. In the presented investigations, this is the case for radii bigger than 200 mm, for which the minimum angle can be extrapolated from the laser signals, Figure 6 (left), identifying the angles from which the stationary regime is stable. The results are summarised and correlated to the target and obtained bending radii in Table 2. At this point, the assumed correlation between the bending radius and the laser signal of Equation (1) can be modeled and described by a quadratic law, of the form:

$$R(t) = A \cdot s(t)^2 + B \cdot s(t) + C \tag{3}$$

The matching between the correlation (red) and the experimental data (dashed, black) can be seen in Figure 6 (right).

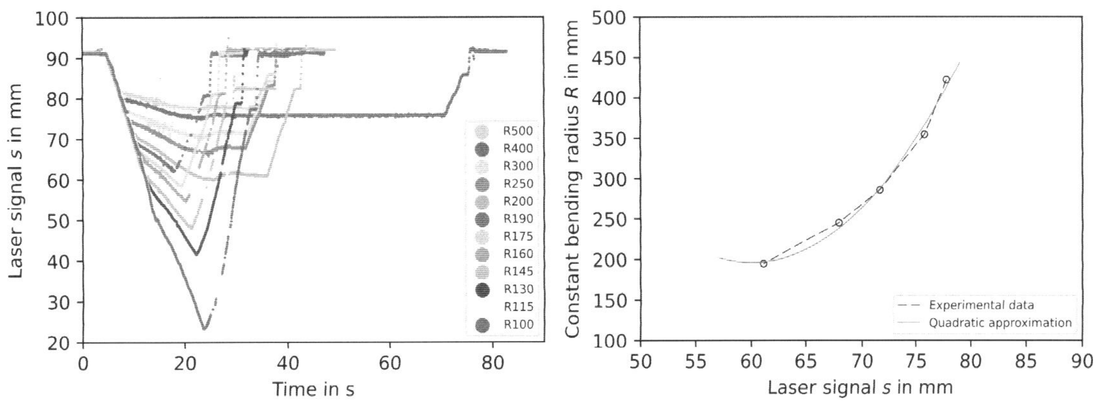

Figure 6. Comparison between the offline and inline measured bending angle considering three repetitions, for severe, moderate and mild bending radii over small, medium and big target angles.

Table 2. Correlation of the laser signal with the target and obtain radius, and identification of the minimum bending angle

R_{tgt} in mm	500	400	300	250	200
R_{obt} in mm	422.0	354.5	286.0	245.1	194.5
s_{tgt} in mm	77.8	75.8	71.6	67.9	61.1
T_{min} in mm	78	76	72	68	61

6.2. Performance of the inline bending angle measurement

Referring to the inline acquisition of the bending angle, the results show a very good agreement with the offline measurement, Figure 7. With a maximum standard deviation of 0.56 deg and a maximum angle difference between the inline and offline measurement of 0.47 deg, it can be concluded that the proposed strategy is suitable for the inline monitoring of the bending angle in free-form bending.

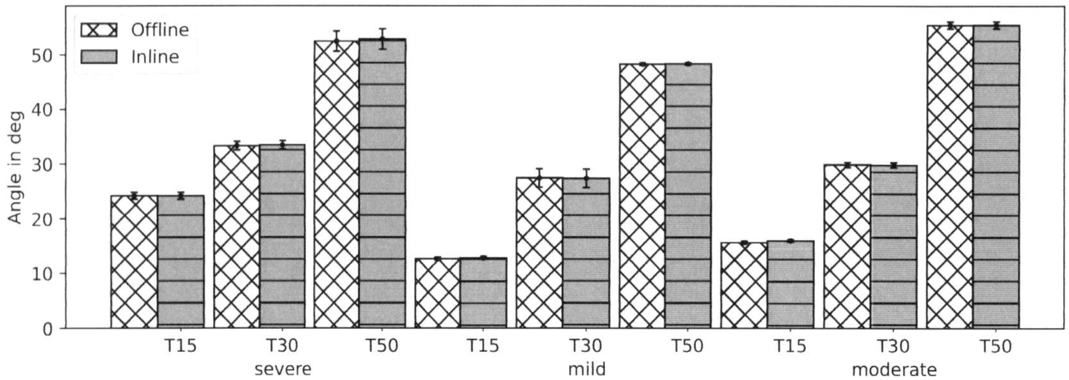

Figure 7. Comparison between the offline and inline measured bending angle considering three repetitions, for severe, moderate and mild bending radii over small, medium and big target angles.

7. Conclusions and outlook

The presented work showed the development and successful application of an inline measurement system for free-form bending. In order to monitor the bending radius, an indirect measurement based on a fixed laser displacement sensor has been employed. The measurements shown a very high reproducibility and allowed to distinguish different phases during the the radius evolution. In this way, a minimum bending angle can be identified, which is required to achieve a full development of the target radius. The proposed system for the inline measurement of the bending angle is based on the tracking of discrete points by employing an IR-camera. From the vector generated by the displacement of two collinear points the bending angle can be reconstructed and observed in time. Also in this case the measurements show a very high reproducibility and are in perfect agreement with the offline results, showing a maximum discrepancy of 0.47 deg. The presented strategy confirms that an inline monitoring of the free-form bending process is possible and can be based on indirect laser displacement measurement as well as camera tracking devices. The obtained results constitute the background for the future development of a intelligent free-form bending machine, which can be monitored in real time and eventually be set up as a closed-loop process.

References

[1] Makoto Murata. "Tube Bending of New Generation by MOS Bending Machine". In: *1st International Tube and Profile Bending Conference*. Ed. by TU Dortmund. Institut für Umformtechnik und Leichtbau, 2011, pp. 9–18.

[2] Bernd Engel, Christoph Gerlach, and Stefan Cordes. "Biegemomentabschätzung des Dornbiegeverfahrens". PhD thesis. Siegen: Universität Siegen, 2008. URL: www.UTFscience.de.

[3] Peter Gantner. "The Characterisation of the Free-Bending Technique". Thesis (Ph.D.) Glasgow Caledonian University, 2008.

[4] J. M. Allwood et al. "Closed-loop control of product properties in metal forming". In: *CIRP Annals* 65.2 (2016), pp. 573–596. ISSN: 00078506. DOI: 10.1016/j.cirp.2016.06.002.

[5] Enrico Simonetto, Andrea Ghiotti, and Stefania Bruschi. "In-process measurement of springback in tube rotary draw bending". In: *The International Journal of Advanced Manufacturing Technology* 112.9-10 (2021), pp. 2485–2496. ISSN: 0268-3768. DOI: 10.1007/s00170-020-06453-w.

[6] Enrico Simonetto, Andrea Ghiotti, and Stefania Bruschi. "Dynamic detection of tubes wrinkling in three roll push bending". In: *Procedia Engineering* 207 (2017), pp. 2316–2321. DOI: 10.1016/j.proeng.2017.10.1001.

[7] Taekwang Ha et al. "In-line Springback Measurement for Tube Bending Using a Laser System". In: *Procedia Manufacturing* 47 (2020), pp. 766–773. ISSN: 23519789. DOI: 10.1016/j.promfg.2020.04.233.

[8] S. Chatti, U. Dirksen, and M. Kleiner. "Optimization of the Design and Manufacturing Process of Bent Profiles". In: *Journal of the Mechanical Behavior of Materials* 15.6 (2004), pp. 437–444. ISSN: 0334-8938. DOI: 10.1515/JMBM.2004.15.6.437.

[9] Christian Löbbe et al. "Closed loop springback control in progressive die bending by induction heating". In: *International Journal of Precision Engineering and Manufacturing* 16.12 (2015), pp. 2441–2449. ISSN: 2234-7593. DOI: 10.1007/s12541-015-0314-8.

[10] W. F. Pan, T. R. Wang, and C. M. Hsu. "A Curvature-ovalization measurement apparatus for circular tubes under cyclic bending". In: *Experimental Mechanics* 38.2 (1998), pp. 99–102. ISSN: 0014-4851. DOI: 10.1007/BF02321651.

[11] Sebastian Groth, Bernd Engel, and Peter Frohn. "Approach to a manufacture-oriented modeling of bent tubes depending on the curvature distribution during three-roll-push-bending". In: *Proceedings of the 21st International ESAFORM Conference on material forming*. AIP Conference Proceedings. Author(s), 2018, p. 110006. DOI: 10.1063/1.5034963.

[12] Sebastian Groth, Bernd Engel, and Kay Langhammer. "Algorithm for the quantitative description of freeform bend tubes produced by the three-roll-push-bending process". In: *Production Engineering* 12.3-4 (2018), pp. 517–524. ISSN: 0944-6524. DOI: 10.1007/s11740-018-0795-2.

[13] GOM GmbH. *ZEISS T-SCAN*. Ed. by GOM GmbH. 2020. URL: https://www.handsonmetrology.com/de/products/t-scan-10-20.

[14] Micro-Epsilon. *optoNCDT 1420 data-sheet*. Ed. by Micro-Epsilon. 2023. URL: https://www.micro-epsilon.com/displacement-position-sensors/laser-sensor/optoNCDT_1420_basic/.

[15] NDI. *Polaris Vicra data-sheet*. Ed. by NDI. 2023. URL: https://www.ndigital.com/optical-measurement-technology/polaris-vicra/.

IOP Conf. Series: Materials Science and Engineering **1307** (2024) 012037 doi:10.1088/1757-899X/1307/1/012037

Creating a Virtual Shadow of the Manufacturing of Automotive Components

A Barlo[1*], O Aeddula[1], A R Chezan[2], J Pilthammar[1,3], M Sigvant[1,4]

[1]Dept. of Mechanical Engineering, Blekinge Institute of Technology, Karlskrona, Sweden
[2]TATA Steel R&D, P.O. Box 10.000, 1970 CA IJmuiden, The Netherlands
[3]Volvo Cars Dept. 81013 Manufacturing Technology Development, Olofström, Sweden
[4]AutoForm Engineering Sweden AB, Olofström, Sweden

E-mail: Alexander.Barlo@bth.se

Abstract. Within the automotive industry, there is an increasing demand for a paradigm shift in terms of which materials are used for the manufacturing of the automotive body. Global climate goals are forcing a rapid adaption of new, advanced, sustainable material grades such as the fossil free steels and materials containing higher scrap content. With the introduction of these new and untested materials, methods for accounting for variation in material properties are needed directly in the press lines.

The following study will focus on creating an initial virtual shadow of the manufacturing of a Volvo XC90 inner door panel through the application of Artificial Neural Networks (ANN). The virtual shadow differs from the concept of the digital twin by only being a virtual representation of the production line, with training data generated exclusively by numerical simulations, and having no automated communication with the physical press line control system. The virtual shadow can be used as an assistance to the press line operators to see how different press line settings and material parameter variations will impact the quality of the stamped component. The study aims to validate the virtual shadow through accurate predictions of the material draw-in measured in the physical press line.

1. Introduction

Within the automotive industry, there is an increasing demand for a paradigm shift in terms of which materials are used for the manufacturing of an automotive body. Global climate challenges are forcing a rapid adaption of new, advanced, sustainable material grades such as the fossil free steels and materials containing higher scrap content, where especially the latter is suspected to increase the in-coil variation of material parameters [1]. The in-coil variation of material parameters of existing materials already forces manufacturing engineers to perform conservative evaluations of Finite Element simulations, where the Forming Limit Curve (FLC) is often lowered by a 10% offset or more to be on the safe side [2]. With this reduction in FLC, the material is underutilized and an optimal process design cannot be achieved. Early attempts to combat this challenge was made by Sigvant & Carleer [3] and Sigvant et al. [4] running large stochastic simulations with focus on variations of material and process parameters to replicate issues that presented themselves in production of automotive body components, but did not show up in the Finite Element simulations used for the process design. These studies added another dimension to the discussion of scatter - that of process parameters such as friction and clamping force. The big issue that presents itself with running large stochastic simulation models is that it is currently only manufacturing CAE engineers that create and use the data, but the data is not available on the production floor for the operators to use directly where it is most needed. Therefore, in order to account for in-coil material parameter variation, and drifts in the process window, this study aims to create an early virtual shadow of a sheet metal forming process using an Artificial Neural Network (ANN), that can be used by the operators on the production floor.

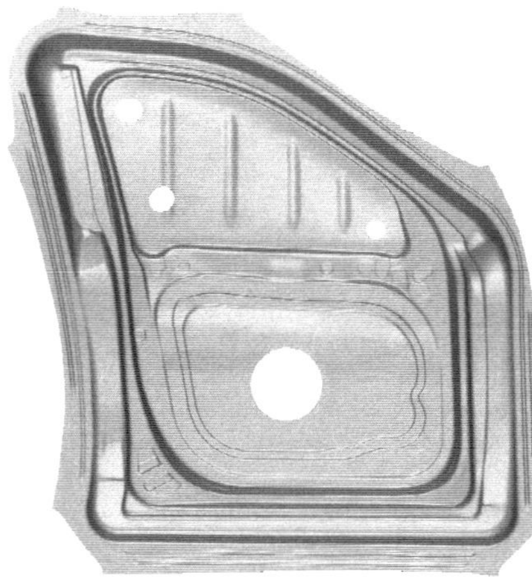

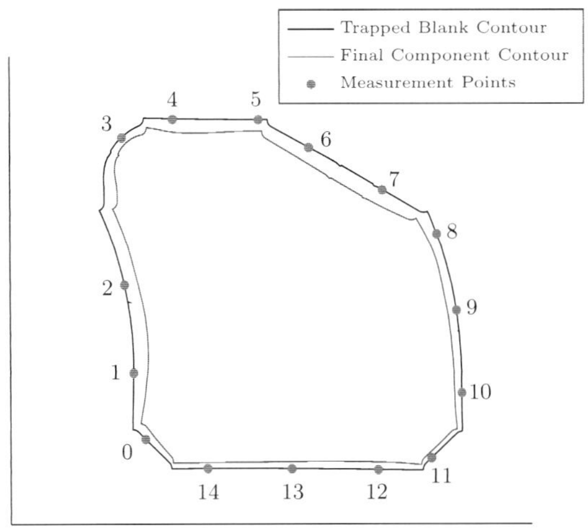

Figure 1. Volvo XC90 Front Door after first forming operation (left), and illustration of trapped blank and final component contours displayed with the 15 measurement points used in the study (right).

The study presented in this paper will focus on the first forming operation of the Volvo XC90 front door (inner) component. The focus will be on an accurate prediction of the draw-in of the trapped blank in 15 predefined points. The geometry of the component after the first forming operation can be seen in Figure 1 (left) and the 15 predefined points can be seen on Figure 1 (right). The ANN will be trained on data obtained from Finite Element simulations and later validated through measurements from the production line.

2. Data Generation

To train the ANN, data were generated through stochastic simulations using the AutoForm sigma-module. A total of five different parameters were chosen to vary to align with the needs specified in the previous section: the initial yield strength (σ_0), the tensile strength (R_m), the average Lankford coefficient (r_m), the lubrication amount, and the cushion force. Since the material used for the component is a VDA239 CR4 mild steel, the base Finite Element model was build using the Vegter 2017 material model. The Full Vegter material model was disregarded as it does not allow for variation of the average Lankford coefficient in the sigma-module. The base model was set up to run with a stroke rate of 14.00 [1/min], which is why the strain rate sensitive hardening behaviour of the material was considered. The hardening curves and yield surface for the CR4 mild steel can be found in Figure 2.

To investigate the impact of friction, the lubrication amount on the sheet was varied, and not the friction coefficient itself. This is due to the fact that a standard Coulomb friction model is not used, but a pressure, velocity, and strain dependent friction model is applied. The behaviour of the friction is modelled in the TriboForm software, where scans of the actual sheet and tool surfaces are used as input. These surfaces are illustrated in Figure 3 (left). In the

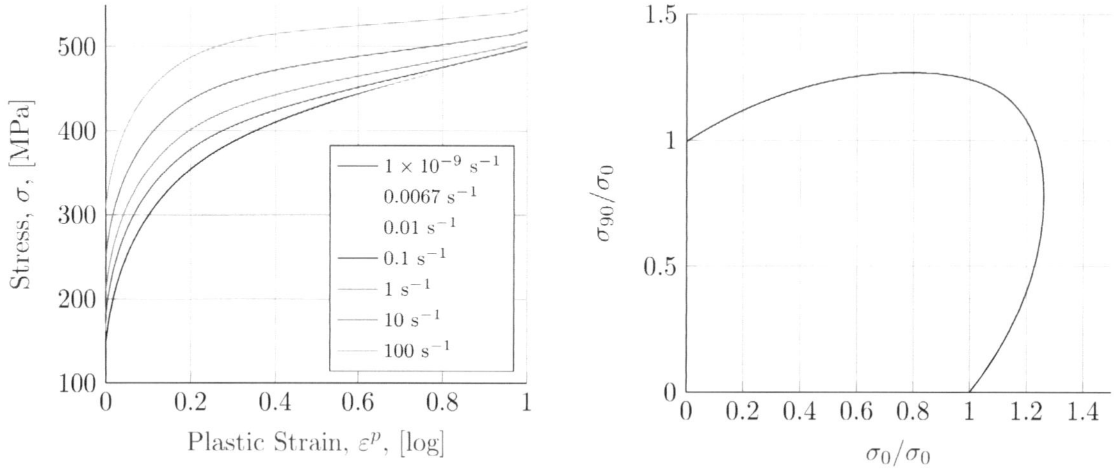

Figure 2. Strain rate sensitive hardening curves (left) and yield surface (right) material inputs of the VDA239 CR4 mild steel.

AutoForm model, the sheet and tool surfaces were modelled with a roughness of 1.30 and 1.00 [μm] respectively. Comparing the applied friction model to the friction conditions applied by Sigvant & Carleer [3] (Figure 3 (right)) it can be observed that the coefficient of friction can extend past the interval of Coulomb friction ($\mu_{c,lower}$ = 0.008 and $\mu_{c,upper}$ = 0.12) depending on the contact pressure of the tools and the sliding velocity of the blank. Adding that to the variation in lubrication amount, a higher accuracy in the modelling of the friction behaviour is expected [5].

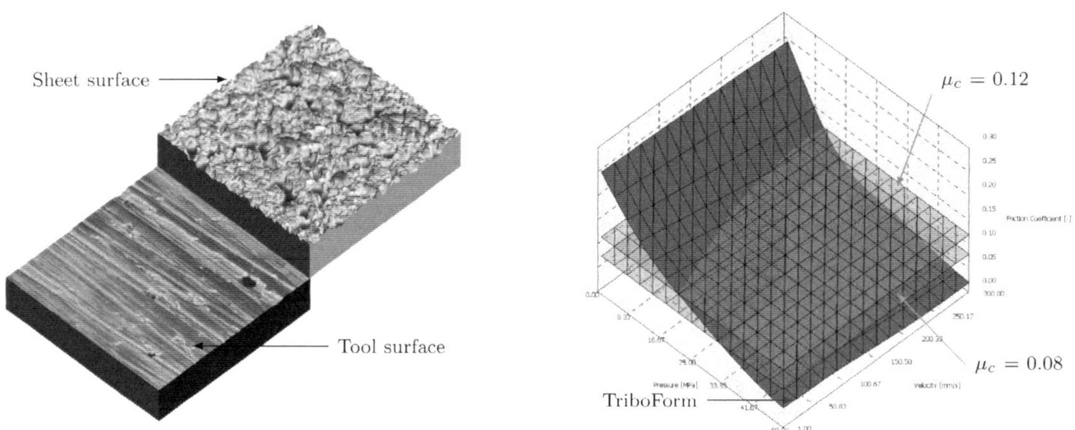

Figure 3. Tribology system used for the base model. Illustration of the sheet and tool surfaces (left), and a comparison of the TriboForm friction coefficient as a function of pressure and velocity (blue) and standard Coulomb friction (green/orange) (right).

To determine the ranges which within the material parameters (σ_0, R_m, and r_m) can vary, the VDA 239-100 standard [6] was consulted. The material used for the component being a CR4 mild steel, the acceptable ranges within the VDA standard were used as the upper and lower limits of the variation. The selected values can be found in Table 1.

For the determination of the ranges of the process parameters (lubrication amount and cushion force), a more observation-based approach was taken. For the lubrication amount, measurements of the oil film on a full coil with a strip length of approximately 1550 [m] was observed. The measurement for the upper side of the coil is presented in Figure 4, where it can be seen that the lubrication amount varies between 1.00 and 2.50 [g/m^2]. In the modelling of the base model, it was decided to use a uniformly distributed lubrication amount, why a value of 1.50 [g/m^2] was chosen. For the variation, the extreme values reaching 2.50 [g/m^2] was disregarded and the variation was chosen to be $\pm$ 0.5 [g/m^2].

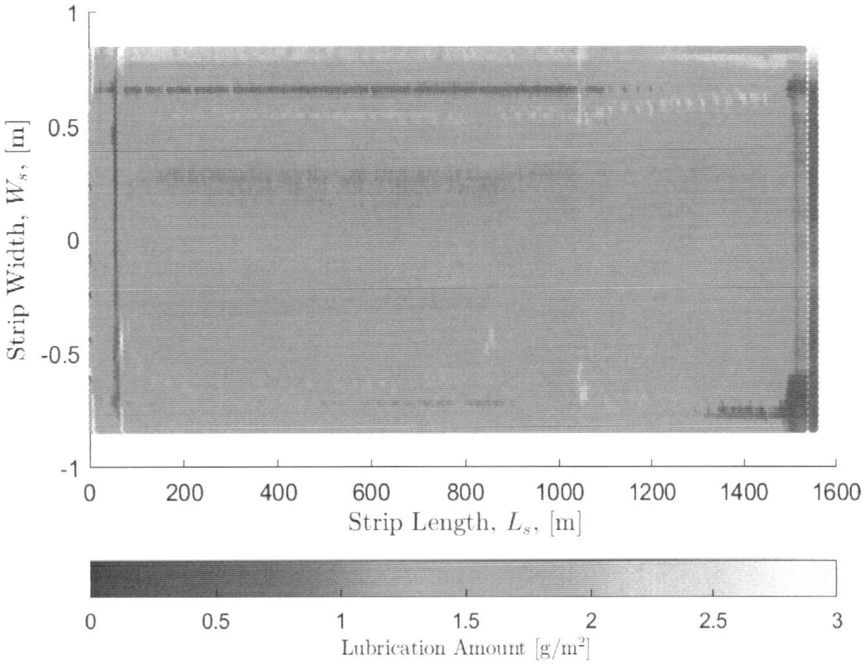

Figure 4. In-line measurement of lubrication amount for full coil of CR4 mild steel used for the XC90 Front Door (inner) component.

For the range of the cushion force, observations from running production were made. The applied cushion force for the batch made from the coil presented in Figure 3 was monitored, and a range between 1200 and 1800 [kN] was determined. Subsequent simulations were run to ensure that no failure occurred when applying the maximum and minimum cushion force values. All the ranges for the parameter variation of the stochastic simulation can be found in Table 1. The stochastic simulation was set to run 500 realizations, with the cushion force and lubrication amount being controllable variables, and the initial yield stress, tensile strength, and average Lankford coefficient being uncontrollable variables. Furthermore, a positive interaction (85%) was set between the initial yield stress and tensile strength to ensure that when initial yield stress increases, the tensile strength will not decrease.

To validate the Artificial Neural Network trained on the simulation data, measurements of three components from a production line were made. The three different components were stamped from the same coil (i.e. same material parameters and lubrication) with a variation in the cushion force used (values presented in Table 2). The measurements were taken after the first forming operation, with the press running in single-stroke mode to avoid temperature effects in the tooling. Figure 5(a) presents the complete blank outline for the three measurements, and

IOP Conf. Series: Materials Science and Engineering **1307** (2024) 012037 doi:10.1088/1757-899X/1307/1/012037

Figure 5(b) presents the draw-in values in the 15 measurement points for the three measurements.

Parameter	**Min**	**Average**	**Max**	σ
Initial yield stress (σ_0) [MPa]	140	160	180	±20
Tensile stress (R_m) [MPa]	270	300	330	±30
Avg. Lankford coefficient (r_m) [-]	1.6	2.1	2.6	±0.5
Lubrication Amount [g/m^2]	1.00	1.50	2.00	±0.5
Cushion Force (F_c) [kN]	1200	1500	1800	±300

Table 1. Parameter variation used in the stochastic simulation.

Measurement	σ_0 **[MPa]**	$\mathbf{R_m}$ **[MPa]**	$\mathbf{r_m}$ **[-]**	**Lubrication [g/m^2]**	$\mathbf{F_c}$ **[kN]**
Measurement 1	147.9	290.7	2.03	1.00	1200
Measurement 2	147.9	290.7	2.03	1.00	1500
Measurement 3	147.9	290.7	2.03	1.00	1800

Table 2. Values used in production for the three measured components.

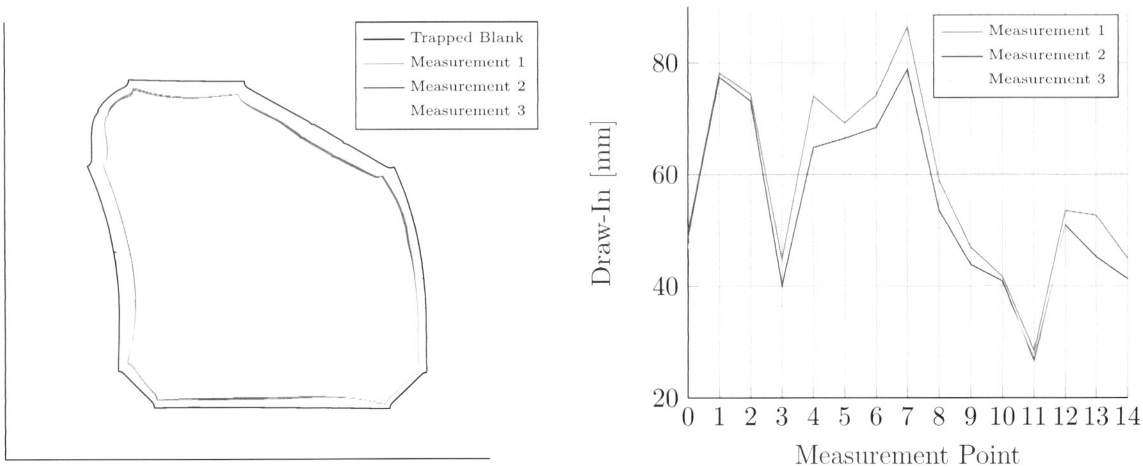

(a) Full component outline measurements from production run.

(b) Draw-in values of the 15 measurement points.

Figure 5. Results of three draw-in measurements made in production.

3. Methodology

An artificial neural network approach was employed in this research to predict draw-in values for the 15 measurement points. The ability of neural networks to learn and predict output values from the input data makes them a powerful tool to solve numerous real-world applications [7]. The neural network architecture consists of three fundamental components: an input layer, hidden layers, and an output layer, all functioning within a feedforward framework. The neural network architecture was modeled to predict 15 measurement points based on five input parameters. The primary objective was to construct a neural network model proficient in capturing the complex relationships between these input parameters and the output measurement points. The neural network architecture can be seen in Figure 6.

3.1. Data Preprocessing
Prior to model training, the data set underwent essential pre-processing steps. This includes partitioning the data set into input and output variables and standardizing the data through the normalization process. The 'min-max' scaling method was employed for normalization, leading to data values confined within the 0 and 1 range. This normalization procedure was instrumental in ensuring consistent scaling across all the parameters, thus enhancing the models robustness.

3.2. Data Splitting
The dataset was methodically divided into two distinct subsets to facilitate model development and evaluation:

- **Training Set:** This segment comprises 80% of the dataset and was exclusively used to model training. During this phase, the neural network model acquired the ability to understand and model the complex relationships existing between the input parameters and the output measurement points.
- **Testing Set:** The remaining 20% of the data served as a benchmark for evaluating the model's performance. This evaluation process follows the principle of "out-of-sample" validation, offering insights into the model's predictive accuracy when applied to previously unseen data.

3.3. Out-of-sample Evaluation
To evaluate the models performance, standard evaluation metrics were utilized in the form of the Mean Squared Error (MSE) and R-squarred (R^2) values. The MSE measures the average magnitude of prediction errors. It is a measure of the average squared difference between predicted and actual values. Lower MSE values indicate better predictive accuracy [8], and is defined as:

$$MSE = \frac{1}{n} \cdot \sum (y_{pred} - y_{true})^2 \tag{1}$$

where n is the number of samples, y_{pred} is the predicted value, and y_{true} is the actual value. The R^2 quantifies the proportion of variance in the measurement points that is explained by the model. A higher R^2 indicates a better fit to the data [9] and is defined as:

$$R^2 = 1 - \frac{\sum (y_{true} - y_{pred})^2}{\sum (y_{true} - y_{mean})^2} \tag{2}$$

where y_{pred} is the predicted value, y_{true} is the actual value, and y_{mean} is the mean value.

3.4. Validation with Experimental Data set
To further assess the model's prediction ability to new, unseen data, an independent experimental data set was introduced. This data set incorporated different input values entirely distinct from the training and testing data sets. The pre-trained neural network model was applied to predict the 15 output variables for these new input values.

3.5. Evaluation Metrics
The performance of the model was evaluated using two metrics: R-squared and MSE. These metrics were calculated for both the testing data set and the experimental data set, as depicted in Figure 7. The assessment outcomes provided insights into the model's capacity to predict the 15 output measurements accurately and its adaptability to new, previously unseen data.

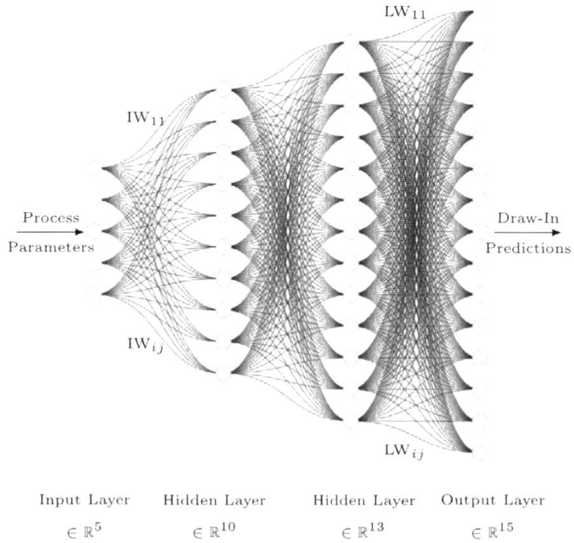

Figure 6. Artificial neural network architecture.

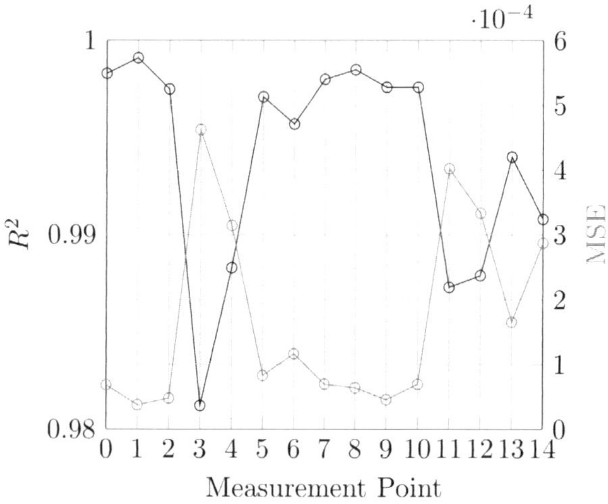

Figure 7. Artificial neural network evaluation metrics.

Figure 8 display the models predictions for measurement points for three random observations based on the testing dataset and validation dataset. These visual representations showcase the model's predictive capabilities for the testing and validation datasets. The model demonstrated favorable performance when applied to new, un-encountered data (out-of-sample accuracy). Figure 9 presented the models predictions of the experimental observations. Even though a good performance of the trained model is seen for the test and validation data, variations were observed for the predictions of the experimental observations.

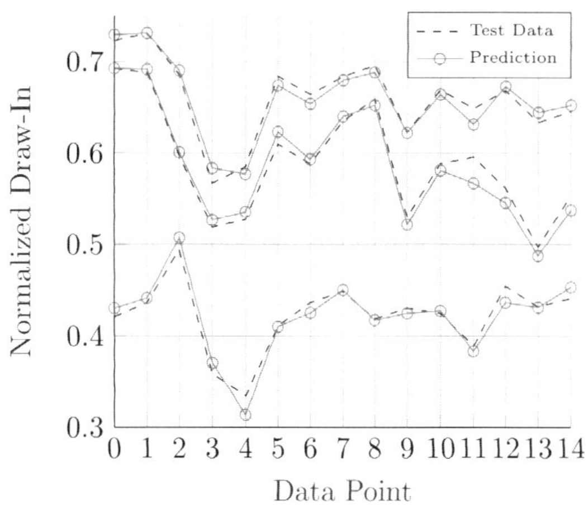

Figure 8. Testing of the trained Artificial Neural Network using the allocated training data.

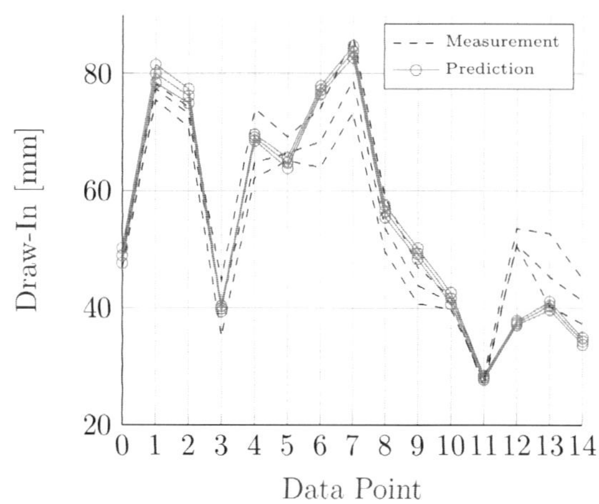

Figure 9. Prediction of draw-in for the three measurements compared to actual measurements.

4. Discussion

The results presented in Figure 9 shows a discrepancy between the predicted and measured values in draw-in. For the validation, the only parameter varied was the cushion force, indicating the model to be relatively insensitive to this parameter. Observing Figure 8 the test of the ANN showed an acceptable accuracy why the root cause for the discrepancy is most likely to be found in the generated data used for training.

An area assumed to be the cause of the discrepancy is the tribology model. A decision was made to set up the stochastic simulation with a uniform lubrication amount over the blank. The largest discrepancies are seen between points 4-7 and 12-14. These points are aligned with the top and bottom edges of the component, which could indicate a migration of lubrication towards these areas (slight indication of this is seen in Figure 4). It is assumed that by increasing the model complexity with respect to the lubrication distribution, a more accurate prediction can be made.

5. Conclusion

The presented study aimed to create a virtual shadow of the Volvo XC90 front door (inner) with focus on accurately predicting the draw-in in 15 points along the trapped blank during the first forming operation. A feedforward Artificial Neural Network (ANN) was chosen as the model base, with the training data being generated from a stochastic Finite Element simulation with 500 realizations varying three material parameters (σ_0, R_m, and r_m) and two process parameters (lubrication amount and cushion Force). The ANN was successfully trained using the simulated data and presented a good accuracy during testing. Three physical measurements with variations in the cushion force only were used to validate the ANN. The ANN was not able to accurately predict the draw-in in all 15 points for the three measurements, and showed particular bad performance around the points located along the top and bottom of the component. This discrepancy between model and measurement is expected to be caused by the assumption of uniform lubrication distribution across the blank used for the stochastic simulation, thereby disregarding the oil migration occurring in reality.

Acknowledgments

Open Access funding was provided by Blekinge Institute of Technology. This study was also funded by VINNOVA in the EUREKA Smart program (grant number 2021-03144).

References

[1] Chezan, A R *et al.* 2023 Material variability effects on automotive part production process *IOP Conf. Ser.: Mater. Sci. Eng.* **1284**, pp. 012037.
[2] Wiebenga, J H *et al.* 2014 Effect of material scatter on the plastic behaviour and stretchability in sheet metal forming *J. Mater. Process. Technol.* **214**(2), pp. 238-252.
[3] Sigvant, M & Carleer, B 2006 Influence on Simulation Results from Material and Process Scatter *In: Proceedings of the IDDRG 2006 Conference, Porto, Portugal.*
[4] Sigvant, M *et al.* 2007 Industrial Experiences of Stochastic Simulations of Sheet Metal Forming *In: Proceedings of the FTF 2007 Conference, Zurich, Switzerland.*
[5] Sigvant, M *et al.* 2016 Friction and lubrication modeling in sheet metal forming simulations of a Volvo XC90 inner door *IOP Conf. Ser.: Mater. Sci. Eng.* **159**, pp. 012021.
[6] Verband der Automobilindustrie (2021) VDA 239-100 - Sheet Steel for Cold Forming.
[7] da Silva, I N *et al.* 2017 Artificial Neural Networks - A Practical Course *Springer International Publishing.*
[8] Witten, I H *et al.* 2011 Data mining: practical machine learning tools and techniques *In: Morgan Kaufmann series in data management systems* 3rd edn.
[9] Kurz-Kim, J -R & Loretan, M 2007 A Note on the Coefficient of Determination in Models with Infinite Variance Variables *FRB International Finance Discussion Paper No. 895.*

IOP Conf. Series: Materials Science and Engineering **1307** (2024) 012038 doi:10.1088/1757-899X/1307/1/012038

Development of a digital material shadow for the press hardening route of medium manganese steel

Karl J. Tilly[1]*, Christian Idzik[1] and David Bailly[1]

[1] *Institute of Metal Forming, RWTH Aachen University, Aachen, Germany*

**karl.tilly@ibf.rwth-aachen.de*

Abstract. Press-hardened ultra-high strength steel parts are widely used in the automotive sector for their lightweight and safety advantages. Medium-Manganese Steels (MMnS) are being explored as an alternative to boron-manganese steels due to their high strength and ductility after quenching, achieved at lower annealing temperatures thus reducing energy usage and carbon emissions. However, industrial adoption of MMnS is hindered by challenging processing requirements, e.g. in cold-rolling and press hardening.
To expedite and improve the process development, data-driven decisions based on process parameters hold promise. Establishing a link between process data and the final produced part necessitates the development of a framework for a Digital Material Shadow (DMS).
This paper investigates the development of a DMS framework for the cold rolling and press hardening process chain. In conjunction with conventional data acquisition methods employed for cold rolling, novel data acquisition techniques are introduced specifically tailored for press hardening, ensuring the comprehensive availability of relevant data. Moreover, a data pipeline is implemented to enable automatic processing, visualization, and analysis of process data. To facilitate seamless data linkage across processes in the DMS, an ID-system is introduced. Finally, the developed framework's validity is demonstrated by creating a DMS for press-hardened MMnS parts, showcasing its potential for practical applications.

1. Introduction

The increasing use of advanced high strength steel parts in the automotive industry is driven by the objective of vehicle weight reduction in order to reduce energy consumption and further improve passive passenger safety. Currently, Medium-Manganese Steels (MMnS) are being examined as a potential alternative to boron manganese steels, such as Mn22B5, for body-in-white applications. MMnS are an appealing option as they offer a combination of high strength and ductility. Remarkably, these properties are achieved at lower annealing temperatures compared to conventional boron manganese steels, reducing energy consumption and CO2 emissions. However, due to the high cost of processing, MMnS faces barriers to widespread industrial adoption. These challenges, particularly in processes such as cold rolling and press hardening, have impeded the smooth integration of MMnS into the manufacturing sector. A data-driven approach, based on a thorough understanding of how process parameters affect material and final product properties, appears to be a promising solution to accelerate the development of MMnS processing by press hardening. Central to this approach is the need to establish a strong link between process data and the final properties of the produced parts. This requires the development of a framework for a Digital Material Shadow (DMS). DMS contain reduced, aggregated data to address specific material issues. The initial step is to implement a tracking system to match process data along the process chain with the resulting parts. Next, a data pipeline that collects, processes and aggregates all data from all processes across the process chain, allowing for real-time decision support.

This paper aims to establish these two steps towards a framework for DMS for the press hardening process chain. For this, an ID system is introduced which allows material tracking along the process chain, so that the entire production history of an individual press-hardened component is available. The process data is cleaned, processed and aggregated using an implemented data pipeline which allows detailed analyses and predictions in real time. This is illustrated in this paper by the prediction of the final hardness of the press-hardened part and the occurrence of cracks with the help of process data.

2. State of the Art

2.1. Processing of Medium Manganese Steels

The utilization of Medium-Manganese Steels (MMnS) in body-in-white applications represents a promising avenue, particularly in the context of crash-worthiness. This is attributable to their unique combination of high tensile strength and exceptional ductility, which underlines their potential to enhance automotive safety [1]. Additionally, MMnS holds the prospect of reducing energy consumption during production, primarily due to their lower austenization temperature in comparison to conventional boron-manganese steels [2]. These advantageous mechanical properties are rooted in the intricate fine-grained multi-phase microstructure inherent to MMnS. However, achieving this microstructure often necessitates the implementation of complex multi-step heat treatment procedures, such as Quench and Partitioning (Q&P) [3]. Consequently, this adds a layer of complexity and required precision to the processing of MMnS, especially concerning aspects like soaking times, tool temperature and cooling rates during the quenching process [4].

Mishra et al. [5] looked into the effect of hot- and cold rolling with subsequent annealing on the austenite morphology and mechanical properties of the material. The cold rolling route showed a recrystallized microstructure which leads to yield point elongation, suggesting that hot – and cold rolling parameters have a significant influence on the final mechanical properties and microstructure.

2.2. Digital Material Twin and Shadow

For several years, digitalization concepts and data-driven methods have been extensively utilized to enhance production processes. Two distinct concepts have recently surfaced in this regard: the digital twin and the digital shadow. These topics have been the subject of numerous academic publications. Nevertheless, digital twin and shadow are often used interchangeably [6]. Bergs et al. [7] present detailed definitions of both concepts in the context of manufacturing technology. The authors state that digital twins and digital shadows exist for physical assets. However, a digital shadow reflects only one specific aspect as it is an assignment of reduced and aggregated data, whereas a digital twin is a virtual representation of a physical asset that is consistently updated [8]. Digital shadows are customised to support real-time decision-making in specific domains. Unlike digital twins, there may be several digital shadows for a single physical asset [9].

This paper focuses on the experimental processing of newly developed materials throughout metal forming processes. Specific concepts for digital twins and shadows for materials science have also been introduced in the literature. For example, Kalidindi et al. [10] proposed a multi-scale concept for a digital twin for materials that aims to accelerate the innovation in materials science. Wang et al. [11] have defined a digital material shadow, which can capture and display the pivotal characteristics of the material within a production chain. Based on that, Wesselmecking et al. [12] further developed the concept and implemented it in the press hardening of medium manganese steel. However, their emphasis lies primarily on the digitalization of press hardening tools. Additionally, their analysis is focused to the press hardening alone, without considering the preceding rolling processes that influence the material.

2.3. *Data Processing*

As previously stated, the focus of this paper is on the collection and analysis of data in the area of cold rolling as well as press hardening and correlate the process parameters with the resulting product quality, in this use case represented by the final hardness. With increasing quality requirements and digital advances, there has been a growing number of publications related to data analysis for forming technology, particularly focusing on hot and cold rolling. Most publications address quality issues that current models cannot explain sufficiently, or non-linear relationships that lack complete understanding. The primary aim of data processing and analysis is to identify relevant parameters and influences [13].

The typical steps for data processing and analysis are data collection, storage, data integration and transformation as well as cleaning, feature selection and finally extracting knowledge by applying classification or regression approaches [14]. One early example of data analysis in cold rolling is the work by Wu et al. [15] who employed data analysis for refining rolling parameters. In their multivariable data analysis for cold rolling, Takami et al. [16] found, that the incoming and outgoing thicknesses, forces, back and forward tension, coil speed at entry and exit, flatness in the middle and at the edges of strips, eccentricity and roll gap positions are the most relevant measurable rolling parameters. Besides that, the authors found that also differences in the material properties and possible introduced damage during the hot rolling processes influences the final product quality.

In the context of press hardening, ascertaining the cooling rate during the process presents a multifaceted challenge due to the unavailability of direct temperature measurements through conventional means. In industrial applications, process monitoring conventionally involves temperature measurements taken on the sheet before and after pressing, as well as monitoring the sheet's placement within the press. In response to this challenge within a multi-step press hardening process, Martschin et al. have introduced a soft sensor utilizing dynamic mode decomposition as a methodology for estimating temperature distribution [17]. Conversely, Li et al. have proposed an alternative approach by determining the interfacial heat transfer coefficient through indirect measurement using a purpose-built testing apparatus equipped with an infrared thermometer and interface pressure sensor [18]. The employment of indirect temperature measurements assumes significant importance in order to facilitate the identification and analysis of essential heat transfer dynamics during the press hardening process.

This brief literature review highlights the critical importance of data processing and analysis in meeting current and future challenges. However, it is noticeable that most of the work done so far focuses on only one process. Data from several processes are rarely processed and analysed together. Specifically, for the press hardening route, a DMS is not yet introduced.

3. Materials and Methods

This chapter describes the material and the processes in more detail. The process chain of interest, including its parameters and measurements, shown in Figure 1. At the beginning, a square-sectioned MMnS billet of 100 mm is cast. Chemically analyzed results of the billet are provided in Table 1, obtained from a sample taken after casting. Following this, the billet is heated to 1200 °C and hot rolled to a 3 mm thickness. To ascertain the effect of process variations on material properties during cold rolling and to allow for experimental processing, the slabs were split into several shorter sheets.

Table 1. Chemical Composition of used alloy.

Fe	C	Mn	Si	Al	Cr	P	S	N
bal.	0.30	4.99	1.55	0.004	0.04	0.005	0.003	0.004

Next, all of the hot-rolled sheets were annealed for a duration of 2 hours at a temperature of 600 °C. This intermediate annealing between hot and cold rolling is essential due to martensite phase of the material and formation of carbides in the material. Otherwise, there is a significant possibility that material failure may result due to the occurrence of cracking during cold rolling. The parameters for annealing were determined on the basis of previous cold rolling trials.

	Cold rolling	Austenitization	Press hardening	Partitioning
Process-parameters	Height-reduction Pass number	Temperature Time	Drawing depth Blankholder force Tool temperature	Temperature Time
Measured data	Force, Torque Rolling speed Hardness	-	Force Displacement Temperature	Hardness

Figure 1. Process chain and its used parameters as well as measured data.

After annealing, the material is descaled and cold rolled in 4 passes from 3 mm to about 1.5 mm using a cold rolling mill with a quarto setup. During cold rolling the force, torque and the rolling speed are measured. In addition, the sheet thickness is measured manually at several points after each pass as well as the hardness (HV1) of the cold rolled sheets is determined.

The final stage of the process chain is the press hardening which needs a defined heat treatment before and after the process, following the Q&P principle. In order to achieve this, the sheets were initially cut into the desired shape. These sheets were then austenitization heat treated in a radiation furnace at 850°C for ten minutes and manually inserted into a Lauffer deep drawing press. The depth of the deep-drawing process remained steady at 16.2 mm. To affect the cooling process following formation, a heated press hardening tool was used, which was separately adjusted to tool temperatures of 120°C, 150°C, and 180°C. After forming, the parts were die-quenched for 30 seconds in the closed tool and manually transferred into a convection oven at 400 °C for 5 minutes. The tool temperature during hot stamping and die quenching was continuously measured by 6 thermocouples. The thermocouples are installed with a 2.5 mm distance to the surface of the die in 3 different positions on each arm of the cross-die, see Figure 2. The final hardness (HV1) was subsequently measured after the specimens cooled to room temperature.

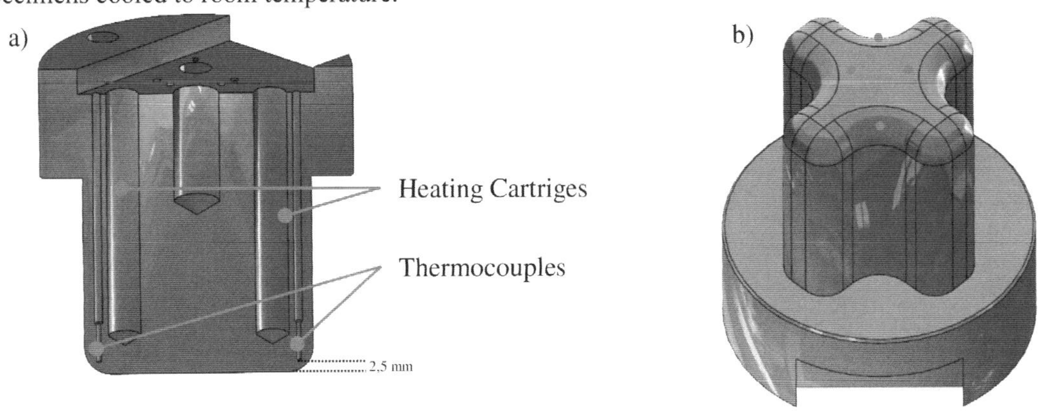

Figure 2. a) Sectional view of used hot-stamping tool b) temperature measurement positions.

4. Results

This chapter outlines the implemented initial steps to develop a framework for DMS. First, an ID system is described, which automates the aggregation of process data from the process chain onto a single sheet that pertains to the final press-hardened part. Following this, the implemented data pipeline is described, and the outcomes are presented, scrutinized, and assessed. Finally, an exemplary demonstration of the potential of the presented approach through the application of regression and classification is shown.

4.1. ID System

In order to automatically match the process data from cold rolling with the data from the press hardening process a simple but informative ID system was established. The ID system consists of several digits that are clearly assigned to a certain characteristic/feature. These features are especially applicable to the manufacturing process and pertain on the one hand to the semi-finished product and on the other hand to the finished part. It consists of the chemical composition, the number of the hot rolled slab, an index for the initial geometry and sheet number for the semi-finished sheet. For the finished product, an additional number for the press-hardened part is added and one number depending on the applied final heat treatment. Figure 3 shows the structure of the developed ID system.

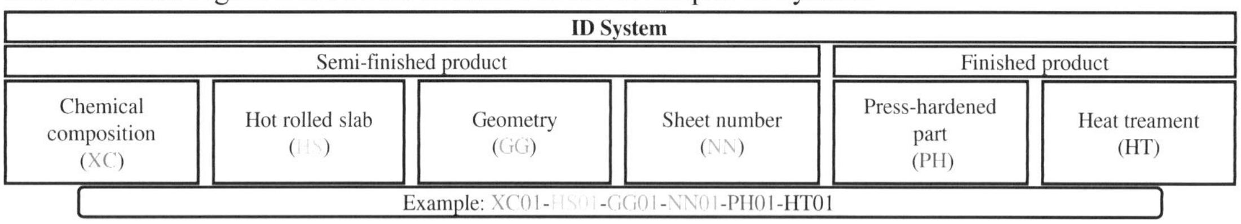

Figure 3. Structure of the ID-System

4.2. Workflow of the data pipeline

Next, automatic data processing is described and explained. As mentioned in the previous chapter, typical data processing consists of the following steps: Data collection, data cleaning/preparation, data processing, visualization and the final data analysis. These procedures were employed for cold rolling and press hardening using Python. Figure 4 shows the steps detailed for cold rolling and press hardening. In contrast to many other publications, the data analysis is not limited to one separate processes, but combines the data from several processes, thus allowing correlations between process parameters and the resulting material properties of a preceding process on the final product properties.

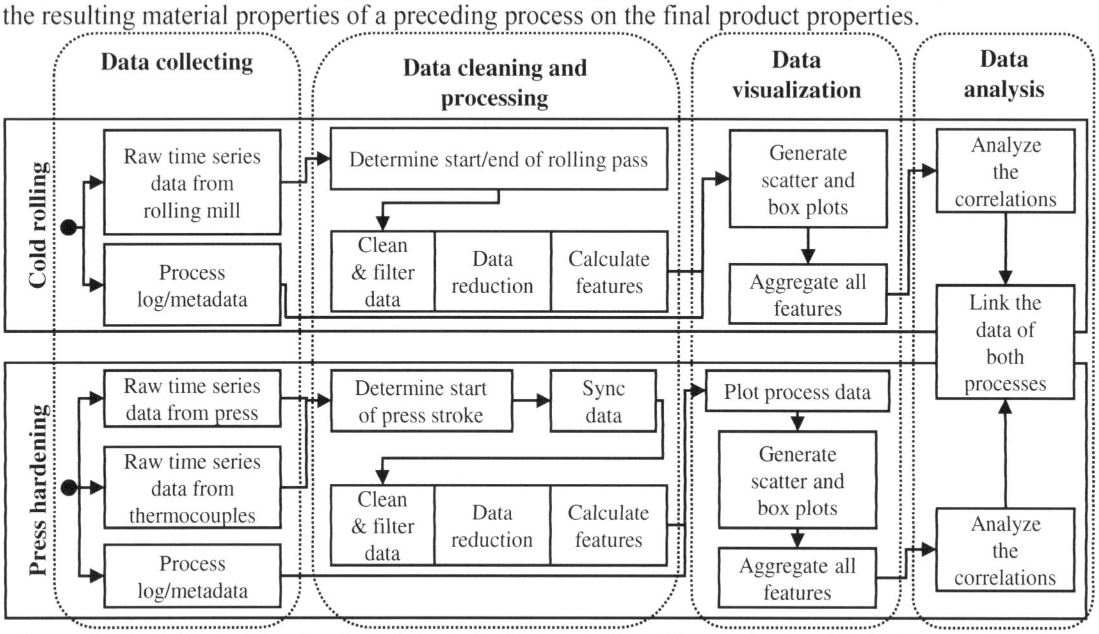

Figure 4. Workflow of the developed data pipeline for cold rolling and press hardening process.

For both cold rolling and press hardening, the fundamental steps in the implemented data pipeline are the same. Data is initially collected, followed by automatic determination of the start and end of the processes, in order to isolate relevant process data. An illustrative example of such process data is provided in Figure 5 a, for the cold rolling process. Next, outliers and measurement errors are eliminated from the process data. Then, the time series data is reduced by calculating features such as the mean rolling force. These features are saved alongside the time-series data for comprehensive analysis and are used to automatically generate plots for the entire trial. Figure 5 b demonstrates further analysis, showcasing the mean rolling force of each cold rolled sheet in the trial plotted as a box plot.

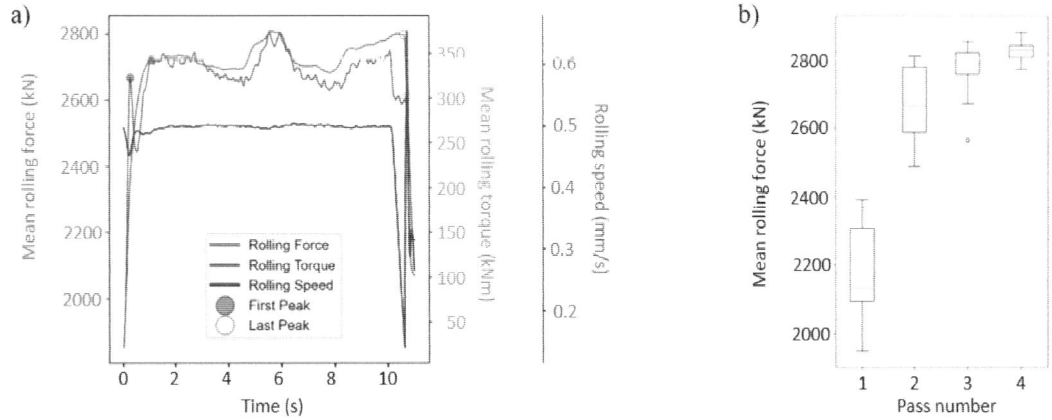

Figure 5. Example of automatically generated images for cold rolling. a): Process data of a pass. b): Aggregation of the mean rolling forces over the 4 cold rolling passes.

To get a better understanding of technical implementation of the data pipeline, the following section provides a detailed explanation of the data pipeline process, exemplified by the process of press hardening. The process data for press hardening is measured by two separate systems. The pressing force and the displacement are stored internally by the control unit of the deep drawing press as a time-series csv-file. The tool temperature measurements by the integrated thermocouples is saved on a thermal measurement system as a time series csv-file. Metadata is logged manually in a xlsx-file on a cloud-based storage system, in which the press data and temperature data files are uploaded. Using a Python script and well-establish libraries like Numpy and Pandas, the data is automatically synchronized by determining the start of the press stroke. After general cleaning and reduction of the data to the relevant time window, the actual holding time during quenching as well as the initial tool temperature is calculated and displayed together with the calculated forming force, displacement, and tool temperature change during and after the hot stamping, as seen in Figure 6.

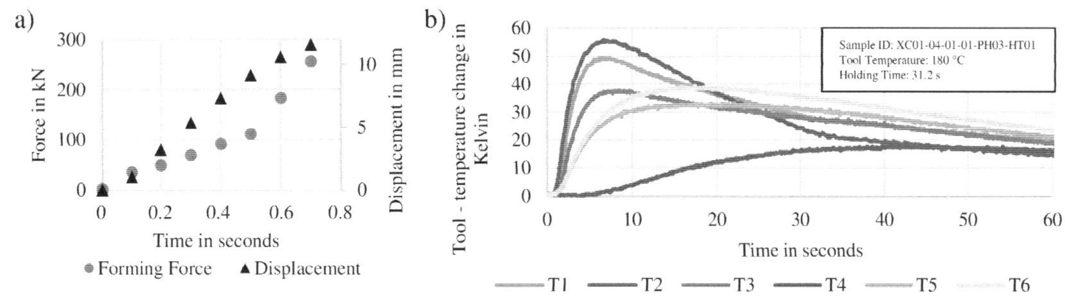

Figure 6. Example of automatically generated data for press hardening; a) forming force and displacement over process time b) temperature change of die for the first minute after start of the stroke.

Additionally, all features shown in Table 2 are automatically extracted within seconds, linked with the corresponding ID and saved for all parts in one xlsx-file. Typical and often used relationships, e.g. distribution of mean rolling force by pass number are directly visualized. Furthermore, the data can be used as a basis for customizing a DMS in order to solve specific issues. Two possible applications of these extracted features are shown in the following chapter.

Table 2. Overview of automatically extracted features.

Cold Rolling	Rolling Force, avg. Rolling Force, max. Rolling Force	Rolling Torque, avg. Rolling Torque, max. Rolling Torque	Thickness reduction, Hardness
Press Hardening	Displacement, Force, max. Forming Force, Holding time	Tool Temperature, max. cum. Temperature increase, avg. cum. Temperature increase, Rate of cum. Temperature increase	Cracks, Hardness

4.3. Exemplary demonstration of the results of the implemented data pipeline

Alongside the previously mentioned processing, aggregation and visualisation of process data, both regression and classification algorithms were utilised to showcase the capabilities of the developed data pipeline. The classification as well as the regression were carried out with algorithms provided by MathWorks Matlab Toolbox Classification Learner and Regression Learner. The aggregated process data from which both algorithms were trained, was taken from the process chain of press-hardened parts presented in chapter 3. In addition, the hardness measurements were used, too.

To illustrate the classification, an algorithm was used to detect the presence of cracks in the press-hardened part using process data. For the training, the press-hardened parts were categorised manually into the "No Cracks" or "Cracks" class (observed cracks) and that information was forwarded to the algorithms. After training, a fine tree algorithm was able to correctly classify (predicted cracks) almost 86% of the parts based on the tool temperature, the maximum force, and three features of the temperature measurement. The result is shown in Figure 7 a, all parts with "No Cracks" were correctly assigned (100%). However, 28.6 % of the parts with "Cracks" were incorrectly assigned. The initial findings are satisfactory, indicating chance for further improvement with additional (simulation) data.

Beside the classification example, regression algorithms were trained to predict the hardness (HV1) of the press-hardened parts. To train the algorithms only the hardness measurements from press-hardened parts without cracks were used in order to use consistent data. The hardness after the press hardening was predicted using the tool temperature, the cold-rolled sheet hardness, the maximal measured force and exactly like in the crack classification example three temperature features. The best result is shown in Figure 7 b. A clear trend can be recognised, although the predictions still deviate from the measured values. Similar to the classification example, it is also clear here that there is potential for further improvements. Presently, there are some manual elements that undermine the precision of data-driven approaches, such as the manual loading and unloading of the component into and out of the press and the manual measurement of hardness. Considering these factors, both outcomes seem optimistic because it becomes clear that correlations can be recognised despite manual influences and a manageable amount of data, thus showing potential for deeper understanding of material behaviour in combination with numerical simulations.

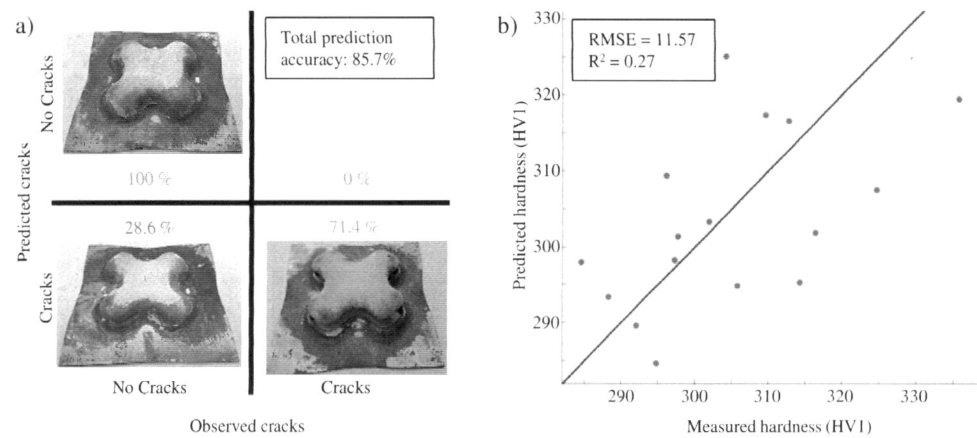

Figure 7. Results of parameter correlation and prediction on a) crack probability and b) hardness.

5. Summary and outlook

This paper presents a data processing framework for press hardening of MMnS, encompassing cold-rolling and hot-stamping with an integrated identification system, linking data across processes and sensor systems. The developed framework enables the usage of DMS for deeper understanding and discovering of complex relationships between process parameters and material properties of the final part. The framework enhances understanding of complex process-material relationships and is exemplified by statistically correlating crack probability and hardness with process data.

In future research, the data processed can be utilised to conduct numerical simulations which yield valuable information that may be otherwise impossible to measure, e.g. cooling rate during press hardening. This data and information can be used to produce a digital twin, accurately detailing each element and allowing for thorough analysis. Combining this principle with fast material models could enable new possibilities of process monitoring by online calculation of material properties based on real-time process data.

Acknowledgements

Funded by the Deutsche Forschungsgemeinschaft (DFG, German Research Foundation) under Germany's Excellence Strategy – EXC-2023 Internet of Production – 390621612.

We thank SeifAllah Mostafa Abdel Hamid Hedeiwy Mohamed for supporting us in developing the ID system and the data pipeline.

References

[1] Rosenstock D, Banik J, Gerber Tand Myslowicki S 2019 *IOP Conf. Ser.: Mater. Sci. Eng.* **651** 12040.
[2] Tong C, Rong Q, Yardley V A, Li X, Luo J, Zhu Gand Shi Z 2020 *Metals* **10** 1652.
[3] Cooman B deand Speer J 2006 *steel research int.* **77** 634.
[4] Blankart C, Emmrich R, Bähren Land Krupp U 2023 *steel research int.*
[5] Mishra G, Chandan A Kand Kundu S 2017 *Materials Science and Engineering: A* **701** 319.
[6] Sepasgozar S M E 2021 *Buildings* **11** 151.
[7] Bergs T, Gierlings S, Auerbach T, Klink A, Schraknepper Dand Augspurger T 2021 *Procedia CIRP* **101** 81.
[8] Zhou G, Zhang C, Li Z, Ding Kand Wang C 2020 *International Journal of Production Research* **58** 1034.
[9] Brecher C, Schuh G, van der Aalst W, Jarke M, Piller F T and Padberg M eds 2023 *Internet of Production* (Cham: Springer International Publishing).
[10] Kalidindi S R, Buzzy M, Boyce B Land Dingreville R 2022 *Front. Mater.* **9**.
[11] Wang J, Wesselmecking S, Krupp Uand Ackermann M 2022 *Data Guided Materials Processing - Digital Material Twin and Digital Material Shadow.*
[12] Wesselmecking S *et al.* 2023 *Internet of Production* eds C Brecher, G Schuh, W van der Aalst, M Jarke, F T Piller and M Padberg (Cham: Springer International Publishing) pp 1–16.
[13] Wang X-Y, Gao Z-Yand Xin Y-L 2023 *SN Appl. Sci.* **5**.
[14] Chen Z, Liu Y, Valera-Medina Aand Robinson F 2019 *Procedia CIRP* **81** 453.
[15] Wu S-X, Zhao X, Shao H-Hand Ren D-X 2004 Cold rolling process data analysis based on SVM *Proceedings of 2004 International Conference on Machine Learning and Cybernetics (IEEE Cat. No.04EX826)* (IEEE) pp 792–5.
[16] Takami K M, Mahmoudi J, Dahlquist Eand Lindenmo M 2011 *Int J Adv Manuf Technol* **54** 553.
[17] Martschin J, Wrobel M, Grodotzki J, Meurer Tand Tekkaya A E 2023 *Automot. Innov.* **6** 352.
[18] Li Y, Li S, He J, Chen Yand Yue L 2019 *Applied Thermal Engineering* **152** 865.

Methodic Approach for the Simulation-Based, Elastic Die Surface Compensation by Creating a FE-model of the Car Body Press Predicated on Spotting Images via Neural Networks

F Göltl[1]*, A Birkert[1], F Harst[2] and N C Stache[2]

[1] Center for Metal Forming and Car Body Manufacturing, Faculty of Engineering, Heilbronn University of Applied Sciences, Heilbronn, Germany

[2] Center for Machine Learning, Faculty of Engineering, Heilbronn University of Applied Sciences, Heilbronn, Germany

* florian.goeltl@hs-heilbronn.de

Abstract. In recent years, the push for streamlined vehicle development and the adoption of higher-strength materials to reduce the car body weight has been prominent. However, this shift results in greater elastic deformations in press and tool components due to the elevated yield stress of these materials. The die active face is subject to the compliance of the body press and die, which influences the manufacturability and dimensional accuracy of the sheet metal part. In practice, manual die spotting is required, relying on empirical data and subjective criteria for acceptance. This paper presents a new methodological approach for the development of the car body press's FE model in structural and forming coupled simulations. To achieve this, a series of measurements quantifies the pressure between tool and sheet metal. This information is fed into a neural network along with the die spotting images as training data. The network evaluates subsequent spotting images, assigning pressure values to distinct tool sections. Based on the die pressure distribution in the process, a virtual press model is created with the help of a newly developed objective function in topology optimization. This model aims to significantly enhance simulation-based die surface compensation, effectively reducing the die spotting process.

1. Introduction

The need for faster vehicle development processes has existed for decades now, and with the introduction of higher strength materials in car body engineering, the process forces are increasing. The total cost of tryout accounts for approximately 25% or more of tooling costs and offers significant potential for cost reduction [1]. The elastic behavior of the press and tool plays an increasingly important role in the result of the forming process. The simplification to ideally rigid tools and presses is therefore obsolete in the FE simulation of car body drawing processes. Multiple works have shown that the press and tools have a significant influence [2-13]. Penter and Großmann et.al. [5][6][7] focused specifically on the blankholders and die cushions and proposed replacement models. Further work has investigated different approaches to reducing model complexity in forming simulations[9][10][12][13]. While the tools can be accurately analyzed in terms of structural mechanics, little is known about the press. Complete modelling is complex and computationally expensive. Pilthammar has presented a method to create a substitute model based on measurement results [14] using topology optimization [15]. Despite advances in FE modelling, tryout is still required and spotting patterns are a mandatory acceptance criterion. An 80% load-bearing percentage for the blankholder closure is a common quality requirement that is currently only estimated by visual inspection. Feedback into the FE models is not yet standard practice. However, Essig has shown that image information from spotting patterns [16] can be fed back into FE models and that a significant increase in prediction quality can be achieved by adapting these models to the real contact areas [17], even under the assumption of an ideally rigid press and die (mathematical model of blankholder compliance available). Zabala et al. have investigated the correlation between paste distribution and pressure values for small, simple rectangular faces and proposed an algorithm for the relative evaluation of spotting patterns [18].

To date, there has been no approach to derive a corresponding explicit pressure distribution directly from the spotting images and feed it back into the FE-supported car body development process.

2. From Die Spotting Images to a Car Body Press's FE-Model with CNN and Topology Optimization

The spotting pattern in deep drawing results from the sheet thickness distribution and force transfer in the press-to-die and die-to-blank processes. Force-displacement measurements show that presses behave linear-elastically during forming [19], allowing to assess the press's structural behavior from pressure distribution during key points in time like blankholder closure and bottom dead center.

Creating a finite element (FE) model of the car body press initially yields a different pressure distribution than reality. Topology optimization helps to align the FE model's structural behavior with the real press, minimizing the deviation in pressure distribution. This optimized substitute model facilitates efficient coupling of the press, die, and forming, ultimately reducing die-spotting effort. In the following chapters, an innovative approach is presented to capture the aforementioned required pressure distributions with convolutional neural networks (CNN) and to introduce a new objective function in topology optimization to generate a substitute model of the press in the context of reducing tryout. The new procedure is demonstrated in Figure 1.

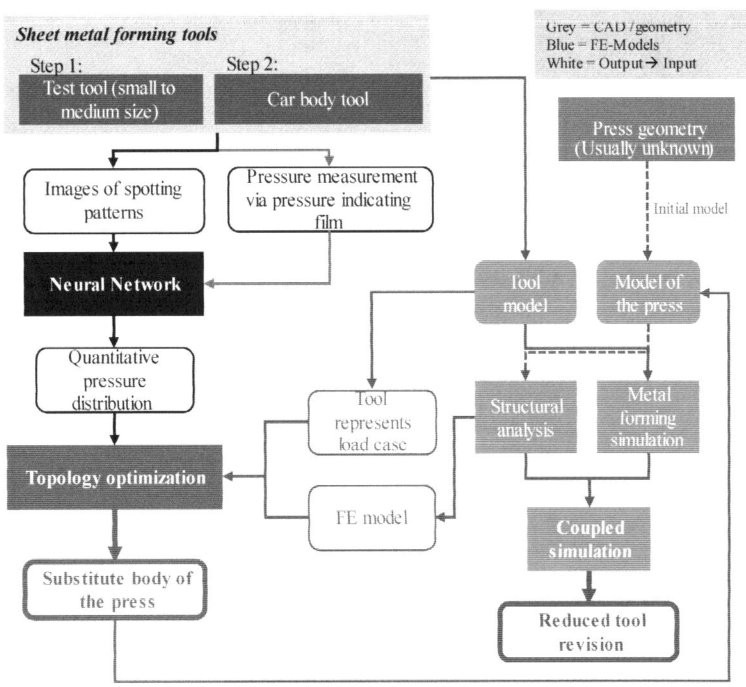

Figure 1. Process to reduce tool tryout by evaluating images of spotting patterns with CNN to get a pressure distribution for the topology optimization to create a substitute model

2.1. Reproducable Die Spotting Images and their Evaluation

In tryout the spotting paste is applied with paint rollers or brushes in a stroking and dabbing manner. For better comparability, a plastic roller was used in the following tests and used several times in advance so that the surface is saturated with color. The utilized paste is Diamant Spotting Paste Blue. The blanks are evenly colored over the entire surface and weighed on a precision scale before and after. On the basis of this, the ratio of ink quantity to surface area was determined and, based on initial practical applications in tryout, is 7.5 - 8.3 g/m². Additional measurements in day-to-day tryout are planned to identify the full range of paste quantity.

The quantification of the measured pressure can be obtained as a mean of the whole active surface during load or with the help of pressure indicating film. The latter was used for a pressure range from 0.2 to 50 MPa. FujiFilm's PreScale film set consists of a two-component film with a total thickness of approximately 0.18 mm and an accuracy of +/- 10%. The pressure measurement range is divided by four different film types (LLLW, LLW, LW, MW). Therefore, all measurements were carried out with several film types to cover the entire range. The individual films per measurement were computationally superimposed in position and value. First test measurements on a car body tool with the film showed that double-curved surfaces lead, as expected in the corner radii, to a strong wrinkling of the film due to the undevelopable geometry. Therefore, only single-curved and flat surfaces can be analysed with the films. The photographs of the blank surface and die surfaces are consistently captured from a "top-down" viewpoint, ensuring that the camera's orientation to the spotting surface remains perpendicular on flat surfaces. Curved surfaces of car body dies are observed from either the ram direction or the opposite direction. A mobile measuring arm with a swiveling head is used to record the spotting pattern and ink residues on the punch, blankholder and die when installed in the press. To reduce reflections on the surfaces of the colored blank and metal tools, an illumination with several indirect diffuse light sources is employed (Figure 2). If necessary, a reference measuring object can be employed to normalize the recorded images. The measuring object has predetermined values for length and color.

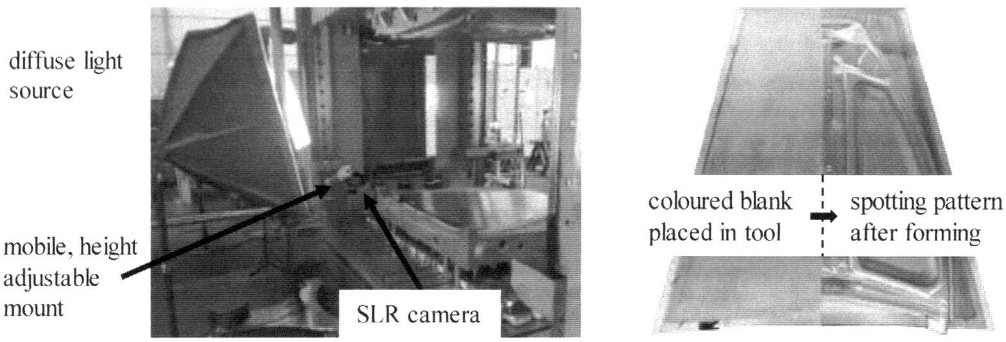

Figure 2. Measuring System for Photographing Spotting Patterns on the workpiece and tool

To achieve the most precise results in the low to medium pressure range, the experiments were conducted using a tensile and compression testing machine. As instruments, two sets of compression discs (D_1=112.7 mm and D_2=210 mm) with circular shape and precision-ground contact surfaces were utilized. The conducted tests reach from 0.4 to 8.5 MPa.

The second test tool is a small deep-drawing tool (341 mm x 248 mm x 117 mm) that was constructed for a tray utilizing a pair of blankholders. One of the holders, by default, is rigid for the given dimensions, while the other has significant inserts to alter its rigidity. The measurements are solely taken within the blankholder area using traditional spotting patterns and pressure indicating films.

The spotting pattern test plan involves evaluating two different blankholder geometries, testing three process forces (F_{ram} 100/130/180 kN and $F_{cushion}$ 21/50/100 kN) and capturing data at two to three specific points in time. Each combination is measured with the films LLW, LW and MW, allowing a pressure range of 0.5 to 50 MPa to be recorded. A total of 54 results were obtained from the 3 x 3 x 2 combinations

IOP Conf. Series: Materials Science and Engineering **1307** (2024) 012039 doi:10.1088/1757-899X/1307/1/012039

with 3 films, with 12 of them excluded. It was observed that shortly (0,6 mm) before bottom dead center did not exhibit any significant influence at the highest loads and was not considered for the other two load sets. One set of results for the massive blankholder geometry is shown in Figure 3. For reasons of space, the spotting pattern solely appears on the sheet metal. However, the die side has also been photographed.

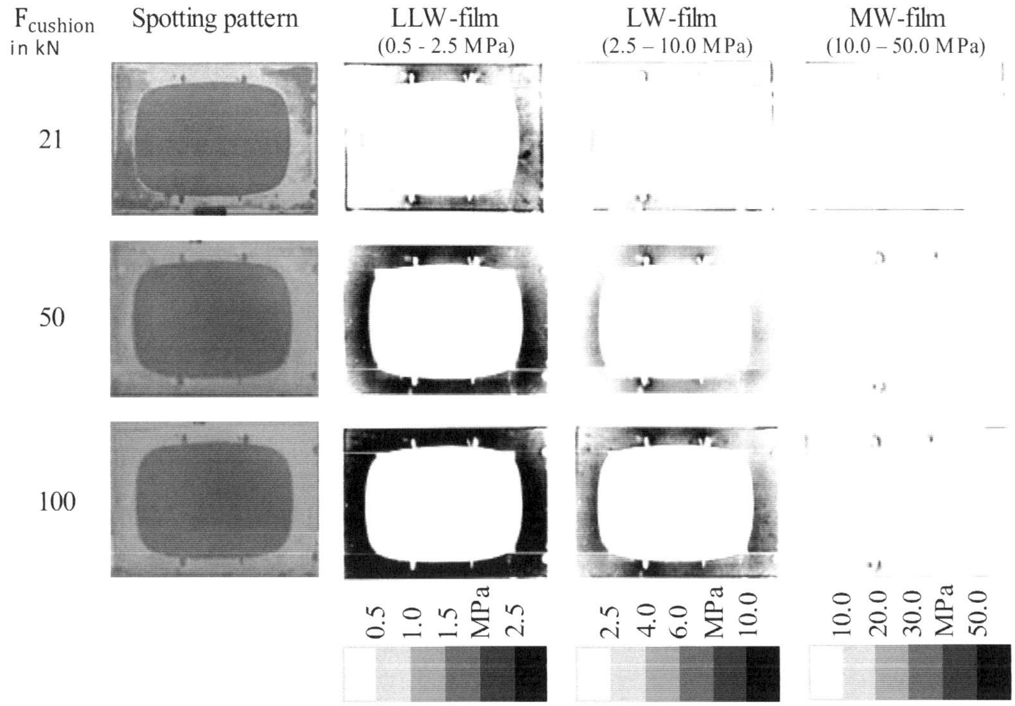

Figure 3. Set of results for the massive blankholder geometry at blankholder closure; spotting pattern on workpiece and evaluation of different pressure indicating films

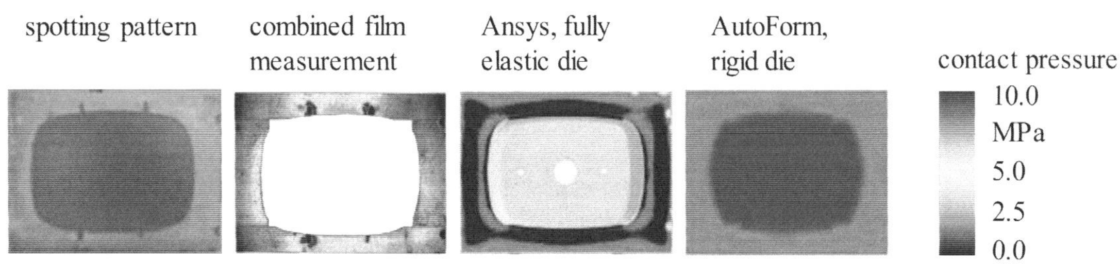

Figure 4. Spotting pattern at blankholder closure with a cushion force of 50 kN compared to contact pressure of the computed pressure distribution from pressure indicating films, the Ansys fully elastic die model and the AutoForm model (from left to right)

The necessity for elastic die and press models is evident from the subsequent results. Using the FE-softwares AutoForm and Ansys, the contact situation can be replicated, and the associated pressure distribution can be assessed. The active surfaces of the tools were 3D scanned and fed back into the CAD models. The AutoForm simulation works with rigid active surfaces and a blankholder model that takes limited elasticity into account. This limitation is readily discernible when compared to the pressure distributions of fully elastic die models in Ansys. The press is assumed to be perfectly rigid in both models, resulting in the expected differences between measurement and simulation outcomes, as evidenced in Figure 4.

From these tests the designation of spotting patterns is categorized by pressure range: low pressure (0-3 MPa), medium pressure (3-8 MPa), high pressure (8-12 MPa), very high pressure (12-50 MPa), and extreme pressure (50+ MPa). In the high-pressure range, typically exceeding 10 MPa, the color is fully displaced in the aftermath of visual inspection.

Spotting images and measured pressure distribution provide input data for a neural network that utilises this result combination to forecast the pressure values of a spotting pattern from a photograph.

2.2. Convolutional Neural Network Design and Pressure Distribution

To assess the requirement for a machine learning algorithm, a correlation analysis was conducted between sections of a spotting pattern extracted from a photograph and the corresponding pressure measurements, using principal component analysis for both the image and frequency domains. Nevertheless, as traditional methods did not reveal any strong correlation between the pressure and the spotting pattern, a neural network was used to determine the possibility of predicting a pressure value from the spotting pattern.

The initial iteration of the neural network has multiple convolutional layers and fully connected layers, culminating in a sole output neuron. The first step in this version requires the CNN to produce the average contact pressure as an absolute value (in MPa) across a two-dimensional segment of the image. To accomplish this, a spotting pattern photograph must be divided into segments and processed individually.

In order to determine the overall ability to deduce the pressure value from a spotting pattern, half of each photograph was employed to form the training data, whilst the remaining half was designated for validation and testing purposes. By adopting this approach, the likelihood of test data and training data overlapping was eliminated, as well as other factors that might vary between individual measurements and complicate the learning process. Four measurements were conducted for the purpose of generating training data, which were then segmented to yield 20,983 data points for training, 11,032 for validation, and 5,435 for testing.

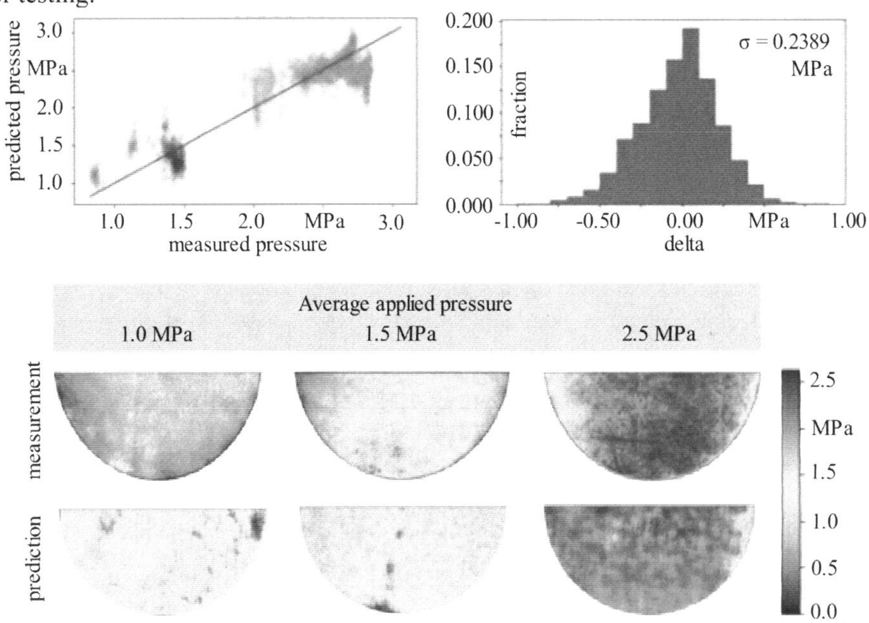

Figure 5. Measured versus predicted pressure values by the CNN and its corresponding value distribution (upper half) and the correlating images (lower half)

A simplified technique of hyperparameter optimization (parameter sweeps) was implemented to identify neural network configurations of high performance. To ensure comparability, uniform datasets of measurement data for pressures ranging from 0.5 to 3 MPa were utilized to train each network.

The preliminary findings on a limited training dataset demonstrate the potential to acquire a correlation between the absolute pressure values and the segments of spotting pattern photographs ($R^2 = 0.84$). Nevertheless, it is evident from Figure 5 that there is a discernible disparity between the measured and predicted pressure values. The regression fit for the lower pressure values is notably poorer than for the higher pressure regions. This is likely due to the limited amount of training data currently available, which will be augmented in future developments. The individual measurements are represented as individually colored dots, and it is apparent that clusters of data points are formed for each measurement, suggesting a greater overlap between pressure ranges is required. Additionally, there is an inequitable distribution of low-range and high-range pressure data, highlighting the necessity for a larger quantity of training data especially in the former.

2.3. Tool pressure as objective function in Topology Optimization

For a selected die (later for several), the pressure distributions measured from the spotting images are to be used as objectives of a topology optimization. The result of the optimization is a virtual press that behaves structurally similar to the real press. The structural model of the press is initially approximated. The most simple model for the ram is a cuboid which acts as the base model for the optimization including load, support (defined via element and node sets) and design space.

A new objective function has been developed for topology optimization, which determines the equilibrium between is- and target forces. For the optimization a certain timestamp is picked in the forming process, representing the moment of the measured spotting image. According to own measurements and literature values, the behavior of a press is approximately linear-elastic in the force range from blankholder closure to the lower forming point. By cutting free the system of press, die and workpiece at the contact between die and workpiece surface, the system is reduced to press and tool. The is-forces at the free surface are known from the spotting patterns and their evaluation. The first contact zone between workpiece and die is resolved with regard to nonlinear behavior. The contact behavior between press and die is assumed to be linear with minimal to no movement between them. It is simplified to bonded contact behavior, which eliminates the remaining nonlinearities in the FE model. This makes it possible to apply the SIMP (Solid Isotropic Material with Penalization) material law. The result of the optimization is a pure numerical calculation model for the press. A continuous distribution of the design variable between 0 and 1 is also possible, in contrast to the usual application of SIMP. This increases the possible solution range of the optimization; the exact advantages and disadvantages in this application case must be investigated in the further project progress.

The data derived from the spotting image (x-, y-coordinates and pressure value) is converted into nodal load vectors with respect to the transformation rule for the measurement and FE data sets. This transformed data is allocated to the surface load elements located on the active surface and is subsequently linearly averaged. When reading the input files, the optimization software XCARAT then automatically generates a nodal force from the pressure loads.

The reaction forces in the model as a result of the applied target forces are determined with the aid of artificially introduced fixed constraints on the die's active surface.

The goal of this optimization is to minimize the deviation of the calculated pressure distribution from the measured pressure distribution or the respective nodal forces by adjusting the element stiffness of the virtual press. The objective function f_{LS} results in a scalar value and is defined as follows:

$$f_{LS}(x,u(x)) = \left[{n_{NF}}^{-1} * \sum_{i=0}^{n_{NF}} \Delta f_i(x,u(x))^2 \right]^{0,5} \tag{1}$$

whereas:

$$\Delta f_i(x,u(x)) = \left| f_{Die_{i,is}}(x,u(x)) \right| - \left| f_{SP_{target}} \right| \tag{2}$$

n_{NF}: number of nodes that experience a pressure load
f_{Die}: vector of the is-forces (support forces) on the inside of the die
f_{SP}: vector of target nodal forces on the inside of the die, determined from spotting patterns
Δf_i: difference between the values of the two force vectors

The objective function (1) and its total derivative were analytically determined and implemented into XCARAT with the help of FEMopt Studios GmbH. The validation of the calculated gradient $\frac{d\, f_{LS}(x,u(x))}{dx_j}$ was carried out with the Finite Difference Method (FDM). The comparison of the calculated PyCarat (Python interface of XCARAT) gradients with the gradients determined using FDM is shown in Figure 6. The relative error is less than 0.23% for any design variable in this academic model.

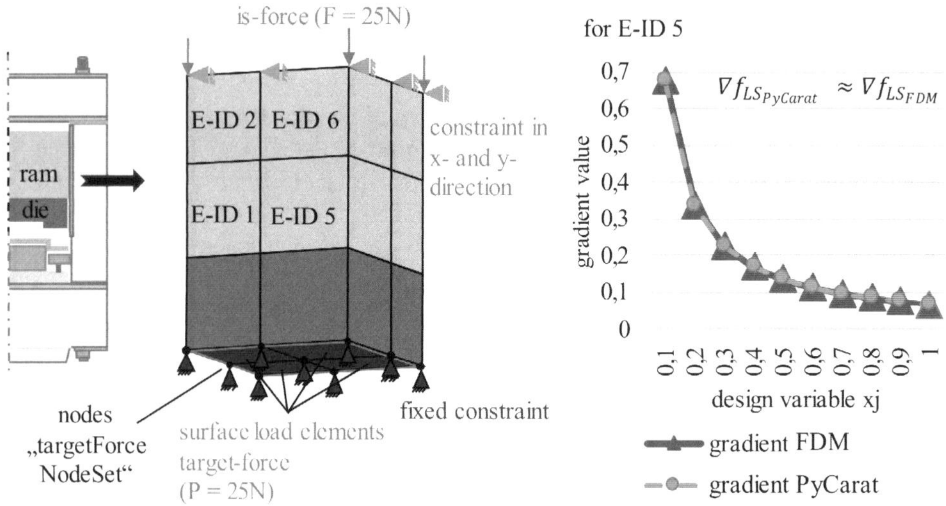

Figure 6. Schematic representation of the car body press (l.) with its reduction to the model for topology optimization, simplified example with few elements to test the new objective function (m.) and validation of the objective function with the use of FDM (r.)

3. Discussion & Conclusion

The published methodology demonstrates a novel approach to quantifying spotting images through measured values and their processing in CNNs. While the prediction quality has been acceptable in limited cases, it remains subject to a small dataset, which is gradually expanding. In addition to paste quantity, pressure, and ambient conditions, metrological factors such as light exposure are being further investigated as parameters for prediction quality. Following completion of laboratory evaluations, the complexity of the analyzed spotting images will be incrementally increased. The CNN-derived pressure distributions can be utilized as target forces for topological optimization of a body using the novel target function (force equilibrium at the contact point of the workpiece tool) in basic examples. The required simplifications, in the form of linearizing the model, have been presented and the new objective function verified. Further advancement involves demonstrating the process in more intricate applications to characterize the structural behavior of a press.

Acknowledgements

This research received funding within the framework of the research project BW1_1435/02 “Fingerprint Press Identification”, which was funded by the Invest BW (VwV Invest BW - Innovation II) program of the federal state Baden-Württemberg, Germany. The authors would like to thank the above-mentioned institutions for providing the financial resources, the companies FMF-WWF Tool and Prototype Construction GmbH and FEMopt Studios GmbH for their technical and material support.

Reference

[1] Birkert A R, Haage S and Straub M 2013 Umformtechnische Herstellung komplexer Karosserieteile. Auslegung von Ziehanlagen *(Berlin, Heidelberg: Springer Berlin Heidelberg)*

[2] Pilthammar J, Sigvant M, Kao-Walter S 2016 Including die and press deformations in sheet metal forming simulations. *Journal of Physics: Conference Series. 734. 032036. 10.1088/1742-6596/734/3/032036.*

[3] Pilthammar J, Sigvant M, Islam M, Schill M, Sjöblom S, Sjöblom V, Lind M 2023 An overview of Methods for Simulating Sheet Metal Forming with Elastic Dies. *IOP Conference Series: Materials Science and Engineering. 1284. 012054. 10.1088/1757-899X/1284/1/012054.*

[4] Pilthammar J, Sigvant M, Islam M, Schill M, Sjöblom S, Sjöblom V, Lind M 2023 Three Industrial Cases of Sheet Metal Forming Simulations with Elastic Dies. *IOP Conference Series: Materials Science and Engineering. 1284. 012055. 10.1088/1757-899X/1284/1/012055.*

[5] Penter L, Ihlenfeldt S, Schenke C, Schulze T, Weber J 2016 Modelltiefe zur Inbetriebnahme von Umformwerkzeugen auf Pressen mit hydraulischen Mehrpunktziehkissen. *Conference: EFB Kolloquium Blechverarbeitung*

[6] Großmann K, Penter L, Schenke C-C, Hardtmann A 2013 Systematische Berücksichtigung der Ziehkissenanlage von Umformpressen in FE-Prozesssimulationen. *Conference: Sächsische Fachtagung für Umformtechnik, Dresden, Volume: 20*

[7] Großmann K, Penter L, Schenke C-C, Hardtmann A, Weber J, Lohse H, Schulze T 2014 FE FORMING MODELS INCLUDING PRESS BEHAVIOR ALLOW FOR REALISTIC COMPUTATION OF BLANKHOLDER FORCES. *Conference: IDDRG, Paris*

[8] Großmann K, Wiemer H, Hardtmann A, Penter L, Kriechenbauer S 2009 Adjusting the Contact Surface of Forming Tools in Order to Compensate for Elastic Deformations during the Process *Conference: 7th European LS-DYNA Conference, Salzburg*

[9] Lingbeek RA 2008 Virtual tool reworking. New strategies in die design using finite elementforming simulations *Enschede, The Netherlands, ISBN 978-90-77172-38-4*

[10] Haufe A, Roll K and Bogon P 2008 Sheet Metal Forming with Elastic Tools in LS-DYNA *Numisheet 2008*

[11] Müller P et al 2019 Experimental analysis of the elastic boundary conditions of press machines for modelling the deep-drawing process *The International Journal of AdvancedManufacturing Technology*

[12] Zgoll F et.al. 2021 Virtual die spotting. Advanced setup for coupling of forming and structure simulation *IOP Conf. Series: Materials Science and Engineering*

[13] Neto D M, Coër J, Oliveira M C, Alves J L, Manach P Y and Menezes L F 2016 Numerical analysis on the elastic deformation of the tools in sheet metal forming processes. *International Journal of Solids and Structures*

[14] Pilthammar J, Sigvant M, Hansson M, Palsson E, Rutgersson W 2017 Characterizing the Elastic Behavior of a Press Table through Topology Optimization. *Journal of Physics: Conference Series. 896. 012068. 10.1088/1742-6596/896/1/012068.*

[15] Bendsøe M P, Sigmund O 2004 Topology Optimization: Theory, Methods, and Applications *Springer-Verlag Berlin Heidelberg 2004*

[16] Essig P et al 2019 Digital process support in toolmaking by using optical metrology *IOP Conf. Ser.: Mater. Sci. Eng. 651 012026*

[17] Essig P, Liewald M, Hol J 2021 Implementation of Real Contact Areas into Sheet metal Forming Simulations using Digital Spotting Images. *IOP Conference Series: Materials Science and Engineering. 1157. 012025. 10.1088/1757-899X/1157/1/012025.*

[18] Zabala A, Llavori I, Sáenz de Argandoña E and Mendiguren J 2020 Towards the automation of the die spotting process: Contact blue pattern decryption *Journal of Manufacturing Processes, Volume 58, 2020, Pages 1285-1296, ISSN 1526-6125,.DOI:10.1016/j.jmapro.2020.09.022.*

[19] Doege E, Behrens B-A 2016 Handbuch Umformtechnik *Springer Vieweg Berlin, Heidelberg*

Optimizing Reverse-Engineered Finite Element Models for Accurate Predictions of Experimental Measurements

A R Chezan[1], T Dhawale[1], E H Atzema[1,2], A Barlo[3], O Aeddula[3], J Pilthammar[3,4], M Sigvant[3,4] and N A J Langerak[1]

[1]Tata Steel R&D, P.O. Box 10.000, 1970 CA IJmuiden, The Netherlands
[2]Faculty of Engineering, University of Twente, P.O. Box 217, 7500 AE Enschede, The Netherlands.
[3]Department of Mechanical Engineering, Blekinge Institute of Technology, Karlskrona, Sweden
[4]Volvo Cars Dept. 81013 Manufacturing Technology Development, Olofström, Sweden

E-mail: toni.chezan@tatasteeleurope.com

Abstract. This study investigates the challenges of reverse engineering in finite element modelling of sheet metal forming, specifically for the Volvo XC90 front door inner component. Advanced models incorporating anisotropic behaviour of steel and non-linear friction are compared against actual real-world measurements. The methodology involves simplifying complex continuous parameters into more manageable representative data sets and assessing model accuracy under both uniform and varied blank holder force settings, guided by measured contact pressure distributions. Although the results indicate an improvement in accuracy, they underscore the need for additional methodological improvements and more accurate replication of tooling effects to enhance the fidelity and effectiveness of these models.

1. Introduction

Accurate finite element modelling has emerged as a strategy to enhance cost-efficiency and streamline the development of forming tools and series production within the automotive sector. Achieving significant improvement in overall accuracy involves using advanced models that describe the anisotropic behaviour of steel sheet metal [1] and incorporate non-linear friction models [2, 3]. To verify model accuracy, it is essential to compare simulation predictions with actual measurements of drawn-in and strain distributions in parts produced under typical series production conditions [4].

Despite these advancements, a significant challenge remains in the field of reverse engineering: the creation of realistic models that accurately replicate existing processes. While incorporating the geometries of existing stamping tools through 3D surface scans is a relatively direct task, accurately capturing the elastic behaviour reported in actual tools and press structures adds a layer of complexity [5]. Current efforts focus on developing methods that efficiently simulate these elastic effects [6, 7, 8], though the prevalent approach still relies on models assuming rigid tooling. This approach necessitates adjustments, particularly in modelling of the restraining effect exerted by the blank holder, as well as the simulation of material stretching in the punch area of the tool set. This is usually achieved through a series of detailed optimization iterations. These iterations are crucial for aligning model predictions

IOP Conf. Series: Materials Science and Engineering **1307** (2024) 012040 doi:10.1088/1757-899X/1307/1/012040

with actual measurements, thereby achieving models with sufficient accuracy to quantitatively describe real-world observations.

In industrial sheet metal forming it is observed that simulations align well with experimental outcomes when considering specific settings of the blank holder force. However, when the range of evaluated blank holder force settings is limited, predictions for series production scenarios outside this calibrated range may require extrapolation. To enhance the model prediction accuracy and reliability, it is essential to incorporate a wider array of blank holder force settings into these assessments. Furthermore, when experiments are conducted on industrial parts, the resulting volume of data is substantial. Consequently, an effective data reduction process is imperative, facilitating both qualitative and quantitative analyses without losing critical information.

This paper studies the first forming operation of the Volvo XC90 front door inner component [4]. The proposed approach aims to simplify the complex continuous parameters describing the part draw-in and part holes geometry modifications into a more manageable representative data set. The model accuracy is evaluated by comparing the model prediction with measurements performed on parts pressed in the production line at various blank holder force settings. Elastic tooling effects are approximated by adjusting the blank holder contract pressure distribution with information from measurements on trapped blanks. The reduced data set enables quantitative comparisons of experimental and model sensitivities to blank holder force adjustments. Pair correlation analysis reveals particularities of the experimental data and of the methodology to model the blank holder contact pressure.

2. Materials, experimental methods and finite element models

The material used in this paper is a 0.7 mm zinc coated mild steel CR4-ZM grade according to VDA 239-100 [9]. Mechanical properties were used according to data from Tata Steel Aurora® Online material data base [10]. ZM represents the MagiZinc Auto zinc coating containing additional magnesium and aluminium. The blanks were oiled with Anticorit PLS100T [11].

The finite element model was reverse engineered from measurements of an existing automotive part manufacturing process. The surfaces of the tools of the finite element model of the door inner automotive panel were based on 3D tool surface digitization using a HandySCAN 700 from Creaform. The trapped blank and draw panels were stamped at the stamping plant of Volvo Cars in Olofström. The model incorporates the tooling kinematics of the used mechanical press. The draw-in measurements were performed using the same HandySCAN 700. The tribology system was modelled using the TriboForm friction model. All finite element simulations were performed using AutoForm^{plus} R10.

The pressure distribution was measured using a Fujifilm Prescale pressure measurement foil with a pressure range between 0.5 and 2.5 MPa.

3. Results

Figure 1 illustrates qualitatively the contact pressure distribution status of the trapped blank, a stage in the forming operation where the blank holder presses against the die under nominal force. At this moment of the forming operation the draw beads are fully engaged with the blank, yet the punch has not made contact. The forming tools utilised in this study are dimensionally precise and the 3D alignment of the blank holder and die surfaces ensures a consistent uniform gap that results in a uniform pressure distribution, as predicted by the FE model in figure 1b, hereinafter referred to as "FEM uniform". Experimental observations indicate some regions in the blank holder that exhibit partial contact as demonstrated in figure 1a. A qualitative mapping of the experimental contact pressure distribution using the "non-bearing area" option in AutoForm was realized. The contact pressure distribution of this second model (hereafter referred to as "FEM mapped") is provided in figure 1c.

Figure 2a outlines the XY plane positions of the equalizer blocks of the die set used to stamp the part. For each block, points on the trapped blank outer edge were defined as the local closest point to the centre of the block. Draw-in was calculated as the distance in a perpendicular direction from the trapped blank edge selected point towards the formed part edge (figure 2b). For the analysis in this paper, the draw-in values corresponding to these 16 points on the trapped blank were considered. The blank has four holes of different initial sizes. They were labelled with capital letters in figure 2a. In the final formed part, the hole shape and dimension change. The average diameter of the deformed hole and the absolute relative displacement of the hole centre relative to the undeformed hole centre, were determined.

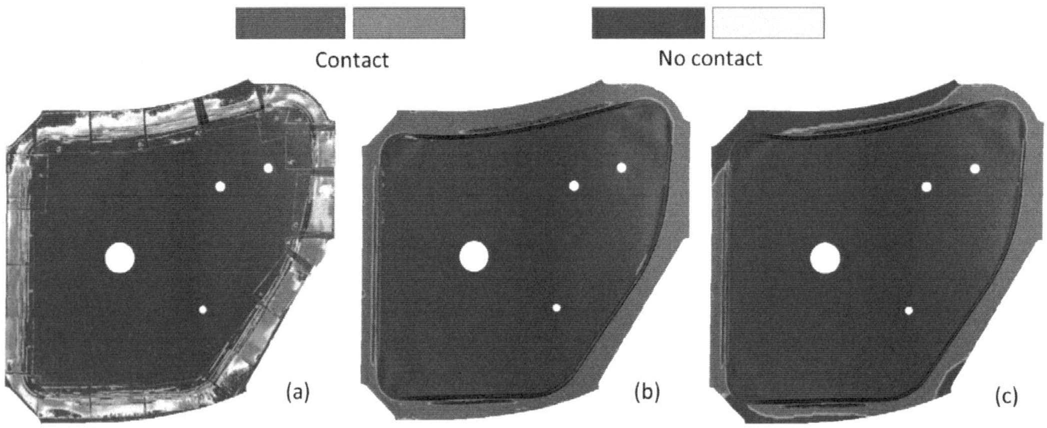

Figure 1. Qualitative visualisations of the trapped blank contact pressure distribution corresponding to (a) experimental observation, (b) FE model prediction with a uniform gap (termed "FEM uniform"), and (c) FE model with experimentally mapped contact pressures (termed "FEM mapped).

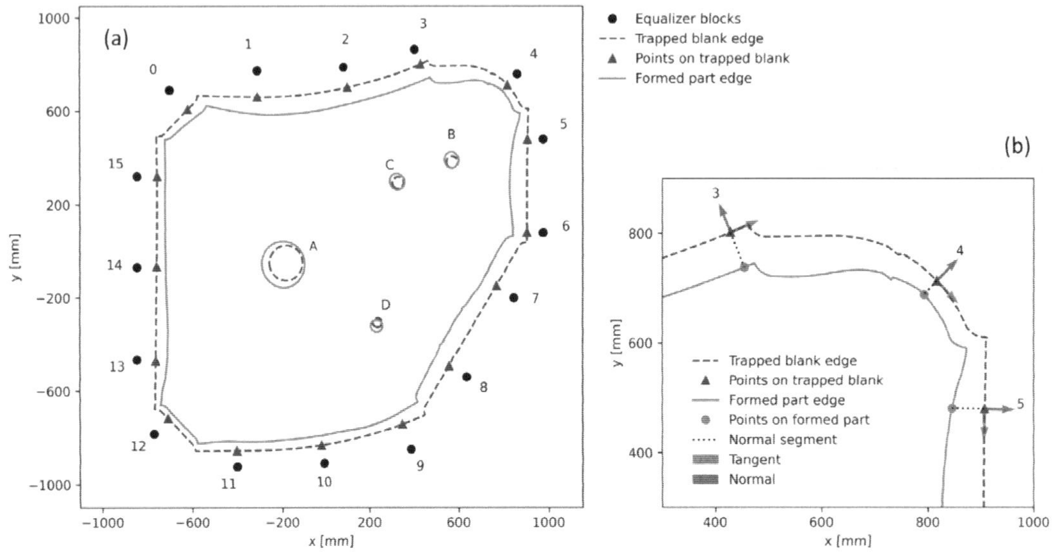

Figure 2. (a) Projection on the XY plane of the equalizer blocks, selected reference points and experimentally measured contours of the trapped blank and formed part (BHF 1800kN). (b) Detail illustrating the position of the normal segments used to characterize the local draw-in.

Examples of measured and FE element models predictions of several points on the trapped blank edge and parameters of the holes are presented in figure 3. The title of the subfigures indicates with a number the locations along the trapped blank edges, respectively, with letters combination, the parameters of the holes as labelled in figure 2a. The average diameter of hole A is labelled as A_d, etc. The relative displacement of the hole centre of hole A is labelled as A_c, etc. Increasing the blank holder force results in higher restraining of the material flow into the die cavity and consequently the draw-in decreases and the diameters of the holes increases. The expected trend of the blank holder effect on draw-in and hole diameters can be observed in both experimental and numerical generated data in figure 3. The locations 3, 7 and 13 shows a well-defined decreases rate of the draw-in for both experiments and simulations, however the absolute values and the rate of decrease can be significantly different. The FEM mapped model predicts well the absolute draw-in values and the draw-in rate change at locations 3 and 13, while the prediction accuracy of the FEM uniform model at these two locations is significantly lower. The two models predict well the size of the holes are blank holder force up to 1500 kN. At higher blank holder force settings, the models predict significantly larger average holed diameters as compared to experimental measurements. Hole positions exhibit a varied response; experimental results show negligible effects of blank holder force on hole centre positioning, while both models predict noticeable shifts.

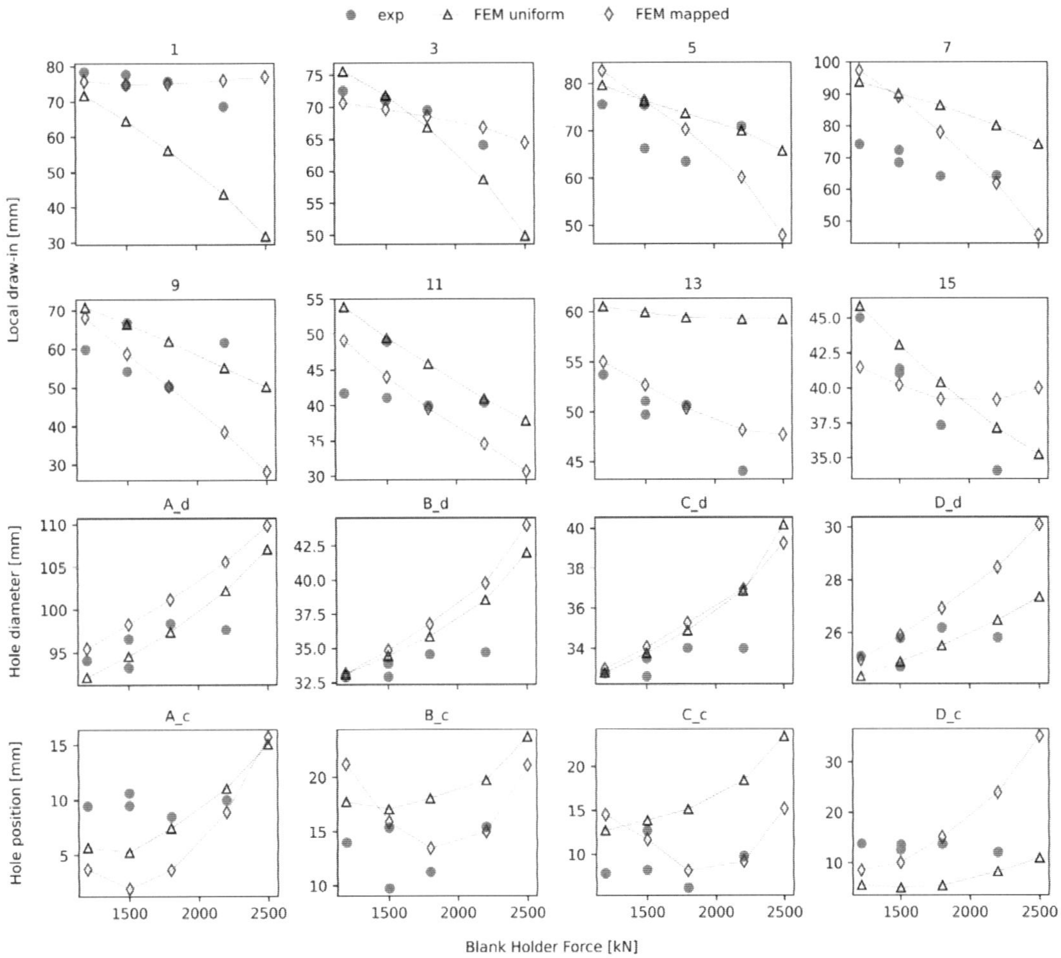

Figure 3. Examples of measurements and simulation results for local draw-in at several locations along the part outer contours and parameters of the holes as specified in figure 2. The figure titles indicate the corresponding locations along the outer contours and the average diameter (e.g. A_d. etc.) respectively hole centre relative displacement (e.g. A_c, etc.) of the four holes.

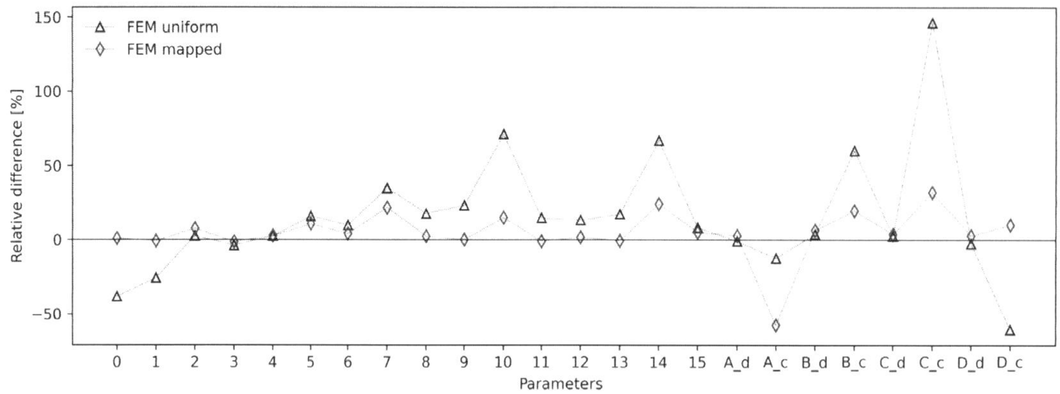

Figure 4. Relative differences between experiments and model predictions at 1800 kN blank holder force setting for the two FE models.

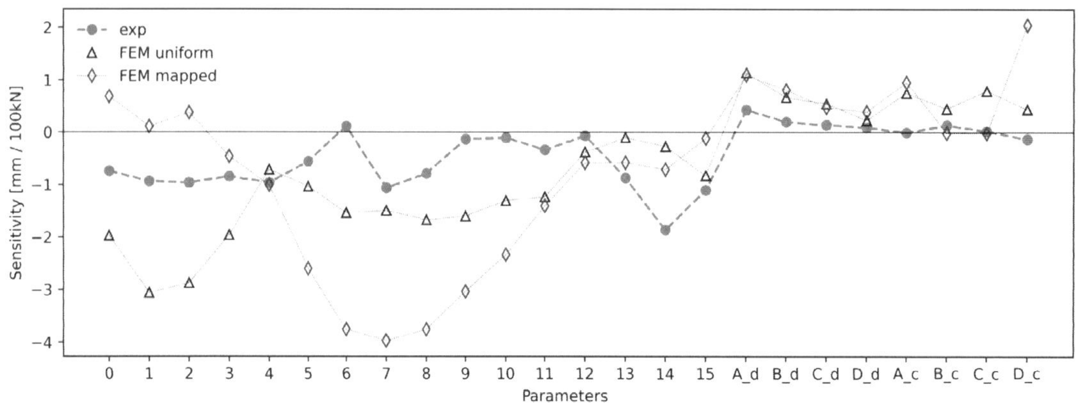

Figure 5. Experimentally measured and FE models predictions of the sensitivity to blank holder force variation for the draw-in parameters and parameters of the holes, relative positions and average diameters expressed as mm effect for 100 kN blank holder force variation.

Figure 4 summarizes the predictive accuracy of the two reverse-engineered models, at a blank holder force setting equal to 1800 kN, by presenting the relative deviations from experimental results in percentage terms for each parameter. The FEM Mapped model delivers more accurate draw-in predictions compared to the FEM uniform model. The accuracy of both models is similar for average hole diameters, but predictions for hole centre shifts are less precise, although the FEM mapped model performs better.

A linear regression analysis was used to gauge the sensitivity to blank holder force variations of the draw-in and the hole parameters. For experimental data a blank holder interval between 1200 and 2200 kN was used. For the FE models 2500 kN was used for the upper limit of the blank holder force interval. In figure 5 the sensitivities of the parameters experimentally measured and calculated using FEM are presented as millimetre changes for 100 kN blank holder force adjustment. For most parameters the experimental sensitivity is smaller as compared to model predictions. Some regions of the pressed parts (locations 6, 9, 10, 11 and 12) and the diameter and locations of the deformed holes are almost insensitive to blank holder force variations. The rocker side of the pressed parts (locations 12, 14 and 15) are more sensitive as compared to the predictions of the models. In the b-pillar area (locations 0 to 3) the FEM uniform model predicts a higher sensitivity as compared to FEM mapped model. In the window frame and a-pillar side (locations 5 to 11) the FEM mapped model predicts higher sensitivity as compared to FEM uniform model.

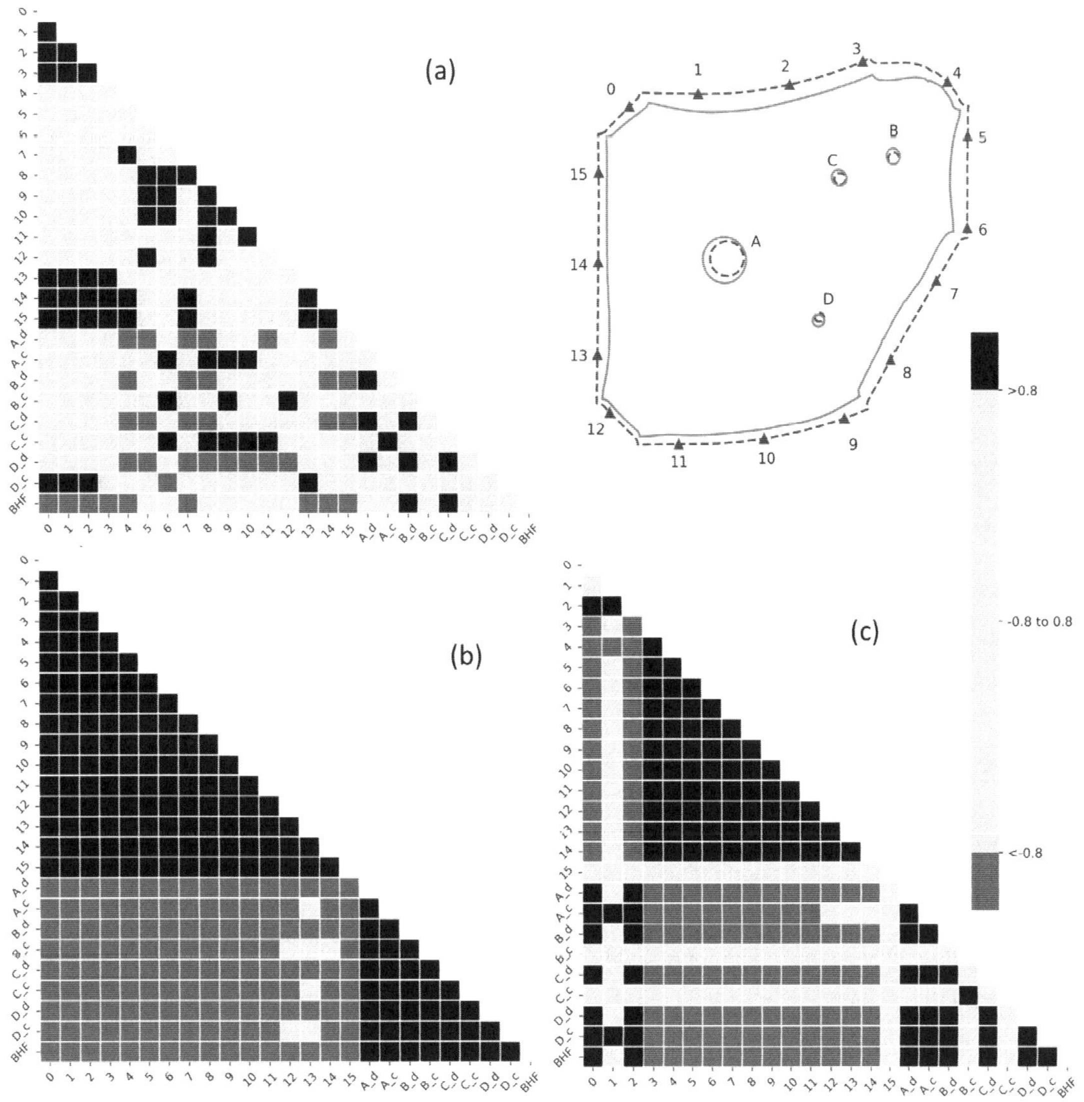

Figure 6. Pair correlation plots of the draw-in, parameters of the holes and blank holder force (BHF) for: (a) experiments, (b) FEM uniform model and (c) FEM mapped model.

The pair correlation plots of the experimental and numerical data are presented in figure 6. The global control parameter of the process, the blank holder force was also considered in the data set (labelled as BHF in figure 6). Strong correlation (above 0.8 and below -0.8) are highlighted in black or dark grey, while moderate and weak correlations appear in light grey. Experimental data (Figure 6a) reveal a heterogeneous correlation pattern, with some parameters like the 0 to 3, 13 to 15 locations, and diameters of holes B and D strongly linked to BHF. A significant number of the considered parameters are relatively weak correlated with the blank holder force and to each other (the light grey) areas in figure 6a. The FEM uniform model (figure 6b) show a very strong pair correlation between almost all considered parameters. A reduced pair correlation is observed between the position of the holes (e.g., A_c, etc.) and some draw-in parameters (12, 13 or 14). The FEM mapped model exhibits a more nuanced correlation pattern as compared to FEM uniform model. Most of the draw in parameters are strongly linked to the blank holder force and consequently to each other. Parameters that are not influenced by the blank holder force (e.g., draw-in locations 1 and 15 or the position of the holes B and C) are also weaker correlated with the rest of the parameters.

4. Discussion

The results presented in figure 2 and figure 3 illustrate a practical methodology to extract a manageable representative set of parameters from the continuous functions representing draw-in variations and hole edge modifications. The selected draw-in locations, adjacent to the positions of the equalizer blocks in the tool, are most affected by the common practice of shimming in press shops. These locations display the highest sensitivity to adjustments in local restraint. The irregular shape of holes in the deformed parts, differing from perfect circles was approximated by two parameters, an average diameter increase and the relative position of the corresponding centre. This reduced set of parameters provides a good estimate of the average in-plane dimensions of the deformed part. The reverse engineering effort is primarily aimed at predicting this set of parameters. Accurate prediction is a critical step before analysing plastic strain distribution and comparing strain measurements to simulations. For both draw-in and hole diameter, as well as position predictions, further refinement is needed to enhance accuracy (an example of accuracy status is illustrated in figure 4), which is essential for qualitative strain comparison of measured strains and strains predicted by the models.

Incorporating measured contact pressure distribution (figure 1) into the analysis alters the restraint on material flow and improves accuracy, though further refinement is needed for a closer alignment between measurements and simulations. The discrepancy between experimental draw-in sensitivity and model predictions (figure 5) necessitates additional examination of methods to emulate the elastic effects of actual tooling, which is typically present in real-world forming tools but absent in rigid virtual tooling models. While the models can capture major trends, finer interpolation or empirical scaling may be necessary to reconcile measured and predicted values. It is interestingly to note that the consistently lower sensitivity to draw-in variations observed in practice suggests that the actual process might be less affected by fluctuations than the models imply.

Pair correlation analysis indicates significant differences between finite element model outcomes and experimental data. In the case of the finite element model featuring a uniform contact pressure distribution, there is almost a complete correlation, as shown in figure 6b. This suggests a high level of predictability across various parameters, indicating that a single parameter might be sufficient to predict the overall behaviour. However, this stands in stark contrast to what is observed experimentally (figure 6a), where the correlations are considerably weaker. The diminished correlations in the experimental data suggest the necessity of monitoring multiple parameters, possibly even all, to gain a comprehensive understanding of the forming process. Incorporating more realistic pressure distributions into the finite element model results in a decrease in correlation strength, as seen in figure 6c. Despite these adjustments, the model's outcomes still significantly differ from the experimental observations, indicating a need for further refinement of the finite element model to more accurately mirror the experimental data.

The data discussed are based on a limited set of experimental observations which were gathered using panels from the production line. To enhance the understanding of the actual process, more extensive experimental data is required. This could be achieved by employing an optical monitoring system that operates non-intrusively during series production. Collecting such data is crucial for the advancement of monitoring and control strategies, both empirical and model-based, within the process control field.

5. Conclusions

In this paper, challenges associated with the reverse engineering of an existing sheet metal forming process for serial production of automotive parts are highlighted. A methodology has been introduced for simplifying complex continuous parameters, which enhances the efficiency of data analysis. The limited accuracy of reverse-engineered models under relatively large variations of blank holder force settings has been revealed. The incorporation of measured contact pressure distributions into the model qualitatively enhances its overall accuracy. A more realistic contact pressure distribution has led to a weakening in correlation strength, aligning more closely with experimental observations. The emphasis on accuracy improvement underscores the need for realistic contact pressure data to enhance predictive abilities.

However, key discrepancies in experimental model predictions, particularly in draw-in sensitivity and pair correlations, have been identified. Several strategies could address these in the short term. One approach involves enhancing local restraining through iterative tuning using available FEM capabilities for location-specific adjustments. Additionally, creating distinct models for interpolation of smaller blank holder intervals could achieve greater precision in data analysis, aligning more closely with serial production practices. Empirical scaling offers another short-term solution, allowing adjustments based on observed data and practical outcomes.

In the long term, the development of methodologies to accurately replicate the elastic effects observed in actual tooling is identified as a crucial need. These effects, common in real-world forming tools, are challenging to approximate with current rigid tooling models and are labour-intensive. This underscores the necessity for innovation in tooling model development to more closely mirror real-world conditions.

A more realistic guidance for improving the model necessitates additional experimental data. The importance of gathering this data extends beyond model enhancement; it deepens the understanding of the forming process as a whole. This understanding is crucial for advancing the development of process monitoring and control strategies, ultimately leading to more efficient, safer, and cost-effective automotive manufacturing processes.

References

[1] Abspoel M, Scholting M E, Lansbergen M and Neelis B M 2017 Accurate anisotropic material modelling using only tensile tests for hot and cold forming *J. Phys.: Conf. Ser.* **896** 012049
[2] Hol J 2013 Multi-scale friction modeling for sheet metal forming, https://doi.org/10.3990/1.9789077172988
[3] Sigvant M, Pilthammar J, Hol J, Wiebenga J H, Chezan T, Carleer B and van den Boogard A H 2018 Friction in Sheet Metal Forming Simulations: Modelling of New Sheet Metal Coatings and Lubricants *IOP Conf. Ser.: Mater. Sci. Eng.* **418** 012093
[4] Tatipala S, Pilthammar J, Sigvant M, Wall J and Johansson C M Introductory study of sheet metal forming simulations to evaluate process robustness 2018 *IOP Conf. Ser.: Mater. Sci. Eng.* **418** 012111
[5] Pilthammar J, Sigvant M and Kao-Walter S 2018 Introduction of elastic die deformations in sheet metal forming simulations, *International Journal of Solids and Structures* 151 p76-90
[6] Pilthammar J, Wall J and Sigvant M 2017 Framework for Simulation-Driven Design of Stamping Dies Considering Elastic Die and Press Deformations, *PROCEEDINGS OF ESAFORM 2017 / [ed] Brabazon D.,Ul Ahad I.,Naher S., American Institute of Physics (AIP)* Vol. 1896
[7] Abbasi F, Sarasua A, Trinidad J, Otegi N, de Argandoña E S and Galdos L 2021 Substitutive Press-Bolster and Press-Ram Models for the Virtual Estimation of Stamping-Tool Cambering *Materials*, **15(1)** 279
[8] Pilthammar J et al. Three Industrial Cases of Sheet Metal Forming Simulations with Elastic Dies 2023 *IOP Conf. Ser.: Mater. Sci. Eng.* **1284** 012055
[9] VDA 239-100 (05/2016)
[10] www.tatasteeleurope.com
[11] www.fuchs.com

Experimental Measurement of the Stress Gradient in a Bent Sheet

Audrie Corral[1], Eric Colllie[1], Zhaojun Lu[1], Grayson Womble[1], and Chester J. Van Tyne[1,2]

[1]Department of Materials Science and Engineering, Virginia Polytechnic Institute and State University, Blacksburg, VA 24060, USA

[2]Department of Metallurgical and Materials Engineering, Colorado School of Mines, Golden, CO 80401, USA

E-mail: cvantyne@mines.edu

Abstract. Stress gradients were experimentally measured in a 316 stainless steel sheet after bending with a press brake. Two different tool radii and three different bend angles were used to produce plastically bent sheets. Microhardness tests were performed across the thickness of the bent sheet, and a correlation between microhardness and strength was experimentally determined and was used to obtain the stress gradient. The position of the experimentally determined neutral plane correlated well with theory. Theoretical calculations of the stress gradient were made using a power law constitutive equation along with a simple mechanics analysis of plastic bending. Four of the six bending experimental conditions had reasonable correlation with the theoretical calculations. Ideas for improving this experimental method are discussed.

1. Introduction

During stamping operations, sheet metal is often bent and undergoes springback [1] when released from the die. In order to determine the amount of springback, the stress states present through the sheet after the bending process need to be known, as well as the unloading modulus. Often the stresses through the thickness of the sheet are not well modeled, especially in membrane models. The purpose of this brief study was to experimentally measure the stresses through the thickness of a bent sheet.

2. Experimental Methods

The test material was a 316 stainless steel 6.35 mm thick sheet. This material is a single-phase austenitic steel and has reasonable strain hardening. Test strips of 127 mm by 12.7 mm were machined from the sheet. The initial hardness of the sheet was 170 ± 8 HK. The test strips were bent on a press break using two different tool radii (9.53 mm and 6.35 mm) to three different bend angles (90°, 115°, and 135°). Figure 1 shows a schematic of the press break bending operation [2].

The bent regions were cut from the bent sample, mounted in Bakelite, and polished to 1000 grit. One hundred Knoop microhardness tests were performed across the thickness of the bent region. The distance between the indentations was approximately 0.06 mm. To prevent overlap of the indentation fields, two adjacent traverses were made across the thickness of the sheet. After the second traverse, the sample was re-polished and a second series of microhardness tests was performed.

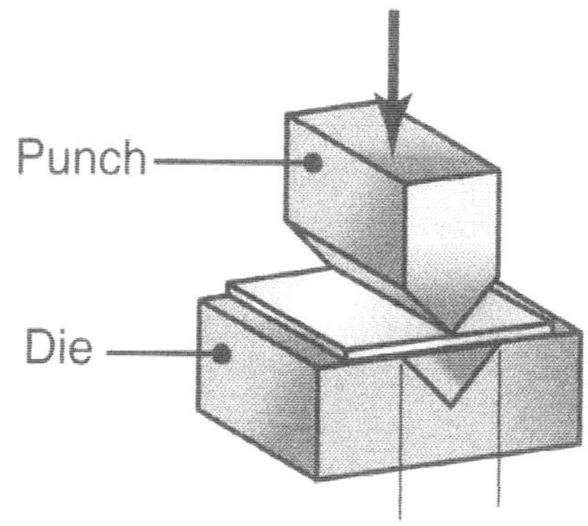

Figure 1 Schematic of the press brake process used to bend the sheet.

To obtain a correlation between the microhardness values and the stress in the steel, a set of additional tests was performed. A standard tensile test was performed to determine the simple power law hardening parameters. Additional tensile tests were performed in a stepped manner. The sample was strained to a known amount, removed from the tensile testing machine, and five Knoop microhardness tests were made. It was returned to the tensile testing machine and strained further followed by a microhardness tests. This process was repeated nine times. A correlation between stress and hardness was obtained. This method is similar to the one used by Pavlina and Van Tyne [3]. From the correlation between stress and hardness, the microhardness test values through the thickness of the sheet were converted to stress values to give the stress gradient through the thickness of the bent sheet.

3. Results

Figure 2 shows the hardness profiles for the six different test conditions. Please note that there were two series of tests through the thickness of the sheet for each condition. The position of the neutral plane was determined by an empirical equation [4].

Figure 3 shows the engineering stress-strain curve for the 316 stainless steel.

Figure 4 gives the experimentally measured correlation between true stress and Knoop hardness value for 316 stainless steel.

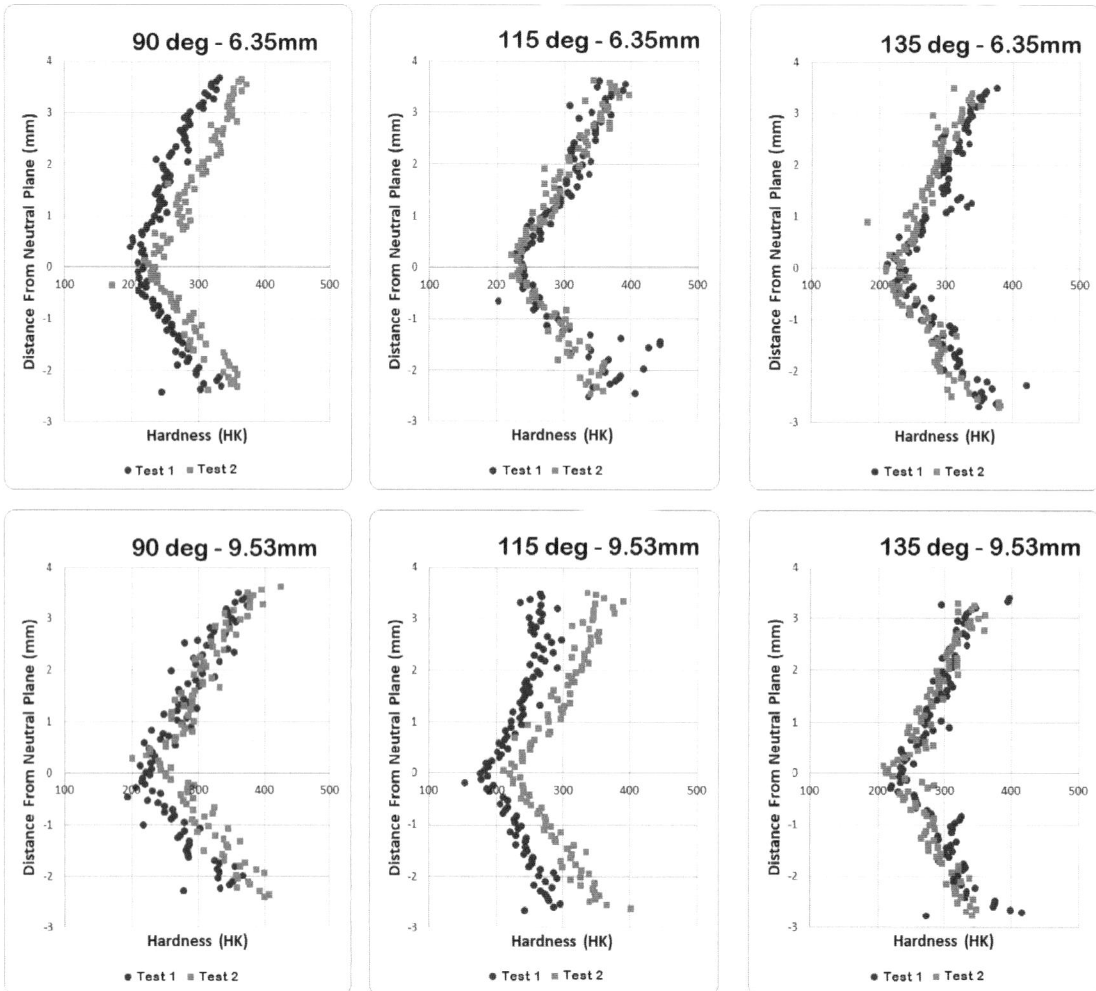

Figure 2. Through thickness hardness profiles for 316 stainless steel bent to various bend angles using two different tool radii. Bend angle and tool radius are shown at the top of each figure.

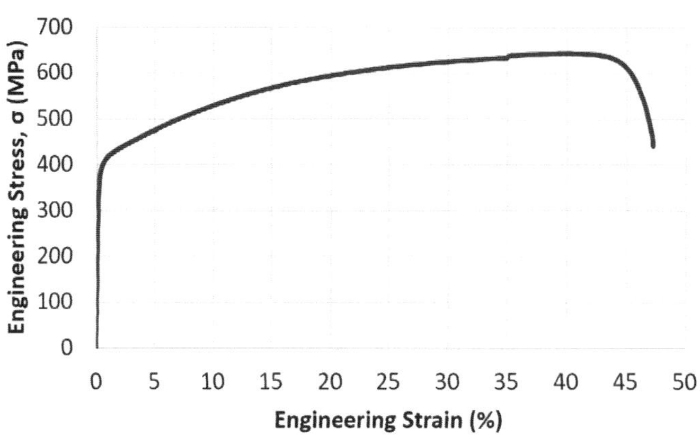

Figure 3. Stress strain curve for 316 stainless steel.

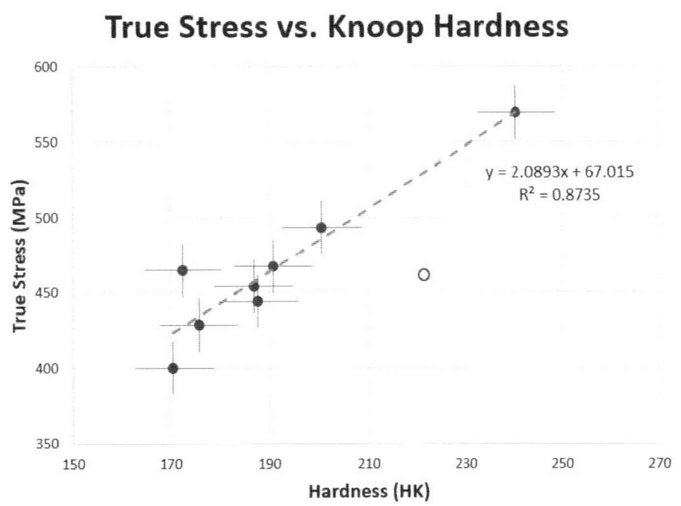

Figure 4. Correlation between microhardness and stress for 316 stainless steel.

4. Analysis of results

Given the stress-strain behavior shown in Figure 3, it was possible to determine the parameters for the power law hardening model. The power law hardening model is

$$\sigma = K\varepsilon_{eff}^{n} \tag{1}$$

where K is the strength coefficient, n is the hardening exponent, and ε_{eff} is the effective strain. The values found for the 316 stainless steel were $K = 1238$ MPa and $n = 0.32$. These values compare reasonably well with values reported in the literature of $K = 1255$ MPa and $n = 0.359$ [5].

From the radius of curvature, ρ, and the distance from the neutral plane, y, the bending strain, ε_b, can be determined as [6]

$$\varepsilon_b = \ln\left(1 + \frac{y}{\rho}\right) \tag{2}$$

For the plane-strain bending the bending strain can be converted to effective strain as

$$\varepsilon_{eff} = \frac{2}{\sqrt{3}}\varepsilon_b \tag{3}$$

By using the power law equation, a theoretical stress gradient through the thickness of the sheet can be calculated.

Figure 5 shows the experimental hardness data converted to stress as a function of the distance from the neutral plane. The theoretical stresses calculated from the power law equation are also shown.

It should be noted that the stresses that have been determined are the flow stress of the material at each point through the cross section of the sheet.

IOP Conf. Series: Materials Science and Engineering **1307** (2024) 012041 doi:10.1088/1757-899X/1307/1/012041

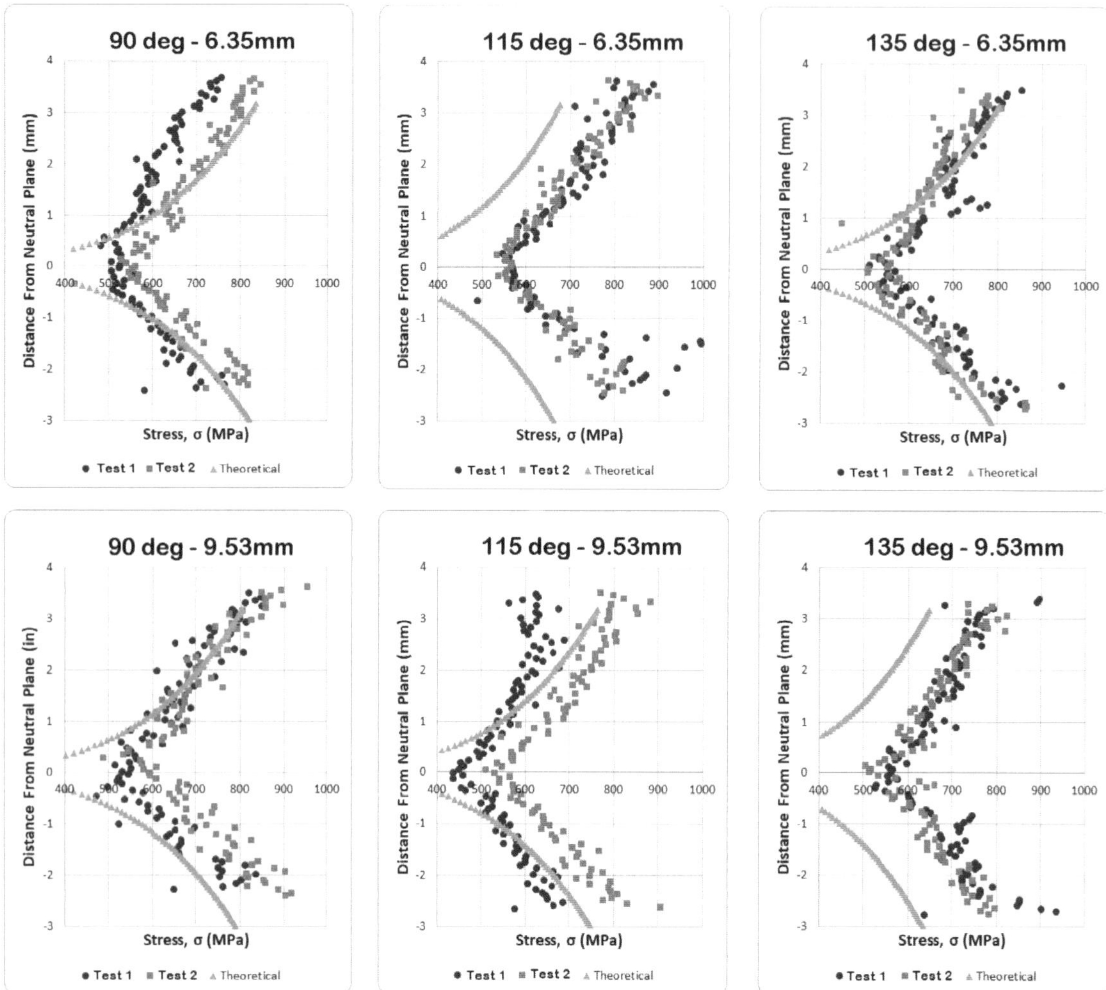

Figure 5. Through thickness stress profiles for 316 stainless steel bent to various bend angles using two different tool radii. Experimental points are calculated from the stress-hardness correlation. The theoretical profile is obtained from the power law calculation. Bend angle and tool radius are shown at the top of each figure.

5. Discussion

The experimentally determined stress gradients, as shown in Figure 5, exhibit the type of shape that would be expected. During bending, the inner region of the sheet is compressed plastically and the outer region is stretched plastically. These regions of plastic deformation show an increase in the stress level. The stress level increases from the neutral plane to the inner surface and from the neutral plane to the outer surface. These are expected trends and the data confirm this behavior.

In all cases, the neutral plane coincides with the minimum in the stress gradient. This is consistent with the fact that the neutral plane is the plane that is neither compressed nor stretched, so there is no plastic deformation. The stress value at the neutral planes is approximately 500 MPa, which is about 20% higher than the yield stress of 400 MPa shown in Figure 2. Part of the reason for this discrepancy is the bending deformation is in plane strain not uniaxial tension

Although the power law model is a simplistic model, it does a reasonable job of matching the trends in the experimental data. Since the model does not demonstrate any elastic behavior, the region around the neutral plane will have the most discrepancy from the experimental data, since it is the region where there is a minimal amount of plastic deformation. The model fits well in four of the six test conditions, but the reason for the two data sets with discrepancies is not known.

It is obvious that the microhardness data has a fair amount of scatter. The grain size of the 316 stainless steel was about 50 microns; it is likely that the tests were often of a single grain region rather than from a polycrystalline conglomerate. Since hardness will vary with crystal orientation, some scatter is to be expected.

Although a good amount of effort was exerted in determining the correlation curve between hardness and stress, the data shown in Figure 3 is not as good as desired. The uncertainty ranges for both hardness and stress are fairly large and the regression equation shows only an adequate fit of the data. This is one area of the method that needs the greatest improvement.

6. Summary

An experimental method has been developed for determining the stress gradient through the thickness of a bent sheet. The experimental stress values have the expected trends, have consistency with the neutral plane, and, in a number of cases, are consistent with a simple power law model for bending. The method described does need further refinement, especially in the correlation between hardness and stress, which is a necessary step in converting the hardness measurements into stress values.

7. Acknowledgements

This work was part of a capstone senior design project in the Material Science and Engineering program at Virginia Tech. The expertise and help from Profs. Staley and Burgoyne were instrumental in bringing this project to completion.

References

[1] R. H. Wagoner and M. Li, Springback, ASM Handbook, Volume 14B: Metalworking: Sheet Forming, ASM International, 2006, p 733–755, doi: 10.31399/asm.hb.v14b.a0005131

[2] S. Kalpakjian and S.R. Smith, Manufacturing Engineering and Technology, 6th ed. Prentice Hall, 2010, p. 400.

[3] E.J. Pavlina and C.J. Van Tyne, Correlation of Yield Strength and Tensile Strength with Hardness for Steels, Journal of Materials Engineering and Performance, 2008, 17(6), p 888–893, doi: 10.1007/s11665-008-9225-5

[4] https://www.cmrp.com/blog/plate-rolling/the-neutral-access-and-considerations-for-bending-plate.html

[5] M. Y. Demeri and S. Lampman, Forming of Carbon Steels, ASM Handbook, Volume 14B:Metalworking: Sheet Forming, ASM International, 2006, p 495–529, doi:10.31399/asm.hb.v14b.a0005162

[6] Z. Marciniak, J. Duncan and S. Hu, Mechanics of sheet metal forming, 2nd ed. Oxford:Butterworth-Heinemann, 2002, p. 82-107.

IOP Conf. Series: Materials Science and Engineering **1307** (2024) 012042 doi:10.1088/1757-899X/1307/1/012042

New test rig for biaxial and plane strain states on uniaxial testing machines

Edgar Marker[1], Lorenz Maier[1], Maximilian Gruber[1], Wolfram Volk[1] and Christoph Hartmann[1]

[1] Chair of Metal Forming and Casting, TUM School of Engineering and Design, Technical University of Munich, Walther-Meißner-Str. 4, 85748 Garching, Germany

edgar.marker@tum.de

Abstract. The increasing interest in hydrogen technology and storage systems has also significant influence on material characterization. Storage solutions such as cryo-compressed hydrogen (CcH2) or compressed gaseous hydrogen (CGH2) use high pressure to improve the energy density. The pressurized hydrogen is stored in a pressure vessel. This leads to various strain states in the vessel, which are difficult to reproduce with standard testing equipment, such as a uniaxial testing machine. This contribution presents a new test rig for uniaxial testing machines that is able to generate a biaxial load on a cruciform specimen. A hinged structure converts the uniaxial load into a biaxial load. Different attachments allow both equi-biaxial and plane strain states. In addition, the test rig provides access to both sides of the specimen, extending the possibilities for testing under load, such as optical measurement, specimen cooling or permeability measurement. Finally, experiments show the functionality of the test rig and the desired strain states in the specimen.

1. Introduction

The renunciation of fossil fuels and energy has major technological and economic implications. New solutions for energy supply and storage need to be found and developed. Mobility applications are particularly challenging in this regard. Limitations of electric drives with batteries, such as low range and long charging cycles, prevent the use of this technology in certain applications, such as heavy-duty transport or aviation [1]. Other technologies are therefore being investigated. One example of such a technology is hydrogen as an energy carrier. However, hydrogen has drawbacks that need to be addressed. A major problem is its low volumetric energy density compared to fossil fuels [2]. Several technologies are being developed to overcome this problem. Possible approaches include pressurizing the hydrogen up to 700 bar or cooling the hydrogen down to 20 K to increase the energy density [3]. The cryo-compressed hydrogen (CcH2) technology uses a combination of high pressure and low temperature to further improve the energy density [4]. Due to the geometry of most pressure vessels, they are subject to a wide range of strain states, including equi-biaxial strain and plane strain. Therefore, material characterization for these strain states preferably with thermo-mechanical load is required [5]. There are well-established experiments for testing both biaxial and plane strain [6–8]. However, they often require specialized equipment such as biaxial testing machines or sheet metal testing machines, which are complex and expensive. The availability of these machines is therefore limited. On the other hand, uniaxial testing machines are very common in most laboratories. This paper presents a new test rig that allows cruciform specimens to be tested on uniaxial testing machines with consideration of additional capabilities for other testing requirements.

2. State of the Art

There are a variety of concepts and designs for converting uniaxial tension into biaxial tension. Their common goal is to enable biaxial testing on uniaxial testing machines. These concepts can be divided into three classes of solutions. The first class uses hinged structures. The second class of solutions is based on sliding surfaces. Finally, the third class of solutions uses different approaches such as rotating elements or cables. These solutions have been analysed to some extent by the scientific community [9,10]. Nevertheless a brief summary of existing designs is required.

2.1. Hinged structures

Ferron et al. developed a hinged structure to allow biaxial testing of cruciform specimens on uniaxial testing machines [11]. In this design, the vertical and horizontal axes are connected by a cross joint out of the specimen plane. Tasan et al. used a similar design for in-plane biaxial loading up to failure of cruciform specimens [12]. A more sophisticated mechanism for free adjustment of stress ratios has been proposed by Bhatnagar et al. [13]. In contrast to the design of Ferron et al., the structure lies completely in the specimen plane. It combines a hinged structure and sliding elements to achieve the desired stress ratios. Barroso et al. transform a compressive load to induce tension in a cruciform specimen [14]. The specimen plane is perpendicular to the load axis of the uniaxial testing machine. A further development of this test rig also allows tension-compression loading [15].

2.2. Sliding surfaces

Albertini proposed a test arrangement for quasi-static biaxial testing using levers moved by a thrust plate [16]. The mechanism is not specifically designed for use in a uniaxial testing machine but could be adapted. Lebedev and Muzyka published a design that allows biaxial tension in cruciform-like specimens on uniaxial testing machines [17]. It uses rigid in-plane elements on which the grips can slide. Lee and Chien presented a test rig that uses triangular sliding surfaces to couple the vertical and horizontal axis of the test apparatus [18]. A similar design is presented by Puente Medellín et al. [19]. Zhao et al developed a test rig that uses four individual sliding surfaces for each arm of the cruciform specimen [20]. When a top plate is pushed down, all four arms move. Dhara et al. designed a test rig for strain path changes [21]. The surfaces are designed to delay the application of load in one axis, resulting in a strain path change.

2.3. Other solutions

Hanabusa et al. presented a test rig using a rotating plate with links to convert uniaxial loading into biaxial loading [22]. The same approach was used by Shao et al [23]. Merklein and Biasutti presented a new design where the specimen clamps are mounted on an upper plate with a translational degree of freedom. At the same time, the clamps are connected to a fixed lower plate by cables. When the upper plate is moved upwards, a tensile force is applied to the specimen [24]. Although this design is considered a stand-alone solution, it could be adapted for uniaxial tensile testing machines.

3. Design of the new test rig

This chapter deals with the design process of the new test rig. Firstly, the requirements for the new design had to be derived. Secondly, a concept had to be developed based on the requirements. And finally, the design had to be worked out.

3.1. Requirements

The proposed test rig solution is based on three experimental challenges. Ultimately, the test rig will be used for material testing under biaxial and plane strain conditions, under cryogenic conditions, and for simultaneous permeation measurement. Although these experiments are not yet possible, they had to be

IOP Conf. Series: Materials Science and Engineering **1307** (2024) 012042 doi:10.1088/1757-899X/1307/1/012042

considered in the design of the test rig. Therefore, the requirements for the new design were derived from these main objectives. It is obvious that two requirements are the ability to obtain both biaxial and plane strain in the specimen. In addition, the effort required to switch between the two strain states should be manageable. This means that the change should be possible when mounted in the uniaxial testing machine. Due to the other objectives of cooling and permeation measurement, there should be an interface for additional test equipment. This interface should ensure the correct position of the additional test equipment in the center of the specimen. The permeation measurement makes another requirement necessary. The test rig must provide access to both sides of the specimen. Finally, the test rig should be designed for a variety of materials such as steel, aluminium and carbon fiber reinforced plastics (CFRP). The latter in particular requires high forces. Therefore, the test rig should be designed for a maximum total force of 400 kN.

3.2. Concept Design

The above requirements preclude the use of most existing test rigs. Therefore, a new design had to be developed. First of all, a mechanism had to be found to convert uniaxial motion or load into biaxial motion or load. According to the state of the art, most concepts are based on a hinged structure or sliding surfaces. Due to the high total force of 400 kN and the resulting high surface pressure, sliding surfaces are a poor solution in this case. For this reason, a hinged structure was preferred. The simplest solution uses four interconnected hinges to couple a vertical with a horizontal movement. A schematic diagram with constraints from the uniaxial testing machine is shown in **Figure *1***.

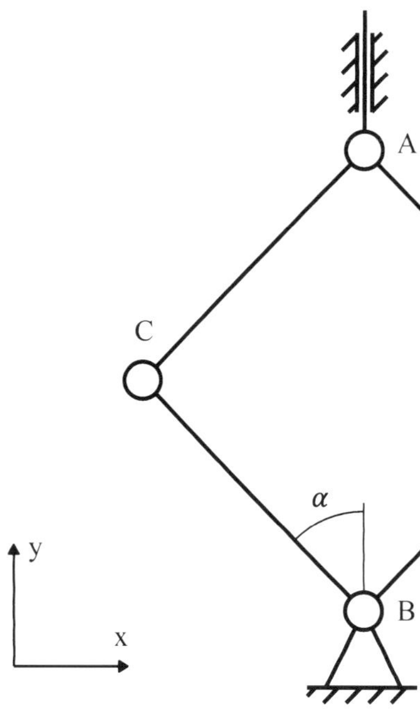

Figure 1. Schematic diagram with constraints from a uniaxial testing machine

The shown nomenclature applies for the whole paper. Points A to D represent the hinges in the structure. They are connected by rods of length l. The uniaxial testing machine provides a fixed base and a traverse which can move in the y-direction. Point B is connected to the base of the machine and is therefore fixed. Point A is connected to the moving traverse. The positions of points C and D can be calculated by knowing the distance between A and B and the length l. This concept has an inherent advantage. The y coordinate of points C and D must be exactly halfway between A and B. This allows the correct positioning of the two grips and additional equipment when point A (traverse) moves upwards.

In this state, the structure is not suitable for biaxial tension. As points A and B move away from each other, points C and D move towards each other. To solve this problem, points C and D must be connected to the opposite grip. This will give the desired behavior of synchronous movement of the four grips away from each other.

The same mechanism can also be used to achieve a plane strain state. Only the connection to the grips has to be adapted. Instead of connecting points C and D to the opposite grips, the distance between the horizontal grips is fixed. This restricts strain in horizontal direction while still allowing A and B to move away from each other. Points C and D are only used to position the grips exactly halfway between A and B. A long hole can provide the necessary degree of freedom (DoF) in x-direction. **Figure *2*** shows a CAD model of both biaxial and plane strain setup.

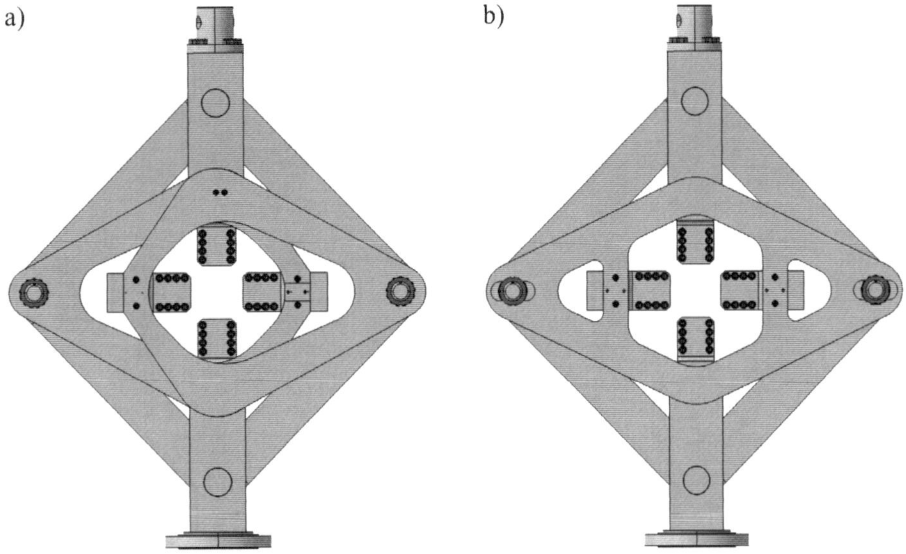

Figure 2. CAD model of the test rig with biaxial (a) and plane strain (b) setup

It can be seen how the points C and D are coupled to the grips in the biaxial and the plane strain setups.

A slight disadvantage of this concept is a non-linear motion characteristic. The deviation from an optimum, i.e. linear motion, is shown in **Figure *3***. The origin represents an angle of $\alpha = 45°$.

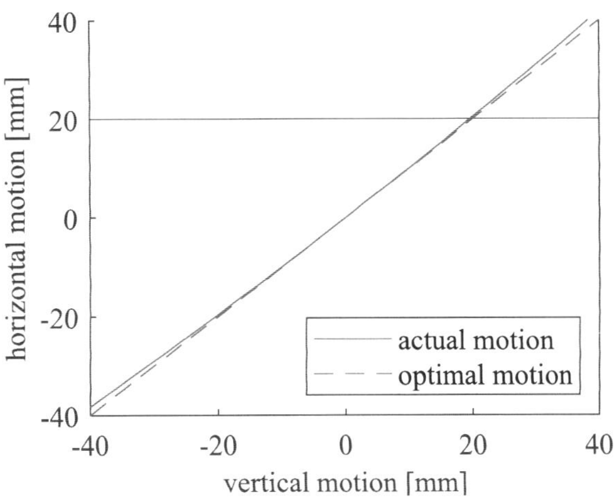

Figure 3. Comparison of actual test rig motion and optimal motion

IOP Conf. Series: Materials Science and Engineering **1307** (2024) 012042 doi:10.1088/1757-899X/1307/1/012042

3.3. Final design

The final design of the test rig was created using CATIA Version 5-6 Release 2014 CAD Software. The basis of the design is the hinged structure from the concept. The points A and B are connected to the vertical grips, which also act as connectors to the uniaxial testing machine. Adapters on the upper and lower grips allow this test rig to be used in different machines. In this case, the test rig was mounted in a 200 kN universal testing machine from Zwick Roell´s Allround line. The base structure without any mounted setup, the plane strain setup and the biaxial setup can be seen in **Figure *4***.

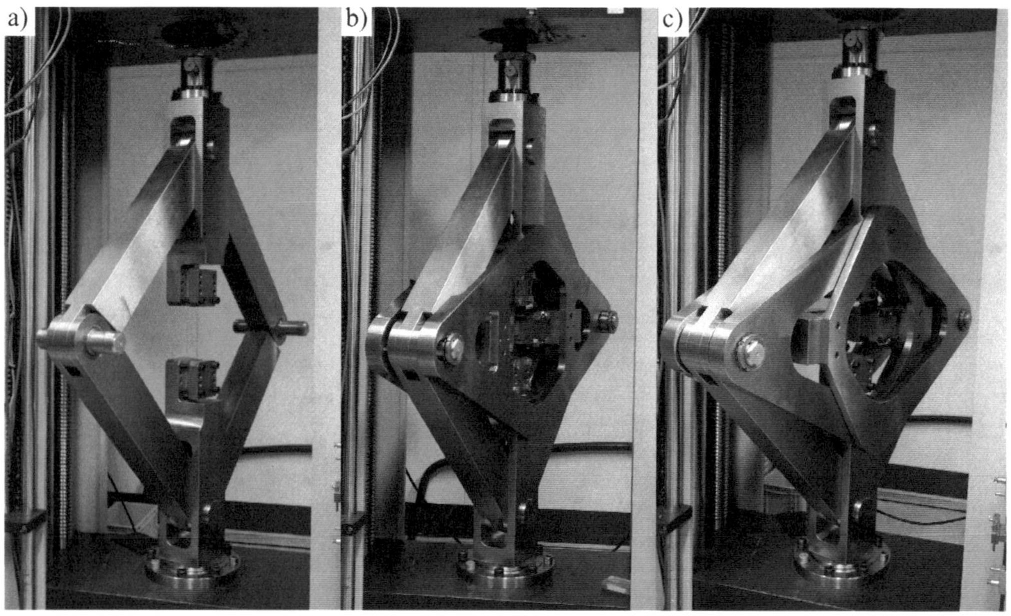

Figure 4. Base structure without setup (a), plane strain setup (b) and biaxial setup (c)

The base structure remains the same for both setups and therefore does not need to be disassembled. The axles at points C and D are designed to allow both the biaxial setup as well as the plane strain setup to be mounted from the side. This meets the requirement for manageable effort to chang the setup. The setup is rotated by 20 degrees inside the uniaxial testing machine to improve accessibility.

A special design for the coupling of points C and D to the specimen grips for both the biaxial and the plane strain setups ensures access to both sides of the specimen. As shown in ***Figure 4a***, there is an outer coupling (D) and an inner coupling (C). They can move against each other to allow biaxial movement of all four grips. In addition to their connection to the Points C and D, they are guided between each other by a linear guide to ensure a purely horizontal movement.

The grips are geometrically fixed on both setups to ensure correct positioning. They allow specimens to be clamped by means of eight M12 screws per grip. The available clamping surface is 50 mm x 89 mm and is textured. This large surface area reduces surface pressure and protects materials such as CFRP from damage in the clamping area.

Threaded holes are provided in both setups for mounting additional equipment. For the biaxial setup a horizontal DoF must be taken into account to allow the movement of the grips without moving the additional equipment.

The dimensioning of the parts was mainly done by explicit finite element (FE) simulation. The software used was ABAQUS/CAE 2022. The simulations were performed for a Young's modulus of 210,000 MPa and a Poisson´s ratio of 0.3. Simulations were run for each structural part of the test rig. As expected, the coupling of the horizontal axis and the axles of the hinges were the most stressed parts.

In particular, the geometry of the coupling had to be adapted to meet the requirements of a maximum total force of 400 kN. Eventually the stresses were sufficiently reduced.

4. Proof of concept

Cruciform specimens were used to demonstrate the ability of the test rig to induce biaxial and plane strain in the specimen. The specimens were laser cut from the commonly used deep drawing steel DC04. The geometry was taken from the ISO/FDIS 16842. The specimens were gritted prior to the experiment in order to measure the plastic deformation afterwards. **Figure 5** shows the specimens before and after the experiment with both setups.

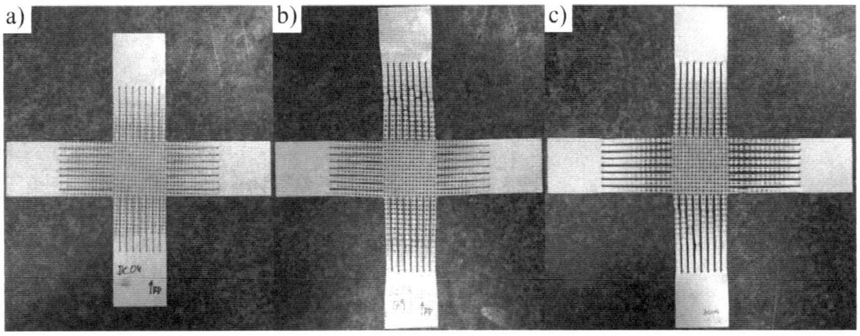

Figure 5. Specimen before (a), after plane strain (b) and after biaxial (c) experiment

All experiments were carried out with the rolling direction oriented in the load direction of the machine. Measurements were made using the GOM ARGUS system. It uses the initial grid geometry to calculate the major and minor strain after deformation. Three experiments were carried out for both setups. The results are shown in **Figure 6**.

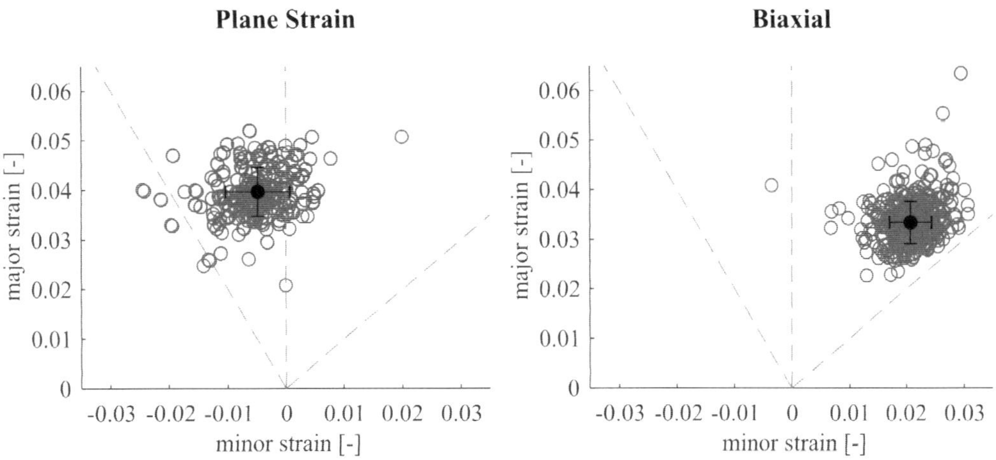

Figure 6. Results of the strain measurement with GOM ARGUS

The blue circles show the major and minor strain of each selected grid point. The black circle represents the mean value of all measurements and is shown with the standard deviation for major and minor strain. All points within a 20 mm diameter circle at the center of the specimen were selected.

The results show a slight deviation from the desired strain states. A possible explanation is the anisotropic behavior of the chosen material and the measurement after failure of the specimen. Overall, it can be concluded that the test rig allows cruciform specimens to be tested under both quasi plane strain and quasi biaxial conditions.

IOP Conf. Series: Materials Science and Engineering **1307** (2024) 012042 doi:10.1088/1757-899X/1307/1/012042

5. Conclusion and outlook

The results show that the test rig can be successfully used in a uniaxial testing machine to allow plane strain and biaxial testing of cruciform specimens. Overall, this test rig provides a simple and inexpensive alternative to complex biaxial testing machines. It also has key features that differentiate it from similar test rigs, such as the easy change of setups, easy two-sided accessibility and its robust design. This makes it suitable for high loads and for testing of strong materials such as CFRP.

The test rig meets the special requirements arising from future testing needs such as permeation measurement of CFRP under load or cooling of specimens. The consideration of such tasks will allow easy integration of additional test equipment. Especially, the ability to access the specimen from both sides expands the possibilities for testing under load.

Nevertheless, further developments of the test rig are possible. At present, the only force measurement is provided by the uniaxial testing machine itself. This can be solved by indirect force measurement at the grips. Strain gauges can be used to correlate a strain at specific positions with the force applied. However, further investigation is required to implement such force measurement.

References

[1] Klell M, Eichlseder H and Trattner A 2018 *Wasserstoff in der Fahrzeugtechnik* (Wiesbaden: Springer Fachmedien Wiesbaden)
[2] Stetson N T, McWhorter S and Ahn C C 2016 *Compendium of Hydrogen Energy* vol 35 (Elsevier) pp 3–25
[3] Yang M, Hunger R, Berrettoni S, Sprecher B and Wang B 2023 *Clean Energy* **7** 190
[4] Xu Z, Zhao N, Hillmansen S, Roberts C and Yan Y 2022 *Energies* **15** 6467
[5] Reimer A, Hartmann C, Norz R, Sturm P and Volk W 2022 *NUMISHEET 2022* vol 86 eds K Inal, J Levesque, M Worswick and C Butcher (Cham: Springer International Publishing) pp 885–97
[6] Marciniak Z and Kuczyński K 1967 *International Journal of Mechanical Sciences* **9** 609
[7] Nakajima K, Kikuma T and Hasuka K 1968 *Yawata Tech Report* 8514
[8] Kohl D and Merklein M 2022 *KEM* **926** 947
[9] Hannon A and Tiernan P 2008 *Journal of Materials Processing Technology* **198** 1
[10] Andrusca L, Doroftei I, Barsanescu P D and Goanta V 2014 *AMM* **658** 3
[11] Wolfenden A, Ferron G and Makinde A 1988 *J. Test. Eval.* **16** 253
[12] Tasan C C, Hoefnagels JPM, Quaak G and Geers MGD 2008 *11th International Congress and Exhibition on Experimental and Applied Mechanics 2008* 1729
[13] Bhatnagar N, Bhardwaj R, Selvakumar P and Brieu M 2007 *Polymer Testing* **26** 154
[14] Barroso A, Correa E, Freire J, Pérez M D and París F 2012 *15th European Conference on Composite Materials: Composites at Venice, ECCM 2012*
[15] Barroso A, Correa E, Freire J and París F 2018 *Exp Mech* **58** 49
[16] Albertini C, Labibes K, Montagnani M, Pizzinato E V, Solomos G and Viaccoz B 2000 *J. Phys. IV France* **10** Pr9-161-Pr9-166
[17] Lebedev A A and Muzyka N R 1980 *A Device for Testing Flat Crosslike Specimens by Biaxial Tension*
[18] Lee R S and Chien T W 2014 *KEM* **626** 275
[19] Puente Medellín L F, Ramírez Elías V A, Balvantín García A d J, Vázquez Gómez P I and La Diosdado De Peña J A 2019 *J Braz. Soc. Mech. Sci. Eng.* **41** 2319
[20] Zhao K, Chen L, Xiao R, Ding Z and Zhou L 2019 *J Mater Sci* **54** 7231
[21] Dhara S, Taylor S, Figiel Ł, Hughes D, Shollock B and Hazra S 2022 *NUMISHEET 2022* vol 24 eds K Inal, J Levesque, M Worswick and C Butcher (Cham: Springer International Publishing) pp 287–300
[22] Hanabusa Y, Takizawa H and Kuwabara T 2014 *J. Japan Inst. Light Metals* **64** 142
[23] Shao Z, Li N, Lin J and Dean T A 2016 *Exp Mech* **56** 1489
[24] Merklein M and Biasutti M 2013 *Journal of Materials Processing Technology* **213** 939

Validation of pneumatic crack detection in hole expansion test

A Leonhardt[1], T Haas[1] and A Kunke[1]

[1] Chemnitz University of Technology, Institute for Machine Tools and Production Processes, Professorship for Forming and Joining, Chemnitz, Germany

E-mail: andre.leonhardt@mb.tu-chemnitz.de

Abstract. According to the standard ISO 16630, the hole expansion test (HET) has to be stopped at the instant when the first through-thickness crack appears at the edge of the hole. The detection of the crack is challenging, since the maximum opening width should not exceed 0.1 mm at the inner diameter according to the newest version of the standard from 2017. A pneumatic crack detection has been introduced in which a pressure drop is used to identify the through-thickness crack. In the present paper, this method for stopping the hole expansion test is validated. It was tested for steels with different edge crack sensitivities and compared to stopping with force drop method. Therefore, the curves of pressure and force have been analysed. The influence of hole insertion method (shear cutting or abrasive water jet cutting) as well as specific hole expansion parameters (punch speed, air pressure) have been investigated. Besides the expanded diameters to evaluate the edge crack sensitivity, the opening width of the edge cracks were analysed to determine the suitability of the crack detection methods for different hole expansion ratios (HER).

1. Introduction

A frequently recurring cause of rejects in cold sheet metal forming of high-strength steels are the so-called edge cracks which originate from the damaged cut edge during forming. In particular, high-strength multiphase steels, such as dual-phase or complex-phase steels, have a high edge crack sensitivity. The reason for this can be found in the special microstructure of these steels, which at the same time determines the advantageous application properties. However, other sheet metal materials also show a lower forming capacity at cut edges. The material is pre-damaged by the hardening associated with cold forming during the shear cutting of the blanks. If this edge is subjected to high tensile stresses within the subsequent forming process like deep drawing, micro or macro cracks will occur. For this reason, the edge crack sensitivity describes a local material-specific forming property, which cannot be determined with conventional testing methods like the tensile test or the Marciniak or Nakajima test [1]. The only standardized method for determining this parameter is the hole expansion test (HET) according to ISO 16630 [2]. A circular hole is inserted into a sheet metal specimen by shear cutting. This is followed by continuous expansion of the hole with a conical punch until at least one crack passes through the entire thickness of the sheet. The so-called hole expansion ratio (HER) λ as a characteristic value for the edge crack sensitivity is determined with the following equation

$$\lambda\ [\%] = \frac{D_1 - D_0}{D_0} \cdot 100 \quad (1)$$

in which D_0 is the initial and D_1 is the expanded diameter of the hole. One major issue of the test procedure is the stopping of the test. According to the standard, the test engineer is supposed to observe/monitor the test. As soon as a crack appears at the edge of the hole (through the entire thickness of the sheet), the test should be stopped. Despite low feed rates of the hole expansion punch of maximum 1 mm/s, the resulting stop of the test is affected by a high subjective influence of the test technician. In addition, the maximum permitted crack width of 0.1 mm at the inner diameter of the expanded hole cannot be reliably reproduced. This procedure also excludes the possibility of automation and possible use in series production. A Europe-wide interlaboratory test showed that the results of different test centres were not comparable with each other [3]. Therefore, major efforts have been made in recent years by research and industry to develop alternative crack detection methods for the HET [4]. The following variants were used:

- Force measurement: Detection of force drop during crack formation [5]
- Video capturing/digital image processing (DIP): Recording and evaluation of the complete test as well as crack detection by test engineer or software during or after the test [6-7]
- Digital image correlation (DIC): Crack detection by evaluation of displacements and strains [8]

In [9], a new crack detection method was introduced using a drop of air pressure. Instead of an open die, a closed pressure vessel was used. The initial pressure inside of the vessel was 6 bar which was measured by a pressure sensor. The test was stopped manually as soon as an air pressure drop of 0.03 bar occurred. This decrease in air pressure was taken from the sensor display by the test engineer. The functionality of the new stopping method was proven by HET with different metals and hole insertion parameters. The results were compared to optical crack detection performed by an experienced test engineer. The HET with pneumatic crack detection showed less pronounced through-thickness cracks up to a hole expansion ratio λ of approx. 35% and comparable results in the range of $\lambda \leq 100\%$. In the present paper, the pneumatic crack detection for HET is validated. Hence, the pressure sensor signal was implemented into the control of the testing machine to provide an automatic stop of the test. The pressure curves were recorded and analysed. Due to the adaption of the standard, the cracks were evaluated in terms of their width at the inner diameter. The results were compared to the stopping of the HET using a drop of force.

2. Methods and materials

2.1. Test setup

The experimental tests were conducted in the universal testing machine inspekt 150 kN from Hegewald & Peschke. A two-column frame is installed in the testing machine containing the hole expansion punch, the vessel die and the blank holder. The force sensor is implemented between the upper plate of the tool and the traverse of the machine. The pneumatic sensor is connected to the vessel die. The signals of both sensors are processed in the machine control. At the beginning of the test, the sheet metal with an inserted pre-hole is positioned on the die. The blank holder is placed above the sample without having contact due to spacers. The punch is inserted into the hole until the sample is centered and sealed. The movement of the punch is stopped to remove the spacers and to apply the blank holder force by screws. In case of pneumatic or combined crack detection, the compressed air line is opened until the required pressure is reached and then closed again. Subsequently, the punch movement is continued for hole expansion. Simultaneously the crack detection for drop of pressure and/or drop of force is activated. The test is conducted until the chosen crack detection method generates a stopping signal. The principle layout of the testing device and procedure for the HET with pneumatic crack detection is shown in [9].

2.2. Experimental studies

Two types of steel were used for validation of the pneumatic crack detection. The hot-dip galvanized dual-phase steel 1.0943 (HCT780XD; s = 2.0 mm) is a widely used advanced high strength steel (AHSS) in automotive industry. It combines a high strength level and sufficient formability at room temperature due to its microscopic composition of a soft ferritic matrix with embedded hard martensite. Dual-phase steel is a representative with high edge cracking sensitivity, primarily because of the differences in hardness of the two phases. The austenitic stainless steel 1.4404 (AISI 316L; s = 1.5 mm) shows lower edge crack sensitivity and is used in applications where high corrosion resistance is needed. The initial hole insertion was realized by shear cutting (SC) on the one hand and abrasive water jet cutting (AWJC) on the other hand. The diameter of the pre-hole was set to 10 mm according to the standard. For shear cutting, a punch diameter of 10.0 mm and punch velocity of 10 mm/s were applied. Despite the standard, the cutting clearance was set to 10% because of the sheet thicknesses and available tool dimensions. Therefore, the cutting die diameter was 10.3 mm for the 1.4404 and 10.4 mm for the 1.0943. The shear cutting was performed in a separate column-guided tool to ensure the alignment of the punch relative to the shear cutting die. The natural mineral abrasive Garnet with a mesh size of 120 was used for abrasive water jet cutting (i.e. size range of 125 – 300 µm). The water pressure was 3,800 bar, the feed rate was set between 500 – 600 mm/min and the mass flow rate of the abrasive media was adjusted to 350 g/min. The diameter of the jet is 0.76 mm and the focus distance between workpiece and focal tube is 2.7 mm. The plan for the series of experiments are summarized in Table 1. The standard parameter for the hole expansion test were a punch speed of 0.05 mm/s and an initial air pressure of ca. 2 bar. A conical punch with a diameter of 55 mm and a tip angle of 60° was applied. The values for the crack detection methods were determined experimentally. The test was stopped when the air pressure dropped by 0.0045 bar (within 5 data points). As the drop in air pressure was detected automatically using the control of the testing machine, smaller stopping values were applied compared to [9]. A decrease of the force value between 0.01 and 0.1 kN was used to stop the HET with the force drop method.

Table 1. Parameter sets for the test series.

Nr.	Material	Hole insertion	Crack detection method	Variable parameter
3.1	1.0943 / 1.4404	Shear cutting / abrasive water jet cutting	Pneumatic	Material / hole insertion method
3.2	1.0943	Abrasive water jet cutting	Pneumatic	Initial air pressure / hole expansion punch speed
3.3	1.0943 / 1.4404	Shear cutting / abrasive water jet cutting	Force / combined	Crack detection method

Beside the choice of steel material, the edge crack sensitivity was also influenced by the hole insertion method. Hence, the pneumatic crack detection was tested for different edge crack sensitivities which generally increases when shear cutting is used instead of abrasive water jet cutting (test series 3.1). The influence of the initial air pressure as well as the punch velocity on the crack detection method and the HER were investigated in test series 3.2. The air pressure was examined in the range of 1 bar to 6 bar and the punch velocity between 0.01 mm/s and 1.00 mm/s. In 3.3, the results for the pneumatic crack detection were compared to the test stop by a defined drop of force and a combined crack detection method was introduced. For calculation of the hole expansion ratio, the initial diameter after hole insertion as well as the final diameter after hole expansion were measured with the 3D microscope VHX-600 from Keyence. The width of the crack at the inner diameter of the expanded hole were recorded to evaluate the stopping method according to the demands of the standard. The maximum crack width was captured for every sample with the minimum distance between the two crack flanks being measured.

3. Results and discussion

3.1. Verification of pneumatic crack detection

The pneumatic crack detection was tested for the chosen steel materials. Four different HER levels were realized by using shear cutting and abrasive water jet cutting. The results for HER and the widths of the cracks are summarized in Table 2. A mean HER of 11.88% for shear cutting and 25.42% for abrasive water jet cutting were calculated for the steel 1.0943. The crack widths met the requirement according to the standard and were below 0.1 mm. For the 1.4404, the crack widths exceeded the demanded value by a multiple. Due to the higher HER of 1.4404, more material adheres to the conical punch. Consequently, the crack must grow to a greater depth in order for the compressed air to be released.

Table 2: Hole expansion ratios and crack widths for pneumatic crack detection.

Material	Hole insertion	Nr.	Hole expansion ratio (%)	Crack width (mm)
1.0943	Shear cutting	1	13.06	0.03
		2	12.16	0.02
		3	10.42	0.06
	Abrasive water jet cutting	1	26.13	0.04
		2	22.93	0.04
		3	27.19	0.02
1.4404	Shear cutting	1	85.50	0.67
		2	98.30	0.76
	Abrasive water jet cutting	1	not determined	> 1.00
		2	not determined	> 1.00

The crack characteristics are shown in Figure 1 marking through-thickness cracks with white arrows. A lot of cracks are present for the shear cut pre-hole in case of the steel 1.0943 (Figure 1(a)). These are distributed along the whole circumference of the hole. In contrast, there are only few cracks (1 – 3) if the pre-hole was inserted by abrasive water jet cutting (Figure 1(b)). This also applies to the steel 1.4404 where two cracks are visible with one being more pronounced (Figure 1(d)). Besides one pronounced crack, there are many cracks or neckings in case of the shear cut pre-hole for 1.4404 (Figure 1(c)).

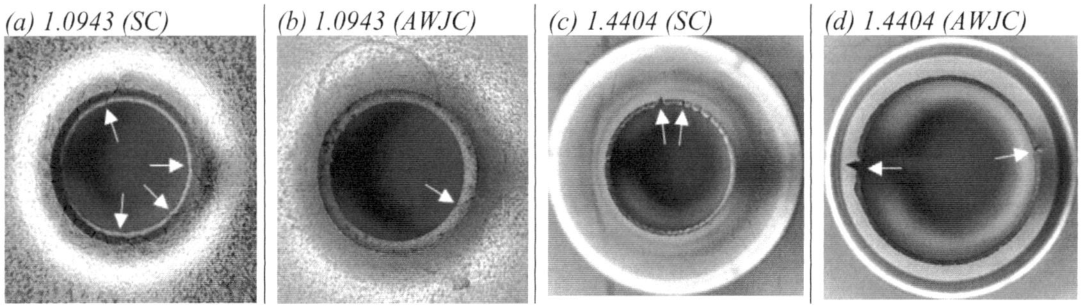

Figure 1: Crack characteristics after HET of 1.0943 and 1.4404 with pneumatic crack detection.

The recorded air pressure and force curves were analysed to check the pneumatic crack detection (Figure 2). It is apparent in all curves that the air pressure value increases in addition the hole expansion force due to the reduction of the volume inside the vessel die. In Figure 2(a) the typical curves for the 1.0943 with a shear cut pre-hole can be seen. The drop of the air pressure curve is recognisable just before the stop of the test. The drop of air pressure does not cause a force drop and is not directly connected with it. A drop in the force curve before the air pressure drop is evident in Figure 2(b) where abrasive water jet cutting of 1.0943 resulted in a higher HER of the 1.0943. This can also be found in test number 3 but not in number 1. A more pronounced decline of the force curve can be found for the hole expansion test of 1.4404 with pneumatic crack detection independent of the pre-hole insertion method. It can be assumed that pneumatic crack detection is suitable for HER up to approx. 50% using the equipment according to the standard. A force drop is detected close or before the decline of the air pressure curve for HER > 25%. This drop in force is more distinctive and occurs sooner than the drop in air pressure for increasing HER. This is also related to the characteristics and number of the edge cracks. A remarkable drop in force is likely to occur for a smaller number of cracks as well as for more pronounced cracks. As it can be seen from the different diagrams, both stopping methods do not influence each other.

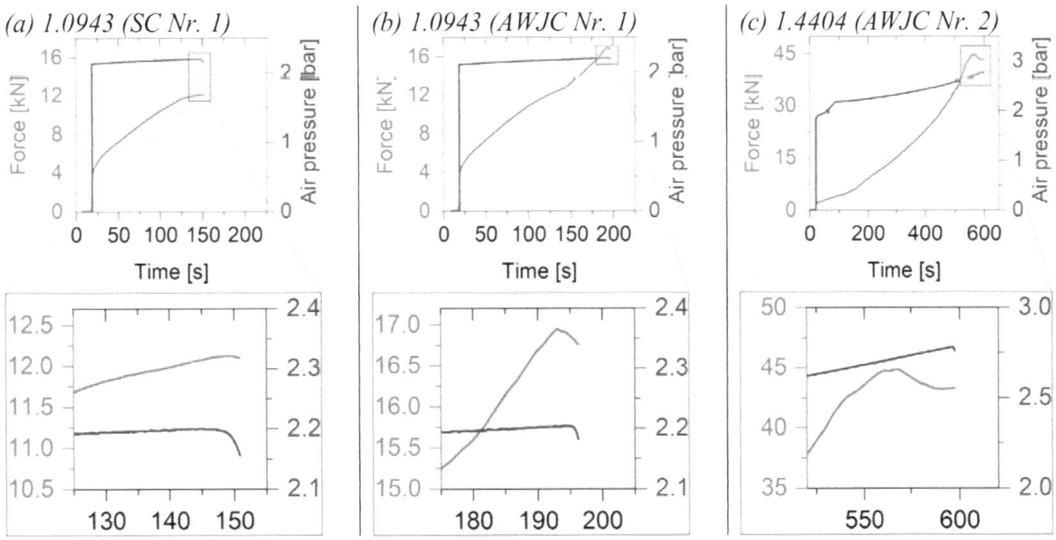

Figure 2: Pressure and force curves for HET of 1.0943 and 1.4404 with pneumatic crack detection.

3.2. Influence of pressure value and hole expansion punch speed

Hole expansion tests of 1.0943 with an abrasive water jet cut pre-hole were conducted to investigate the influence of the initial pressure value as well as the hole expansion punch speed. The results are presented in Figure 3. No effect of the initial air pressure was determined for the experiments between 1 and 6 bar (Figure 3(a)). The mean HER for the different pressures vary within the standard deviation. Crack widths less than 0.1 mm were realized for all the different starting pressure values except for 1 bar. Crack width values above 0.1 mm as well as non through-thickness cracks were obtained for the pneumatic crack detection at 1 bar initial pressure. Due to this variation regarding the edge cracks, 2 bar is proposed as the initial pressure value.

In contrast, an influence of the punch speed on the HER is shown in Figure 3(b). The HER increases with rising velocity in the range between 0.01 mm/s and 1.00 mm/s. The reason for this is the time between the start of the pressure drop and the stop of the test due to reaching the critical pressure drop value. If the velocity of the punch is higher, it will cover a longer distance and cause a greater expansion of the hole. Besides the HER, this effect can be seen from the crack widths which also increase for higher speeds. The crack width is below 0.1 mm for velocities up to 0.1 mm/s and increases to above 0.2 mm for 0.5 and 1.0 mm/s. The value 0.05 mm/s was chosen as standard punch speed which was the highest velocity that met the crack width requirements. The adaption of the stop value is an approach to be able to use higher speeds.

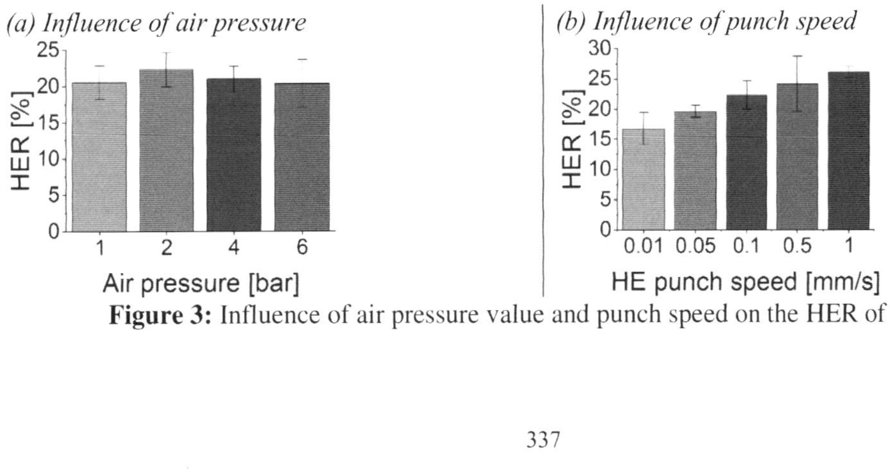

Figure 3: Influence of air pressure value and punch speed on the HER of 1.0943.

3.3. Comparison of crack detection by drop of air pressure and force

As shown in 3.1, a drop of the expansion force can be found close or even before the pressure drop for higher HER. In addition, the pressure and force curves do not influence each other. For this reason, the crack detection by force and a combined crack detection were investigated for the steel 1.0943 (Table 3). Through-thickness cracks with widths less than 0.1 mm were found for all experiments. The comparison of the HER for all three crack detection methods shows that the values are in a close range independent of the stopping approach. It is noticeable that the hole expansion test with combined crack detection is stopped by drop of air pressure for smaller HER (in case of shear cutting) and by drop of force for larger HER (in case of abrasive water jet cutting).

Table 3: Results of the HET of 1.0943 using force drop and combined crack detection method.

Crack detection method	Hole insertion	Nr.	Crack detection by	Hole expansion ratio (%)	Crack width (mm)
Force drop	Shear cutting	1	Force drop	11.95	0.07
		2	Force drop	10.47	0.04
		3	Force drop	12.96	0.07
	Abrasive water jet cutting	1	Force drop	17.79	0.06
		2	Force drop	29.56	0.03
		3	Force drop	21.11	0.04
Combined	Shear cutting	1	Pressure drop	9.77	0.07
		2	Pressure drop	13.27	0.07
		3	Pressure drop	13.76	0.08
	Abrasive water jet cutting	1	Force drop	30.59	0.03
		2	Force drop	30.77	0.05
		3	Force drop	26.58	0.06

All recorded curves for force and air pressure were analysed. Two representative curves for the hole expansion of 1.0943 with shear cut pre-hole (Figure 4(a)) and abrasive water jet cut pre-hole (Figure 4(b)) are shown using the combined crack detection. A drop of air pressure causes the stop of the HET for smaller HER in case of the pre-hole insertion with shear cutting (Figure 4(a)). No drop in the force curve is recognisable here. In Figure 4(b), the crack is detected by a force drop without any evidence of a decline in the air pressure curve. Both results prove that the curves are not influenced by each other. In addition to these two cases, there are also curves in which both drops occur and the more pronounced one ultimately leads to test stop.

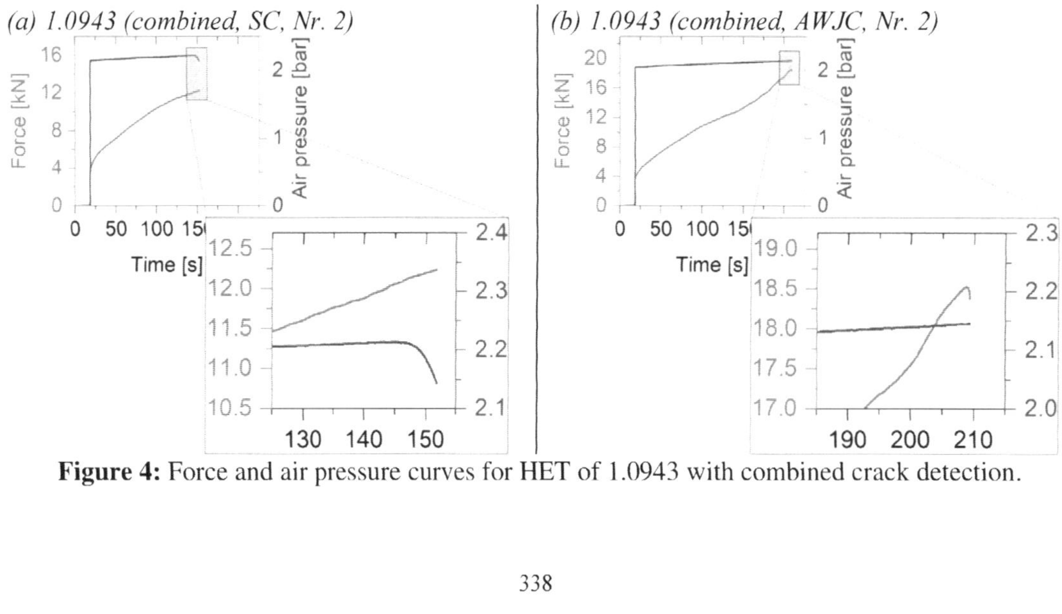

Figure 4: Force and air pressure curves for HET of 1.0943 with combined crack detection.

Finally, the force drop method was investigated for the HET of 1.4404 (Table 4) as a decline in the force curve (or a local minimum) was detected before the pressure drop as shown in 3.1. All tests were stopped by a force drop which was smaller compared to the decline in Figure 2(c). Two selected force curves are shown in Figure 5.

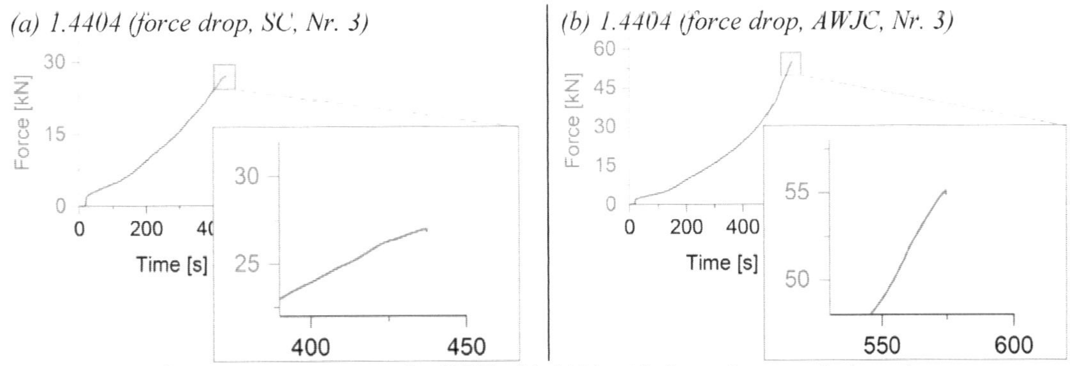

Figure 5: Force curves for HET of 1.4404 with force drop crack detection.

Although the tests were stopped within a close range of crack widths, all of them exceeded the specification of the standard. This is justified with the applied force sensor and its position within the test set up. An adapted sensor for the application and force range in combination with a position closer to the actual forming process (e.g. above the hole expansion punch) is an approach for improved results regarding the crack width.

Table 4: Results of the hole expansion tests of 1.4404 using force drop crack detection method.

Crack detection method	Hole insertion	Nr.	Crack detection by	Hole expansion ratio (%)	Crack width (mm)
Force drop	Shear cutting	1	Force drop	122.36	0.44
		2	Force drop	106.47	0.34
		3	Force drop	111.89	0.24
		4	Force drop	104.20	0.42
	Abrasive water jet cutting	1	Force drop	151.09	0.40
		2	Force drop	164.33	0.14
		3	Force drop	159.20	0.14
		4	Force drop	138.56	0.83
		5	Force drop	150.34	0.17

4. Conclusion

Due to the adapted version of the ISO 16630 from 2017, more demanding requirements have to be achieved regarding the HET. Besides the determination of the HER, the crack widths have to be analysed. This sets higher requirements regarding the crack detection method to stop the test. In this paper, the crack detection by air pressure drop was validated for different HER and compared to the detection method by a drop in expansion force. The pressure drop method has proved to be suitable for HER up to approx. 50%. All test were stopped with maximum crack widths below 0.1 mm within this range of the HER. The method showed no dependency on the initial air pressure which is suggested to be set to 2 bar for the presented test setup. In addition, the crack width requirements were not met using a punch speed above 0.1 mm/s. Higher velocities can be used if the sensor sensitivity and the critical pressure value are adapted. An alternative crack detection is recommended for higher HER if the parameter for the test set up are taken from the standard. Besides the common variants applying video capturing or DIC, the force drop method is a promising approach for higher HER if a suitable force sensor and sensor positioning is used. The application of a combined crack detection method using both pressure and force drop is an universal approach to cover a large HER range as the two methods do not influence each other. Both variants are implemented easily and cost-effectively. Due to the possibility of automation, the combined crack detection is suitable for serial testing without operator influence. The next steps include the improvement of test setup to widen the range of testable parameter. Both detection methods will also be tested for alternative punch geometries. Besides the conical punch specified by the standard, forming punches with a hemispherical or a flat-bottom tip were already used for hole expansion tests which were derived from the tools applied for the determination of the Forming Limit Curve according to Nakajima and Marciniak. Finally, larger test series for different steel and other metallic materials will be conducted.

Acknowledgments
The authors would like to thank the German Federal Ministry for Economic Affairs and Climate Action (BMWK) for funding the research project "Entwicklung von Prüfverfahren und -technik zur Ermittlung relevanter Eigenschaften höherfester Stahlwerkstoffe für ein ressourceneffizientes Produkt- und Prozessdesign – effiTEST".

References
[1] Schneider M, Geffert A, Peshekhodov I et al 2015 Overview and comparison of various test methods to determine formability of a sheet metal cut-edge and approaches to the test results application in forming analysis *Materialwissenschaft und Werkstofftechnik* 46(12) pp 1196–1217
[2] N N 2017 ISO 16630:2017: Metallic Materials – Sheet and Strip – Hole Expanding Test *International Organization for Standardization* (Geneva, Switzerland)
[3] Atzema E, Borsutzki M, Braun M et al 2011 Ergebnisse eines europäischen Ringversuchs zum Lochaufweitungsversuch nach ISO 16630 *Tagung Werkstoffprüfung* (Berlin)
[4] Schneider M 2019 Ermittlung des Formänderungsvermögens schergeschnittener Kanten von hochfesten Stahlblechen *Berichte aus dem IFUM* 8/2018 (TEWISS-Verlag)
[5] Krempaszky C, Larour, P, Freudenthaler J et al 2014 Towards More Efficient Hole Expansion Testing *Proceedings of IDDRG 2014 Conference* (Paris, France)
[6] Chiriac C, Chen G 2008 Formability Characterization of AHSS-Digital Camera Based Hole Expansion Test Development *Proceedings of IDDRG 2008 Conference* (Olofström, Sweden) pp 16–18
[7] Cruz D J, Amaral R L, Santos A D et al 2023 Application of Digital Image Processing Techniques to Detect Through-Thickness Crack in Hole Expansion Test *Metals* 13(1197)
[8] Chen X, Yang L, Chirac C et al 2011 Measurement of Strain Distribution for Hole Expansion with Digital Image Correlation (DIC) System *SAE* (Warrendale, PA, USA)
[9] Leonhardt A, Kräusel V, Paar U 2019 Automated hole expansion test with pneumatic crack detection *IOP Conf Ser Mater Sci Eng* 480(012026)

IOP Conf. Series: Materials Science and Engineering **1307** (2024) 012044 doi:10.1088/1757-899X/1307/1/012044

Influence of Hole Preparation Technique on Stretch Flangeability of Aluminium Alloys

Vinay Kumar[1*], D Ravi Kumar[1]

[1]Department of Mechanical Engineering, Indian Institute of Technology Delhi, New Delhi, India-110016

Email*: vinay.kumar@mech.iitd.ac.in

Abstract. Application of aluminium alloys has increased for light-weighting the sheet metal parts in manufacturing of automobiles. Formability is important in sheet metal forming of aluminium alloys. In the assessment of formability of sheet metals, stretch flangeability or edge formability is an important aspect in application such as flanges or plunged rims. The stretch flangeability refers to the ability to avoid cracking during hole expansion and it is evaluated by performing the hole expansion test. Besides material properties, hole making process also influences the stretch flangeability of aluminium alloys. In this work, the stretch flangeability of two Al-Mg alloys has been investigated through hole expansion tests by using three different hole making techniques namely punching, water jet cutting and laser cutting. The stretch flangeability of aluminium alloys has been compared with that of a high strength steel (Dual phase steel of grade DP600) which is more commonly used steel for stretch flanging applications.

1. Introduction

The use of aluminium alloys is continuously growing in automobile industry due to the increasing necessity for lightweighting. Though aluminium alloys possess high strength to weight ratio for manufacturing intricate sheet metal components for vehicles. Al-Mg alloys of 5xxx series are the most significant for applications in automobile, aeronautics, and marine industries where strength and formability of sheet metals are important[1].

In the estimation of overall formability of sheet metals, stretch flangeability is one of the key attributes. In a stretch flanging operation, it is the indicator of local formability near the hole's edge. The hole expansion ratio (HER) is commonly used to quantify the stretch flangeability of sheet metals. It is determined by a hole expansion test in which a circular hole in a blank is expanded by a conical punch until failure. The increase in hole diameter to the initial hole size when edge fractures show through thickness is known as the hole expansion ratio [2] and is relevant as a relative "local formability" characteristic for edge-cracking assessment[3]. A schematic of hole expansion experimental test setup is shown in Fig. 1.

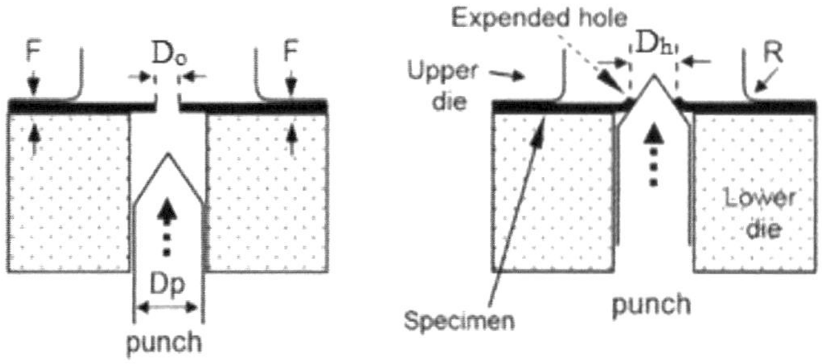

Figure 1. Schematic diagram of hole expansion test

The hole expansion ratio $\boldsymbol{\lambda}$ is generally stated in percentage by Equation 1:

$$\lambda(\%) = \left[\frac{D_h - D_0}{D_0}\right] * 100 \tag{1}$$

where D_0 is the initial hole diameter and D_h is the average hole diameter when the through thickness crack occurs.

Over the period, there have been a several studies[3–5] to investigate the stretch flangeability of primarily steels and to a limited extent on aluminium alloys. In hole expansion tests, stretch flangeability of Al alloys is dependent on several variables, including sheet thickness, burr location, hole preparation method, and punch geometry [3,4,6,7]. Good edge conditions can enhance the HER of the sheet. But they certainly do not reduce the HER alterations because of the material variability.

HER is influenced by the surface condition of the hole edges. Hance[8] studied the effect of hole making technique on HER of advanced high strength steel by using different techniques such as punching (piercing), machining, water jet cutting, and laser cutting and found out punching resulted in lower HER because of the work hardening at sheared edge and presence of micro voids in the shear affected zone whereas water jet and laser cutting give better HER results. Many other researchers [7–11] have used different techniques (punching, water jet cutting, laser cutting, hole drilling, drilling + reaming and wire cutting) to improve hole expansion ratio for different types of steels and investigate the effect of hole making method on stretch flangeability. But there have been very limited studies on stretch flangeability of aluminium alloys.

So, this work mainly focuses on the influence of hole preparation technique on stretch flangeability of two Al-Mg alloys (AA5083 and AA5754) and their behavior is compared with DP600 steel to explore the feasibility of replacing steel with these alloys intended for lightweighting of automotive components where stretch flangeability is very important. Three different techniques are used for preparing holes are punching, waterjet cutting and laser cutting for hole expansion test (HET). The hole expansion test has also been simulated in FEA and the predicted HER has been compared with experimental results.

2. Alloy composition and properties

5xxx series aluminium alloys have good mechanical properties and resistance to corrosion due to which they are used for a variety of applications. In this work, two Al-Mg alloys, AA5083 and AA5754, have been used in annealed condition because of their ability to work harden during plastic deformation and better formability than many other aluminium alloys. For comparison of their stretch flangeability, high strength dual phase steel (DP600) is used due to its wide application in automotive sector for stretch flanging applications.

The tensile properties of the Al-Mg alloy and DP600 steel were determined using uniaxial tension tests according to ASTM E8M specification and the properties are given in Table 1. From true stress and true strain curve, the coefficient in the swift power law of hardening namely strain hardening exponent (n) and strength coefficient(K) values were determined. The plastic strain ratio values (R_0, R_{45}, and R_{90}) that were determined in accordance with the ASTM E517 specification to measure anisotropy and value are shown in Table 2.

Table 1. Tensile properties of the materials used

Grade	Thickness (mm)	YS (MPa)	UTS (MPa)	% Elongation	n	K (MPa)
AA5083	1.75	149.0	286.0	15.1	0.27	541.8
AA5754	1.00	108.5	231.3	24.5	0.30	504.9
DP600	1.40	406.5	648.9	19.1	0.22	1164.1

Table 2. Anisotropic properties of the materials used

	AA5083	AA5754	DP600
R_0	0.705	0.740	0.957
R_{45}	0.898	0.720	0.705
R_{90}	0.751	0.800	0.931

3. Methodology

3.1 Blank preparation for HET

Specimens of dimensions 100 mm × 100 mm were used for determination of stretch flangeability in hole expansion tests. A hole of 10 mm diameter was made at center of specimen by using three different methods- punching (shearing), abrasive waterjet cutting and laser cutting, as explained below:

3.1.1 Punching. According to ISO 16630:2017[2], punching is the recommended method for preparing the hole in the blank for hole expansion test. So, the holes were made by punching on a hydro-pneumatic punching press of 15-ton capacity using a cylindrical punch size of 10 mm diameter. The die diameter was decided by providing a punch-die clearance of 2% of sheet thickness. The clearance is one of the major parameters that affect the burr formation in a punching process. Due to the burr formed during shearing, the test samples were placed on the die with burr side on the top (burr facing away from the conical punch) during the hole expansion test[7].

3.1.2 Water jet cutting. An abrasive water jet cutting method involves cutting materials by employing a high-pressure water jet along with an abrasive slurry to facilitate erosion [12]. Water jet pressure and cutting speed are the two main process parameters that depend on the material being cut. The details of the parameters used for abrasive water jet cutting of 10 mm diameter hole in the HET samples are shown in table 3.

Table 3. Parameters used in Water Jet Cutting

Parameter	Name/value
Abrasive Grit	Grade 80 HP Garnet Grit
Water Jet Pressure	250-300 MPa
Nozzle Diameter	1 mm
Cutting Speed	0.7-0.9 m/min

3.1.3 Laser cutting. Laser cutting is a process that utilizes a laser to cut a variety of materials into desired shapes by harnessing the intense focus of the cutting beam through the application of energy and high-purity technical gas[13]. In laser cutting, laser power and cutting speed are the major process parameters. In the present work, a 4000W Industrial CO_2 laser was used for cutting the hole in HET specimens. A cutting speed of 5-6 m/min was used and nitrogen gas was used for reducing oxidation and heat generation.

3.2 Hole expansion test

Using the blanks prepared from all the three techniques, the hole expansion tests were done as per ISO standard 16630: 2017 [2]. The tests were conducted on a double-action hydraulic press of 100-ton capacity. A 50 mm diameter conical punch with a 60° cone angle was used in the experiments as illustrated in Fig. 2. The experiments were conducted with a constant punch speed of 0.5 mm/sec and a blank holding force of 40 kN. During the hole expansion tests, the samples were placed with the burr surface on the top [7]. The tests were stopped upon the detection of a crack at the edge, identified through a camera positioned above the die setup. The final hole size (D_h) at the appearance of first crack was determined by an image analyzer tool using the images captured during the test.

3.3 Prediction of HER by simulation

The stretch flangeability of Al-Mg alloys and DP600 steel was predicted by numerical simulation of HET using Dynaform software with LS-Dyna solver. The tools (die, blank holder, and conical punch) and the blank were modeled in SolidWorks with the dimensions used in the HET experiments. The assembled model is shown in Fig. 3. To secure the blank between the dies, a blank holding force of 40 kN was applied, and a constant punch velocity of 100 mm/min was defined in the simulations. The blank was meshed using Belytschko-Tsay shell elements with a size of 2.0 mm and five integration points. To improve the accuracy of prediction, the mesh was refined using the refining mesh option, with a maximum refinement level set to 5. The contact between the tool surface and the blank was specified using the 'forming one way surface to surface' option, and a constant friction coefficient of 0.125 was defined at the tool-blank interface.

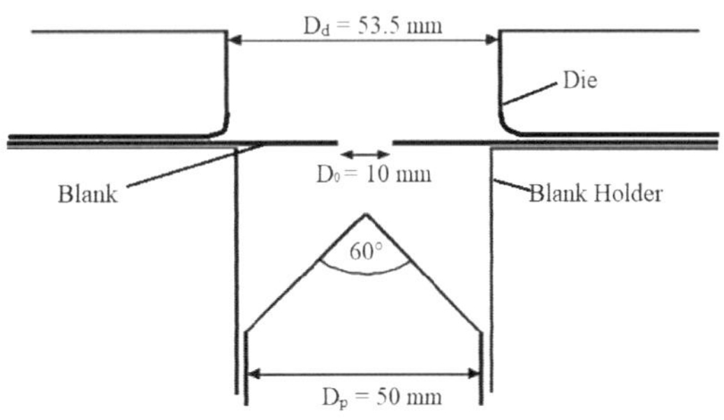

Figure 2. Dimensions of tools used in HET

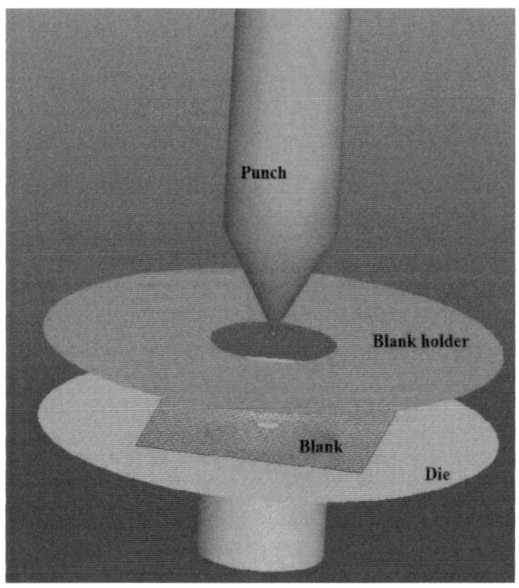

Figure 3. Model of tools and blank for HET simulation

In the material model, the yielding behavior of the alloy was represented using the 3-parameter Barlat's 89 yield criterion. The yield exponent (m) was set to 8, as recommended for FCC (face-centred cubic) materials[14]. The yield locus was plotted using the anisotropic coefficients calculated with the help of the plastic strain ratio R values R_0, R_{45} and R_{90}. To define the hardening curve in simulation, the experimentally determined strain hardening exponent (n) and strength coefficient(K) were given as input from table 1 of material properties. As an example, the hardening curve and the yield surface of AA5083 alloy are shown in Fig.4.

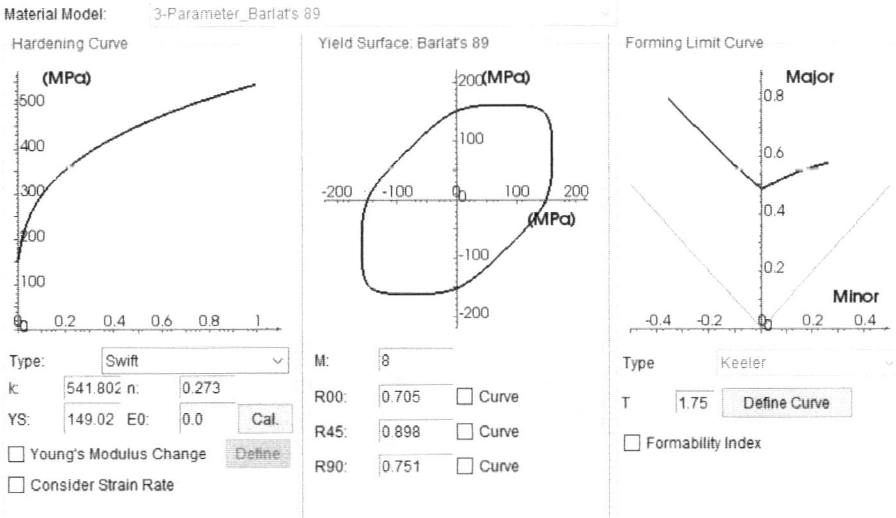

Figure 4. Material Model of AA5083 for blank in simulation

The forming limit curve was used to predict the failure in simulation is also shown in Fig. 4. The predicted results from the simulation of HER for all three materials were compared with experimental results.

4. Results and discussion

4.1 HET simulation results

The failure prediction in the hole expansion test (HET) simulation of AA5083 alloy is shown in Fig. 5. The figure illustrates the initiation of failure as the strains in the tension-compression region exceeded the limit strains on the left-hand side of the Forming Limit Curve (FLC). At the point of failure on the hole edge, the state of stress is nearly uniaxial tension. The predicted values of hole expansion ratio (HER) are 66.8 % and 68.8 % for AA5083 and AA5754 alloy, respectively. It is predicted that both these alloys have higher stretch flangeability than DP600 steel as the predicted HER value for DP600 steel is 57.8 %.

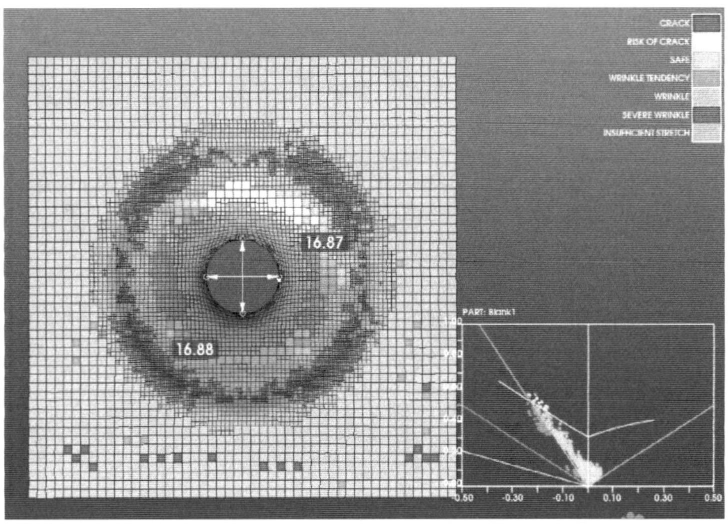

Figure 5. Prediction of failure in HET simulation using AA5083

IOP Conf. Series: Materials Science and Engineering **1307** (2024) 012044 doi:10.1088/1757-899X/1307/1/012044

4.2 Experimental results of HER

The experimentally determined average HER values of the two aluminium alloys (AA5083 and AA5754) and DP600 steel are shown in Table 4 for all three hole making techniques (punching, water jet cutting (WJC) and laser cutting (LC)). The predicted HER results from simulation are also shown for comparison with experimental results.

Table 4. Comparison of HER value from experiment and simulation

Grade	HER% Punching	HER % WJC	HER% LC	HER% Simulation
AA5083-O	45.8 ± 1.6	49.5 ± 2.4	62.7 ± 4.6	66.8
AA5754-O	67.4 ± 3.8	49.9 ± 3.9	82.6 ± 3.6	68.8
DP600	41.1 ± 3.5	53.5 ± 7.9	75.1 ± 3.0	57.8

The tested specimens of all the three materials with the three different hole making techniques are shown in Fig. 7.

(a) 5083-Punching

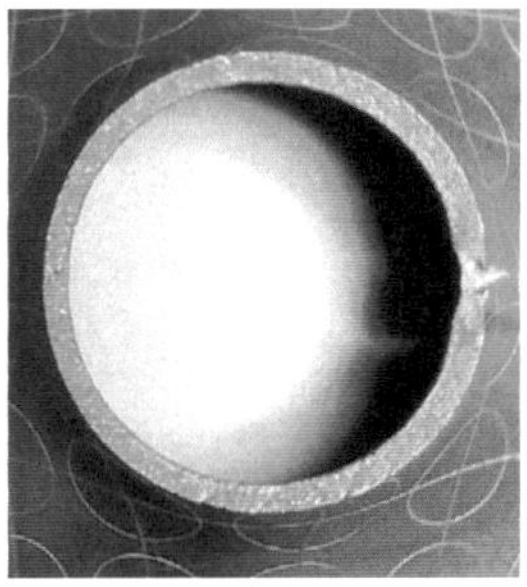
(b) 5083-WJC

(c) 5083-LC

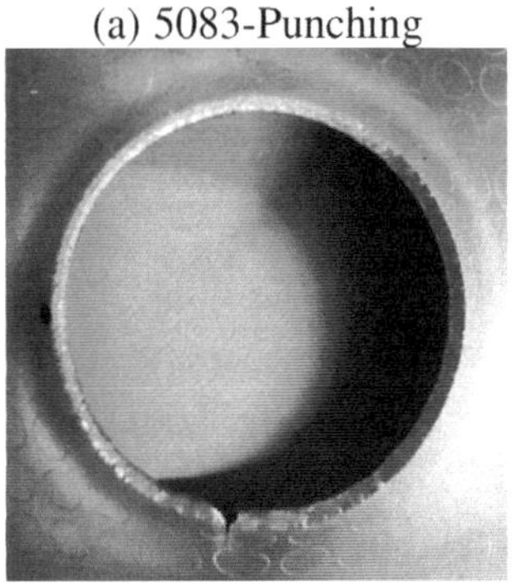
(d) 5754-Punching

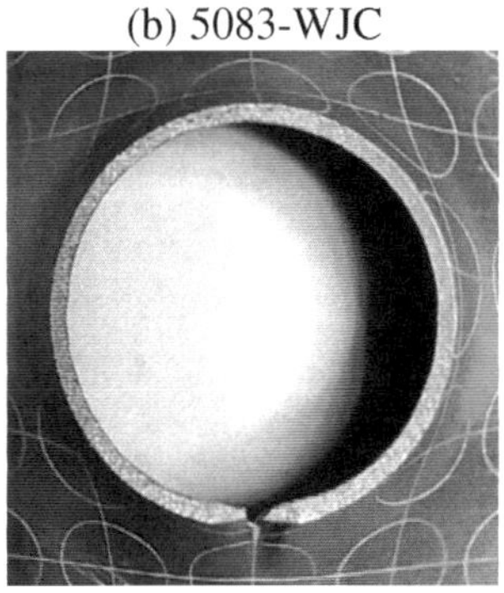
(e) 5754-WJC

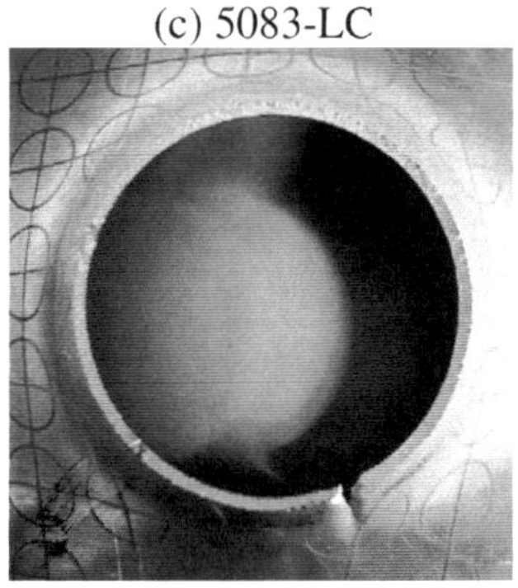
(f) 5754-LC

(g) DP600-Punching

(h) DP600-WJC

(i) DP600-LC

Figure 6. Tested specimens of (a-c) AA5083, (d-f) AA5754 and (g-i) DP600 using three hole making techniques (Punching, WJC and LC)

4.2.1 Effect of hole making technique. It can be clearly observed from table 4 that specimens in which the hole was prepared by laser cutting resulted in much higher HER values than in the case of the other two hole making methods (WJC and punching) for both the aluminium alloys. An improvement of 36.8% and 22.5% has been obtained with LC when compared to punching in the case of AA5083 and AA5754, respectively. Though a marginal improvement in HER 8.1% has been observed with WJC when compared to punching in the case of AA5083 alloy, the HER decreased in the case of AA5754. It could be due to presence of a small notch generated due to WJC process which was clearly seen after the hole was prepared in all the materials. Though WJC produced a better quality hole edge than punching, the presence of small notch resulted in early cracking at the location of the notch in all the materials shown in Fig.6 (b), (e) and (h) and hence resulted in lower HER than in the case of laser cutting. However, in the case of DP600 steel, WJC resulted in an improvement of 30.2% when compared to punching despite the presence of a very small notch. It could be due to the high strength of DP600 steel which resisted the early failure at the notch which was observed in case of the Aluminium alloys.
In case of punching, which is a standard method to prepare the blank for HET, the large shear deformation near the sheared edge causes significant work hardening of material decreasing ductility of the material locally and increasing hardness. Significant increase in local hardness near the sheared edge due to punching can be clearly seen from microhardness measurement presented in Fig. 7 for all the three hole making methods. Shearing also induces a no. of voids and microcracks on the sheared edge. This results in early onset of cracks at multiple location during hole expansion test which can be observed in Fig. 6 (a), (d) and (g).

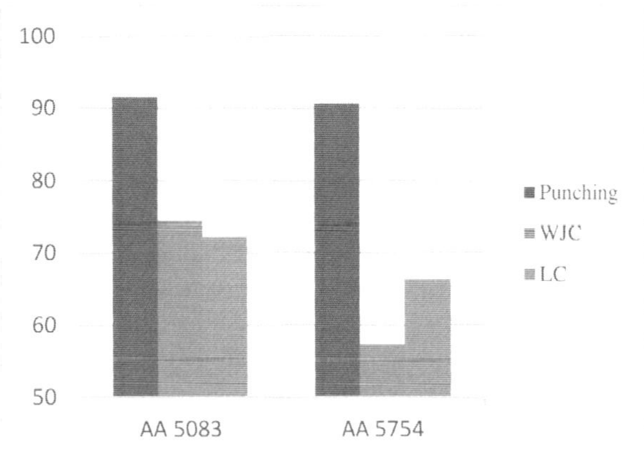

Figure 7. Vickers Micro Hardness measurement results near hole edge

Whereas cracking at only one location can be observed in the case of hole made by WJC and LC. Overall, hole making by laser cutting resulted in maximum HER in all the materials due to the excellent quality of the hole edge with no severe plastic deformation and burr at the edge (as in case of punching) and no presence of a small notch (as in case of WJC). Slight softening of the hole edge due to heating and cooling could also have contributed to improvement in HER.
It can also be observed that HER of AA5754 alloy is higher than that of DP600 steel when hole was made by both punching and LC. It clearly shows that between the two alloys investigated in the present work, AA5754 has better stretch flangeability than DP600 steel making it suitable for lightweight applications in automotive industry where stretch flangeability is an important requirement and strength is not a major concern.
4.2.2 Comparison of experimental and simulation HER results. The predicted HER values from simulation agreed well with the experimental values in general. Since, the edge quality of the hole depends on the hole making process, variation in experimental value could be observed. However, in simulation the effect of edge quality and modification of local mechanical properties of the material due to hole making process have not been accounted for. Therefore, some discrepancies have been found

between the predicted and experimental values. However, it can be observed that the predicted values of HER of the two aluminium alloys are higher than that of DP600 steel in simulation. It is consistent with the experimental observation for both alloys in the case of punching and only for AA5754 in the case of LC. It is due to higher ductility and work hardening ability of these alloys (indicated by higher n values) when compared to DP600 steel.

5. Conclusions

In this work, the following are the main findings of this investigation:

1. Hole making by laser cutting in HET test specimens demonstrated significantly higher hole expansion ratios for both aluminium alloys (AA5083 and AA5754) and DP600 steel compared to punching and water jet cutting. It can be attributed to the excellent quality of the hole edge, absence of severe work hardening, burrs, or notches near the hole edge.
2. Water jet cutting showed a marginal HER improvement in HER for AA5083 but a decrease in the case of AA5754 compared to punching. Despite producing better hole edges, the presence of a small notch led to early cracking in all the materials, impacting HER values.
3. Punching resulted in lower HER values due to severe shear deformation near the hole edge. This induced work hardening, reduced ductility, and increased local hardness, contributing to early crack initiation at multiple locations during hole expansion tests.
4. DP600 steel exhibited improved HER with water jet cutting despite the presence of a small notch, highlighting the influence of material strength on the effectiveness of hole-making methods. AA5754 alloy consistently demonstrated higher HER values than DP600 steel, indicating better stretch flangeability and suitability for lightweight applications in the automotive industry.
5. Predicted HER values from simulation agreed well with the experimental values in general. The observed discrepancies could be due to the fact that the edge quality of the hole has not been accounted for in the simulation. Predicted HER values of aluminium alloys are higher than that of DP600 steel in simulations also.

Acknowledgement

The author would like to acknowledge the Ministry of Human Resource Development (MHRD), Government of India for receiving the assistantship for this research work.

References

[1] Kleiner M, Geiger M and Klaus A 2003 *CIRP Ann Manuf Technol* **52** pp 521–42
[2] ISO 16630:2017 *Metallic materials-Sheet and strip-Hole expanding test*
[3] Hance B M 2017 *SAE Int J Engines* **10** pp247–57
[4] Stanton M, Bhattacharya R, Dargue I, Aylmore R and Williams G 2011 *AIP Conference Proceedings* vol **1353** pp 1488–93
[5] Pathak N, Butcher C and Worswick M 2016 *J Mater Eng Perform* **25** pp 4919–32
[6] Li J, Wang H and Yan D-W 2016 *Int conference AMTIA*
[7] Paul S K 2020 *Materialia* **9** p.100566
[8] Hance B M, Comstock R J and Scherrer D K 2013 *SAE Technical Papers* vol 2 (SAE
[9] Balisetty V, Chakkingal U and Venugopal S Evaluation of stretch flangeability of dual-phase steels by hole expansion test
[10] Karelova A, Krempaszky C, Werner E, Hebesberger T and Pichler A 2009 Steel research international 80(1) pp 71-77ns
[11] Karelova A, Krempaszky C, Werner E, Hebesberger T and Pichler A 2007 *Proc. MS&T* pp. 159–169
[12] Wang J and Wong W C K 1999 *International Journal of Machine Tools and Manufacture* **39(6)** pp.855-870
[13] Krajcarz D 2014 *Procedia Engineering* **69** pp 838–43
[14] Barlat F and Lian K 1989 *Int J Plast,* **5(1***)* pp 51-66

IOP Conf. Series: Materials Science and Engineering **1307** (2024) 012045 doi:10.1088/1757-899X/1307/1/012045

Wrinkling Determination in Sheet Metal Loaded by Shear Stresses

J.P.G. Magrinho[1*], J.A.O. Santos[1] and M.B. Silva[1]

[1]IDMEC, Instituto Superior Técnico, Lisbon, Portugal

[*]joao.magrinho@tecnico.ulisboa.pt

Abstract. Lightweight concept approach in sheet metal forming requires, among others, thickness reduction and geometry complexity. This combination dramatically promotes the wrinkling occurrence, which can be considered as one of main failure mode that toolmakers must consider. The sheet metal wrinkling determination research usually considers the wrinkling in flange and conical area of a cylindrical cup. For complex geometries the loadings are more diverse than in the deep drawing of a cylindrical cup involving shear stress states. In this context, the present work proposes the application of two different wrinkling determination methodologies in an experimental sheet metal specimen loaded by shear in an aluminum alloy. One of the methodologies involves geometrical parameters to trigger the onset of wrinkling and the other methodology is physically based and considers the evolutions of the in-plane minor strain and its strain rate. The experimental formability test requires a dedicated clamping apparatus to allow its kinematics to be similar to a tensile test. Results allowed the identification of the stress state that triggered the wrinkling occurrence, and the determination of the critical strains that allow plotting the wrinkling limit curve of the sheet material.

1. Introduction

In a world constantly pushing the boundaries of innovation, the quest for efficiency and performance takes a central stage, and nothing embodies this pursuit quite like the paradigm of lightweight design. As we navigate an era defined by technological breakthroughs and sustainability imperatives, the concept of shedding excess weight has transcended mere physicality. Beyond the realm of materials and structures, the essence of lightweight extends into a profound redefinition of industries, from transportation to manufacturing [1].

In sheet metal forming, thickness reduction and geometry complexity are the key factors to ensure lightweight systems. However, this combination increases the propension for wrinkling failure involving complex stress states. Hence, the identification of the failure limits is of major importance when designing a sheet metal forming process [2].

In this sense, wrinkling is of major relevance in sheet forming process, despite being the least explored phenomenon by the scientific community. Due to the fact that it is influenced by different aspects such as the material, the geometry, the contact conditions, and the stress state [3]. Considering that complex geometries involve shear stress states it is crucial to extend the wrinkling study to shear stress state.

Du *et al.* [4] studied failure by wrinkling on sheet metal specimens under a shear stress state made from 304 stainless steel. The onset of wrinkling was defined as the instant when bifurcation occurs in the strain loading path of the integration points defined through the thickness of the sheet metal. Tang *et al.* [5] also studied wrinkling in specimens subject to shear stresses. However, the specimens developed by the authors were asymmetric under offset loading.

In a previous work [6] the authors proposed a new experimental methodology physically based to determine the onset of wrinkling by means of the in-plane minor true strain and compared the results with the commonly used experimental methodology [7]. In the present work, the authors considered the shear tests developed by Du *et al.* [4] and applies the new experimental methodology to determine the onset of wrinkling under shear stress state. The new experimental methodology approach was shown as appropriate to determine the wrinkling limit curve (WLC) for a specific material.

2. Materials and Methods

2.1. *Mechanical Characterization.*

Mechanical characterization of the commercial AA1050-O aluminum sheets with 1.5 mm thickness used in the investigation was carried out by means of tensile tests on an INSTRON 5900 universal testing machine. The tests were carried out at room temperature and in accordance to ASTM standard E8/E8M-16 [8]. The tensile test specimens were machined out from the supplied sheets at 0, 45 and 90° with respect to the rolling direction (RD) to evaluate the influence of anisotropy.

The mechanical properties obtained from the tensile tests are presented in Table 1, such as the modulus of elasticity, E, the yield stress, σ_y, the ultimate tensile stress, σ_{uts}, the elongation at break, A, and the planar anisotropy coefficient, r.

Table 1. Mechanical properties of aluminum AA1050-O sheet with a thickness of 1.5 mm obtained from tensile tests [6].

Specimen Direction	**Modulus of elasticity *E*** (GPa)	**Yield strength** σ_y (MPa)	**Ultimate strength** σ_{uts} (MPa)	**Elongation at break *A*** (%)	**Anisotropy coefficient *r***	**K** (MPa)	**n**
0°	78.63	24.78	72.74	45.98	1.08	136.94	0.27
45°	79.40	25.64	82.12	52.86	3.20	159.07	0.30
90°	78.36	24.11	71.48	47.26	0.81	134.44	0.27
Average	**78.95**	**25.04**	**77.11**	**49.74**	**2.08**	**147.38**	**0.29**

The average values included in Table 1 were determined as follows (where x denotes a mechanical property),

$$\bar{x} = \frac{x_0 + 2x_{45} + x_{90}}{4} \quad (1)$$

The anisotropy coefficient obtained from the tensile tests for sheets aligned at 45° with the rolling direction is considerably higher than those at 0°, and 90°, which is a phenomenon that was previously identified by Sharma et al. [9] in sheets due to the manufacturing of the sheets by means of accumulative roll bonding.

The average stress-strain curve can be approximated by the following Ludwik-Hollomon's equation. The strength coefficient, K, and work hardening exponent, n, are presented in Table 1.

$$\bar{\sigma} = 147.38\bar{\varepsilon}^{0.29} \quad (2)$$

2.2. *Experimental Plan*

2.2.1. Shear Test Specimens. Table 2 presents a summary of the shear test specimens that were used in the characterization of the WLC of the aluminium sheet and at least three repetitions were made for each dimension of the two different geometries. The shear tests were performed at room temperature and in quasi-static operating conditions with a constant velocity of 5 mm/min to ensure a quasi-static condition.

IOP Conf. Series: Materials Science and Engineering **1307** (2024) 012045 doi:10.1088/1757-899X/1307/1/012045

Table 2. Summary of the experimental work plan for the shear tests.

Geometry	Thickness (mm)	L (mm)	Identification
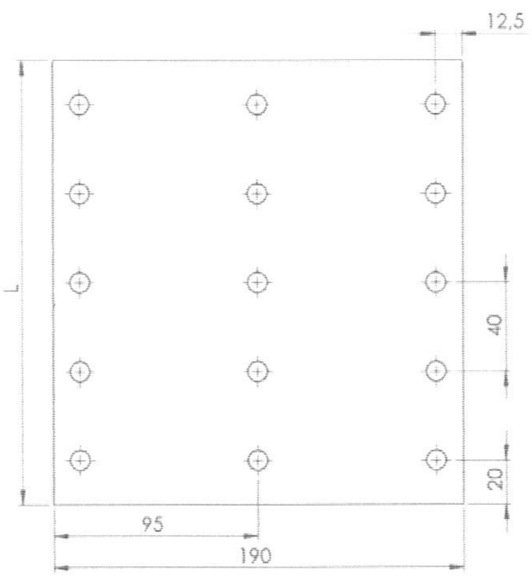 (mm)	1.5	160	S160
		180	S180

The specimens were cut out from the supplied sheets considering the rolling direction longitudinally in the specimens.

The geometrical dimensions used in the experimental work plan are similar to the dimensions used by Du et al. [4]. To ease the identification of each set of specimens in the manuscript, the identification is done as 'SX', where X can be 180, or 160, depending on whether the length of the specimen, L, is 180 mm, or 160 mm, respectively.

2.2.2. Experimental Apparatus. The tool required to induce shear stresses in the specimens is shown in Figure 1(a) and it corresponds to an adaptation of the design proposed by Du et al. [4]. A constant gauge width of 57.5 mm was used to ensure the same shear zone width for the entire set of specimens.

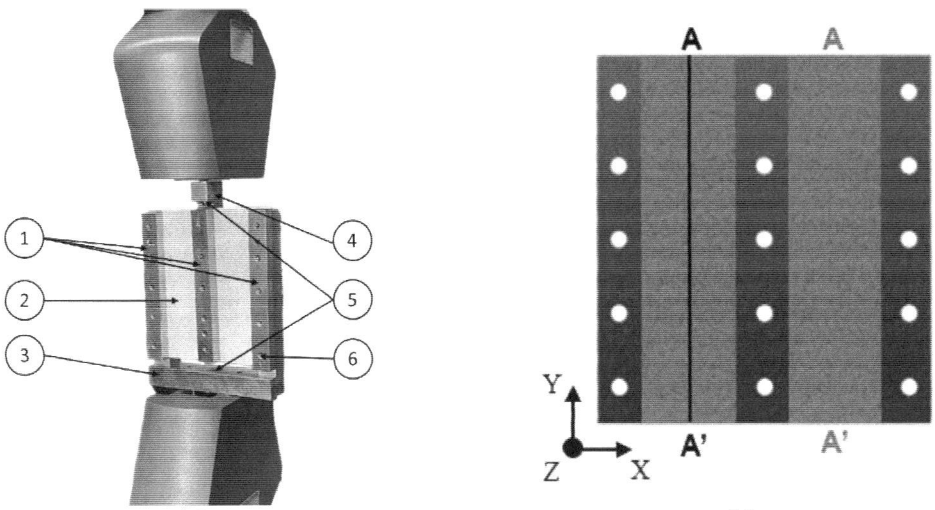

(a) (b)

Figure 1. (a) Elements of the tool apparatus for the shear tests: 1. removable blocks, 2. shear specimen, 3. fixed body, 4 central fixed part, 5. rods to fix the tool, and 6. bolts. **(b)** Schematic representation of the shear test specimen with identification of section AA'.

IOP Conf. Series: Materials Science and Engineering **1307** (2024) 012045 doi:10.1088/1757-899X/1307/1/012045

To fix the tool to the universal testing machine a rod is used, represented as number 5 in Figure 1(a). It has an M10 thread to connect the rod to the tool and a rough surface on the opposite end to prevent sliding when fixing the rod in the grip of the universal testing machine.

To measure the 3D displacements and the deformation of the shear test specimens a DIC system was used (Dantec Dynamics Q400 3D). The data acquisition of the DIC system was set to 10 Hz. The surface of the specimens was sprayed with a stochastic black speckle pattern on a uniform background previously painted in white (Figure 1(b)) and the test specimens were illuminated with one spotlight.

The DIC system is equipped with 2 cameras with 6 megapixels of resolution and 50.2 mm of focal lenses. An aperture of f/11 was selected to increase the field depth. The frequency of image acquirement was set to 10 frames per second and the correlation algorithm was performed with the INSTRA 4D software. A facet size of 13 px with a spacing grid of 7 px was considered.

2.3. *Onset of Wrinkling Determination*

The present work proposes the application of two different wrinkling determination methodologies in the experimental sheet metal specimen loaded by shear. Due to the geometry of the shear test specimen, there is a need to differentiate the left-hand side (LHS) and the right-hand side (RHS) of the specimen, as represented in Figure 1b. Therefore, each methodology is applied on both sides of the specimen.

Methodology A. The principle behind the position-dependent methodology is the direct analysis of the Z displacement, along the shear test specimen in the vertical direction (section AA' in Figure 1(b)). In the shear test specimens, the wrinkle is not symmetric, and for this reason, the highest point can occur in a random coordinate along section AA'.

Figure 2(a) shows the profile of section AA' at the initial, critical, and final stages of the experimental tests. The following step consists in plotting the wrinkle amplitude as a function of time, as shown in Figure 2(b). The onset of wrinkling is defined when the wrinkle amplitude reaches a critical value. The strain loading path and the critical strain pairs (major strain and minor strain at the onset of wrinkling instant) can be plotted in the principal strain space, as shown in Figure 2(c).

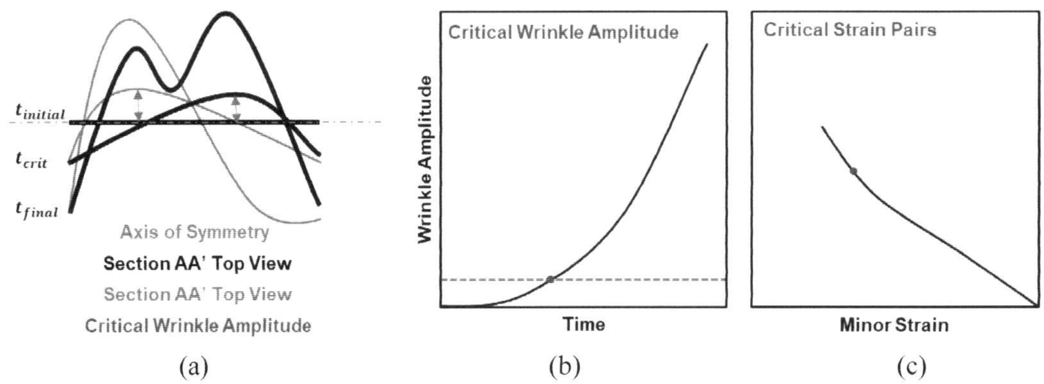

Figure 2. Methodology A to characterize the onset of wrinkling: **(a)** Identification of wrinkling amplitude; **(b)** Evolution of the wrinkling amplitude over time; **(c)** Representation of the strain loading paths and identification of the critical strains in principal strain space.

Methodology B. The physically based methodology is based on the direct analysis of the in-plane minor strain in the measuring area (section AA'), as shown in Figure 1(b). Figure 3(a) represents the evolution of the minor strain and strain rate of the point located in the section AA' with the larger Z-displacement. The critical instant is identified when the minor strain rate reaches a minimum corresponding to the wrinkling triggering. The strain loading path and the critical strains can be plotted in the principal strain space, as shown in Figure 3(b).

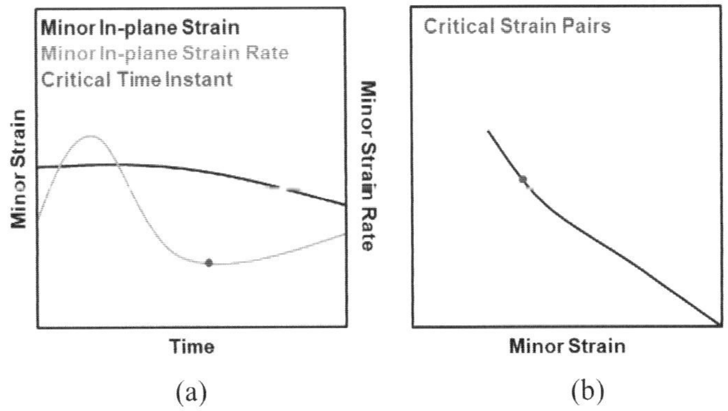

Figure 3. Methodology B to characterize the onset of wrinkling: **(a)** evolution of the in-plane minor strain and strain rate; **(b)** Representation of the strain loading paths and identification of the critical strains in principal strain space.

The critical strain pairs obtained from both methodologies can be used to draw the WLC. Methodology A is a position–dependent approach because it relies on the analysis of displacements and methodology B is a time–dependent approach because it is based on the analysis of in-plane minor strain rates. Although the above-mentioned methodologies have different grounds, both are physically based methodologies to detect the onset of plastic instability and determine the WLC of sheet metal forming processes.

3. Results and Discussion

The force-displacement curves of the shear test specimens measured by the universal testing machine are presented in Figure 4. The curves allow verifying the repeatability within specimens with the same geometric dimensions.

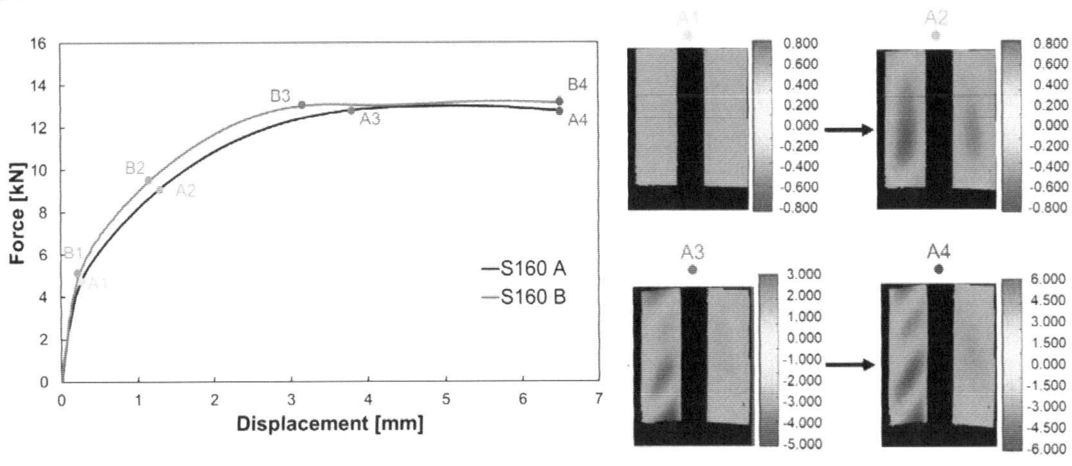

Figure 4. Evolution of the force throughout the experimental test for the S160 specimens with the respective points that show the Z-displacement at the respective instant in time.

For the S160 specimens, the evolution of the force throughout the experimental test continuously increases. Therefore, four instants throughout the force-displacement curves of these specimens were considered. Instant A1 and A2 reveal that at an early stage of the experimental test, the wrinkle develops uniformly throughout the length of the specimen and aligned with the rolling direction of the sheet metal.

However, a second stage occurs between instant A2 and A3, where the wrinkle rotates 45° clockwise and splits into two separate wrinkles, while also increasing in amplitude, as seen in point A3 of Figure 4. The last point, A4, shows that the wrinkles remained in this position and only increased in amplitude, thus, the boundary between the two wrinkles is further noticeable on the same specimen. Similar results were also obtained for the S180 specimens.

3.1. Methodology A

Methodology A considers the onset of wrinkling to occur when the wrinkle amplitude is greater than 5% of the sheet metal thickness. This pre-set percentage on the sheet metal thickness is represented in Figure 5(a) by a dashed-black-line. The total Z-displacement evolution through time of the LHS and RHS for the S160 specimen is also represented in Figure 5(b).

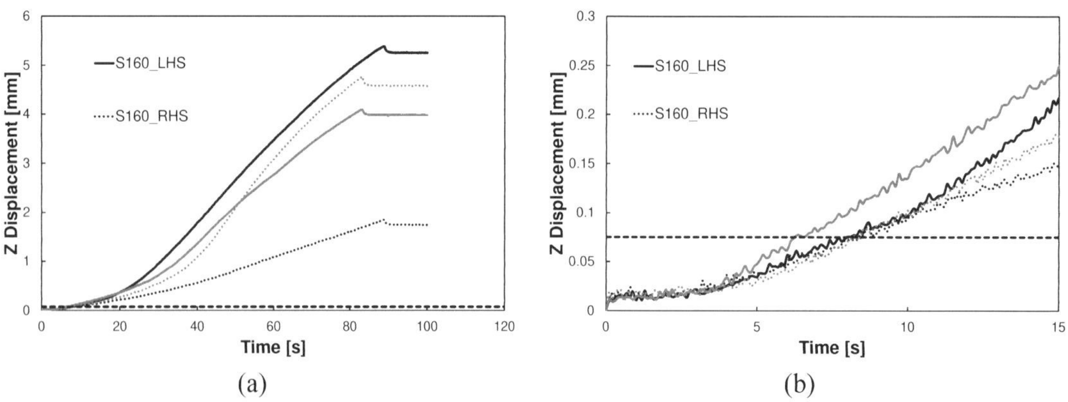

Figure 5. Z-displacement over time of the LHS and RHS for the S160 specimens: **(a)** macro and **(b)** micro view.

The S160 specimens show a similar behavior of wrinkle height progression during the experimental test, despite the difference in the rate at which the wrinkle amplitude develops between specimens, and between the LHS and RHS of the same specimen, as seen in Figure 5(a). Nonetheless, the Z-displacement registered in these specimens monotonically increases during the experimental test, and the maximum wrinkle height can occur on either side, LHS or RHS.

Proceeding to a more micro-oriented viewpoint of the Z-displacement registered by the S160 specimens, presented in Figure 5(b), it is possible to identify the instant where the onset of wrinkling defined by methodology A occurs. As seen in Figure 5(b), the Z-displacement evolution through time of the S160 specimens presents similar evolutions for both sides of the specimen, thus sustaining repeatability within the experimental tests performed with the specimens with a length of 160 mm.

3.2. Methodology B

The physically-based methodology assesses the evolution of the in-plane minor strain rate at the points located in Figure 6(b). These points correspond to the local of section AA' that first initiates the Z-displacement and reaches a maximum. Section AA' is in the center of the gauge width because is the local of the specimen with the biggest Z-displacement at the final stage of the shear test. The in-plane minor strain and in-plane minor strain rate evolution for the LHS and RHS of the S160 is presented in Figure 6(a). The onset of wrinkling is defined as the instant in time when the in-plane minor strain rate reaches a minimum, highlighted by the vertical lines in Figure 6(a). Additionally, the in-plane minor strain distribution throughout the LHS and RHS of the S160 specimen at the onset of wrinkling established by methodology B is shown in Figure 6(b).

To assess if this methodology is independent from which side the specimen wrinkles, this methodology was successfully implemented in a coordinate of section AA' that registered the lowest point in the Z-displacement.

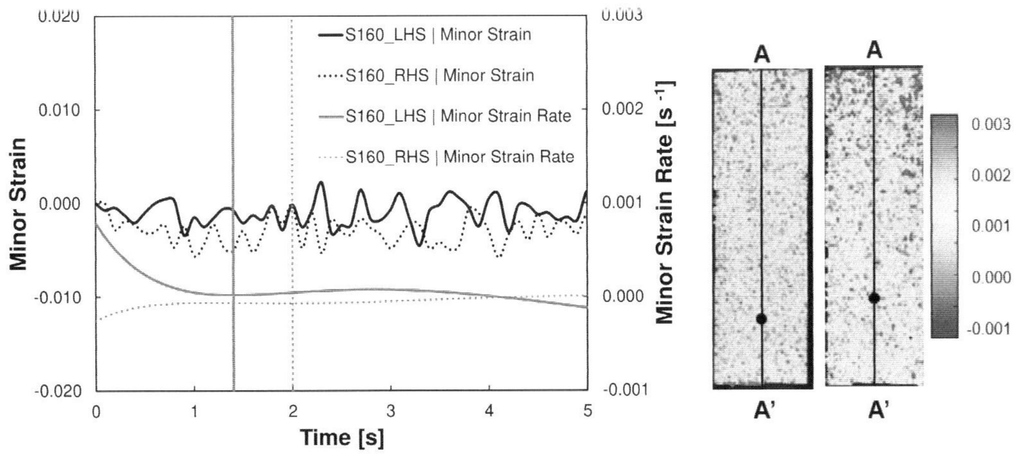

Figure 6. (a) In-plane minor strain and strain rate evolution over time for specimen S160 and **(b)** in-plane minor strain distribution at the onset of wrinkling according to methodology B.

3.3. Wrinkling Limit Curve

The experimental results obtained from applying methodologies A and B to the shear specimens are present in Table 3, where the average for the critical instant of time is presented along with the critical strain pairs. An accuracy of 0.002 in the measured strains was ensured by the DIC system.

Table 3. Experimental results for the onset of wrinkling determined by methodology A and B for the shear test specimens.

Specimen	Side	Methodology A			Methodology B		
		Critical time instant	Major strain	Minor strain	Critical time instant	Major strain	Minor strain
S160	LHS	7.3	0.00460	-0.004400	2.4	0.00279	-0.00100
	RHS	8.4	0.00470	-0.0036	2.1	0.00122	-0.00063
S180	LHS	8.2	0.00335	-0.00377	2.5	0.00194	-0.00311
	RHS	10.4	0.00417	-0.00343	1.8	0.00226	-0.00048

The critical time instants in Table 3 reveal that, according to methodology A, the onset of wrinkling occurs sooner in the LHS than in the RHS, for both the S160 and S180 specimens. For methodology B, the onset of wrinkling occurs at similar time instants in the LHS than in the RHS, for both the S160 and S180 specimens.

Figure 7 represents the critical strain pairs obtained from applying both methodologies to the shear test specimens. Two WLCs were defined, one for each methodology. The formability limits by wrinkling provided by methodology A and B are obtained by applying a linear approximation to the black and grey markers, respectively.

The critical strain pairs obtained from applying methodology A reveal a substantially higher degree of scattering when compared to those obtained by methodology B. In addition to this, the WLC obtained from methodology B has a slope of −0.65, indicating a slope corresponding to uniaxial compression in the principal strain space. The WLC from methodology A has a slope of −0.91, which corresponds to a slope of pure shear in the principal strain space.

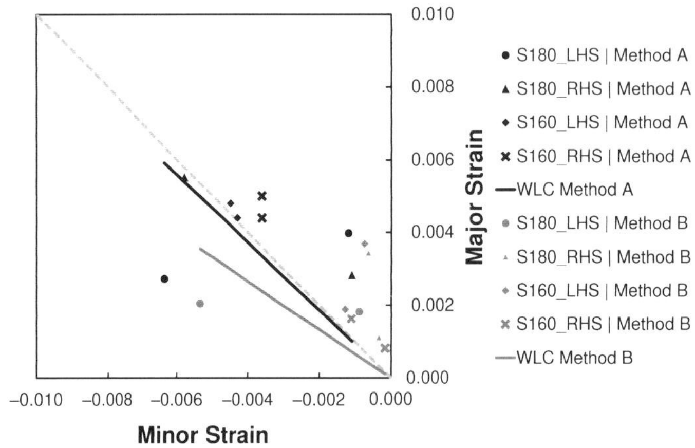

Figure 7. Formability limits by wrinkling determined by methodology A and B for a AA1050-O aluminium sheet with 1.5 mm thickness.

4. Conclusions

The two methodologies allowed to successfully obtain the critical strain pairs at the onset of wrinkling in the shear test specimens necessary to determine the wrinkling limit curve for a specific material.

The onset of wrinkling defined by methodology A occurs for higher critical strain values when compared to those obtained at the onset of wrinkling defined by methodology B. In addition, the onset of wrinkling defined by methodology A occurs at a stage in wrinkle development when the rate of wrinkle amplitude increase is significant.

References

[1] Kleiner, M.; Geiger, M.; Klaus, A. Manufacturing of Lightweight Components by Metal Forming. CIRP Ann. 2003, 52, 521–542.

[2] Martins, P.A.F.; Bay, N.; Tekkaya, A.E. Atkins AG. Characterization of fracture loci in metal forming. Int. J. Mech. Sci. 2014, 83, 112–123.

[3] Kasaei, M.M.; Moslemi Naeini, H.; Liaghat, G.H.; Silva, C.M.A.; Silva, M.B.; Martins, P.A.F. Revisiting the wrinkling limits in flexible roll forming. J. Strain Anal. 2015, 50, 528–541.

[4] Du, B., Wang, X., Wang, J., Yang, H., Chen, X., & Yang, Z. (2022). The establishment of critical wrinkling judgment line for shear wrinkling instability of thin plate and its influencing factors. Thin-Walled Structures, 109226. https://doi.org/10.1016/j.tws.2022.109226.

[5] Tang, H., Wen, T., Hong, J., Zhou, Y., & Zhang, L. (2022). Analysis of shear stress wrinkling of asymmetric sheet specimen under offset loading. Journal of Mechanical Science and Technology, (3), 1451–1457. https://doi.org/10.1007/s12206-022-0232-5.

[6] Santos JAO, Magrinho JPG, Silva MB. Physically-Based Methodology for the Characterization of Wrinkling Limit Curve Validated by Yoshida Tests. Metals. 2023; 13(4):746. https://doi.org/10.3390/met13040746.

[7] Du, B.; Xie, J.; Li, H.; Zhao, C.; Zhang, X.; Yuan, X. Determining factors affecting sheet metal plastic wrinkling in response to nonuniform tension using wrinkling limit diagrams. Thin-Walled Struct. 2020, 147, 106535.

[8] ASTM Standard E8/E8M. Standard Test Methods for Tension Testing of Metallic Materials. West Conshohocken, USA: ASTM International. 2016. https://doi.org/10.1520/E0008_E0008M-16A.

[9] Sharma, S.; Singh, R.; Kumar, S. Mechanical Anisotropy of Aluminium AA1050 and Aluminium Alloy AA6016 produced by Accumulative Roll Bonding. Int. J. Innov. Eng. Technol. 2016, 7, 350–356.

IOP Conf. Series: Materials Science and Engineering **1307** (2024) 012046 doi:10.1088/1757-899X/1307/1/012046

The effect of elastic and plastic stresses on the electrical resistivity of conductive materials

F. Steinlehner[1*], M. Ott[1], B. Stiegeler[1], L. Maier[1], Q. Nguyen[1] and W. Volk[1]

[1] Chair of Metal Forming and Casting, Technical University of Munich, Walther-Meissner-Strasse 4, D-85748 Garching, Germany

* Florian.Steinlehner@tum.de

Abstract. The mobility sector has been responsible for large emissions of harmful greenhouse gases worldwide for years. E-mobility can contribute to a reduction of these emissions. For a faster spread and acceptance of electric vehicles, the achievable range and thus the efficiency of the individual components in the drive and charging train is essential. In addition to the battery and the electric machine, losses in the cables and bus bars are playing an increasingly important role. The cross-sectional shape, material and manufacturing technology of these conductive materials depends on the voltage and power to be transmitted and may still change in the future, especially in the field of fast charging and increased voltage.
The aim of this work is to investigate the influence of the mechanical stress condition of copper, brass and aluminium materials on their electrical conductivity. The conductivity is measured by means of 4-wire measurements under defined tensile and compressive loads. The effects of elastic and plastic strains are taken into account. The findings can be used for an optimized manufacturing strategy for conductive components for future applications.

1. Introduction

In order to achieve sustainability goals in general, but especially in the automotive sector, efficient generation, provision and transportation of energy is necessary. Electromobility is characterized by a high level of efficiency when converting electrical energy stored in batteries into mechanical energy through the electric motor. Compared with efforts to achieve compact and high-capacity batteries, storage with high energy density and efficient energy utilization in the electric motor, the focus has not yet been on the efficiency of the current carrying structures. [1]

In addition to the achievable range, fast charging of cars is crucial for the acceptance of the future transport transition. Very high power levels are achieved with fast charging using DC current. This leads to an increase in the required cross-sections, which also reduces the flexibility of the current-carrying cables and conductors. Where flexible cables were often used in the past, it is now necessary to use rigid power conductors manufactured by forming technology. Similar to current collectors and bus bars, there is a change in electrical properties in the formed areas, e.g. sharp radii. [1]

For the realization of structural current-conducting components, different forming processes are used, which lead to a different residual stress state in the formed area of the manufactured product. The aim of this work is to investigate typical current conductor materials with respect to the change in electrical resistivity as a function of the degree of forming. The electrical resistivity is measured using a four-wire measurement on round specimens for uniaxial tension.

2. State of the Art

According to equation (1), for an electrical conductor material with isotropic properties, the electrical resistivity ρ is defined as the quotient of the magnitude of the electric field E by the magnitude of the current density J [2]:

$$\rho = \frac{E}{J} \tag{1}$$

Elastic and plastic strains influence the value of ρ. While an increase in electrical resistivity with increasing strain was observed for most of the metallic materials examined, for example, nickel shows a decrease at low strains [3]. Grain boundaries [4] and dislocation density [5] were identified as influencing variables responsible for the change in ρ at the microstructure level. ρ can also be affected by the heat treatment history [6]. Furthermore, ρ varies with temperature T, whereby the following linearized relationship in equation (2) with the temperature coefficient α can be derived in the range around a reference temperature T_0:

$$\rho(T) = \rho(T_0) * (1 + \alpha\,(T - T_0)) \tag{2}$$

The electrical resistivity ρ can be determined by measuring the electrical resistance R on a known geometry. For a cylindrical conductor of length l, cross-section A and homogeneous current density J, the equation (3) follows:

$$R = \rho\,\frac{l}{A} \tag{3}$$

For determining small values of R, the four-wire measurement method is established. For the measurement of electrical steel sheets, this method is specified in more detail in DIN EN IEC 60404-13 [7]. During the measurement, the sample is contacted at four points. An electric current is applied to the two outer contact points, causing a homogeneous current density in between. At the two remaining contact points, located between the outer contact points, the potential difference U due to the electrical resistance is sensed. By using two separate pairs of connecting electrodes, the wiring and contact resistance are eliminated [2]. As a result, the electrical resistance R according to Ohm's law is quantified.

3. Experimental Setup

Strain controlled tensile and compression tests were performed on an Allround Line 150 kN tensile-compression testing machine from ZwickRoell GmbH, Ulm, Germany. The change in the cross-section of the specimens was recorded using a stereo DIC system, GOM Aramis SRX from Carl Zeiss GOM Metrology GmbH, Braunschweig, Germany. For the four-wire measurement a Keithly 2400 SourceMeter was used as the current source and a Keithly 2182A nanovoltmeter from Keithley Instruments SRL, Cleveland, Ohio, USA was used to record the voltage. The temperature of the samples was recorded during the tests using a PT-1000 sensor. The experimental setup is shown in Figure 1. The materials listed in Table 1 are typically used in electrical conductors and cover a broad variety of electrical conductivity:

Table 1. Investigated materials.

	Cu-ETP CW004A	**AlMgSi0.5 EN AW-6060**	**Brass CuZn39Pb3, EN12164**	**Free-cutting steel, 1.0715**
electrical conductivity	58.58 Ms/m	28-30 Ms/m	15 MS/m	1.71 MS/m

Figure 1. Experimental setup.

The electrical resistivity is a material constant that depends on the temperature and the residual stress state. The four-wire measurement is used to determine the electrical resistivity. In the multi-conductor measurement, the resistance is measured using two voltage taps U1 and U2 at a defined distance from the resistor while current flows through the conductor. The current flows between the two external conductors I1 and I2. To maintain the distance between the contacting pins, a rigid mounting device is used. The schematic structure of the four-wire measurement is explained in Figure 2:

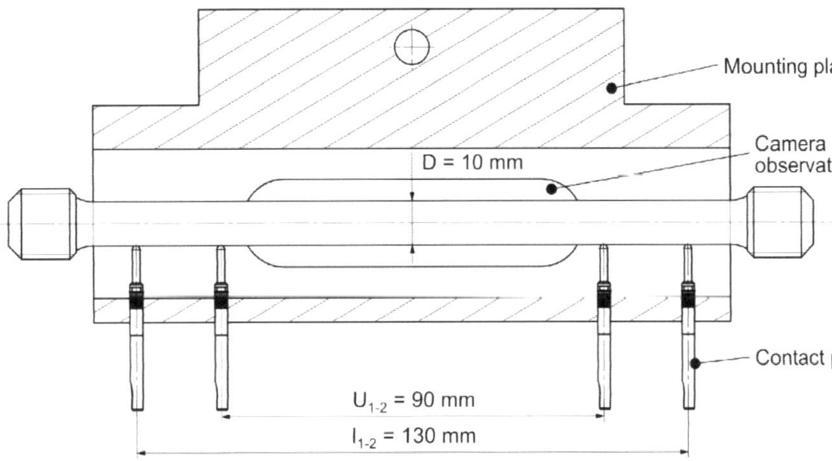

Figure 2. Specimen and contacting system for the four-wire measurement.

The specific resistance ρ is calculated according to the following equation (4) from the measured voltage $U_{1\text{-}2}$, the flowing current $I_{1\text{-}2}$, the current cross-sectional area A and the linear distance of the voltage measurement:

$$\rho = \frac{U_{1-2}}{I_{1-2}} * \frac{A}{l_U} \tag{4}$$

The main test parameters and the holding times for the cyclic tensile and compression tests are summarized in the following table 2:

Table 2. Experimental and process parameter.

Initial Specimen Diameter	Distance length I1 – I2	Distance length U1 – U2	Ambient temperature	Holding Time hysteresis cycle	Hysteresis control
10 mm	130 mm	90 mm	20 °C	10 s	strain-rate

An electrical simulation was carried out to ensure a uniform distribution of the electrical distribution for the voltage measurement area. The simulation was performed using Abaqus 2022, Dassault Systèmes, Johnston, RI, USA. An electrical current was applied to two contact points at a distance of 130 mm, analogue to the test setup. For a distance of 90 mm between the voltage measuring points, a homogeneous distribution of the EPG (electrical potential gradient) is maintained as shown in Figure 3.

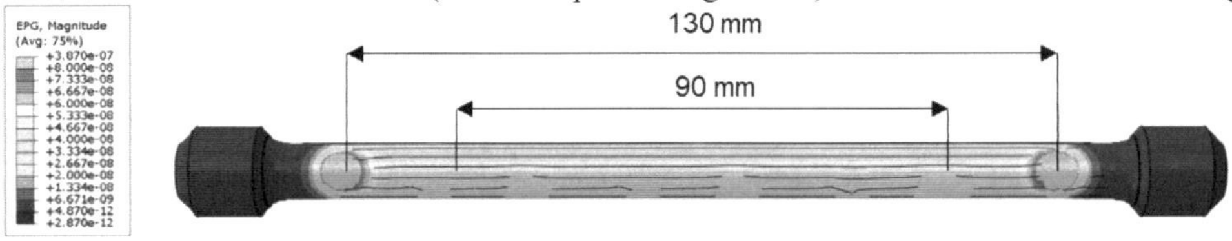

Figure 3. Distribution of the EPG (el. potential gradient) for the four-wire measurement.

4. Results

Strain-controlled cyclic tensile tests were performed to determine the electrical resistivity as a function of the degree of deformation. In order to achieve a high resolution in the area of elastic deformation, the strain per cycle was chosen to be lower. After the start of the flow, the elongation steps per cycle were increased. The temperature was recorded for all tests, whereby no significant sample heating occurred due to the current flow or material deformation. The temperature change during the tests was within the measurement accuracy range of ± 0.1 °C for all materials.

4.1. Results for Steel

For the steel material Figure 4 shows the force curve over the increasing elongation and the resulting relative change in the specific resistance per cycle is shown in Figure 5. The strain-controlled hysteresis had an increase of 0.005 mm/cycle until the point of yielding and 0.05 mm/cycle afterwards:

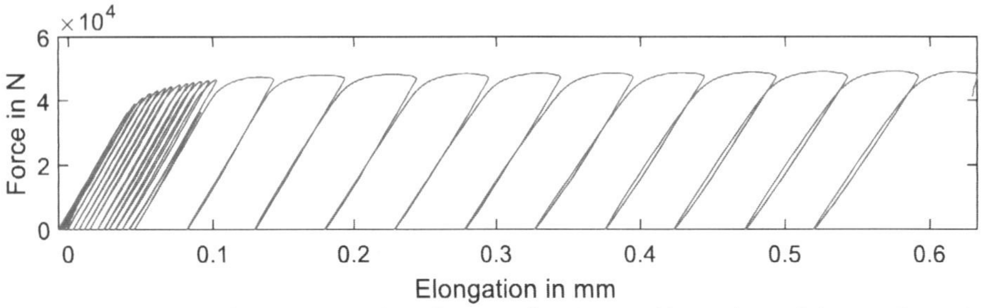

Figure 4. Hysteresis of the force-elongation diagram for the tensile testing of the steel specimen.

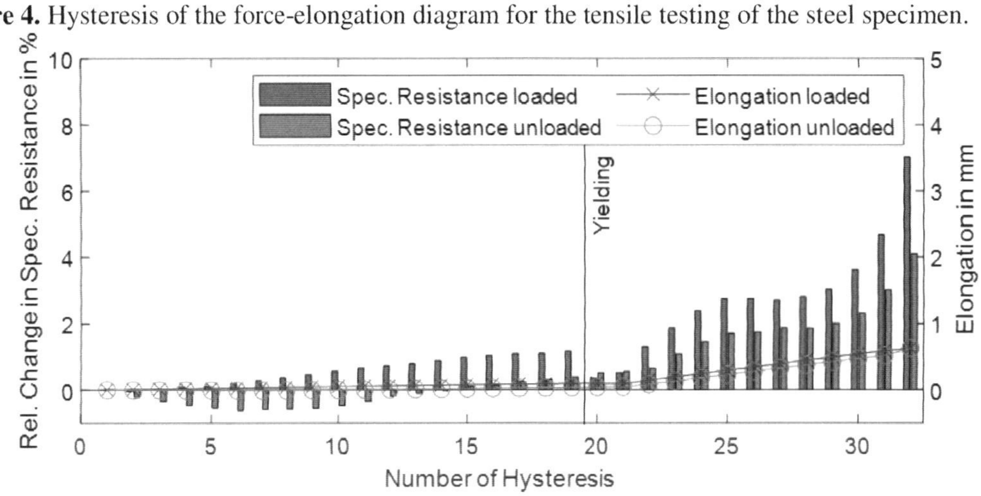

Figure 5. Relative change in specific resistance and elongation for tensile testing Cu-ETP for hysteresis cycle.

The specific resistance for steel increases steadily with increasing uniaxial tensile load. With the onset of yielding, there is a slight decrease in the specific resistance, which then increases until failure. As expected, the breakpoints show higher resistance values in the loaded state of hysteresis. The initial electrical resistivity used for the relative change of the steel specimen is $149.58*10^{-3}\,\frac{\Omega * mm^2}{m}$.

4.2. Results for Copper (Cu-ETP)

Figure 6 shows the force curve for each hysteresis cycle over the increasing elongation of the copper material and the resulting relative change in the specific resistance per cycle is shown in Figure 7. The increase of the strain was chosen to be 0.005 mm/cycle before yielding and 0.05 mm/cycle after.

The initial electrical resistivity used for the rel. change of the Cu-ETP specimen is $16.55*10^{-3}\,\frac{\Omega * mm^2}{m}$.

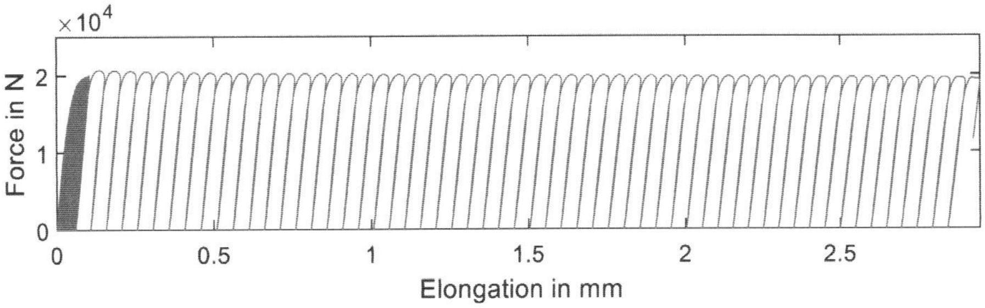

Figue 6. Hysteresis of the force-elongation diagram for the tensile testing of the copper specimen.

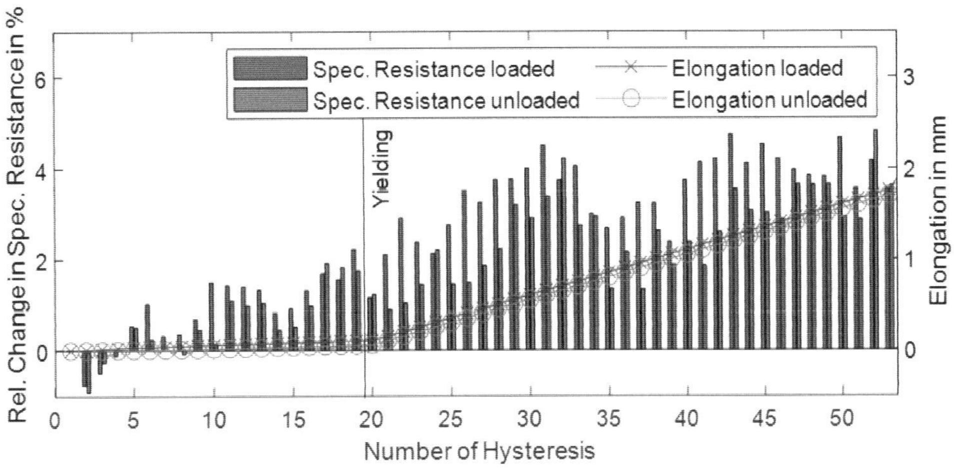

Figure 7. Relative change in specific resistance and elongation for tensile testing of Cu-ETP for hysteresis cycle.

The tested copper specimen shows an increase in the specific electrical resistance with increasing uniaxial tensile load. In the yielding range, there is a slight decrease in resistance for the unloaded and a strong increase during the loaded plateaus of the hysteresis. Due to the high ductility of the copper material, necking only occurs at very high strain values. In the area before necking, a reduction in the specific electrical resistance occurs at an elongation of approx. 1.2 mm before it subsequently increases again.

4.3. Results for Aluminium (EN AW-6060)

For the brass material Figure 10 shows the force curve with each hysteresis cycle over the increasing elongation and the resulting relative change in the specific resistance per cycle is presented in . The strain-controlled hysteresis increased by 0.005 mm/cycle until the point of yielding was passed and

0.05 mm/cycle afterwards. The initial electrical resistivity used for the relative change of the brass specimen is 33.46*10^{-3} $\frac{\Omega * mm^2}{m}$.

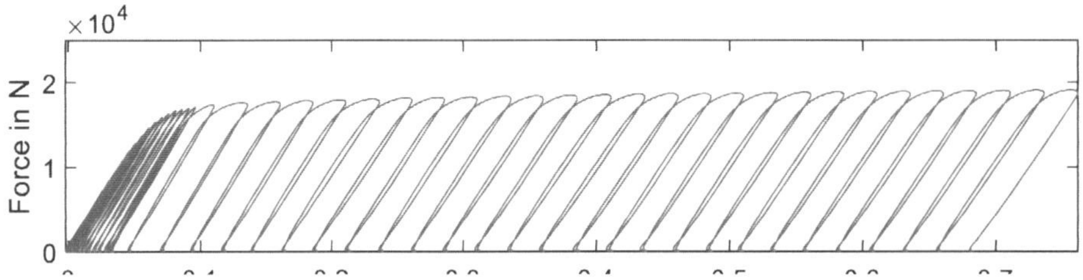

Figure 8. Hysteresis of the force-elongation diagram for tensile testing of the aluminium specimen.

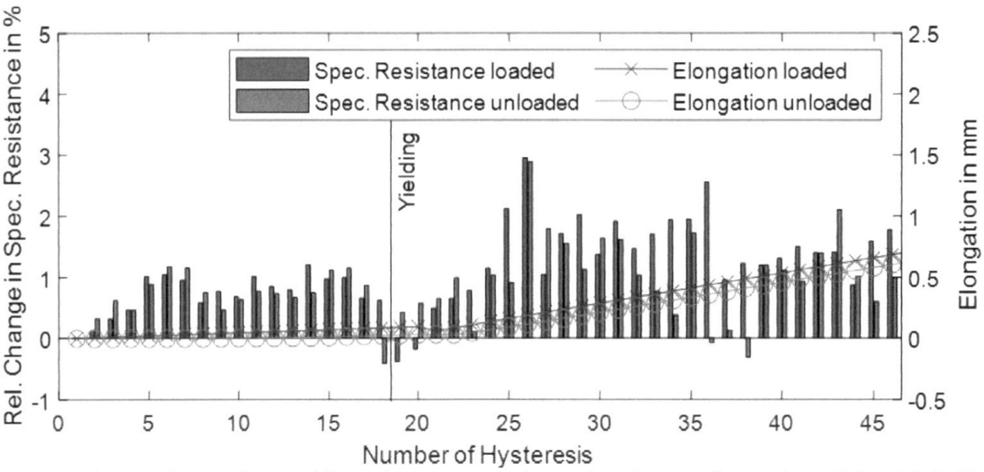

Figure 9. Relative change in specific resistance and elongation for tensile testing of the AlMgSi0.5 material for hysteresis cycle.

The aluminium specimen has the smallest change in electrical resistivity of all tested samples. While there is an increase of 1 % in the area before yielding, the value at the point of yielding reaches the initial value of the resistance. After yielding, there is a very fluctuating behaviour and a strong decrease in the unloaded hysteresis plateaus in the area before constriction. In this area, the uniform reduction in diameter of the tensile specimen is superimposed on the decreasing voltage of the four-wire measurement.

4.4. Results for Brass (CuZn39Pb3, EN12164)

For the brass material Figure 10 shows the force curve with each hysteresis cycle over the increasing elongation and the resulting relative change in the specific resistance per cycle is shown in Figure 11. The strain-controlled hysteresis had an increase of 0.005 mm/cycle until the point of yielding was passed and 0.05 mm/cycle afterwards. The initial electrical resistivity used for the relative change of the brass specimen is 68.22*10^{-3} $\frac{\Omega * mm^2}{m}$.

The course of the specific resistance of the brass tensile specimen shows a steady increase with increasing tensile load. In the area shortly before necking at an elongation of approx. 3 mm, there is a slight decrease in the electrical resistance. After necking began, the four-wire measurement lost contact with the sample, meaning that the measured values after necking could not be utilised.

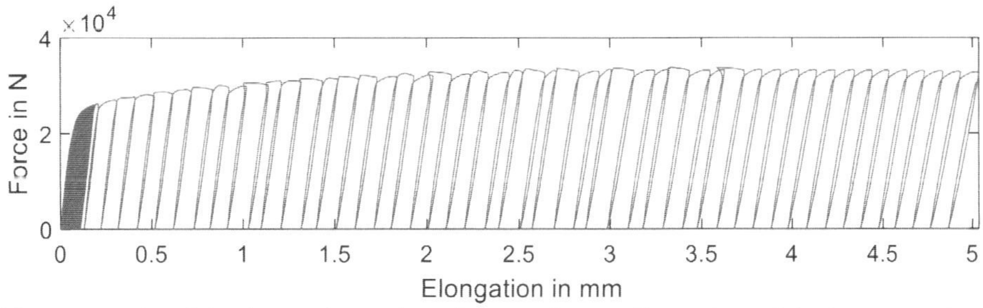

Figure 10. Hysteresis of the force-elongation diagram for tensile testing of the brass specimen.

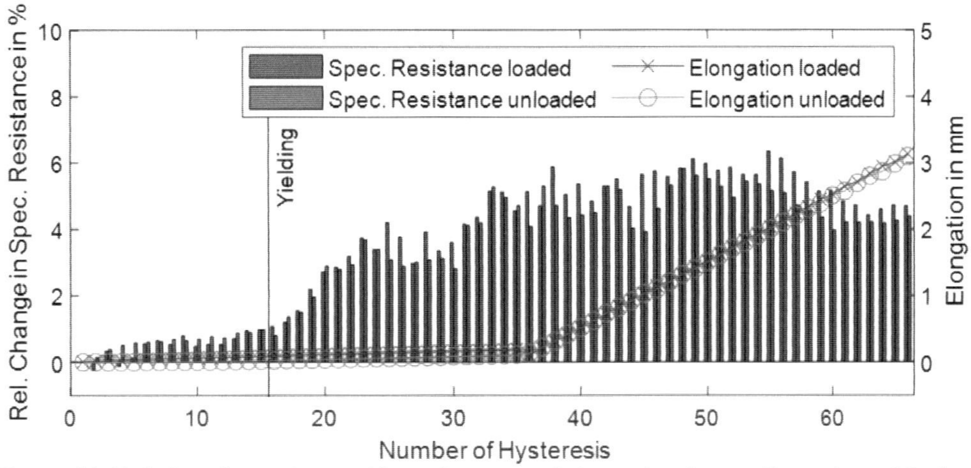

Figure 11. Relative change in specific resistance and elongation for tensile testing of the brass material for hysteresis cycle.

4.5. Results for Brass (CuZn39Pb3, EN12164), compression testing

In addition to the previous tests, a pressure test was performed on a brass specimen. The compression test was carried out with the same test setup as the tensile tests, but the control of the compression was done force controlled, with an increase of 500 N per hysteresis cycle.

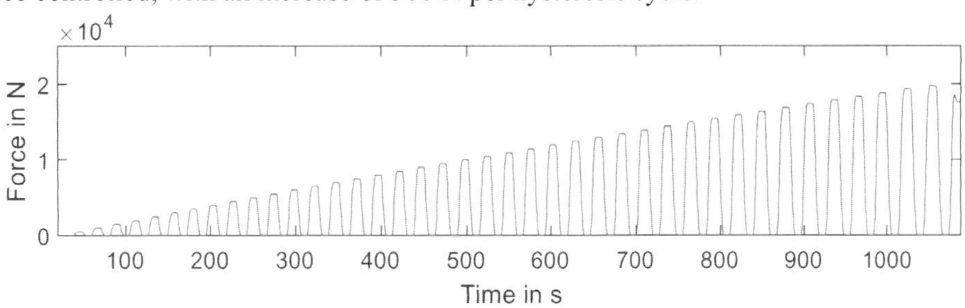

Figure 12. Hysteresis of the force diagram for the compression testing of the brass specimen.

Due to the length of the sample, no significant plastic deformation occurred. The test was stopped when the specimen buckled.

Figure 12 shows the increasing compression force with each hysteresis cycle over time and the resulting relative change in the specific resistance is shown in Figure 13. The initial electrical resistivity used for the relative change of the brass specimen is $68.20*10^{-3}\,\frac{\Omega*mm^2}{m}$. The Force curve indicates that there is no yielding occurring while performing the test. In contrast to the tensile tests, the compression test shows a decrease in the relative specific resistance shortly before the specimen buckles.

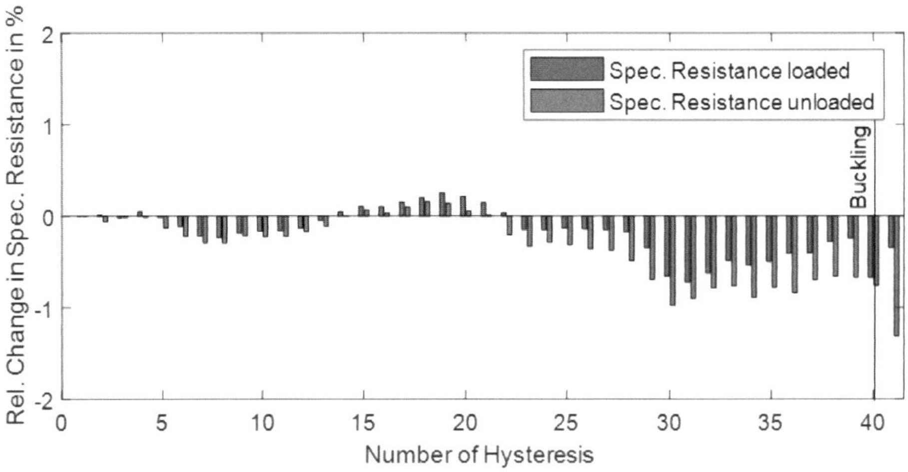

Figure 13. Relative change in specific resistance and elongation for compression testing of the brass material for hysteresis cycle.

5. Conclusion and Outlook

Tensile tests were carried out on steel, aluminium, brass and copper materials. The work shows a change in specific electrical conductivity of up to 8 % due to uniaxial loading. This must be taken into account when designing and dimensioning future conductors. An exemplary compression test for the brass material shows a decrease in the specific resistance due to a uniaxial compressive load.

The investigations of the materials with tensile test provide an initial indication of the sensitivity of the specific resistance. Based on the promising results of the compression test and the potential increase in electrical conductivity, tests with pressure samples will be carried out in further work. As electrical components are usually bent, further work will be executed for bent samples. Production processes with different mechanisms and the resulting residual stress states, such as pressure-superimposed free-form bending or rotational tensile bending, can be used to utilise the advantages and increase the performance of electrical conductors. As the results show that the electrical conductivity decreases with plastic elongation due to tensile stresses, it would be interesting to investigate the possibility of stress relief annealing in greater depth for subsequent studies. Stress relief annealing can reduce residual stresses and possibly improve electrical conductivity.

References

[1] Stanton, N. A. (2018). Advances in Intelligent Systems and Computing: Bd. 597. Advances in Human Aspects of Transportation: Proceedings of the AHFE 2017 International Conference on Human Factors in Transportation.

[2] Miccoli, I., et al. 2015. The 100th anniversary of the four-point probe technique: the role of probe geometries in isotropic and anisotropic systems. *J. Phys.: Condens. Matter* 27, 22.

[3] Kuczynski, G. C. 1954. Effect of Elastic Strain on the Electrical Resistance of Metals. *Phys. Rev.* 94, 1, 61–64.

[4] Bishara, H., et al. 2021. Understanding Grain Boundary Electrical Resistivity in Cu: The Effect of Boundary Structure. *ACS Nano* 15, 10, 16607–16615.

[5] Zürcher, R., et al. 1995. Dislocation resistivity in Cu: dependence of the deviations from Matthiessen's rule on temperature, dislocation density and impurity content. *J. Phys.: Condens. Matter* 7, 18, 3515–3528.

[6] Shakouri, M., et al. 2021. Effect of the heat treatment on the electrical resistivity and magnetization reversal behavior of MnAl alloys. *Materials Science and Engineering: B* 274.

[7] *DIN EN IEC 60404-13:2020-06, Magnetische Werkstoffe_- Teil_13 (IEC_60404-13:2018); Deutsche Fassung EN_IEC_60404-13:2018*. Beuth Verlag GmbH, Berlin.

Reinforcement of thin hot-stamped components by fiber reinforced plastic structures with optimized fatigue strength properties

B-A Behrens[1], **S. Hübner**[1], **P. Müller**[1], **T. Fünfkirchler**[1], **J. Wehmeyer**[1*], **K. Dilger**[2], **S. Hartwig**[2], **C. Gundlach**[2]
[1] Leibniz Universität Hannover, Institute of Forming Technology and Forming Machines, An der Universität 2, 30823, Garbsen, Germany.
[2]Technical University Braunschweig, Institute of Joining and Welding, Braunschweig, Langer Kamp 8, 38106, Germany
* Corresponding contact: +49 511 762 3813, wehmeyer@ifum.uni-hannover.de

Abstract. Multi-material structures in the automotive industry hold great potential for lightweight design, body construction, and functionalization due to their favorable mechanical properties and reduced structural weight. The combination of metal and plastic, in particular, is commonly used to enhance the overall properties of the end product when compared to single-material structures.
This paper describes a process development with a hot-stamping and an extrusion tool. By means of this tool, a thermally assisted extrusion process can be used to join GMT (Glass Mat reinforced Thermoplastics) and 22MnB5 steel in a single process step. Through adhesion, the GMT adheres to the rough surface of the AlSi(aluminum-silicon)-coated 22MnB5. Test components were manufactured and through static tests the influence of process parameters was evaluated. Assuming that the parameters determined for the reference component are already sufficient for a design in the vehicle, the reduction of the steel thickness of the structure from 1.5 mm to 1.2 mm can be recommended on the basis of the results obtained. This is accompanied by a reduction in the mass of the test structure used while maintaining or improving its dynamic and static properties. Further weight savings appear possible through further component and process optimization.

Keywords: Advanced/smart materials, multi-material concepts/solutions, lightweight structures

1. Introduction

People's consumer behaviour clearly indicates a growing emphasis on safety and comfort-related requirements. This trend is observable in various sectors, especially in the automotive industry. The installation of more mechanical and electronic components adds weight to vehicles, necessitating a strong focus on lightweight design in the body-in-white construction. To achieve environmentally and economically optimal outcomes, considerable investments are being directed towards research and development in this regard.

Hence, it is essential to secure weight reduction possibilities by leveraging lightweight materials and integrating them with advanced manufacturing techniques. Broadly speaking, principles of lightweight design can be categorized into three groups: Constructive lightweight design, lightweight material design and conceptual lightweight design. Shape-hardened steels, particularly relevant for crash scenarios, exhibit substantial strength values and possess the ability to absorb significant amounts of energy [1].

Through the integration of metals and fiber-reinforced plastics (FRP) to create hybrid components, the benefits of both material groups can be capitalizing on synergy effects and fully maximizing the potential for lightweight construction. The production of these hybrid components necessitates a tailored handling and forming process to accommodate varying material characteristics and ensure a robust, large-area connection between the two components.

In this publication, a process chain was established to combine form-hardened sheets with additional fiber-reinforced plastic stiffening, all without the use of adhesion promoters. Adhesion promoters refer to organic additives or layers intended to enhance the bonding of FRP reinforcing structures. It's important to note that the AlSi layer of 22MnB5 used is not considered an additive material in the context of an adhesion promoter as this layer is present from the factory.

The objective of this research is to achieve weight reduction by decreasing the sheet thickness of the steel component while optimizing stiffness and necessary crash properties through the reinforcement of FRP.

At first, the research discusses the test die designed for the investigations, featuring interchangeable punches for various hot stamping and impact extrusion components. Afterwards, the implementation of the process chain is outlined. The publication concludes with a comprehensive assessment of the test hybrid components manufactured and monolith parts, summarizing the testing and evaluation processes.

2. State of the Art

2.1. *Fiber-reinforced thermoplastics*

FRP offer a promising avenue to address light weight solutions [2]. In working towards future lightweight construction goals, the utilization of FRP has witnessed a steady rise over recent years, establishing itself as a pivotal element. Semi-finished FRP products can be categorized into diverse types, including long fiber reinforced thermoplastics (LFT), glass mat reinforced thermoplastics (GMT), fiber reinforced unidirectional tapes (UD tapes), and organic sheets [3].

Forming processes are contingenting on the matrix material as this represents the connection to the metal component [4; 5]. In addition to consolidating the matrix materials, alternative joining methods such as clinching can also be employed under high-temperature conditions [6].

Although FRPs are classified as a distinct material class, they fundamentally comprise a composite of multiple materials. Within this composite structure, the individual components have well-defined roles: the high-strength fibers bear the mechanical compressive forces and loads, while the matrix in which the fibers are embedded serves to secure them in place, provide support, and transmit the mechanical stresses to the fibers [7].

2.2. *Joining of hybrid FRP steel components*

Numerous approaches exist for the production of hybrid structures incorporating a combination of FRP and metals. These approaches typically involve the separate manufacturing of the respective components, followed by their subsequent bonding to create the hybrid structure. Recent literature has highlighted the substantial potential for lightweight construction through such hybrid structures, as demonstrated in various prototype parts.

One example of this is at the Technical University of Dresden, where a B-pillar geometry was reinforced using a fiber-reinforced composite (FRC) sheet and glass-fiber-reinforced polyamide. The FRC sheet was bonded to the steel component using an adhesion promoter [8]. In another example [9], a cross-member was produced from thermoplastics, where the components were manufactured independently and later bonded using adhesive solvents. Another noteworthy application of hybrid components can be observed in the roof frame of the model A6 by the company Audi AG. Here, a steel profile with reinforcing ribbing made of plastic was created using injection moulding (see Figure 1). Compared to the previous models, this modification resulted in a weight reduction of 0.3 kg and a significant improvement in force absorption of 200 %, all without increasing costs [10; 11]. Potential applications for components with localized or full-surface reinforcements in motor vehicles include side impact beams, bumper cross members, roof frames, sills, and B-pillars.

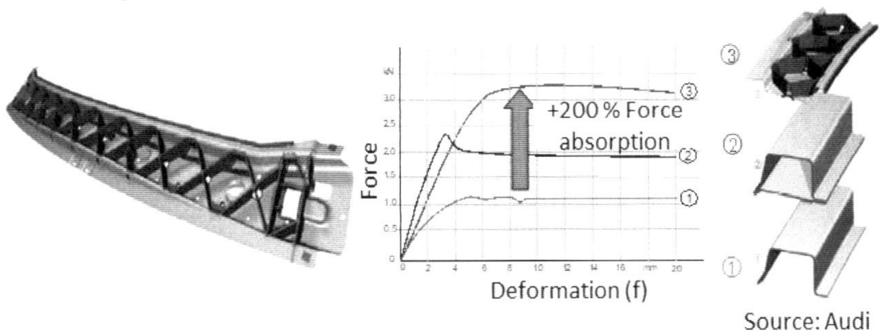

Figure 1. Roof frame as hybrid part in the Audi A6 and improved force absorption according to [11]

Moreover, hybrid lightweight components are detailed in the subsequent patent disclosures: DE102009035777A1[12] elucidates a structural element, predominantly comprising a bending cross member for a motor vehicle connected to vertically oriented energy absorption elements. The connection is established through internal high-pressure forming of the hollow cross-section aluminum profile with a thermoplastic melt (PP or PA reinforced with carbon or aramid fibers), creating a form-fit junction. DE202009012432U1[13] outlines a car body component comprising an outer skin component (preferably aluminum) and a rear-molded FRC joint through form fit, enabling a weight reduction of approximately 60% compared to a pure steel structure. The lightweight component presented in DE102009042272A1 [14], specifically a body component, incorporates a steel or aluminum sheet and a reinforcing structure made of plastic. The metal component is physically bonded to the reinforcing structure via a plastic reinforcing layer. Fibers, preferably glass fibers, can be integrated into both the plastic reinforcing layer and the plastic reinforcing structure. The reinforcing structure is produced either through extrusion followed by subsequent pressing or injection molding.

As evident, multiple joining techniques exist for the creation of hybrid structures integrating fiber-reinforced plastics and metals [15, 16]. This aspect holds immense significance for the performance of the hybrid component, as delamination of the reinforcement structure could result in a significant shift in both static and dynamic performance. In addition to adhesive bonding, the direct fusion bonding of the thermoplastic FRC component to the metal structure by melting its thermoplastic matrix component can offer advantages in terms of time and energy efficiency.

In this study, the bonding of reinforcement structures made of FRP to press-hardened components without necessitating additional adhesion promoters or surface pretreatment, as demonstrated in [17] was explored. This aspect is presently underrepresented in current research literature. The aim was to investigate the introduction and bonding of the GMT rib structure via extrusion molding to enhance the buckling stiffness and buckling behavior of the U-profile, aligning with the principles of constructive lightweight design. The investigation involves testing hybrid demonstrator parts to establish fundamental processing parameters. Subsequently, the results from the demonstrator part are meticulously assessed and compared against a monolithic steel structure. Ultimately, this comparative analysis aids in evaluating the lightweight potential of components manufactured using the approach of the thermal press joining.

3. Process-chain and experimental set-up

3.1. Materials

The metallic adherend consists of press-hardened 22MnB5 steel coated with aluminum silicon (AlSi). The glass mat reinforced thermoplastic (GMT) is produced by Mitsubishi Chemicals and constitutes a polyamide 6 thermoplastic with a 32% tangled fiber content, available in sheet form. Essential material properties are outlined in Table 1 for reference.

Table 1. Material properties

22MnB5	
Yield Strength [MPa]	310 - 550
Tensile strength [MPa]	500 - 700
Material thickness [mm]	1.5
Coating	AlSi

[Volkswagen AG]

GMT (Quadrant EP589-8)	
Matrix material	Polyamide 6
Fibre content [%]	32
Melting temperature [°C]*	220
Elasticity Modulus [MPa]	6620
Tensile strength [MPa]	88

* measured with dynamic difference calorimetry according to DIN EN ISO 11357-2

3.2. Demonstrator part and process route

This study focuses on exploring the benefits of merging steel press-hardening with GMT reinforcement at a demonstrator scale. To achieve this objective, a hybrid demonstrator part specifically tailored for this purpose was conceptualized (refer to Figure 2 a). Subsequently, the manufacturing process involved the integration of a two-stage tooling concept, as depicted in Figure 3 b, to produce the hybrid demonstrator part.

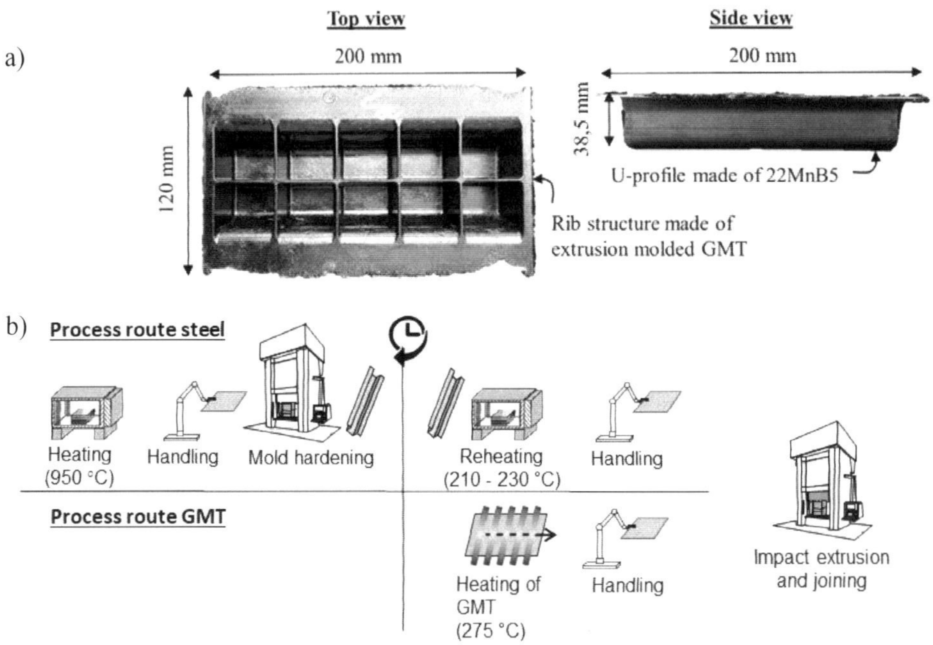

Figure 2. a) Dimensions of the demonstrator part, **b)** Process steps for producing the hybrid demonstrator part

In the initial of the production process of the hybrid demonstrator, a press-hardening tool is used to manufacture the hardened U-profiles using pre-heated 22MnB5 sheets, serving as the foundational components for the subsequent demonstrator parts. The press hardening tool is equipped with cooling channels to ensure rapid sheet cooling. In the second stage, the extrusion molding tool is utilized to introduce GMT ribs. This entails executing a combined extrusion molding and joining process once the formed steel profiles and GMT sheets have been simultaneously heated to their respective temperatures in different furnaces. Figure 3 provides a detailed perspective of the tool's design. To prevent rapid cooling of the GMT material and the reheated 22MnB5 sheets, the tool is heated by three heating cartridges in the lower part and two heating cartridges in the upper part. Through this process chain, a comprehensive surface joint is formed between the press-hardened U-profile and the GMT rib structure. Given that the joint is established through the direct bonding of the molten thermoplastic PA6 matrix of the GMT to the coarse Al-Si-Fe surface of the heated 22MnB5, there is no need for surface treatment or texturing of the U-profile to achieve a reliable bonding strength.

To assess the bonding characteristics between the U-profile and rib structure, as well as the overall performance of the hybrid component, demonstrator parts with varied structural (radius of the ribs) and processing parameters (tool and sheet temperatures) are produced. In terms of structure, the standard configuration employs a metal sheet thickness of 1.5 mm, matching the thickness of a monolithic, press-hardened U-profile without a rib structure, serving as the benchmark state-of-the-art reference. Additionally, a demonstrator version with reduced metal sheet thickness (1.2 mm) was defined to assess the potential for lightweight. In the second stage of the tooling concept, different radii in the foot area of the GMT ribs are achievable by modifying the tool's upper part. Radii of 4.0, and 5.5 mm are considered for sensitivity analysis. Regarding process parameters, the joining force is set to 1750 and 2000 kN, while the holding time in the die is incrementally increased in steps of 10 and 30 seconds. Moreover, the reheat temperature of the 22MnB5 sheets ranges from 260 to 300 °C.

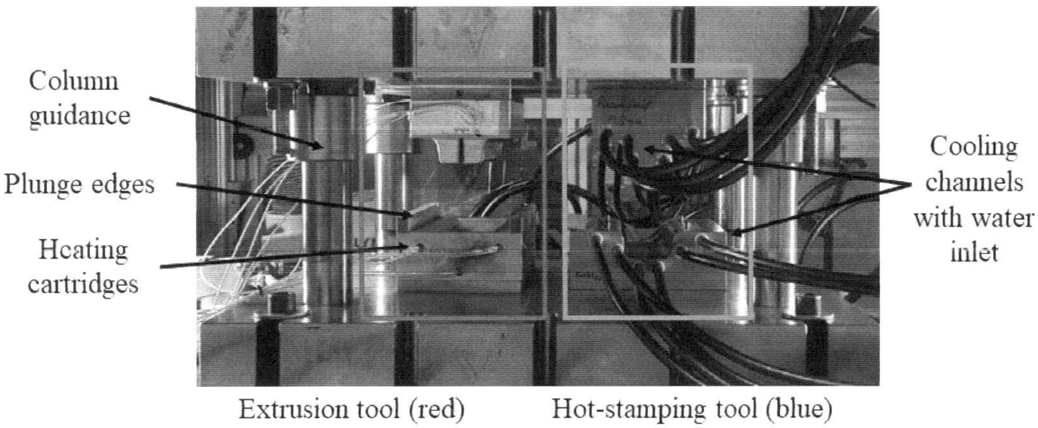

Figure 3. Extrusion tool (left) and hot-stamping tool (right)

4. Testing procedure

In order to facilitate quasi-static testing, test samples were extracted from the designated demonstrators using a wet grinder (ATM Brillant 250). This approach aimes to optimize the utilization of the available components for characterization purposes.

In terms of length and height, these test samples match the dimensions of the demonstrator, measuring 120 mm in length, 38.5 mm in height, and approximately 25 mm in width. The specimen's geometry does not align with any existing standardized geometry.

Quasi-static testing was performed using an Instron universal testing machine (type 5567). A customized 3-point bending test configuration was applied to assess bending stiffness (refer to Figure 4). The specimen was securely clamped between two supports. To prevent the central rib structure from contacting the support plate during full deformation, two identical metal cuboids were positioned

IOP Conf. Series: Materials Science and Engineering **1307** (2024) 012047 doi:10.1088/1757-899X/1307/1/012047

beneath the specimen's ends. Thin plates on the sides facilitated central alignment of the rib structure under the bending specimen, compensating for any manufacturing tolerances in the demonstrator. The bending die had a radius of 6 mm.

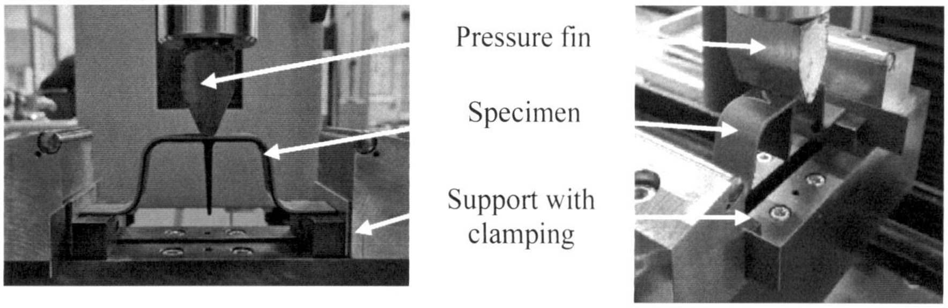

(a) (b)

Figure 4. Test setup for determining the bending stiffness from the front (a) and diagonally (b)

5. Results

5.1. Bending stiffness

The components underwent three-point bending tests. Within those, the bending force up to the point of initial delamination offered valuable insights, particularly regarding the bond strength of the metal and GMT in the shear direction. The delamination of the GMT was detected through a sudden deflection in the negative y-direction in measured force-displacement curves. This relationship is derived from the bending of the demonstrator resulting from a central compressive load.

In Figure 5, the test results are presented in an effect plot. The height of the bars reflects the magnitude of the mean values of the output factor, derived from a series of measurements, each associated with the variations indicated on the x-axis. For instance, for R4.0, the mean value was calculated and plotted across all samples with a rib radius of 4.0 mm. The series of measurements encompass additional variations in parameters; however, these remain consistent for each investigated input factor. This consistency allows for the formation of comparable mean values, such as for R4.0 and R5.5. However, it's worth noting that determining a standard deviation purely related to a specific parameter is challenging using this method, as it gets influenced by interactions with other variations in the measurement series. Hence, the presentation of the standard deviation has been omitted. A comprehensive evaluation of the significance of the results is somewhat limited, and only trends can be elucidated.

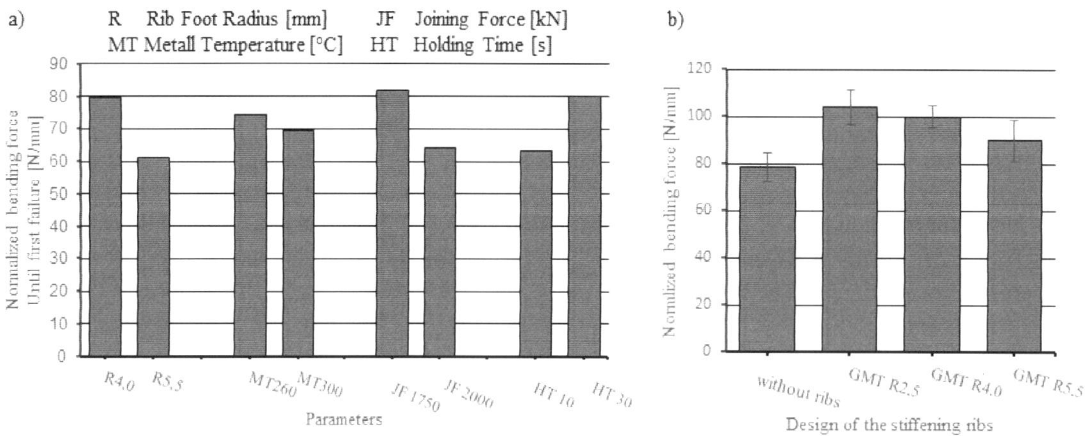

Figure 5. Effect plot for the bending force in relation to the investigated parameters

Basically, the following statements result from Figure 5 a). A smaller radius of the rib foot increases the achievable bending force up to delamination.

The combined effect of metal temperature and elevated GMT molding temperature at 260/75 °C and 300/95 °C does not significantly impact the achievable bending force up to delamination. This can be rationalized by the minimal bending force observed at the rib's base. The GMT matrix, being sufficiently flowable at the lower temperature combination, is able to distribute within the AlSi layer's undercuts over an ample duration, establishing stable mechanical interlocking after solidification. The transition to 300/95 °C does not yield further enhancement of this effect within the current series of measurements.

Regarding the joining force and holding time, they exhibit opposing influences on the attainable bending force. A longer holding time improves undercuts' filling on the surface, thus enhancing the mechanical clamping of the molten GMT matrix. Conversely, a lower value is advantageous for the joining force. One might anticipate the opposite effect, expecting higher force to result in improved interlocking of the GMT and metal at the interface. However, it is conceivable that the combination of high pressures and elevated joining temperatures, significantly surpassing the GMT matrix's melting point, leads to its degradation, thereby reducing the influence of mechanical interlocking.

Figure 5 b) shows a compression of a monolith part with the hybrid parts. The normalized bending force is for the hybrid parts up to 25 % higher compared to a pure steel part.

6. Summary and Conclusion

The primary objective of this study was to assess the bonding behaviour between the press hardened steel 22MnB5 and glass mat reinforced thermoplastic composites through fusion bonding, focusing on a demonstrator scale to showcase its viability for mass production. The newly developed process chain, integrating a two-step tooling setup, facilitated the bonding of GMT to ultra-high-strength steel on a demonstrator scale without relying on adhesion promoters or surface structuring. The natural surface roughness of the AlSi layer on 22MnB5 was sufficient to achieve reasonable bonding strength. Simultaneously, a reinforcing GMT rib structure could be incorporated.

Various process parameters, aligning with those yielding positive outcomes in previous lab-scale experiments (joining force, holding time, and metal temperature), were employed, along with different design parameters (metal sheet thickness, rib foot radius), for a sensitivity analysis. Bending, peel, and impact testing were conducted on the demonstrator part to comprehensively characterize the joint's quality under practically relevant load conditions and determine optimal parameters.

The static experiments revealed that an increase in joining force and holding time tend to negatively impact bending stiffness and bonding strength. Concurrently, rib radius and particularly metal temperature exhibited varying degrees of influence depending on the load direction of the part. Consequently, optimal parameter selection needs to primarily consider the dominant load direction of the part in its final installation position.

Assuming that the parameters determined for the reference component are already adequate for vehicle design, the reduction of the steel structure's thickness from 1.5 mm to 1.2 mm is recommended. This reduction accompanies a decrease in the demonstrator's mass while maintaining or even improving its dynamic and static properties. Further weight savings seem achievable through additional component and process optimization within the realm of constructive lightweight design. Based on these findings, this manufacturing technology provides potential for cost-effective production of hybrid components in volume production.

7. Outlook

However, the reheating of the metal parts after the press hardening process is very energy consuming. In future research a modified process rout, shown in Figure 6 should be investigated. In this process rout the press hardening process will be stopped at a temperature of 250 °C, so that the rest heat of the metal component is still high enough for the GMT extrusion process.

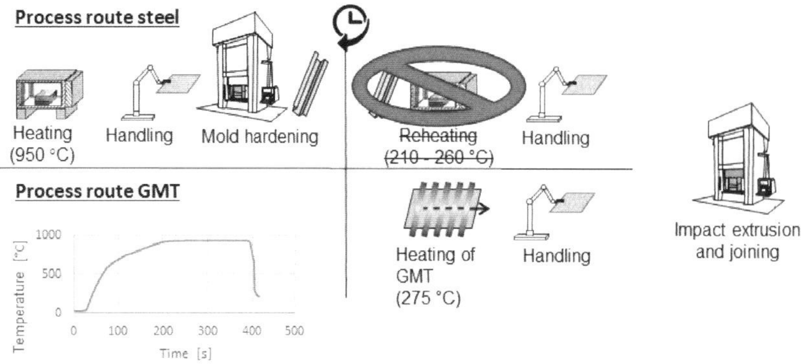

Figure 6. a) New process route, b) Temperature time course of press hardening process cancel at 220 °C

First tests at the Institute of Forming Technology a Machines showed that the 22MnB5 steel has already a hardness of 528 HV10 after cancelling the press hardening process at 220 °C.
With an automated handling system for a fast transfer from the press hardening tool to the extrusion tool and a programmable hydraulic press for a fast change of different forming force and holding time, there is a good possibility to use the residual heat. This will be focus of future studies.

Acknowledgements

The authors thank the Industrial Collaborative Research (IGF) and the Research Association for Automotive Technology (FAT) for the financial support of the research project "Reinforcement of thin mold-hardened components by means of FRP ribbing" (IGF-Code: 19597N).

References

[1] Hoff, C; Untersuchung der Prozesseinflussgrößen beim Presshärten des höchstfesten Vergütungsstahls 22MnB5, Bamberg. Meisenbach, 2007
[2] Behrens B-A, Neumann A 2016 Ermittlung der Prozessgrenzen bei der Umformung von Musterbauteilen aus Faser-Kunststoff/MetallVerbunden; Schlussbericht zu dem IGF-Vorhaben DFG-AiF-Cluster Großserientaugliche Prozessketten für hochintegrierte Bauteile aus hybriden FaserKunststoff/Metall-Verbunden Teilprojekt 4
[3] Friedrich K, Almajid A A 2013 Manufacturing of Advanced Polymer Composites for Automotive Applications. In: Applied Composite Materials 20, pp. 107-128
[4] Behrens B-A, Hübner S, Bonk C, Bohne F, Micke-Camuz M 2017 Development of a combined Process of Organic Sheet forming and GMT Compression Molding. In: Procedia Engineering 207, Cambridge, pp. 101-106
[5] Behrens B-A, Hübner S, Neumann A 2014 Forming sheets of metal and fibre-reinforced plastics to hybrid parts in one deep drawing process. In: Procedia Engineering 81, pp. 1608-1613
[6] Behrens B-A, Hübner S, et al. 2017 Forming and Joining of Carbon-Fiber-Reinforced Thermoplastics and Sheet Metal in One Step. In: Procedia Engineering 183, pp. 227-232
[7] Schürmann H 2007 Konstruieren mit Faser-Kunststoff-Verbunden. Springer-Verlag, Berlin, ISBN 9783540721895
[8] Composites Europe 2014: Wissenschaftler der TU Dresden stellen Neuentwicklungen vor, https://tu-dresden.de/die_tu_dresden/fakultaeten/fakultaet maschinen-wesen/ilk/news/composites_europe_2014
[9] Coulton J, Ringenbach S, Richeton J 2016 Thermoplastic Bumper Beam (Querträger aus thermoplastischen Kunststoffen) Strategien des Karos-seriebaus, Bad Nauheim
[10] Jäschke A, Dajek U 2004 Roof-frame design using hybrid technology, Sonderdruck aus VDI-Tagungsband Nr. 4260, VDI Verlag GmbH, Düs-seldorf, S. 25-45
[11] Goede M, Ferkel H, Stieg J 2005 Dröder, K.: Mischbauweisen Karosse-riekonzepte -Innovationen durch bezahlbaren Leichtbau, Multi-Material Body Concepts – Innovations to Affordable Lightweight Design, 14. Aachener Kolloquium Fahrzeug- und Motorentechnik
[12] Offenlegungsschrift DE102009035777A1 2011 Strukturbauteil sowie Verfahren zur Herstellung eines Strukturbauteils, Daimler AG,
[13] Gebrauchsmusterschrift DE202009012432U1 2010 Karosseriebauteil, EDAG GmbH & Co. KGaA
[14] Offenlegungsschrift DE102009042272A1 2011 Leichtbauteil, Porsche AG
[15] Dean A, Rolfes R, Behrens B-A, et. Al. 2018 Parametric study of hybrid metal-composites clinching joints, Key Engineering Materials,767 KEM, pp. 413–42
[16] Behrens B-A, Dilger K, Lippky, K, Farahmand, E, Micke-Camuz M 2019 Procedia Combined deep drawing and fusion bonding of structural FRP-metal hybrid parts, Manufacturing, pp. 296–304
[17] Lippky K et. AL2014 „online proceedings", 37. Annual Meeting of the Adhesion Society, Bahia Resort, San Diego

Quenching optimization of a hot stamping line for aluminum automotive components

U Ibarretxe[1,2]*, S Garmendia[1], N Otegi[1], U Argarate[2], A Aranburu[3], A Ormaetxea[3], M Carranza[4] and L Galdos[1]

[1] Mechanical and Industrial Production Department, Mondragon Unibertsitatea, Loramendi 4, 20500 Mondragon, Spain
[2] Koniker S. Coop. Technology Centre, Barrio San Andres 20, 20500 Mondragón, Spain
[3] Fagor Arrasate S. Coop., Barrio San Andres 20, 20500 Mondragón, Spain.
[4] Batz S. Coop., Torrea Auzoa 2, 48140 Igorre, Spain.

* uibarretxe@mondragon.edu

Abstract. The rising cost of fuel and harder environmental regulations driving the need for more eco-friendly vehicles have compelled automotive manufacturers to search for innovative lightweight solutions. Consequently, hot stamped aluminum alloys are gaining prominence due to their exceptional specific strength and improved formability compared to cold formed components. Unfortunately, the current state of knowledge in the field lacks the necessary depth to establish a reliable, fast and optimized process for aluminum hot forming. To address this gap, a collaborative effort between Mondragon Unibertsitatea (knowledge provider), Fagor Arrasate (hot stamping line supplier), and Batz (tool manufacturer) has been initiated. The collaborative endeavour aims to conduct semi-industrial trials involving the hot stamping of an authentic automobile bumper employing high-strength 6xxx and 7xxx aluminum alloys. The hot stamping trials have been performed under different in-die quenching times and mechanical properties of the stamped bumpers have been empirically tested. These results have helped to define the optimal quenching time. With that aim, first, a thermophysical and rheological characterization of the aluminum has been performed, followed by the development of a model to predict mechanical properties of aluminum alloys. This predictive model, together with the aluminum material data, has been then integrated into the simulation framework to enable accurate forecasting of mechanical properties and, consequently, the identification of the optimal quenching duration. The results revealed that quenching times as brief as 1 second could be employed while maintaining acceptable mechanical properties in a rapid cycle hot stamping process. These expedited quenching times have the potential to boost production by an impressive 70 % when compared to the conventional hot stamping process applied to the equivalent steel bumper.

1. Introduction

The weight reduction of vehicle parts has been a key priority for meeting the demand of re-ducing fuel consumption and environmental pollution. Therefore, high strength aluminum alloys (HSAA) are emerging due to their high specific strength and collision performance over press hardening steels [1]. Despite their lightweighting potential [2], they are more challenging to form than steels yet. One of the promising alternatives to improve the formability of these aluminum alloys is to increase the forming temperature. On the one hand, warm forming of high strength aluminum alloys seems to solve some of the formability issues, but in exchange of 18% loss of hardness in comparison with the initial material state [3]. On the other hand, hot stamping of aluminum could help carmakers manufacture complex automotive components such as B-pillars made of AA7075-T6 with acceptable mechanical properties [4] and good forming precision [5]. However, the thermal history during the stamping and quenching steps must be taken with special care as it plays a major role on the post form strength of the final component [6].

Therefore, Mondragon Unibertsitatea (Knowledge provider), Fagor Arrasate (Hot Stamping line supplier) and Batz (Tool manufacturer) teamed up to perform semi-industrial trials on hot stamping a real car bumper using high strength aluminum alloys. The behavior of the hot formed components towards mechanical properties have been analyzed within specific thermal history ruled by different process parameters such as the quenching time.

2. Methodology

2.1. Material characterization

The materials selected for this work are two high strength aluminum alloys from the 7xxx and 6xxx series: AA7075-T6 and AA6082-T6. In order to obtain an accurate FEM simulation model, a deep material and process characterization is required to implement them as input parameters. For that aim, the following test and measurements were performed at Mondragon Unibertsitatea laboratories:

- Mechanical characterization: hardness, tensile tests (Table 1) and energy absorption capacity were tested [2].
- Thermophysical characterization between 20 - 550 ºC: density (Dilatometry), specific heat capacity (DSC, Differential scanning calorimetry), Diffusivity (Laser flash method) and conductivity (Table 1).

Table 1. Thermophysical and Mechanical properties of AA7075-T6 and AA6082-T6 at room temperature.

Material	Specific Heat Capacity [J/g·K]	*Density [kg·dm3]*	Conductivity [W/m·K]	*YS [MPa]*	UTS [MPa]	*Hardness [HV3]*
AA7075-T6	0.885	2,794	135,308	440	500	160

- Localized measurement and control of temperature and strain during rheological characterization between 20 - 550 ºC at different strain rates ranging from quasi-static 0.001 s-1 to 10 s-1.
- Temperature dependant Heat Transfer Coefficients (HTC) between aluminum-air and aluminum-dies from 0.5 MPa to 60 MPa pressures.
- Friction coefficients: Hot ring test were performed in dry and wet conditions to obtain the friction coefficients.
- Surface topography characterization of the aluminum sheets was also performed using optical profilometer through interferometry method. The roughness of the reconditioned die was also measured by the replication method using resins.

2.2. Case study and facilities

The stamping trials were done by reengineering a prototyping die used for hot stamping of boron steels. The die was originally built to hot form 2.3 mm thickness press hardening steel (Usibor 1500) into a rear bumper, which is a critical car component for passive safety needs. The aim of the tool reconditioning was to machine the die surface to a nominal gap of 3 mm, being the thickness of the AA7075 and AA6083 aluminum sheets used to form an aluminum bumper fulfilling the car

IOP Conf. Series: Materials Science and Engineering **1307** (2024) 012048 doi:10.1088/1757-899X/1307/1/012048

manufacturer's mechanical and geometrical requirements as well as enabling a high lightweighting of the component.

The reconditioning of the tool consisted in the following steps:

- Clone the old 3D surface corresponding to the 2.3 mm thickness bumper into the new 3 mm one.
- Generate the optimal die surface by means of FEA for hot forming a 3 mm aluminum alloy bumper using Autoform. For that aim, a try-out thermography study in tool-shop and optimization of the surface geometry has been performed to increase the quenching efficacy.
- Machine the old die surface according to the new optimized surface.
- Readapt all die features to match the new geometry (plates, press bench, transfer feeder claws, sheet supporters…).
- Assemble the die and execute the first forming trials for the die spotting with blue paint until desired final quality and contact was reached.

The trials were performed at the hot stamping line inside the industrial facilities of Batz equipped with HOTTEKNIK features. The line consists of a servo-mechanical hot forming press patented by Fagor Arrasate, an automatic feeding system and a multilevel furnace. The line was designed for press hardening of steels. Therefore, feeding system and other peripheral features were adjusted to the needs of aluminum alloys.

2.3. Hot Stamping Line Tuning Aided by FEM Simulation

With the aim of obtaining all input parameters for the FEM simulation model, the following measurements have been performed at Batz facilities:

- The new machined tool surfaces were 3D-Scanned using GOM Atos software to enable the modelling of the reconditioned dies after the machining and die spotting. The scanned surfaces have been directly imported in Autoform, so that the simulation leads to a more accurate contact distribution.
- In addition, press velocity and stroke were measured with a draw-wire displacement sensor installed between upper and lower die surfaces, and real press kinematics were used in the simulation model.

Once all input data was imported in the FEM model, Autoform simulations were used to predict formability issues. As it can be seen in Figure 1, the failure areas were identified using the plastic strain and thinning output of Autoform. During the set-up of the line, initial components were hot formed and cracked in the same critical areas predicted by the simulations, see Figure 2.

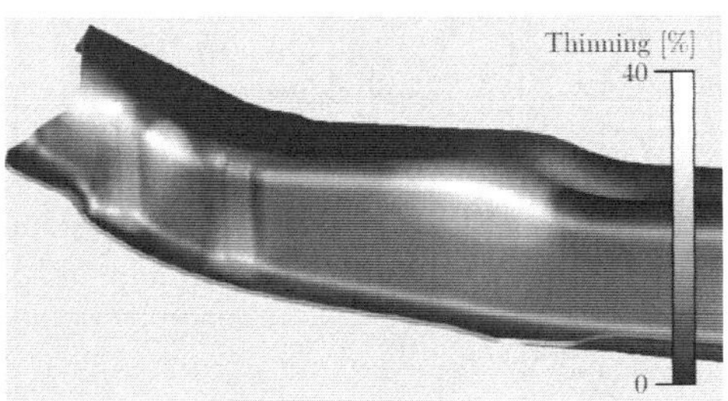

Figure 1. Thinning field output predicted by Autoform.

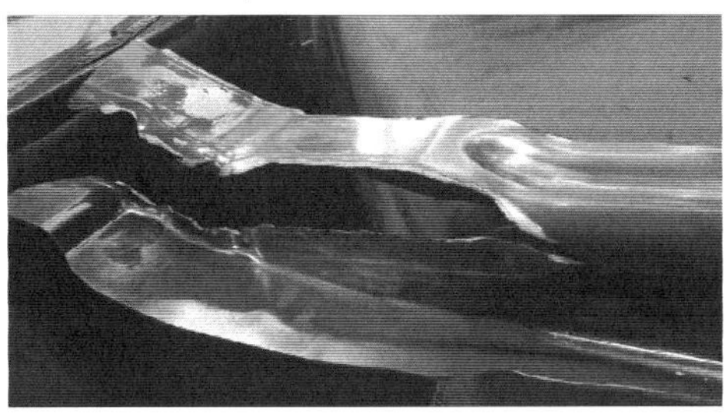

Figure 2. First bumper formed without any optimization.

Process parameters were changed to address formability issues without losing thermal requirements which ensure good mechanical properties and accurate geometry tolerances both at the same time. The most relevant modified parameters were the SHT (Solution Heat Treatment) temperature and soaking time, feeder transfer time, forming speed, lubricant type, as well as lubricant amount and distribution, press forces and quenching times among others. For that aim, background knowledge developed by a PhD study about hot stamping of aluminum alloys at the Mondragon Unibertsitatea was applied. Once the process parameters were determined so that the test parts are successfully formed without cracks, the hot stamping line was ready for the trials and the fine tuning of the quenching time.

2.4. Experimental Set-Up and Program

The trials were fully monitored thermally and cinematically. For tracking the complete thermal cycle of the aluminum blanks the following strategy was followed: as the line configuration makes impossible to use thermocouples attached to the blank due to the transfer of the feeder, high resolution FLIR (Figure 3d & 3e) and OPTRIS (Figure 3b & 3c) thermal cameras were used framing at the press bench and furnace table respectively. Surface emissivity was adjusted for different surface temperatures, viewing angles and blank conditions such as lubricated, dry and coated in order to accurately measure temperatures with a contactless method. The cooling during the in-die quenching step was tracked using contact thermocouples strategically attached around the most critical zones of the lower die (Figure 3a). That way, almost the full thermal cycle of the aluminum can be registered during the hot stamping process from the exit of the furnace to the opening of the die after forming.

The kinematic behaviour of the press was measured using the draw-wire displacement sensor, and the displacement and press forces were exported from the servo-mechanical press directly to validate the FEM simulation model.

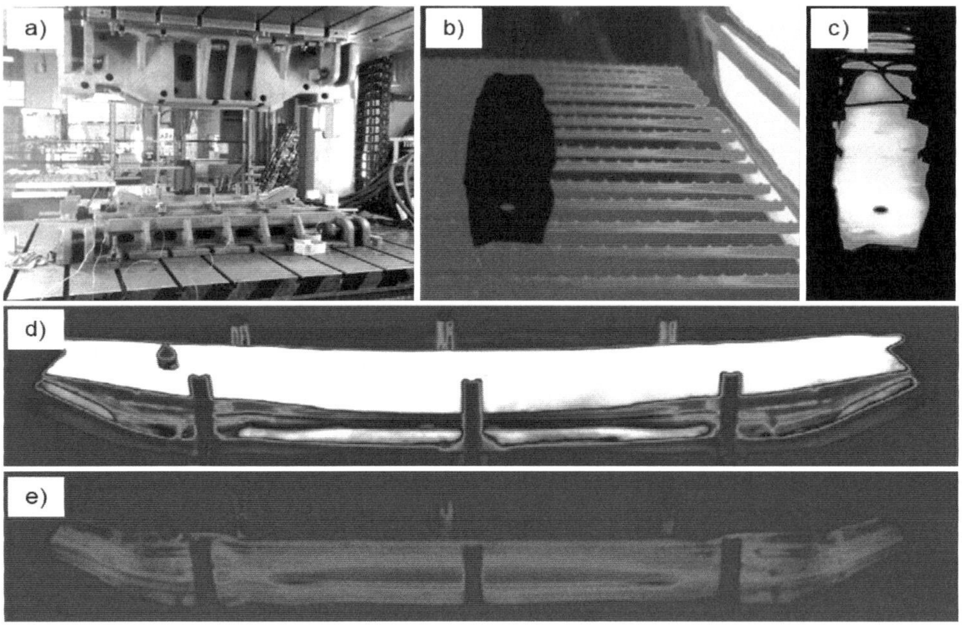

Figure 3. a) Assembled die with the thermocouples attached to its surface, b) and c) OPTRIS thermal camera images of the aluminum sheets before and after the SHT respectively, d) and e) high resolution FLIR thermal camera images of the part before and after forming respectively.

Each forming test was performed as follows:

- The blank was automatically introduced in the furnace by the feeder at the corresponding SHT temperature for each alloy (AA7075 & 6082) for a certain amount of time.
- In the meantime, the die surfaces were sprayed with different lubricants to reduce friction, avoid galling and increase the HTC. The lubricants used for the trials were two non-commercial FUCHS lubricants, which are currently in research phase. These lubricants were

specially designed for hot forming aluminum application, which are totally safe, effective and easy to clean as they do not have dirty components such as graphite particles.

- The blank is quickly transferred by the automatic feeder to the holders above the tool to avoid contact and heat loss between the blank and the dies before forming and quenching.
- The blanks were hot formed and quenched at different quenching times.
- Finally, the parts were age hardened to achieve the desired T6 mechanical properties.

With the aim of optimizing the hot stamping line a wide variety of process parameters were studied, such as the effect of different SHT temperatures and times, different transfer times, press forces, quenching times, lubrication conditions and age hardening types (T6 and T4 tempers). Overall, more than 150 parts were hot formed during the trials with both AA7075 and AA6082 aluminum alloys to optimize a broad range of hot forming parameters and line configurations. In this study, the effect of the quenching time has been analyzed.

2.5. Quenching-time measurement

The quenching time, also known as holding time, is an important variable during the hot stamping process, which essentially determines the cycle time and final mechanical properties of the component when performed properly. In this particular scenario, the quenching time has been defined as the time elapsed with the tool completely closed. The minimum quenching time required to achieve the optimal mechanical properties highly depends on the blank thickness, thermophysical properties and cooling rates for the hardening of the material. The typical quenching times used for PHS are around 10 seconds, although it can be improved by the application of higher conductivity tools [290] and more efficient cooling systems [291]. For instance, the quenching time used during the continuous hot stamping of the original 2.3 mm thickness PHS bumper is 11 s with refrigerated dies, yielding a cycle time of 20 s.

The quenching time for HSAA, however, is totally different from several reasons: on the one hand, longer quenching times would be needed to fully cool the HSAA due to higher specific heat capacity and thicker components in order to match the crash performance requirements of the PHS components. On the other hand, the lower mass, lower quenching temperatures and higher conductivity becomes HSAA in a faster quenchable material, as these last reasons prevail over the previous ones. Therefore, shorter quenching rates than 8 s in comparison to the original PHS bumper have been tested ranging from 7 s to as short as 0 s, where just the forming operation takes place. Due to the press configuration and control functions the exact nominal quenching times have not been achieved. However, they have been accurately measured with the use of the displacement sensor, and for instance, the 0 s nominal quenching time has happened to become a 0.3 s quenching time.

2.6. Mechanical properties characterization

To begin with, three different profiles have been cut out the A, B and C sections, shown in Figure 4. Micro hardness Vickers indentations have been performed along A, B and C profiles to measure the hardness through thickness according to ISO 6507 standard. At least, 10 measurements have been performed per profile to ensure a reliable repeatability of the measurements and for detecting any hardness variation along the profile itself.

Tensile test samples have been cut next to the A, B and C sections location as seen in Figure 4 and tested under the ISO 6892 and ASTM E8/E8M standards. Tensile test sample from location A, TT_A, has been taken from the base, whereas the TT_C has been taken from the flange. Two tensile test samples have been taken from the B section on the base and on the flange named TT_{B1} and TT_{B2} respectively.

IOP Conf. Series: Materials Science and Engineering **1307** (2024) 012048 doi:10.1088/1757-899X/1307/1/012048

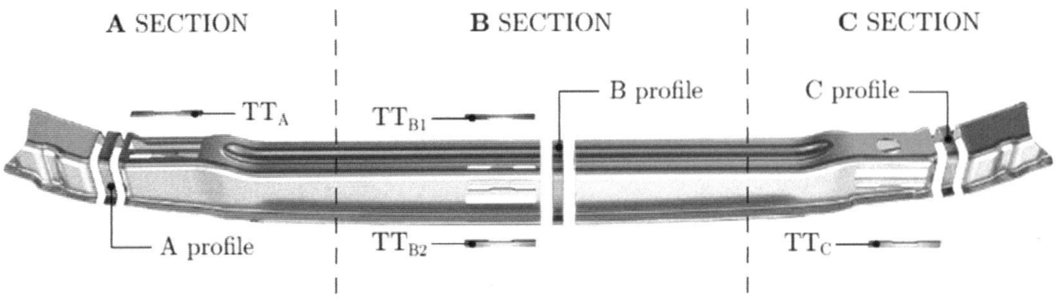

Figure 4. Specimen breakdown.

Even though the forming process itself takes less than 0.5 seconds, and all component locations have been practically quenched at the same time, there are still some differences regarding cooling rates and plastic strains in different sections of the bumper. For instance, the flat base of sections A and C have been first quenched due to the blank-holder, whereas B section is quenched almost once all the tool has been completely closed. This may induce heterogeneities in the bumper and must be also investigated. Anyway, the average of the deviations regarding mechanical properties do not exceed the 7 %, and therefore, the average mechanical properties of each bumper have been analyzed in the following sections for simplification purposes.

3. Results and discussion

It can be observed in Figure 5 and Figure 6 that except from the shortest 0.3 s quenching time, acceptable mechanical properties have been obtained for the remaining quenching times within the nominal threshold, even for as short as 1.3 s quenching times. Moreover, the trials have been performed without the cooling system, so shorter quenching times might be achieved with the proper tooling design. That means, that the cycle times of HSAA in comparison with the PHS could be drastically reduced.

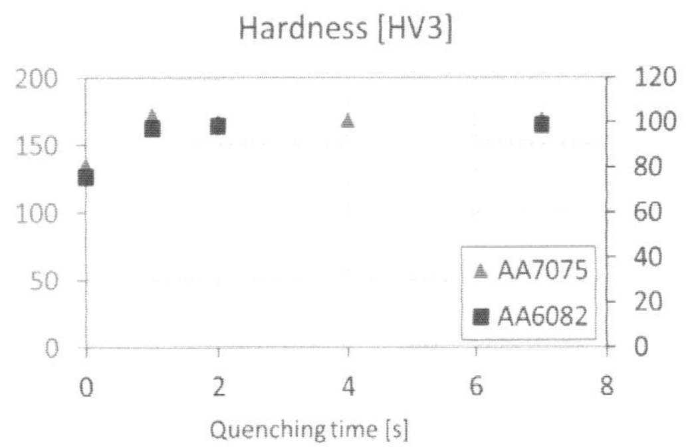

Figure 5. Hardness under different quenching times for both HSAA.

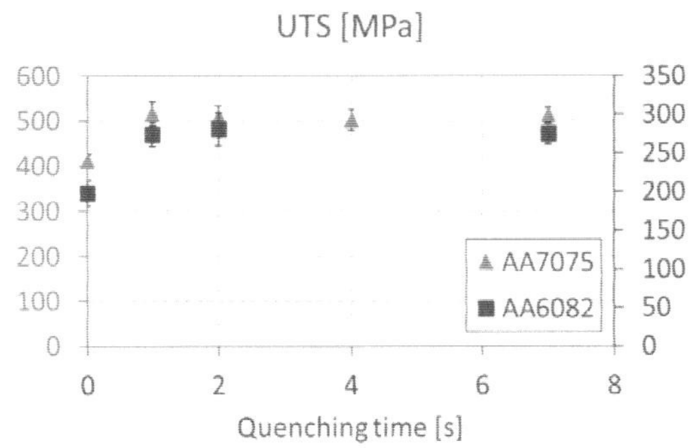

Figure 6. UTS under different quenching times for both HSAA.

According to the thermal history provided by the thermal cameras and in-die thermocouples, sections A and B are subjected to earlier and faster quenching due to the earlier contact of the blank holders, which are next to this section and further away from the middle C section. However, appropriate mechanical properties were achieved for both AA7075-T6 and AA6082-T6 alloys in the whole analysed locations according to the standard values of HV and UTS described in Table 1. Therefore, quenching time, and thus, cycle time can be drastically improved with the use of an optimized tool surface with the corresponding die spotting, even for long cycled simulations and thicker components. An schematic comparison between the cycle times for the original PHS and optimized AA7075 aluminum bumpers can be seen in Figure 7. By reducing the quenching time below 1 s, the total cycle time has been reduced from 18.5 s (3.24 part/min) to 10.5 s (5.71 part/min), which is traduced in a 76 % increase on the productivity.

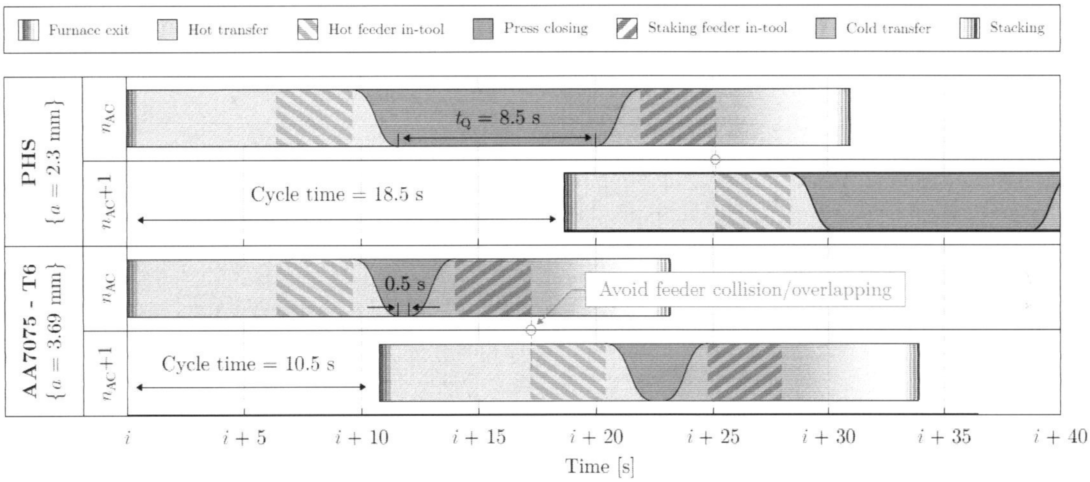

Figure 7. Comparison of cycled times between PHS and AA7075 bumper.

4. Conclusions and Future Work

Based on these results, the following main conclusions can be drawn from this chapter:

- The hot stamping line optimization step aided by FEM simulation plays an important role on the success of the final result. A proper optimization is needed to avoid issues related with formability and furnace temperature, as the radiation furnace is not optimal for heating aluminum blanks.
- In comparison with press hardening steel (PHS) bumpers, a really short quenching time is sufficient to obtain acceptable mechanical properties and geometrical tolerances.
- The reduced cycled times of hot stamping of high strength aluminum alloys (HSAA) components in comparison with PHS, could potentially increase the productivity by around 70 %.

Therefore, high strength aluminum alloys have a great potential to be used for hot stamping operations regarding geometry and mechanical properties. These tests have enabled to write technological guidelines for Fagor, and have allowed the generation of a deep industrial knowledge about hot stamping of high strength aluminum alloys in the partnership. Nowadays, being a trusted technological partner for hot stamping of boron steels, Fagor is also able to offer tailor made hot stamping lines for aluminum alloys equipped with roller or multilevel forced convection furnaces, for low, medium to ultra-high production rates with a production capacity of 600 strokes/hour.

5. Acknowledgements

The authors would like to thank the Ministry of Economy and Competitiveness of the Basque Government for the funding of BIKAINTEK scholarship and BEROALU project under the program Hazitek. The collaboration and technical support of the companies Fagor Arrasate S. Coop. and Batz S. Coop. are also gratefully acknowledged.

6. References

[1] Lee, M. S; Moon, Y. H.: Collision Resistance of a Lightweight Center Pillar Made of AA7075-T6. In: International Journal of Automotive Technology (2021) Vol. 22, No. 4, pp. 853-62.

[2] Ibarretxe, U. et al.: Lightweighting potential of an automotive roof reinforcement using aluminum alloys in place of high strength steels. In: AIP Conference Proceedings 2113 (2019), 170021.

[3] Polak, S. et al.: Warm forming of 7075 aluminum alloys. In: Procedia Engineering (2017), No. 207, pp 2399-2404.

[4] Gronostajski, Z. et al.: Properties of B-pillar made of aluminium 7075 in warm forming process. In: Procedia Manufacturing (2019) No. 27, pp 98-103.

[5] Liu, Y. et al.: Hot Stamping of a B-Pillar Reinforced Panel With 7075 Aluminum Alloy and the Feasibility Study of Short-Time Aging. In: Frontiers in Materials (2021) Vol. 7, 61722.

[6] Liu, J. et al.: Hot Stamping of AA6082 tailor welded blanks for automotive applications. In: Procedia Engineering (2017) No. 207, pp 729-734.

IOP Conf. Series: Materials Science and Engineering **1307** (2024) 012049 doi:10.1088/1757-899X/1307/1/012049

Investigation on effect of high-efficiency solid solution and hot stamping process on microstructure evolution and mechanical properties of high-strength aluminum alloy

Y.L. Wang[1]*, H.C. Geng[1], B. Zhu[1], Y.S. Zhang[1], X.Q. Ren[2] and X.P. Ren[2]

[1]State Key Lab of Materials Processing and Die and Mould Technology, Huazhong University of Science and Technology, Wuhan 430074, China
[2]DongYing Guangda Jinke Robot Co.,Ltd, DongYing 257100 , China

E-mail: wangyilin@hust.edu.cn

Abstract. Aluminum alloy has become an important lightweight material in the automotive industry due to its excellent performance. The development of hot stamping technology has effectively promoted the application of high-strength aluminum alloys in auto parts. However, aluminum alloy hot stamping requires a long time for solid solution and aging heat treatment, which hinders its mass production application in the existing mature hot stamping production line. Therefore, this paper studies the effect of a high-efficiency solid solution and hot stamping process on the microstructure and mechanical property strengthening mechanism of high-strength aluminum alloy. The contact heating and hot stamping experiment device was designed. It was found that the temperature rise rate of the sheet during the contact heating process could reach 44.51 °C/s. The experiment results show that contact heating treatment with a solution temperature of 480 °C and a solution time of 20s can achieve the high-efficient solution treatment of the sheet metal. The high-temperature deformation behavior of 7075-T6 at different temperatures (300-450 °C) and different strain rates (0.01~1/s) under the condition of high-efficiency solid solution rapid heating was studied.

1. Introduction

Although aluminum alloy has the advantages of high specific strength, corrosion resistance, and recyclability, its formability is poor at room temperature due to high deformation resistance and obvious springback. To effectively enhance formability and reduce springback in high-strength aluminum alloy sheets, researchers proposed the use of hot forming and quenching technology [1]. Zheng et al. [2] investigated the effects of transferring aluminum alloy sheets from a heating furnace to the die stage on the mechanical properties of the final part during the hot stamping. Their analysis focused on the temperature-time-properties correlation and the reason for the decline in material performance. Omer et al. [3] conducted a study on the hot stamping process of AA7075 aluminum alloy and highlighted that a cooling rate no less than 56 °C/s is required when quenching the aluminum alloy after solid solution to achieve an optimal solid solution effect.

The hot stamping process enables the use of high-strength aluminum alloy sheets for complex-structured automobile parts. Nevertheless, the lengthy solid solution and aging heat treatment of aluminum alloy pose a hindrance to its mass production application in the current mature hot stamping production line. The processes of hot stamping and quenching occur simultaneously, whereas the solid solution and artificial aging treatments require extensive time, with the aging process particularly requiring several hours or more. The incongruity between the processes makes synchronization challenging, while the lengthy heat treatment results in high energy consumption, contradicting the "double carbon" objective.

During the aluminum alloy hot stamping process, a heating furnace is utilized for solution treatment, with the sheet being heated primarily through radiation heat transfer. A smooth aluminum alloy sheet has an emissivity of approximately 0.09, while a sheet with a rough surface has an emissivity of around 0.28. Due to the lower emissivity, the heating rate during the radiation heating is significantly slowed. To enhance the heating rate of sheet metal, Liu et al. [4] investigated the impact of lubricant spraying on the radiation heating of 6061 and 7075 aluminum alloy sheets. They discovered that applying lubricant enhanced the surface condition, resulting in reduced time required to heat up the sheet during solution treatment. They indicated that spraying boron nitride (BN) or graphite lubricant can lessen the solid solution time to less than 50% of the initial value. Xu et al. [5] employed electric pulses to conduct solution treatment on 7075 aluminum alloy. Their findings indicate that the supersaturation was less than that of traditional solution treatment, despite taking only 220 ms, resulting in smaller second-phase particles and grain sizes after aging. Furthermore, the mechanical properties of the samples are superior to those of regular T6 aluminum alloys. Shao et al. [6] utilized a jet flame arranged in a matrix pattern to perform the solid solution treatment on a thin sheet of AA6082 with a thickness of 1.5 mm. Their experimentation observed that the sheet's heating rate exceeded 10 °C/s. However, they also found that the temperature uniformity was inadequate, and proposed that the configuration of the flame nozzles be arranged reasonably to enhance the temperature distribution across the sheet. Our research indicates that using the contact heating process on a 7075 sheet with a thickness of 2 mm, can achieve solid solution temperature (480 °C) within 15 to 20 seconds [7].

The contact heating method offers high heating efficiency and shape flexibility, making it a low-cost and adaptable option for rapidly heating sheet metals. However, the impact of the increased heating velocity on high-temperature formability of the sheet warrants further investigation. Additionally, the efficacy of short-term heating in ensuring a fully solid-solution of the aluminum alloy sheet and its consequent effects on the material's final properties deserves scrutiny.

2. Experimental Methods

2.1. Experimental Material

The 7075-T6 aluminum alloy, with a thickness of 2 mm, was utilized in this study. The chemical composition of the material can be found in Table 2.1. The sheet was cut into 300 × 150 mm specimens to facilitate experiments.

Table 2.1 Chemical composition of 7075-T6 aluminum alloy (wt. %)

Zn	Mg	Cu	Cr	Fe	Si	Mn	Ti	Zr	Al
5.341	3.131	1.675	0.213	0.176	0.158	0.067	0.039	0.022	Bal.

2.2. High-temperature Tensile Test

The Gleeble-3500 thermal simulation testing machine was used to simulate the temperature history of the contact heating and hot stamping process. High-temperature tensile experiments were then conducted on the heated samples at different strain rates and temperatures. The aim was to analyze the material's behavior during high-temperature deformation under rapid heating conditions. Figure 1 illustrates the test process. The experimental steps included: Heating the sample to 460 °C at a rate of 44.51 °C per second. Continue to heat at a rate of 5 °C per second until the temperature reaches 480 °C. Hold the temperature for 6 seconds to maintain consistency with the contact heating solution treatment process. Following this, proceed to cool the specimens using an air blowing system at a cooling rate of 20 °C/s until they reach the specified tensile test temperatures, including 300 °C, 350 °C, 400 °C, 450 °C. After each cooling stage, stabilize the temperature of the specimens for 2 seconds before conducting any further testing. Finally, the high-temperature tensile experiments should be conducted at stretching rates of 0.01 s^{-1}, 0.1 s^{-1}, and 1 s^{-1} until the sample fractures and then allowed to cool to room temperature.

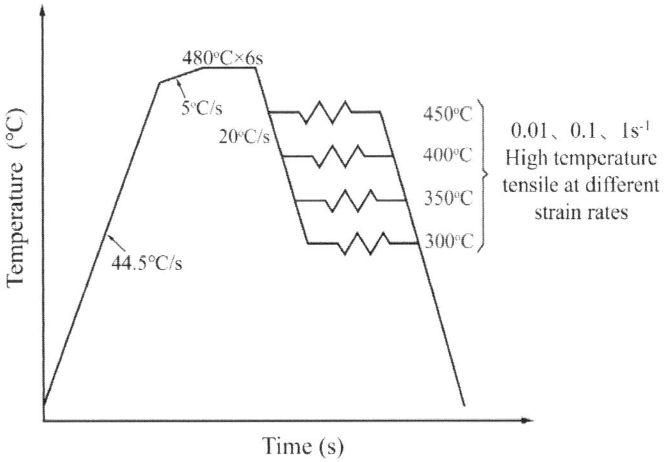

Figure 1 High-temperature tensile experiment flow chart

To replicate the pressure of the contact heating device on the sheet during sample heating, a constant pressure of approximately 564 N (determined by the pressure of the contact heating device) is applied to the sample via the testing machine's chuck, directed toward the center of the sample. However, due to the sheet specimen's limited stiffness and susceptibility to instability and warping when pressure is applied at both ends, an additional fixture was developed to prevent this occurrence, as illustrated in Figure 2a. Figure 2b displays the actual fixture used. A heat-insulating cloth was placed between the clamp and the sample to hinder heat transfer and decrease friction. It is important to avoid clamping the clamp too tightly as it may lead to excessive force, which can ultimately impact the accuracy of the later tensile experiment.

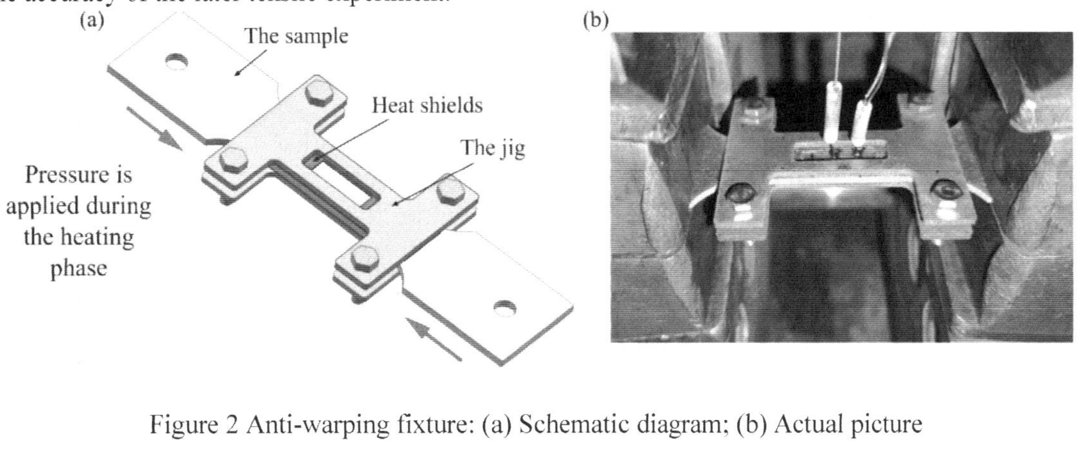

Figure 2 Anti-warping fixture: (a) Schematic diagram; (b) Actual picture

2.3. *Contact heating-hot stamping-aging Test*

The designed contact heating device is shown in Figure 3a and 3b. The experimental device's main body is comprised of two thick metal heating blocks made of QRO 90 Supreme hot work die steel. This material possesses high thermal conductivity, which improves temperature uniformity in contact heating devices. Insulation material surrounds the heating block, and a pressure-resistant insulation board is placed underneath to minimize heat loss and enhance the temperature uniformity within the heating block. The heating rods and K-type thermocouples are methodically positioned inside the block to facilitate temperature adjustment feedback. The total surface area of the heating block is 380 × 290 square millimeters. However, in high-temperature conditions, heat dissipates quickly in the surrounding area of the device, resulting in lower temperature in that area compared to the central area. As a result, the sheet should be placed in the middle area of the heating block when undergoing contact heating. Figure 3c shows the temperature distribution of the surface when the heating temperature is set to 480 oC.

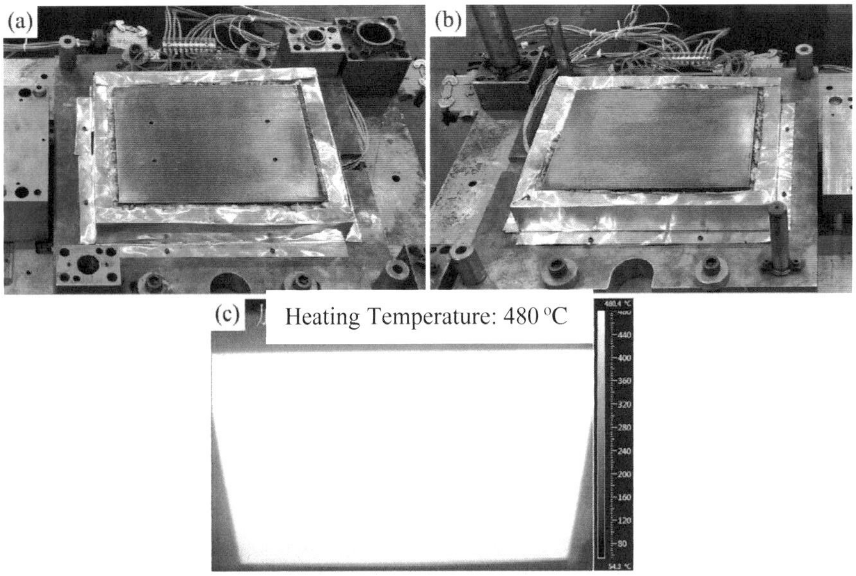

Figure 3 Physical picture of the contact heating module: (a) upper heating block; (b) lower heating block; (c) surface temperature distribution of the contact heating device

Research was conducted on the heating process of aluminum alloy sheets at various pressures (100 kN, 500 kN, 1000 kN) and under different heating temperatures of the contact heating device. The rate of the aluminum alloy sheet heating during contact heating was calculated.

Then, the contact heating-hot stamping experiments, as depicted in Figure 4, were carried out through the following steps: (a) Heating the contact heating apparatus up to the desired temperature of 480 °C and holding it steady for a period to obtain a uniform temperature distribution across the apparatus. (b) Laying the aluminum alloy sheet onto the heated apparatus and close it while simultaneously exerting pressure for 20 seconds to complete solution treatment of the sheet. (c) Transfer the heated aluminum alloy sheet to the cold U-shaped die, then form, quench, and hold for 20 seconds. (d) Transfer the formed parts to a radiation heating furnace for artificial aging treatment at 120 °C/24 h. This will precipitate the second-phase particles in the material to strengthen the parts.

Finally, samples for microstructure observation were processed from the bottom of the formed U-shaped part. The microstructure was then observed using Optical Microscopy (OM) and Scanning Electron Microscopy (SEM), while the recrystallization of materials was analyzed through SEM's backscatter diffraction (Electron Backscatter Diffraction, EBSD) function.

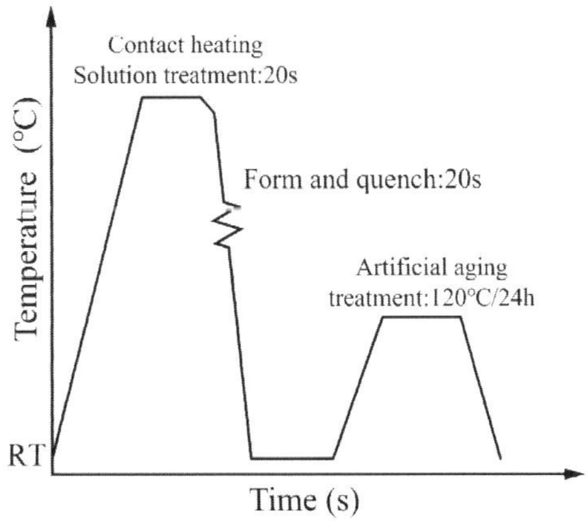

Figure 4 Contact heating-hot stamping experiment flow chart

3. Experimental Results and Analysis

3.1. Analysis of Contact Heating Temperature Rise Process

From the temperature rise curve obtained for the aluminum alloy sheet at a heating temperature of 480 °C, Figure 5 displays the change in the calculated heating rate in response to pressure. It is noticeable that an increase in pressure results in an acceleration of the heating rate. At a pressure of 1000 kN, the heating rate can reach up to 44.51 °C/s. The temperature rise curve measurement results indicate that when the contact heating temperature is adjusted to the solid solution temperature of 7075 aluminum alloy (480 °C) and exposed to varying pressures, the aluminum alloy sheets can attain near the target temperature (480±5 °C) within 15 to 20 seconds.

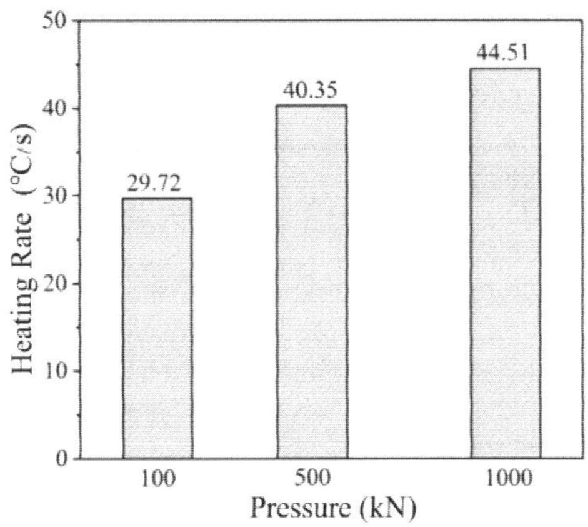

Figure 5 The heating rate of the aluminum alloy sheet changes with pressure when the heating temperature is 480 °C

3.2. Plastic deformation behavior of 7075 under high-temperature conditions

The true stress-strain curves for 7075 aluminum alloy were obtained by stretched it at different temperatures and different strain rates after rapid heating. The obtained curves are shown in Figure 6.

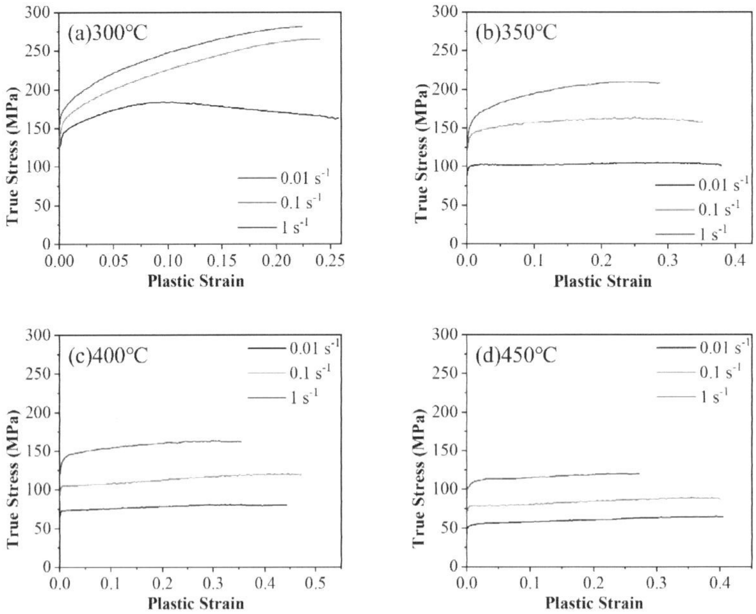

Figure 6 True stress-plastic strain curves of 7075 aluminum alloy after rapid heating at different temperatures and different strain rates: (a) 300 °C; (b) 350 °C; (c) 400 °C; (d) 450 °C

Throughout the process of high-temperature plastic deformation, both the work hardening and dynamic recovery of the material have a simultaneous impact on the flow stress change with plastic strain. At a lower temperature of 300 °C, work hardening dominates, resulting in a significant increases in the material's true stress as plastic strain increase. However, at a higher deformation temperature (≥400 °C), work hardening and dynamic recovery of the material are in a balanced state. As the plastic strain increases during the deformation process, the material's true stress increment is negligible. Of course, when the deformation rate of the material varies, there are differences in the effects of work hardening and dynamic recovery. When the strain rate is high, the work-hardening effect is noticeably stronger. However, when the material is deformed at a higher temperature (450 °C), even if the deformation rate is high (1 s^{-1}), the work-hardening effect is still mostly counterbalanced by the dynamic recovery.

After comparing these results to our prior work under slow heating conditions, we discovered that, even when the deformation temperature and strain rate remain constant, the flow stress of the samples subjected to rapid heating conditions increases to a certain extent. Moreover, the lower the deformation temperature, the greater the rise in flow stress. However, it is evident that both slow and fast heating samples show consistent effects of work hardening and dynamic recovery on the flow stress curve of the material when deformed at varying temperatures and strain rates.

3.3. Effect of contact heating process on microstructure and strengthening mechanism of 7075 aluminum alloy

Figure 7 depicts the distribution of recrystallized structure, sub-grain and deformation structure in various regions of the samples and the as-supplied samples. The supplied 7075-T6 aluminum alloy displays a relatively complete degree of recrystallization, and the majority of the grains within its microstructure are recrystallized. Nonetheless, for the materials after contact heating, a substantial proportion of sub-grain and deformed grains are evident in the microstructure. Especially concerning the microstructure of the surface area after contact heating, there is a significant proportion of deformed grains, resulting in a substantial reduction in grain size. Due to the brief contact heating solution treatment time, the deformed grains do not have sufficient time to transform into recrystallized grains, resulting in a large number of deformed grains. These observation suggest that the contact heating process induces considerable plastic deformation in the metal sheet.

After applying contact heating and hot stamping to the sample, we selected 5 positions from the surface of the sample to the center, and observed the microstructure using an optical microscope. The corresponding positions' grain size was counted. In comparison to the supplied sample's microstructure (with an average grain size is 10.85 μm), the material's grain size decreased after contact heating process. The grain size in surface area of the sample was refined. Most obviously, there is a decrease in the average grain size to 6.28 μm in this area. Additionally, the grain refinement effect is lessened in the central area of the sample's thickness direction to yield an average grain size of 8.85 μm.

The statistical results of the grain sizes clearly demonstrate that the contact heating process refines the material grains, resulting in an enhanced grain boundary strengthening effect. Calculations reveal that after contact heating and hot stamping, the sample's grain boundary strengthening effect provides a strength ranging from 40.34 to 47.89 MPa. In comparison, the supplied sample's grain boundary strengthening effect is only 36.43 MPa.

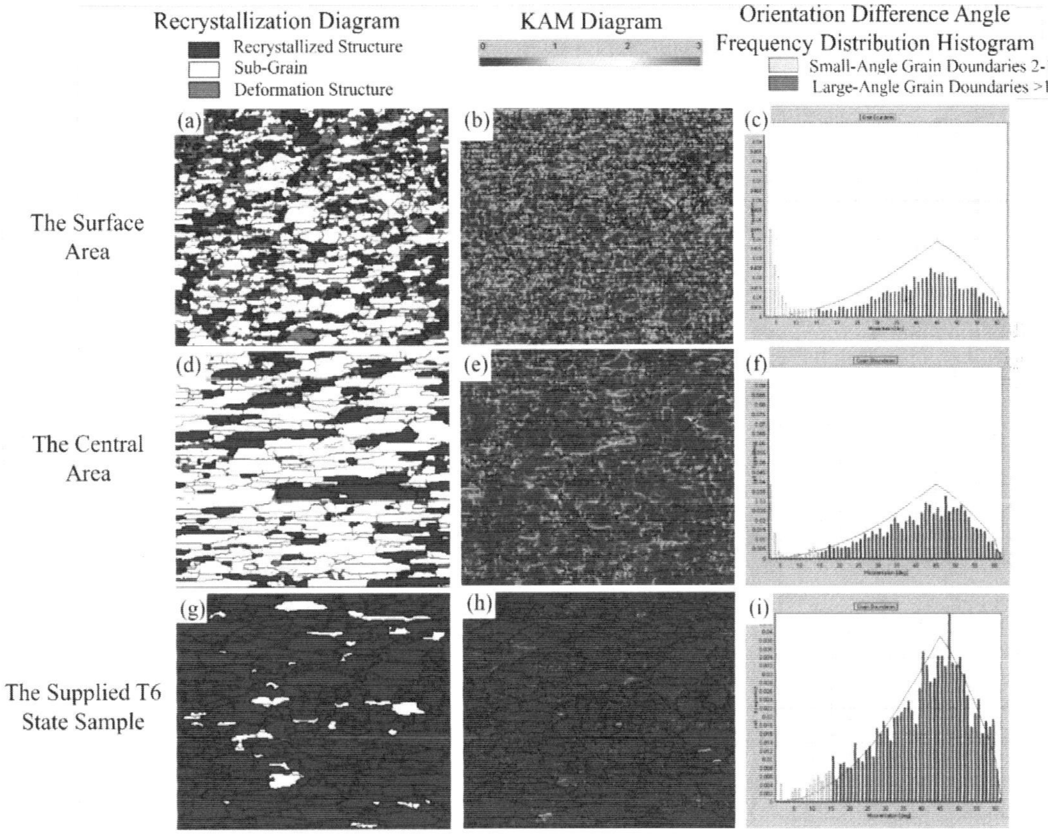

Figure 7 EBSD observation results of contact heating-hot stamping sample and T6 state sample: recrystallization diagram, KAM diagram, orientation difference angle frequency distribution histogram of (a-c) contact heating-hot stamping sample surface and (d-f) middle area and (g-i) T6 state sample.

Through calculations, it was determined that the contact heating-hot stamping sample has a dislocation density of about 1.904×10^{13} m^{-2}, contributing to a mechanical property enhancement of 20.54 MPa through dislocation strengthening. Similarly, for the supplied T6 state sample, the dislocation density was calculated to be around 1.349×10^{13} m^{-2}, with a dislocation strengthening contribution of 17.29 MPa. After contact heating and hot stamping, the dislocation density in the

IOP Conf. Series: Materials Science and Engineering **1307** (2024) 012049 doi:10.1088/1757-899X/1307/1/012049

sample increased slightly increased. However, its enhancement of the dislocation strengthening effect was limited. The occurrence of this phenomenon is believed to be due to the recovery of the contact-heated sample during the longer artificial aging process, resulting in a decrease in the dislocation density in the material.

4. Conclusion

The high-efficiency solid solution treatment for aluminum alloy sheets during the hot stamping process was achieved using the contact heating method. Experimental verification confirmed the feasibility of high-efficiency solid solution. The study investigated the high-temperature deformation behavior of aluminum alloy under rapid heating conditions. Additionally, the study examined the impact of the contact heating and hot stamping process on the microstructure and mechanical properties of the material.

The study examined the high-temperature deformation behavior of the 7075 aluminum alloy under rapid heating conditions at various temperatures (300-450 °C) and strain rates (0.01~1 s^{-1}). It was determined that at lower deformation temperature (300 °C), the material's work hardening dominates, resulting in a significant increase in true stress with increasing plastic strain. Conversely, at higher deformation temperature (≥400 °C), the material's work hardening and dynamic recovery are essentially in equilibrium. The increase in true stress with plastic strain is minimal under these conditions.

A contact heating-hot stamping experimental device was developed to analyze the appropriate contact heating time for achieving efficient solution treatment of 7075 aluminum alloy sheets. Based on the results of mechanical property testing and microstructure observations, the treatment at 480 °C/20 s complete efficient solid solution process. Parts produced using the contact heating process exhibited higher strength than the supplied T6 material. The analysis demonstrates that the values of grain boundary and dislocation strengthening are 40.34 to 47.89 MPa and 20.54 MPa, respectively. These strengthening effects are superior to the supplied T6 state material, resulting in higher final strength of the parts than the T6 state material strength. The utilization of contact heating, in lieu of a radiant furnace, not only reduces the aluminum alloy solution treatment time but also enhances the final part's performance.

References

[1] O. El Fakir, L. Wang, D. Balint, J. P. Dear, J. Lin, T. A. Dean. Numerical study of the solution heat treatment, forming, and in-die quenching (HFQ) process on AA5754. International Journal of Machine Tools and Manufacture, 2014, 87: 39-48

[2] K. Zheng, Y. Dong, J. Zheng, A. Foster, J. Lin, H. Dong, et al. The effect of hot form quench (HFQ) conditions on precipitation and mechanical properties of aluminium alloys. Materials Science and Engineering: A, 2019, 761: 138017

[3] K. Omer, A. Abolhasani, S. Kim, T. Nikdejad, C. Butcher, M. Wells, et al. Process parameters for hot stamping of AA7075 and D-7xxx to achieve high performance aged products. Journal of Materials Processing Technology, 2018, 257: 170-179

[4] Y. Liu, B. Zhu, Y. Wang, S. Li, Y. Zhang. Fast solution heat treatment of high strength aluminum alloy sheets in radiant heating furnace during hot stamping. International Journal of Lightweight Materials and Manufacture, 2020, 3(1): 20-25

[5] X. Xu, Y. Zhao, X. Wang, Y. Zhang, Y. Ning. Effect of rapid solid-solution induced by electropulsing on the microstructure and mechanical properties in 7075 Al alloy. Materials Science and Engineering: A, 2016, 654: 278-281

[6] Z. Shao, J. Jiang, J. Lin. Feasibility study on direct flame impingement heating applied for the solution heat treatment, forming and cold die quenching technique. Journal of Manufacturing Processes, 2018, 36: 398-404

[7] H. Geng, Y. Wang, Z. Wang, Y. Zhang. Investigation on contact heating of aluminum alloy sheets in hot stamping process. Metals, 2019, 9(12): 1341

Study on electric pulse effect on steel DP780 during roll forming

Zhichao Yu[1], Jiahui Chen[1], Han Li[1], Jie Zhou[1,*] and Libo Pan[1,*]

[1] Institute for Intelligent Manufacturing, Jianghan University, Wuhan, China

[*] Correspondence: (Jie Zhou) zhoujie@jhun.edu.cn; (Libo Pan) pan97181@126.com

Abstract. The electrophysical impact on metals introduces a novel optimization avenue for roll forming. This study explores the influence of different electric pulse parameters (peak current and frequency) on the roll forming behavior of DP780 high-strength steel. Through electric pulse-assisted tensile, bending, and roll forming experiments, coupled with microstructure and fracture morphology analyses, the study comprehensively assesses the impact of electric pulses on the mechanical properties and springback mechanism of DP780 sheets. Results indicate that electric pulses reduce deformation resistance and enhance the plasticity of DP780 sheets by repairing defects, promoting dislocation slip, and changing the fracture form of the material. Moreover, electric pulses effectively inhibit springback in roll forming, particularly at larger bending angles and under specific E-pulsing parameters. This is attributed to the promotion of residual stress release by electric pulses, leading to a reduction in springback. In conclusion, electric pulse assistance optimizes the roll forming process of DP780 high-strength steel, evidenced by decreased deformation resistance, improved material plasticity, and effective suppression of springback. These findings open up a new optimization pathway for sheet metal forming.

1. Introduction

Roll forming is a plastic processing and forming method in which metal sheets are rolled through multiple passes with variable cross sections and angles to obtain different cross-sectional shapes. Compared with traditional bending forming, roll forming can solve the problem of continuous variable radius forming of large-radius high-strength parts, however, the ultra-high-strength steel samples obtained by roll forming often fail to meet the requirements of the actual part because of springback, cracks and other defects. Electroplasticity effect[1] refers to the phenomenon that the deformation resistance of material decreases and plasticity increases significantly under the effect of electric current, and the introduction of electroplasticity effect can improve the formability of ultra-high-strength steel.

Jong-sup Leea[2-3] et al. performed uniaxial tensile experiments on DP980 in three directions and utilized isotropic and Yoshida-Uemori hardening for finite element analysis. Roll bending experiments under the same conditions confirmed the superior accuracy of the Yoshida-Uemori hardening model in predicting springback. Junghan Song[4-5] et al. carried out V-bending test on 1180MPa grade steel sheet under pulsed current and found that the angle of springback decreases sharply with the increase of current density and pulse duration. Huanyang Xie[6-8] et al. investigated a method of using electric current to control the springback of the steel sheet after stamping. A new method of using electric pulse to promote the release of residual stresses was investigated, and a model for predicting the springback angle considering electric pulse was proposed and the proposed model was verified by the experimental results. Nguyen Trung Thien[9-11] et al. applied a single pulse of current with a duration shorter than 0.5 s to a specimen, and the flow stress of ultra-high-strength steels decreased instantaneously upon electric current, followed by strain hardening until the specimen necked. Xiaoli Liu[12-15] et al. proposed a mathematical model that considers the variation of Young's modulus and applied it to three-dimensional finite element analysis to simulate the cold rolling forming process. The results showed that springback increases with increasing flange width, sidewall height, roll gap and the distance, decreases with increasing the strip thickness and the web width.

Lots of scholars carried out on researches on roll forming of ultra-high-strength steel however, most of them were mainly focused on the influence of sheet thickness, bending radius, stand spacing and other factors on roll forming. The effect of electric pulse-assisted roll forming on ultra-high-strength steel is not so clear. In this study, the effects of electric pulse on the mechanical properties and springback of DP780 were comprehensively investigated through electric pulse-assisted tensile, bending and roll forming experiments.

2. Experimental materials and methods

For the electric pulse-assisted tensile experiments, DP780 sheets with a thickness of 1.5 mm were used, and the chemical composition was shown in Table 1. The tensile specimen was processed along the rolling direction by wire cutting method, with a marking distance of 15mm×6mm. The tensile experiment was carried out on LCJ-150 electronic tensile and torsion universal testing machine, with a tensile rate of 1mm/min. In the experimental process, SMD-300 numerically controlled bidirectional pulse plating power supply was used to apply a pulse current to the specimen. The output electric pulse was in the form of square wave, and the duty cycle was set at 20%. During the test, the temperature change of the tensile specimen was observed by a temperature measuring device.

Table 1. Chemical composition of DP780 sheets(wt%).

C	Si	Mn	P+S	Al	Cr	Fe
0.16	0.42	2.07	≤0.01	0.053	0.1	Bal.

Electric pulse-assisted V-shaped bending experiments were also carried out in the LCJ-150 electronic tensile and twisting universal testing machine, during V-shaped bending, the upper and lower die holder were made of nylon for insulation, as shown in Figure 1.Bending specimens were created from a 160mm × 30mm rectangular sheet with a thickness of 1.5mm. Experiments were conducted on a stretching machine with a pressure rate of 5mm/min, a bending radius of 2mm, and bending angles of 90°, 120°, and 150°. The duty cycle was set at 40%. The study focused on the impact of varying peak current and frequency on reducing springback and minimizing molding force.

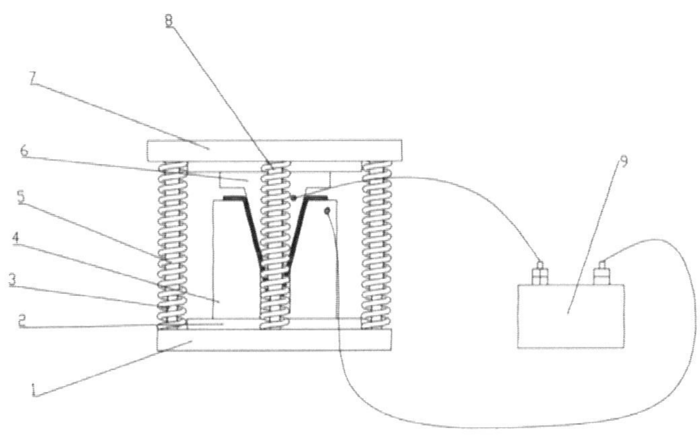

Figure 1. Schematic diagram of V-shaped bending mold. 1-Nylon lower die holder, 2-Pad, 3-Guide pillar, 4-bending concave mold, 5-spring, 6-bending convex mold, 7-nylon upper die holder, 8-guide sleeve, 9-pulse power supply.

The electric pulse assisted roll forming experiment was conducted on the roll former, and the schematic diagram of the roll former is shown in Figure 2. The original sample is rolled by a rolling roller with six increasing angles and a final angle of 120 ° , resulting in the rolled sample. The sheets with length of 1200mm, width of 100mm, and thickness of 2mm were utilized for experiments during the second and fifth passes of the energized work roll. The experiments included a friction factor of 0.15, a roll speed of 10mm/s, and a 60% duty cycle. The study focused on investigating the impact of varying peak current and frequency size on reducing the springback effect.

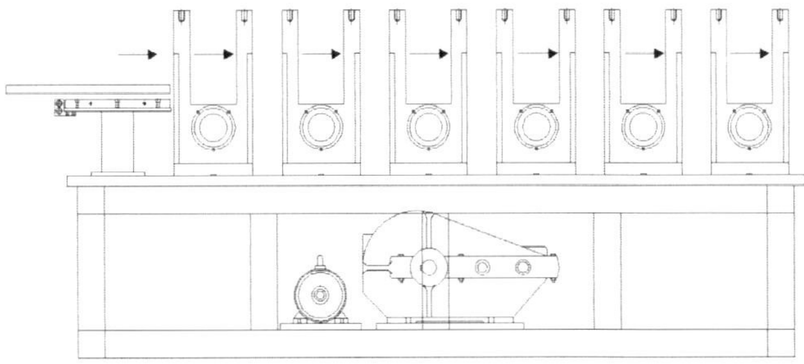

Figure 2. Schematic diagram of roll former.

3. Experimental results

3.1 Electrical pulse-assisted stretching experiment

Figure 3 displays the outcomes of DP780 tensile experiments under varying peak current densities. Stress refers to engineering stress, which is the ratio of the load to the cross-sectional area of sheet. Strain is the difference between the length of the gauge length of sheet after tensile deformation and the initial gauge length, which is divided by the initial gauge length. Following impulse current introduction, DP780 sheet experienced a slight reduction in tensile strength. As peak current density increased, there was a consistent decline in tensile strength, with elongation initially increasing and then decreasing, reaching its maximum at 20 A/mm^2, indicative of a plastic saturation point. In Figure 4, the fracture scanning diagram of DP780 at various peak current densities is presented. DP780 exhibited a plastic fracture mode during electrostatic stretching, characterized by large and densely populated dimples, including numerous small dimples within larger ones, and numerous tearing ridge. Pulsed current resulted in a reduction in dimples density as peak current density increased. At a peak current density of 30 A/mm^2, the tearing ridge in the fractured microstructure disappeared, revealing a laminar and step-like fracture surface. This transition signified a shift from plastic fracture to brittle fracture, leading to a significant decrease in plasticity and tensile strength. Conversely, at peak current densities of 10 A/mm^2 and 20 A/mm^2, the microstructure retained a substantial number of tearing ridge, consistent with a plastic fracture mode. The application of electric pulses induced the repair of defects and facilitates dislocation slip, resulting in a slight increase in DP780 plasticity and a corresponding slight decrease in tensile strength.

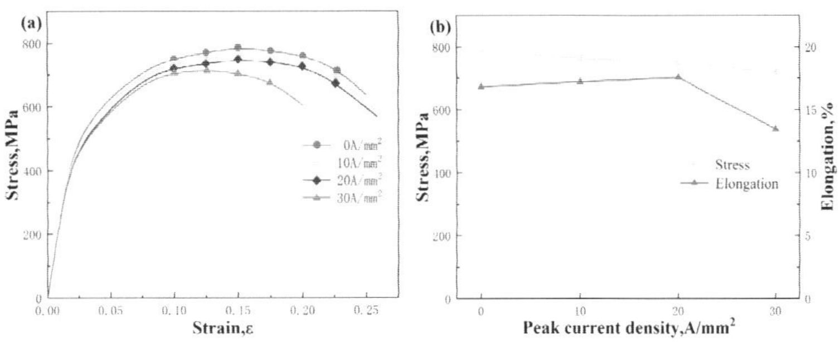

Figure 3. Stress-strain curves (a) vs. tensile strength and elongation (b) of DP780 at different peak current densities.

IOP Conf. Series: Materials Science and Engineering **1307** (2024) 012050 doi:10.1088/1757-899X/1307/1/012050

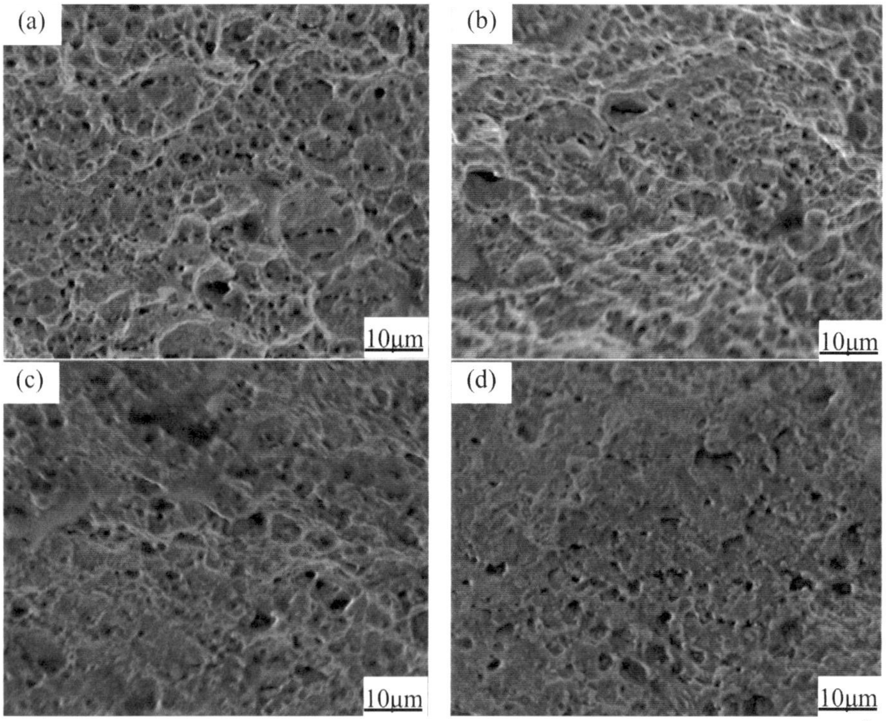

Figure 4. Fracture scan of DP780 at different peak current densities (a) 0 A/mm^2, (b) 10 A/mm^2 100 Hz, (c) 20 A/mm^2 100 Hz, (d) 30 A/mm^2 100 Hz.

The findings from DP780 tensile experiments at varying frequencies are presented in Figure 5. The figure indicated a subtle increase in tensile strength with the rise in frequency. The elongation followed a pattern of initial increase and subsequent decrease, peaking at a frequency of 300 Hz, reaching a plastic saturation point. In Figure 6, the fracture scanning graph for DP780 at different frequencies, under a peak current density of 20 A/mm^2, illustrated ductile fractures. As frequency ranged from 100 Hz to 300 Hz, there was an increase in dimples density and depth, leading to enhance plasticity. Conversely, at a frequency of 500 Hz, dimples density decreased and depth became shallower, resulting in reduced plasticity. These observations highlighted a maximum plasticity threshold for DP780 sheet between 100 Hz and 500 Hz under the same peak current density.

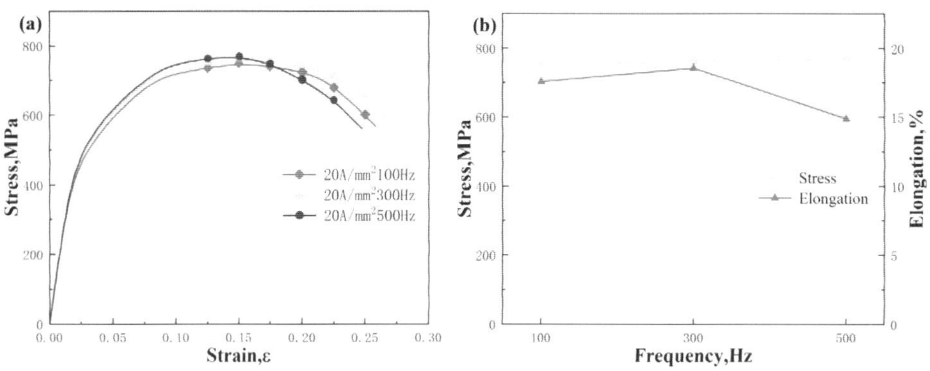

Figure 5. Stress-strain curves (a) vs. tensile strength and elongation (b) of DP780 at different frequencies.

IOP Conf. Series: Materials Science and Engineering **1307** (2024) 012050 doi:10.1088/1757-899X/1307/1/012050

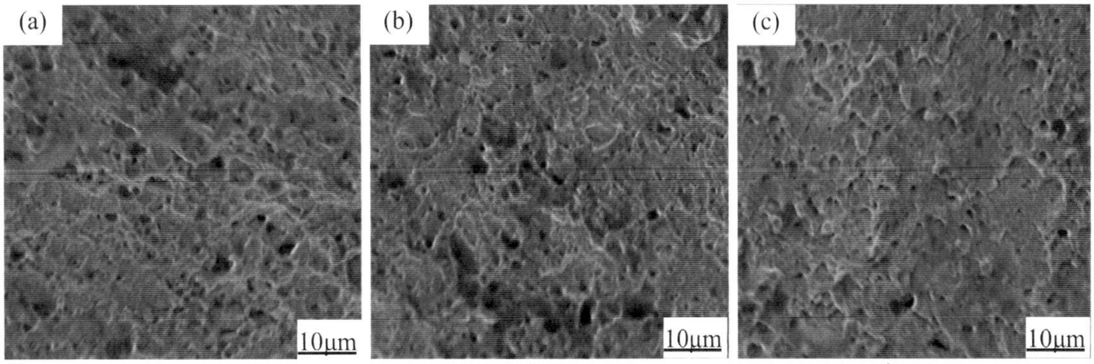

Figure 6. Fracture scan of DP780 at different frequencies (a) 20 A/mm^2 100 Hz, (b) 20 A/mm^2 300 Hz, (c) 20 A/mm^2 500 Hz.

3.2 Electric pulse-assisted V-bending experiment

Figure 7 illustrates the bending forming force variation curves for 120° and 150° V-bending of DP780 under different peak currents and frequencies. For both bending angles, under various peak current and frequency conditions, bending force in the elastic phase rapidly increased with the bending stroke. As bending stroke continued to increase, force changing became minimal, reaching a stable deformation stage. Further increase in stroke lead to the end of the stable deformation stage, marked by a sharp increase in bending force. The application of electric pulse had a negligible effect on bending forming force in the elastic and rapid deformation stages, but it reduced force in the stable deformation stage. At a 120° bending angle, bending force was largely unaffected by peak current. While at a 150° angle, force was minimized at a peak current of 450 A, with maximum force occurring at a frequency of 300 Hz for both angles.

In Figure 8, the data was measured by a measuring instrument with an accuracy of ± 0.1 mm, and springback reduction refers to the difference in the springback angle of DP780 under electric pulse bending compared to non-electrically pulsed bending. Under both bending angles, introduction of pulse current into the DP780 sheets resulted in a decrease in springback angle. As peak current gradually increased, reduction in springback diminished, while an increase in frequency enhanced springback reduction. Thus, smaller peak current and higher frequency proved more effective in inhibiting springback.

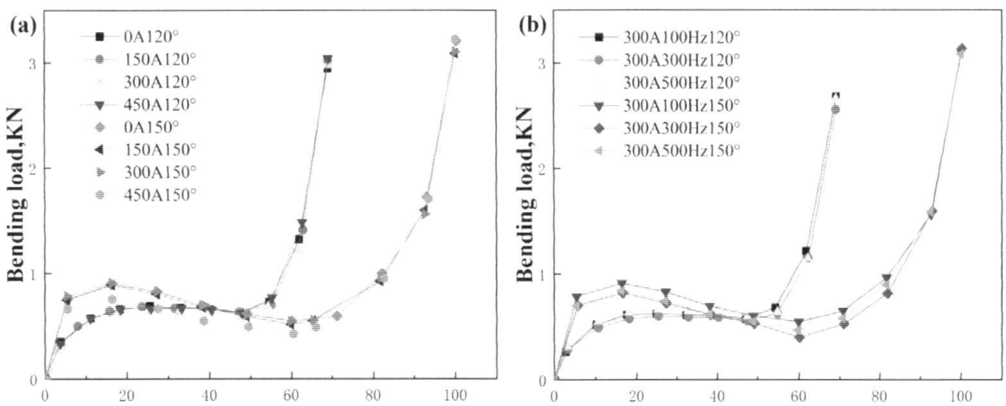

Figure 7. Bending force variation curves of 120° and 150° V-bending DP780 under different peak currents (a) and frequencies (b).

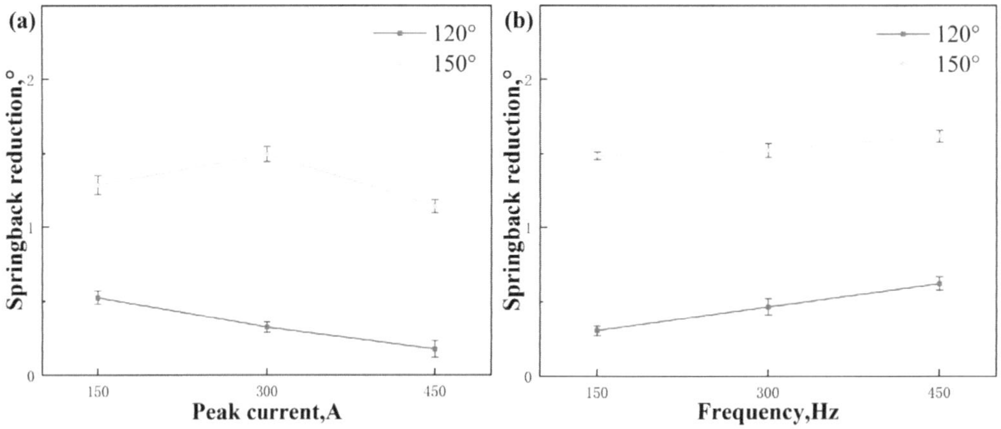

Figure 8. Springback reduction of 120° and 150° V-bend DP780 at different peak currents (a) and frequencies (b).

3.3 Electric pulse-assisted roller pressing experiment

Figure 9 depicts the impact of different peak currents and frequencies on springback reduction in roll forming of DP780 sheet during electric pulse experiments conducted on the 2nd and 5th passes. The electric pulse roll forming effectively reduced springback of the DP780 sheets, and the effect is more pronounced as the peak current increased. The overall reduction in springback is greater in the 5th pass electric pulsed condition than in the 2nd under the same electric pulse parameters. This suggested that applying a rectangular electric pulse at the tail end of roll forming was more effective in suppressing DP780 sheets springback than at the front end. Under 2nd pass electric pulsed conditions, reduction in springback initially decreased and then increased with increasing frequency. Conversely, under 5th pass electric pulsed conditions, reduction in springback initially increased and then decreased. Despite these variations, springback decreased compared to non-electrically pulsed roll forming, indicating that electric pulse roll forming effectively reduced DP780 sheets springback. Notably, for 2nd pass electric pulsed conditions, higher frequencies contributed to better springback reduction, whereas for 5th pass electric pulsed conditions, lower frequencies were more effective in inhibiting springback.

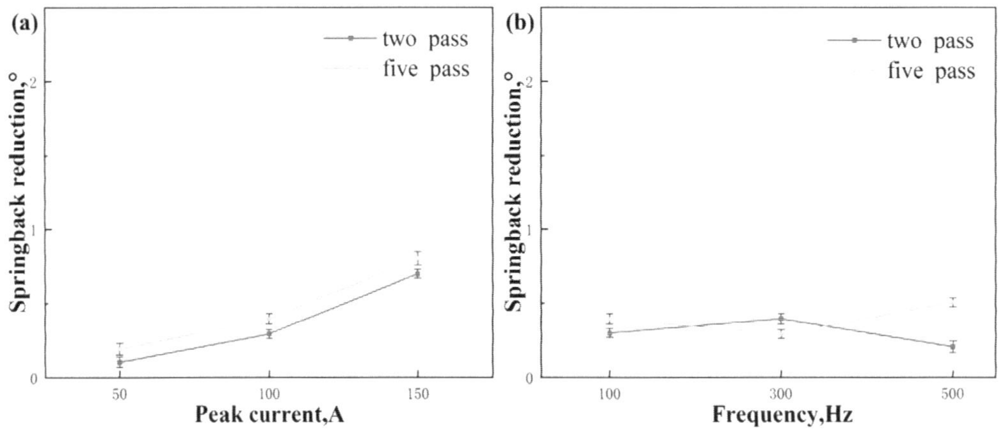

Figure 9. springback reduction at different peak currents (a) and frequencies (b).

4. Discussion

In electric pulse-assisted tensile experiments, there is no notable difference in the engineering stress-strain curves of DP780 at various peak current densities and frequencies. This lack of distinction may be attributed to the existence of a threshold effect in its pure electrophysical nature. Specifically, when peak current density or pulse frequency surpasses a certain value, a distinct pure electrophysical effect becomes apparent. However, the values of the variables in our study remain below this threshold, resulting in insignificant changes in tensile strength and elongation. At the microstructure level, the limited electrophysical effect in DP780 may arise from the abundance of movable dislocations in its ferrite phase. Additionally, the martensite phase contains a high density of dislocations. Plastic deformation initiates in the ferrite phase, and as deformation progresses, the martensite phase becomes involved. When movable dislocations in both phases aggregate at phase junctions or precipitate, high-density dislocation groups form. This impedes dislocation movement significantly. At this stage, the drifting electrons' force on the dislocations has not exceeded the threshold necessary for dislocation movement. Consequently, there is insufficient force to disentangle and release the pinned dislocations, leading to only slight changes in tensile strength and elongation.

In contrast, the enhanced bending properties of DP780 under pulsed current primarily stem from purely electroplastic effect. Force exerted by drifting electrons is directly proportional to peak current, and electroplastic effect strengthens with increasing peak current. Consequently, during the bending process, pulsed current significantly improves the formability of DP780 and reduces bending forming force. This improvement becomes more pronounced with higher peak currents. Additionally, the rise in frequency correlates with an upward trend in DP780 springback reduction, indicating that increased pulse frequency is advantageous for mitigating springback. This is attributed to the higher concentration of pulse current energy introduced into the specimen, resulting in a more efficient elimination of residual stresses and, consequently, better suppression of springback.

In the roll forming process, external force induces deformation, with a portion of released heat contributing to plastic deformation, and the remaining elastic deformation storing energy in the form of residual stresses within the material. These internal stresses primarily manifest as crystal lattice deformation and aberration at the microstructure level, maintaining the material system in a higher-level sub-stable state. Applying electric pulses to remove residual stress in DP780 involves introducing electric energy as external input, elevating the material system beyond the energy potential barrier. This action releases the stored residual stress work, reducing the system to a lower energy state and achieving overall stability. The macroscopic outcome is a reduction in springback, with better inhibition of springback observed as peak current increasing.

5. Conclusion

In this study, electric pulse-assisted experiments were conducted on DP780 ultra-high-strength steel, including tensile, V-bending, and roll pressing tests, yielding the following conclusions:

The pulsed current slightly reduces DP780 tensile strength. With constant frequency, tensile strength decreases with increasing peak current density, while elongation initially increases and then decreases. Conversely, at constant peak current density, tensile strength increases with rising frequency, with a similar trend observed for elongation. Scanning electron microscopy analysis of fractured DP780 specimens reveals that increasing peak current density leads to a gradual reduction in ligament density. Similarly, increasing frequency initially enlarges ligament density before subsequently reducing it, aligning with experimental results. Pulse current effectively reduces DP780 springback during bending, and the reduction is more prominent with larger bending mold angles. The impact of pulse current on bending force is relatively small, but variations in peak current and frequency can marginally decrease bending force. Electric pulses inhibit DP780 roll forming springback by inducing stress relaxation, and as peak current and frequency increase, springback further decreases. Applying electric pulses at the tail end of roll forming passes is more effective in suppressing DP780 sheet springback than at the front end.

Acknowledgements
This work was supported by the Research Fund of Jianghan University (Grant No. 2021yb152, 06750001).

IOP Conf. Series: Materials Science and Engineering **1307** (2024) 012050 doi:10.1088/1757-899X/1307/1/012050

References

[1] Liu X, Lan S, Ni J. Experimental study of electro-plastic effect on advanced high strength steels. Mater Sci Eng, A. 2013; 582: 211-8.

[2] Lee J-s, Choi S-Y, Lee S-k, Song J-h, Noh W, Choi S, et al. Roll forming process of automotive seat rail with 980 DP steel using Yoshida-Uemori kinematic hardening model. Procedia Manufacturing. 2018; 15: 796-803.

[3] Liu X, Cao J, Chai X, Liu J, Zhao R, Kong N. Investigation of forming parameters on springback for ultra high strength steel considering Young's modulus variation in cold roll forming. Journal of Manufacturing Processes. 2017; 29: 289-97.

[4] Song J, Jang I, Gwak S, Noh W, Lee J, Bae G, et al. Effect of pulsed currents on the springback reduction of ultra-high strength steels. Procedia engineering. 2017; 207: 359-64.

[5] Kim M, Bae G, Park N, Song JH. Springback Reduction of Ultra-High-Strength Martensitic Steel Sheet by Electrically Single-Pulsed Current. Materials. 2022; 15: 2373.

[6] Xie H, Dong X, Wang Q, Peng F, Liu K, Wang X, et al. Investigation on transient electrically-assisted stress relaxation of QP980 advanced high strength steel. Mechanics of Materials. 2016; 93: 238-45.

[7] Xie H, Dong X, Liu K, Ai Z, Peng F, Wang Q, et al. Experimental investigation on electroplastic effect of DP980 advanced high strength steel. Mater Sci Eng, A. 2015; 637: 23-8.

[8] Egea AJS, Rojas HAG, Celentano DJ, Travieso-Rodríguez JA, i Fuentes JL. Electroplasticity-assisted bottom bending process. J Mater Process Technol. 2014; 214: 2261-7.

[9] Thien NT, Jeong Y-H, Hong S-T, Kim M-J, Han HN, Lee M-G. Electrically assisted tensile behavior of complex phase ultra-high strength steel. International Journal of Precision Engineering and Manufacturing-Green Technology. 2016; 3: 325-33.

[10] Liu K, Dong X, Xie H, Wu Y, Peng F. Influence of pulsed current on deformation mechanism of AZ31B sheets during tension. J Alloys Compd. 2016; 676: 106-12.

[11] Krishnaswamy H, Kim MJ, Hong S-T, Kim D, Song J-H, Lee M-G, et al. Electroplastic behaviour in an aluminium alloy and dislocation density based modelling. Materials & Design. 2017; 124: 131-42.

[12] Xiaoli L, Jianguo C, Xueting C, Jiang L, Rongguo Z, Ning K, Investigation of forming parameters on springback for ultra high strength steel considering Young's modulus variation in cold roll forming. Journal of Manufacturing Processes. 2017; 29: 289-297.

[13] Li M, Guo D, Li J, Zhu S, Xu C, Li K, et al. Achieving heterogeneous structure in hcp Zr via electroplastic rolling. Mater Sci Eng, A. 2018; 722: 93-8.

[14] Qian L, Zhan L, Zhou B, Zhang X, Liu S, Lv Z. Effects of electroplastic rolling on mechanical properties and microstructure of low-carbon martensitic steel. Mater Sci Eng, A. 2021; 812: 141144.

[15] Jeong K, Jin S-W, Kang S-G, Park J-W, Jeong H-J, Hong S-T, et al. A thermally enhanced recrystallization kinetics of ultra-low carbon steel via electric current treatment. Acta Mater. 2022; 232: 117925.

IOP Conf. Series: Materials Science and Engineering **1307** (2024) 012051 doi:10.1088/1757-899X/1307/1/012051

Process design for flexible T-profile-rolling of tailored thickness stringers with different cross sections

F S Aign[1], J Kilz[1] and P Groche[1]

[1] Institute for Production Engineering and Forming Machines (PtU), Technical University of Darmstadt, Otto-Berndt-Straße 2, 64287 Darmstadt, Germany

e-mail: franziska.aign@ptu.tu-darmstadt.de

Abstract. A new process, called flexible T-profile-rolling, is being developed to produce T-profiles with variable thickness distribution in longitudinal direction by flexible rolling. These tailored thicknesses allow for load-adapted and weight-optimised design of stringer profiles for aircrafts. The new process can replace the chemical milling currently used in the production of stringers and enables a more environmentally friendly and resource-saving production of the thickness changes. In order to achieve uniform thicknesses in the web and flange of the T-profile, a new rolling stand with rolls inclined against the T-profile and varying roll radii was developed. However, the varying roll radii cause different longitudinal strains in the profile cross-section due to the varied contact areas between the roll and the rolling stock, resulting in unwanted profile bowing. This bowing depends on the profile dimensions, as the adjacent roll radii vary over the profile cross-sections.

This paper demonstrates the effects of profile dimensions on deflection through numerical investigations with different web heights, flange widths and thickness reductions. A straightening method is presented to obtain straight profiles as a function of the profile dimensions. The results show that this process can produce almost straight T-profiles with variable thicknesses for different cross section dimensions.

1. Introduction

With the increasing demand for efficient energy and resource utilisation, lightweight construction is gaining importance in the transport sector, machine tool production and architecture [1]. According to Carruth et al., lightweight construction can lead to a reduction in material consumption of 25 to 30 % [2]. An essential principle in lightweight construction is load-adapted design, especially implemented through individual wall thickness gradients in Tailor Rolled Blanks.

Tailor Rolled Blanks are semi-finished sheet products with a local, continuous variation of the sheet thickness in the longitudinal direction via a flexible rolling process with online adjustment of the roll gap. There is no need for reinforcing blanks and less joining elements are necessary [3]. The position, number and shape of the thickness transitions are arbitrary and the continuous process ensures high material utilisation [4].

Flexible rolling is currently mainly used for the production of flat sheets. Aign et al. [5] introduced a new process idea called flexible T-profile rolling, to produce branched T-sections with varying thicknesses. This process could be used, for example, in the manufacture of aircraft stringer sections with thickness variations in the longitudinal direction. Currently, the thickness transition is achieved by chemical etching [6], whereas the new process offers an environmentally friendly alternative.

The flexible T-profile rolling process uses a new roll geometry with two skew rolls arranged at an angle to the T-profile and one cylindrical roll, as shown in figure 1. The simultaneous translational infeed of the two skew rolls causes the thickness variation.

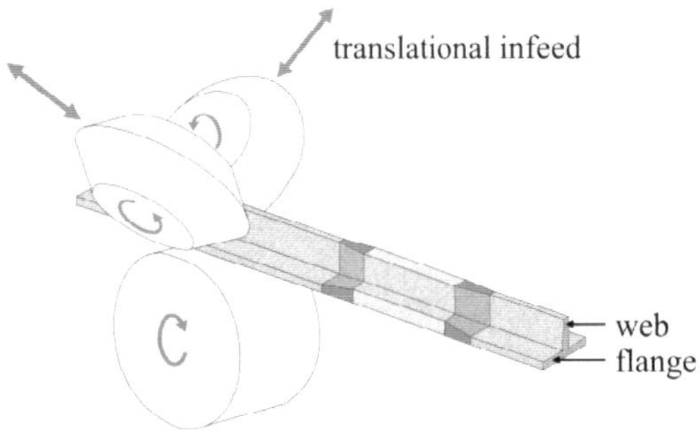

Figure 1. Principle of the new flexible T-profile rolling process according to [7]

The challenge of the new rolling process is the vertical profile bowing caused by the varying radius of the skew roll and the asymmetry of the profile shape. The varying radius of the skew roll results in different contact areas of the roll with the profile and in a difference in longitudinal strain over the height of the web. By choosing a different thickness reduction in the web and flange, the difference in longitudinal strain could be compensated and the profile could thereby be straightened [5]. This idea is also used by Takashima et al. who developed the T-bar universal rolling, a hot rolling process. They were able to minimise the bowing of T-sections by adjusting the thickness reduction in the web and flange and demonstrated a linear relationship [8].

Furthermore, in asymmetric rolling, different roll radii, different roll circumferential speeds or different friction conditions lead to a shift of the neutral points at each side of the plate. This results in shear stresses across the plate thickness, resulting in bowing [9, 10]. There is also evidence from other sources that bowing can be influenced by locally different rolling in the cross section. For example, inhomogeneous longitudinal strain distributions in a profile cause bowing during roll forming [11, 12] and can be straightened by partial rolling [11].

To design the new rolling process with as little bowing as possible, this paper shows parameters that influence bowing and presents a new and simple calculation model for designing the infeed depths in the web and flange. For the first approach, the constant rolled thickness area of the profile was investigated to provide a basis for extending the process to the flexible process.

2. Numerical model of flexible T-profile rolling

The roller geometry consists of two tapered skew rolls inclined at 45°, bearing against the T-profile with a varying radius, and a cylindrical roller. The corner of the T-profile and the skew rolls are rounded with the same radius of 2.5 mm. Only half of the T-profile is simulated using a symmetry plane. The dimensions of the profile can be found in figure 2. The cylindrical roll has a radius r_{CR} of 136 mm and the maximum radius r_0 of the skew roll in the corner, without rounding, is 145 mm. See figure 7 (left) for the radius designations. The profile has a total initial length of 350 mm as the process is stationary once the target thickness has been reached. For sheet feed, all rollers are driven in rotation at a circumferential speed of 0.25 rad/s. The thickness variation in the longitudinal direction is achieved by the simultaneous inline infeed of the skew rolls at the same infeed depth. For the numerical simulation of the process, the software package Marc Mentat is used and the model is structured as in Aign et al. [5]. The rollers are represented as rigid bodies and the profile as a deformable part. The sheet is modelled as a solid using element type 7 as 8-node hexagonal full integration with a length of 1.125 mm, a width of 0.875 mm and 4 elements are arranged in thickness direction.

The thickness reduction of the initial profile is 0.5 mm. Since the web is rolled through the skew rolls from both sides, the infeed of the skew roll in the flange is twice as much as in the web to ensure a consistent infeed depth across the entire cross-section. This means that the skew rolls are fed 0.5 mm in the y-direction and 0.25 mm in the x-direction for the initial profile. The infeed is specified as a linear motion curve. At the start of the simulation, the start of the profile is in the roll gap. After a feed of

20 mm, the infeed starts at a pitch of 1:100 (1 mm thickness change per 100 mm length), analogous to Kopp et al. [4]. Once the target infeed depth has been reached, the profile continues to be rolled with a constant thickness reduction. It is simulated with a Coulomb friction coefficient of 0.1. The material used is Al7075 O with the flow curve shown in figure 2. The flow curve is obtained from flat crush tests according to Becker [13] with a strain rate of 0.1 1/s. The von Mises yield criterion is implemented, as the profile is soft annealed before rolling. Elastic-plastic isotropic material behaviour is presumed with a Young's modulus of 73 GPa, a density of 2.78 g/cm^3 and a Poisson's ratio of 0.3. Since the temperature during cold rolling is close to ambient, temperature effects are ignored [14]. Due to low rolling speeds and strain rates in the process, it is assumed that the influence of the strain rate is low and can be neglected [15]. The assumed strain formulation allows improved modelling of the shear behaviour and to avoid the locking effect, the constant dilatation option is used [5].

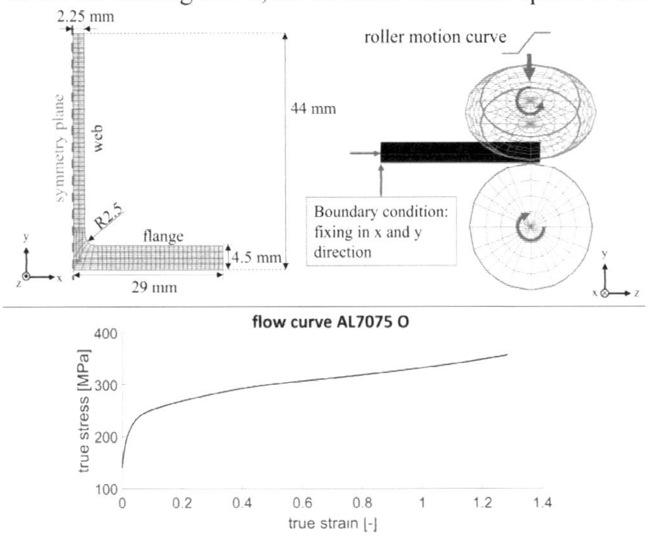

Figure 2. Numerical model of flexible T-profile-rolling according to [5]

3. Dependence of the bowing on the profile cross section

In roll forming, differences in longitudinal strain distributions across the profile cross section, caused by non-uniform longitudinal elongation, lead to unwanted bowing. The amount and direction of bowing follows the variation in strain distribution. The profile bows in the direction of the lower longitudinal strain apparently when the equivalent stress exceeds the yield strength [12, 16].

Non-uniform deformations also occur during the asymmetrical rolling of flat sheets, resulting in different plastic strains in the top and the bottom surface [17]. Mousavi et al. show that the sheet bends towards the surface with less plastic strain (inner area of the bow). Despite the shear zone, no bowing occurred when the normal plastic strain was similar on the top and bottom surface [17].

This bowing behaviour can also be transferred to the flexible T-profile rolling. Due to the varying radius of the skew rolls, see Fig. 6 (left), the contact area of the rolls on the profile varies with the height of the web and therefore the displaced volume. In Aign et al. it was shown that the width deviation is negligible [5]. Due to volume constancy and assuming constant friction conditions and constant peripheral speeds, the change in elongation is caused by the change of the displaced volume. As the roll radius decreases and web height increases, the displaced volume decreases, which reduces the longitudinal elongation above the web height. Consequently, there is a difference in longitudinal strain in the web, as shown in figure 4. It can be seen that for profiles with different web heights, the smallest web height near the corner has the largest longitudinal strain. The profile bows in the direction of the lower longitudinal strain at the top of the web, resulting in a negative bow for the initial profile with a web height of 44 mm, see figure 3.

Based on considerations of how the longitudinal elongation differences in the T-profile arise and influence the bowing, the web height, flange width and thickness reduction were varied in numerical simulations. The variables were investigated using the constant rolled thickness area. Friction factor and roll peripheral speed remain constant. The vertical bowing at the centre of gravity of the profile, see figure 3 (right), was evaluated in the steady state range from 110 mm to 240 mm of the profile length, analogous to Aign et al. [5]. The web height and the flange width were then varied starting from the initial profile.

When varying the web height, it can be observed in figure 3 that the bowing increases continuously with increasing web height from 25.188 mm to 37.313 mm. This confirms the assumption that the longitudinal strain at the top of the web decreases with increasing web height, causing an increasing bowing. This can also be seen in the longitudinal strain distribution over the web height, evaluated in the middle of the web, in figure 4 for the web heights from 19.125 mm to 37.313 mm. For the web height of 19.125 mm, the bowing changes in a positive direction. This may be due to the radius of the skew roll adjacent to the top of the web still being larger than the radius r_{CR}, with the lower elongation now being at the bottom of the flange. Another possible reason could be the modified moment of inertia of the profile. From 37.313 mm onwards, the bowing decreases again, see figure 3. Slippage is assumed to be the cause, whereby the material at the upper edge of the web is dragged along, leading to larger elongation. The longitudinal strain distribution over the web height also shows that the longitudinal strain increases again from approximately 90 % of the web height.

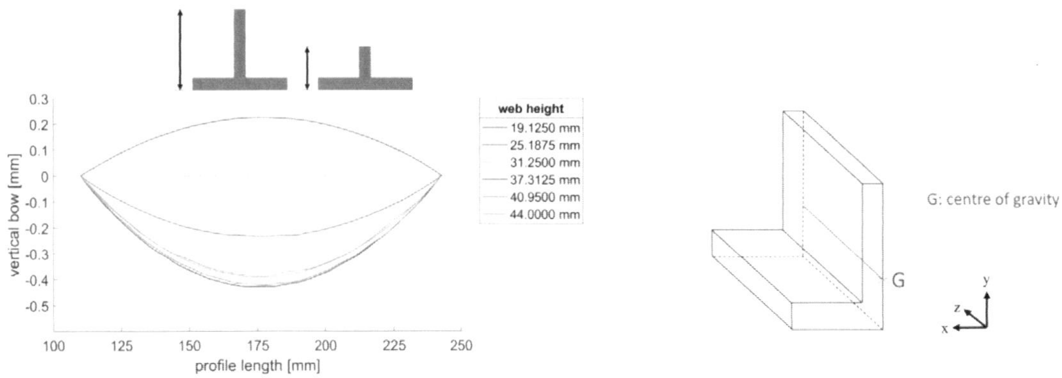

Figure 3. vertical bowing for web height variation and node path for evaluation on profile half according to [5]

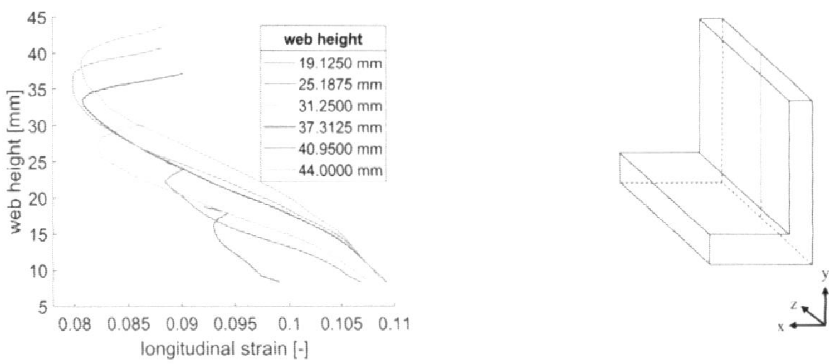

Figure 4. longitudinal strain over web height for web height variation and node path for evaluation according to [5]

The width of the profile's flange was varied between 39.1 mm and 58 mm. The bowing increases as the flange width increases, as seen in figure 5 (left). One possible reason for this is that with a smaller flange width the mean adjacent radius of the skew roll $r_{f,mean}$ is larger than that of the cylindrical roll r_{CR}. This is shown schematically in figure 5 (right) for a very small flange width. Therefore, if the strain

IOP Conf. Series: Materials Science and Engineering **1307** (2024) 012051 doi:10.1088/1757-899X/1307/1/012051

difference in the flange and web were considered separately, the flange deflection could counteract the web deflection. As the flange width increases, the relationship is reversed and the bowing increases. For the initial profile, $r_{f,mean}$ is nearly the same as r_{CR}.

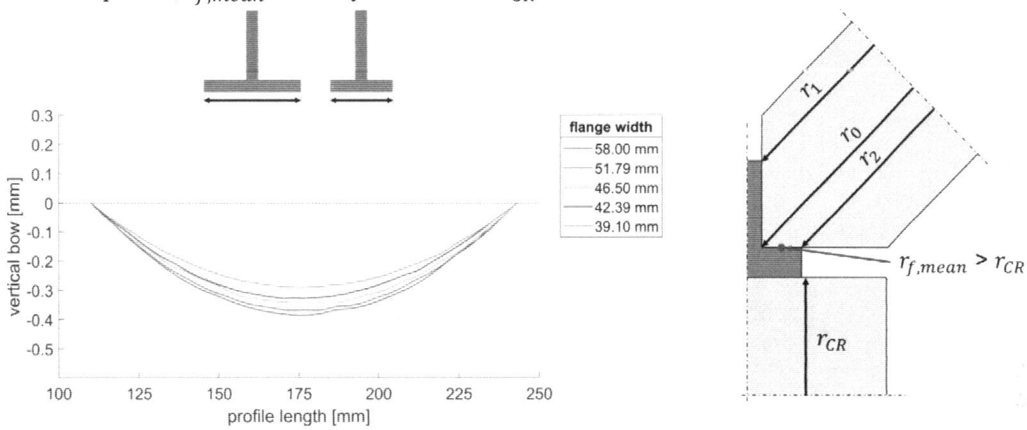

Figure 5. (left) vertical bowing for flange width variation; (right) radius ratio for short flange width

The infeed depth was varied from 0.3 mm to 0.6 mm. As the thickness reduction increases, the bowing increases, see figure 6. A larger infeed depth results in a larger displaced volume, thus introducing a larger difference in longitudinal strain.

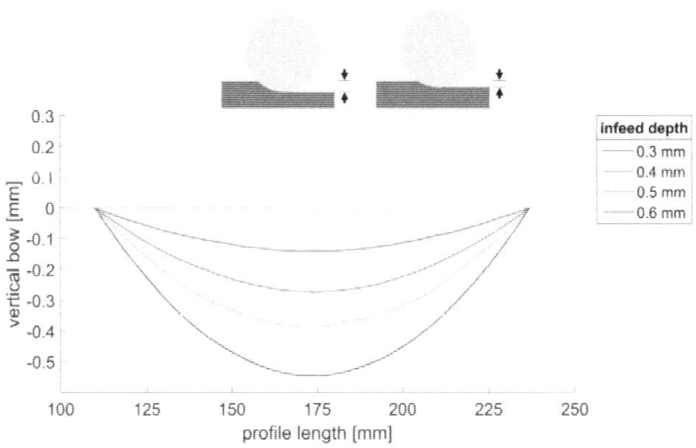

Figure 6. vertical bowing for variation of infeed depth

4. Straightening process design for constant thickness

Based on the investigations of [10] and [16], it is postulated that a profile is considered straight if equal longitudinal strains are present at the top of the web and in the flange, specifically at the locations denoted by blue arrows in figure 7 (left). To achieve equal longitudinal strains in these locations, the displaced volume must be equal as well. This can be accomplished by employing different infeed depths in the web and flange, as demonstrated in [8]. The infeed depth in the flange remains unchanged in the straightening method presented, while that of the web is adjusted. Since both the skew roll and the cylinder roll affect the strains in the flange asymmetrically, they are averaged.Based on these considerations, the following model was developed to calculate the infeed depth required in the web and flange to straighten the profiles with different cross sections for a constant thickness reduction.

4.1. Calculation of the infeed depth difference for straight profiles

First, it is assumed that the infeed depth Δh in the web and flange is equal and corresponds with the intended thickness adjustment. The contact length l_d between the roll and the rolling stock is then calculated according to [18]. r_i represents the radius of the roll at point i.

$$l_d = \sqrt{r_i \cdot \Delta h} \tag{1}$$

As the initial profile has a constant length, the calculation is confined to the displaced area A, which is shown in figure 7 (right) for rolling a flat sheet. The process involves determining the area of the circular section and then subtracting the area of the triangle over area A.

$$A(r_i) = \pi {r_i}^2 \cdot \frac{\alpha}{360°} - \frac{1}{2} l_d \cdot (r_i - \frac{\Delta h}{2}) \tag{2}$$

The angle α of the circle section can be calculated using the small angle approximation acc. to [18].

$$\sin(\alpha) = \frac{l_d}{r} \approx \alpha \tag{3}$$

To determine δ_{web}, the ratio of the displaced area in the web to the initial profile thickness in the web h_{web}, equation 2 requires the radius r_1 of the skew roll located at the top of the web to be inserted. As the roll radius is the same on both sides of the top of the web, multiply the result by 2.

$$\delta_{web} = \frac{2 \cdot \mathrm{A}(r_1)}{h_{web}} \tag{4}$$

To calculate δ_{flange}, the ratio of the displaced area in the flange to the initial profile thickness in the flange h_{flange}, the areas of the adjacent radii of the inclined and cylindrical rolls are averaged. The radius designations can be found in figure 7 (left).

$$\delta_{flange} = \frac{A(r_{CR}) + A(\frac{r_0 + r_2}{2})}{h_{flange}} \tag{5}$$

To calculate the new infeed depth in the web, the related displaced areas in the web and flange are set equal. The infeed depth in the flange remains unchanged and the equation is solved for the new required infeed depth in the web Δh_{new}.

$$\delta_{web}(\Delta h_{new}) = \delta_{flange} \tag{6}$$

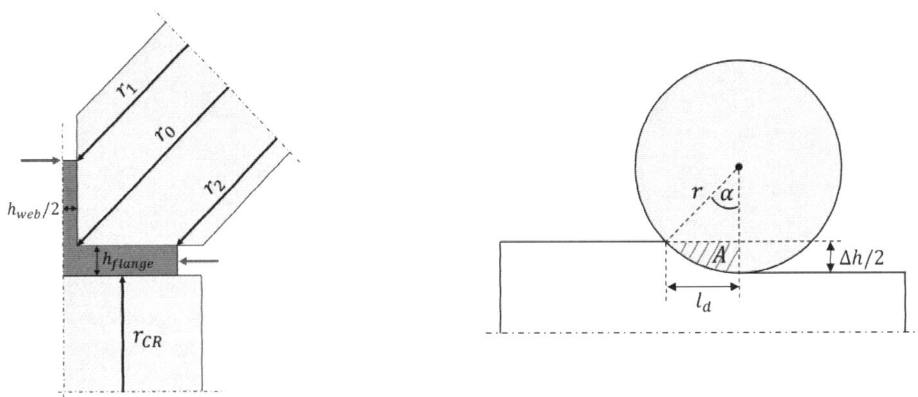

Figure 7. (left) radius designations of the rolls; (right) schematic representation of the displaced area for rolling a flat sheet

4.2. Results of the straightening method

In order to test the straightening method from chapter 4.1, the modified infeed depth in the web was calculated and numerically simulated for selected profiles. For this purpose, one size was randomly selected from each parameter variation of infeed depth, web height and flange width. The vertical bow for the straightened initial profile is shown in figure 8. Table 1 shows the bowing maxima and minima for the other profiles as well as the percentage improvement in absolute bowing.

Table 1. bowing reduction by straightening method

	calculated infeed depth in the web Δh_{new}	min. /max. of unstraightened bow	min. /max. of straightened bow	absolute percentage improvement
initial profile	0.5253 mm	- 0.4543 mm	- 0.0139 mm	97 %
web height 31.25 mm	0.5125 mm	- 0.4603 mm	- 0.1830 mm	60 %
infeed depth 0.4 mm	0.4203 mm	- 0.2731 mm	+ 0.0539 mm	80 %
flange width 46.5 mm	0.5266 mm	- 0.3310 mm	+ 0.0145 mm	96 %

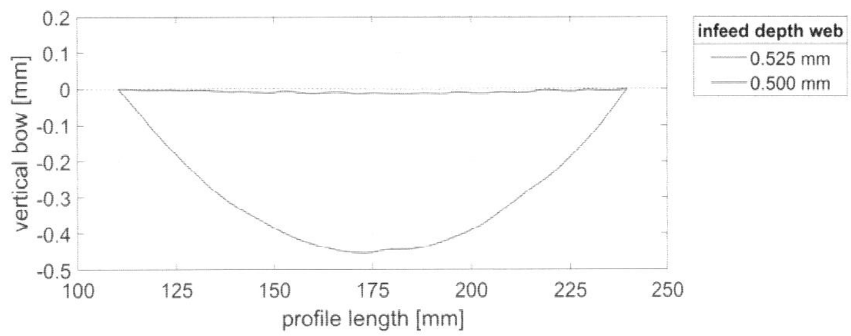

Figure 8. Straightening of the initial profile with calculated infeed depth

In case of the initial profile, the bow could be reduced by 97 %. For the web height of 31.25 mm, on the other hand, the improvement in bowing is only 60 %. The causes could be the slip mentioned in chapter 3 or the changing surface moment of inertia of the profile, which is not taken into account. In addition, the distance of the longitudinal strain from the centre of gravity of the profile could also have an influence as a lever arm for the bending moment resulting from the strain difference.For the infeed depth of 0.4 mm and the flange width of 46.5 mm, the absolute bows were reduced by 80 % and 96 % respectively, see table 1. In the case of the flange width variation, the differences in the unstraightened bows for the varied parameters are already small in figure 5 (left), so the influence on the straightened bows is also likely to be small and very good results can be obtained. Therefore, the direction of the partial bowing in the flange is neglected in this model for the time being.Another factor that has not been taken into account is the corner, which is likely to have less longitudinal strain than the web and flange due to the accumulation of more material and less penetration of the skew rolls. For larger profile widths, it represents only a small percentage of the total rolled area and is therefore neglected.

5. Conclusion and Outlook

In this paper, influences on the bowing that occur during flexible T-profile rolling were investigated and a countermeasure for straightening the profiles was presented. The straightening is accomplished by adjusting infeed depths in the web and flange, which can be approximately calculated by the presented model. The effects of web height, flange width and infeed depth on the bowing during rolling of flexible T-profiles have been shown for the constant rolled thickness area. In summary, the bowing increases with increasing infeed depth and flange width. As the web height increases, the bowing also increases until it decreases slightly. This is probably due to slippage in the roll gap.

Overall, it is shown that the model for designing the adjustment of thickness reduction for straightening can significantly reduce the bowing. This would make it possible to roll almost straight T-profiles with different cross-sections in the future. However, there is still room for improvement.

Experimental verification of the process is required and the model will be extensively validated and extended in the future. In practice, it needs to be shown how additional influences such as friction and circumferential speed of the rolls affect the bowing. The influence of the flexible thickness transition also remains to be investigated; it is assumed that a linear transition must also result in a linear increase in the infeed depth. In addition, the influence of other geometry variables, such as the influence of the initial profile thickness, can be investigated in the future.

6. Acknowledgement

The presented research forms part of the project "MeGro - Flexibler Walzprozess zum Einbringen variabler Wanddickenverläufe in Stringer", which is a component of Lower Saxony's aviation research program and receives funding from "Investitions- und Förderbank Niedersachsen (NBank)". We extend our gratitude to all industry partners, especially AIRBUS Aerostructures GmbH, for their valuable contribution to the project.

References

[1] Kleiner M, Geiger M and Klaus A 2003 Manufacturing of lightweight components by metal forming *CIRP Annals* **52 (2)** 521-542.
[2] Carruth M, Allwood J and Moynihan, M 2011 The technical potential for reducing metal requirements through lightweight product design *Resources, Conservation and Recycling* **57** 48–60
[3] Merklein M, Johannes M, Lechner M and Kuppert A 2014 A review on tailored blanks - production, applications and evaluation *J. Mater. Process. Technol.* **214** 151 - 164
[4] Kopp R, Wiedner C and Meyer A 2005 Flexibly rolled sheet metal and its use in sheet metal forming *Adv. Mat. Res.* **6–8** 81–92
[5] Aign F and Groche P 2023 T-profile-rolling of stringers with tailored thicknesses – challenges and approaches *Production at the Leading Edge of Technology, 13th Congress of the German Academic Association for Production Technology (WGP)* (Freudenstadt: Springer) 569-578
[6] Torng C, Huang C and Chang H 2009 Process control for aerospace chemical milling process *Int. J. Manuf. Technol. Manag.* **18(3)** 308-318
[7] Aign F, Groche P and Wang, T 2021 Flexibles Auswalzen von T-Profilen zur Anwendung im Flugzeugbau *Blechnet* **4** 54-55
[8] Takashima Y and Hiruta T 2012 T-bar universal rolling and its deformation properties *ISIJ International* **52(7)** 1328-1334
[9] Yekta R, Zohoor H and Motekallem A 2018 Analysis of asymmetric cold rolling using finite element method to create flat sheet *Scientia Iranica* **B 25(6)** 3133-3140
[10] Liang H, Di H and Gong D 2013 Analysis of sheet curvature in asymmetrical cold rolling *J. Iron Steel Res. Int.* **20(5)** 34-37
[11] Groche P, Güngör B and Kilz J 2023 Straight roll formed profiles through partial rolling *CIRP Ann. - Manuf. Technol.* **00** 1-4
[12] Halmos G T 2006 *Roll forming handbook* (Toronto, Ontario, Canada: Taylor & Francis Group)
[13] Becker N 1994 *Weiterentwicklung von Verfahren zur Aufnahme von Fließkurven im Bereich hoher Umformgrade (IFU - Berichte aus dem Institut für Umformtechnik der Universität Stuttgart* vol. 122*)* Springer-Verlag Berlin (Heidelberg)
[14] Ghosh S, Li M, Gardiner D 2004 A Computational and Experimental Study of Cold Rolling of Aluminum Alloys With Edge Cracking *J. Manuf. Sci. Eng.* **126(1)** 74-82
[15] Ostermann F 2014 *Anwendungstechnologie Aluminium* Springer Vieweg Berlin (Heidelberg) p 292 f.
[16] Kilz J, Güngör B, Aign F and Groche P 2023 Profile defects caused by inhomogeneous longitudinal strain distribution in roll forming *Int. J. Mater. Form* **16** 36
[17] Mousavi S A A., Ebrahimi S M and Madoliat R 2007 Three dimensional numerical analyses of asymmetric rolling *J. Mater. Process. Technol.* **187–188** 725–729
[18] Kugler H 2009 *Umformtechnik – Umformen metallischer Konstruktionswerkstoffe* Fachbuchverlag Leipzig im Carl Hanser Verlag

IOP Conf. Series: Materials Science and Engineering **1307** (2024) 012052 doi:10.1088/1757-899X/1307/1/012052

Straightening rollformed profiles by controlled partial rolling

J Kilz[1], **S C Arnold**[1], **P Groche**[1]

[1] Institute for Production Engineering and Forming Machines (PtU), Technical University of Darmstadt, Otto-Berndt-Straße 2, 64287 Darmstadt, Germany

E-mail: johannes.kilz@ptu.tu-darmstadt.de

Abstract. Roll forming is a sheet metal forming process in which flat sheets are incrementally formed into a desired profile geometry. The process is characterized by high material utilization and output. Along with these advantages, process-related longitudinal strains occur, causing profile defects such as bowing and twisting of the profile, which require subsequent straightening. Conventional straightening methods use overbending and twisting to eliminate existing profile defects. However, this approach is purely experience-based and iterative due to the complex interactions between the various bending and twisting operations. Therefore, a closed-loop control of the straightening process is currently not implemented. This can be remedied by the novel knowledge-based straightening method of partial rolling.
In this paper, the basic idea of partial rolling will be introduced and the idea of knowledge-based process control will be demonstrated. A simulation model is then presented to test different process control strategies and evaluate their impact on the profile defects. Finally, an outlook is given on how a closed-loop controlled partial rolling process could be set up in a real process.

1. Introduction

Roll forming is a continuous forming process to produce open and closed profile structures with a constant profile cross-section. The profile structure is iteratively formed from flat sheets or strips into the target geometry by successive roll forming tools. In addition to the desired transverse bending of the material, this process causes longitudinal strains due to the three-dimensional forming zone resulting in profile defects such as horizontal and vertical bow and twist [1], as shown in Figure 1. These defects occur particularly in asymmetric profiles due to their asymmetric longitudinal strain distributions. These profile defects require subsequent straightening of the profiles.

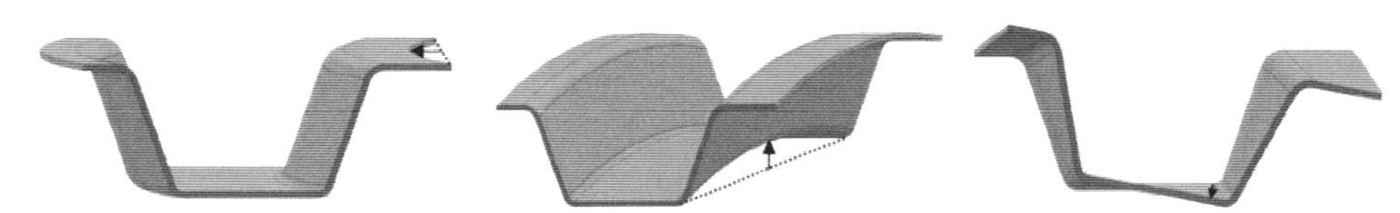

Figure 1. Profile defects in roll forming [2]

For this purpose, straighteners are used at the end of the roll forming line. Various scientific studies show the benefits and limitations of different straightening concepts. Ona et al. [3] investigate various straightening methods such as overbending the profile in horizontal and vertical direction, twisting the profile and combining the various individual methods to straighten asymmetrical U-profiles. They show that complete straightening of all profile defects is only possible by combining the different methods. Most conventionally used straighteners use a combination of these straightening methods. Due to the complex interactions, the straightener is adjusted manually and the straightening success is therefore dependent on the experience and knowledge of the machine operator. A closed-loop control of the straightener has so far only been investigated numerically. Abeyrathna [4] shows that there is a relationship between the forces and torques acting on the forming rolls and the vertical bow of a trapezoidal profile and uses this to control an overbending stand in the simulation. However, this straightening method is limited exclusively to straightening the vertical bow and cannot be transferred to more complex profiles.

Recent studies on the straightening of roll forming profiles have focused on the targeted introduction of additional longitudinal strain to homogenize the inhomogeneous longitudinal strain distribution in the profile cross-section, which is the cause of the profile defects. Wiebenga et al [5] investigate a partial rolling of a vertically bowed symmetrical V-profile by reducing the rolling gap and show that the vertical bow can be significantly reduced. Cha [6] numerically investigates the straightening of asymmetric U-sections by selectively reducing the thickness of the profile web after profiling. The investigations show that by reducing the material thickness in the web region of the U-section, longitudinal strains are introduced into the web of the section, homogenizing the longitudinal strain distribution over the cross-section and reducing vertical bow and twist. This is confirmed by numerical and experimental investigations by Güngör [7]. In another study, Groche et al [2] show that all profile defects can be straightened by optimally adjusting the roll gap of a separate rolling stand for a hat profile.

As a basis for this approach, the longitudinal strain distribution in the profile cross-section is required, which can be calculated by an analytical model of Kilz et al [8] based on the expression of the profile defects or taken from simulation results. Based on the longitudinal strain distribution in the profile cross-section, the additional longitudinal strain required to homogenize the longitudinal strain distribution can be calculated and transformed into a required roll gap of the rolling stand to straighten the profile. Figure 2 shows the principle of the knowledge-based straightening method partial rolling with the initial longitudinal strain, the necessary additional longitudinal strain required for straightening and the result.

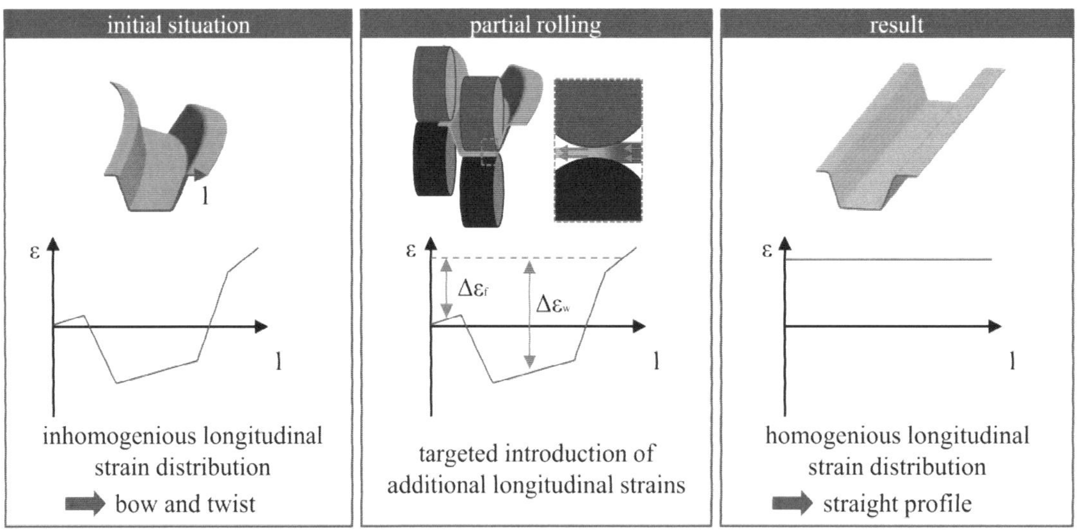

Figure 2. Principle of knowledge-based partial rolling [9]

The roll gap required to introduce a given amount of additional longitudinal strain can be estimated by considering the volume constancy and the expected springback of the material after rolling. Assuming that the ratio of the width of the roll to the compressed length of the material is greater than

IOP Conf. Series: Materials Science and Engineering **1307** (2024) 012052 doi:10.1088/1757-899X/1307/1/012052

20, the change in width of the material can be neglected [10]. Thus, the required roll gap h_i of the cross-section element i is as follows:

$$h_i = \frac{h_0}{(1 + \Delta\varepsilon_i + \varepsilon_{el})} \tag{1}$$

h_0 stands for the initial thickness of the sheet, $\Delta\varepsilon_i$ for the required additional longitudinal strain in a specific part i of the profile cross-section, and ε_{el} for the additionally required strain fraction that springs back elastically and therefore does not lead to plastic longitudinal strains.

In this paper, this context is used to investigate different approaches to control a roll stand for straightening roll forming profiles. Therefore, the roll forming and the partial rolling processes are numerically simulated and user subroutines are used to implement different control strategies.

2. Numerical model

2.1. Simulation setup

The numerical model is set up in Hexagon's Simufact Forming non-linear finite element software. In addition to the predefined simulation tools, the simulation is extended by user-subroutines.

The roll forming model consists of the conventional roll forming stands with six passes and the partial rolling stand, each with a stand distance of 525 mm. Each conventional pass contains two upper rolls and one lower roll. The model is used to simulate the forming process of an asymmetric hat profile, as shown in Figure 3 (a). The asymmetry is achieved by displacing the relative sheet metal to tool position. Following, the right flange is 10 mm longer than the left flange. To reduce computing time, only one quarter of the total roll body is simulated, as Figure 3 (b) shows. Additionally, it is assumed that there is no friction between the sheet metal and the rolls, since the influence of fiction is negligible in roll forming simulations according to [11]. Kinematic inversion is used, meaning that the roll forming system moves along the longitudinal axis over the sheet metal. The translational speed in negative z-direction is 6 m/min, while no rotational speed is set for the rolls.

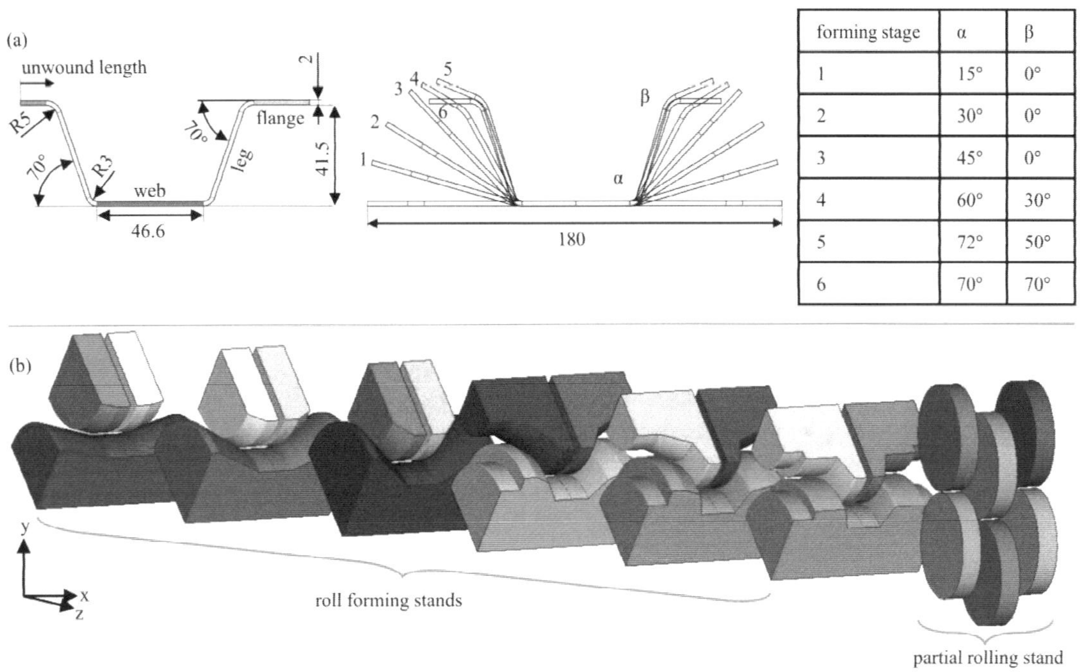

forming stage	α	β
1	15°	0°
2	30°	0°
3	45°	0°
4	60°	30°
5	72°	50°
6	70°	70°

Figure 3. a) flower pattern and profile dimensions [2], b) numerical model

In addition to the roll forming system, a novel partial rolling system is simulated, as shown in Figure 3 (b). The simulated tool is represented by six cylindrical bodies, positioned 525 mm downstream from the roll forming system. The rolls have a diameter of 150 mm and a width of 30 mm. It is important to note that the lower rolls are stationary and only the upper rolls have a degree of freedom to move in y-direction for roll gap adjustment. The rolls are positioned so that one pair of rolls works on the profile's web, one pair on the left flange and one pair on the right flange. Unlike the roll forming system, the rolls for partial rolling are assigned a friction coefficient of 0.15 to determine the primary forming direction. This also requires the definition of a rotational speed which is adapted to the translational speed of the roll forming stands.

For the forming process, a sheet metal with a length of 1998 mm, a width of 180 mm and a thickness of 2 mm is used. Based on previous convergence analyses, the sheet is discretized using hexahedral elements with a constant length of 6 mm, a thickness of 1 mm and a variable width across the cross-section ranging from 1 mm in bending zones to 6 mm in areas that remain flat during forming. The material used is a DC04 steel with material data provided by the Simufact Forming library. Boundary conditions are set in z-direction at two nodes in the front row to prevent the sheet from being pushed through the roll forming system.

The simulation is run with 4000 increments, the simulation time is 60 seconds and the complete simulated way amounts to 6000 mm. This results in a displacement of 1.5 mm per increment.

2.2. User subroutines

As described in section one, the adjustment of the roll gaps is based on the longitudinal strain difference in the profile cross-section. For computational purposes, the profile data is divided into the three different areas that are rolled - profile web, left flange and right flange. This makes it possible to average of the longitudinal strains of the separated profile cross-sections and predict the necessary roll movement required to homogenize the longitudinal strain distribution. After the roll forming process, the longitudinal strains ε_l are evaluated by comparing the current length of the elements with their original length:

$$\varepsilon_l = \frac{l - l_0}{l_0} \tag{2}$$

The length of the current elements l is calculated by:

$$l = \sqrt{(\Delta x_1 - \Delta x_2)^2 + (\Delta y_1 - \Delta y_2)^2 + \left(l_0 + (\Delta z_1 - \Delta z_2)\right)^2}. \tag{3}$$

The delta terms describe the displacement of a node, index 1 represents one node, index 2 represents the neighboring node in z-direction, l_0 describes the original length of the elements. This length comparison is continuously carried out for small longitudinal areas of the sheet.

Using the longitudinal strain, the necessary gap between the rolls can be determined. The highest longitudinal strain is set as the reference value. There will be no adjustment of the rolls in the area of the highest strain. Areas with a lower strain need to be processed by the partial rolling station so that a homogenous level of longitudinal strain is reached. Equation (1) is used to determine the required rolling gap. After calculating the gap, a command is sent to move the different rolls in y-direction in order to adjust the roll gap.

3. Control strategies and results

3.1. Open-loop control

In a first step, an open-loop control is used to control the roll gaps of the partial rolling stand. The input value of the control system is the longitudinal strain distribution over the profile cross-section. It is calculated 200 mm in front of the partial rolling stand, as shown in Figure 4 (a). Depending on the deviations in the measured longitudinal strain, it is determined which pair of partial rolls must be moved to reach a uniform level of strain.

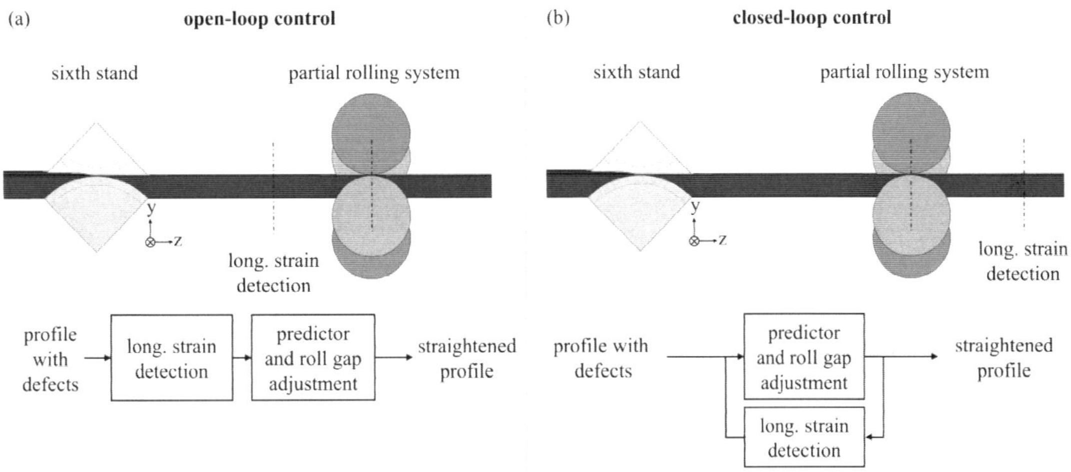

Figure 4. (a) open-loop control, (b) closed-loop control

Due to the spatial distance between strain detection and roll gap, a dead time needs to be considered. Applied to the simulation, the calculated rolling gaps must be saved and the rolls moved in a later increment.

The advantage of using the open-loop control with an evaluation point upstream of the partial rolling stand is that the longitudinal strain distribution is known before the profile enters the partial rolling stand. Thus, the gap between the roll pairs can be predicted in advance but there is no information about remaining strain deviations of the rolled-out part.

Results. To validate the performance of the open-loop control, the partially rolled part is compared to a reference part that has been simulated without partial rolling. The bow and twist reduction are calculated by comparing the bow and twist maxima of the partially rolled and the reference part. As shown in Figure 5, a significant decrease in profile defects in the stationary area between 700 mm and 1400 mm profile length is achieved. Vertical bow is reduced by 80 %, twist by 65 % and horizontal bow by 70 %. This shows that the aim of straightening by partial rolling can be achieved if the longitudinal strain distribution in the cross-section is known. Knowledge-based straightening is therefore possible.

Further evaluations of the results show an increase in plastic deformation in longitudinal direction in the parts that have been rolled by the partial rolling system. The total longitudinal strain distribution of the reference profile and the partial rolled part are shown in Figure 6. To better understand to which part of the profile the respective longitudinal strain distribution belongs, the background color was adapted to the color of the different profile parts in Figure 2 (a). It can be seen that the longitudinal strain distribution of the straightened part is more homogeneous than that of the reference part. According to Groche et al [2], the vertical bow of a hat profile is the result of a difference of the longitudinal strains in the profile web and the profile flanges. As shown in Figure 6, this difference decreased, resulting in a profile with less vertical bow. The remaining horizontal bow of the straightened profile results from a remaining difference in the longitudinal strain in the flanges.

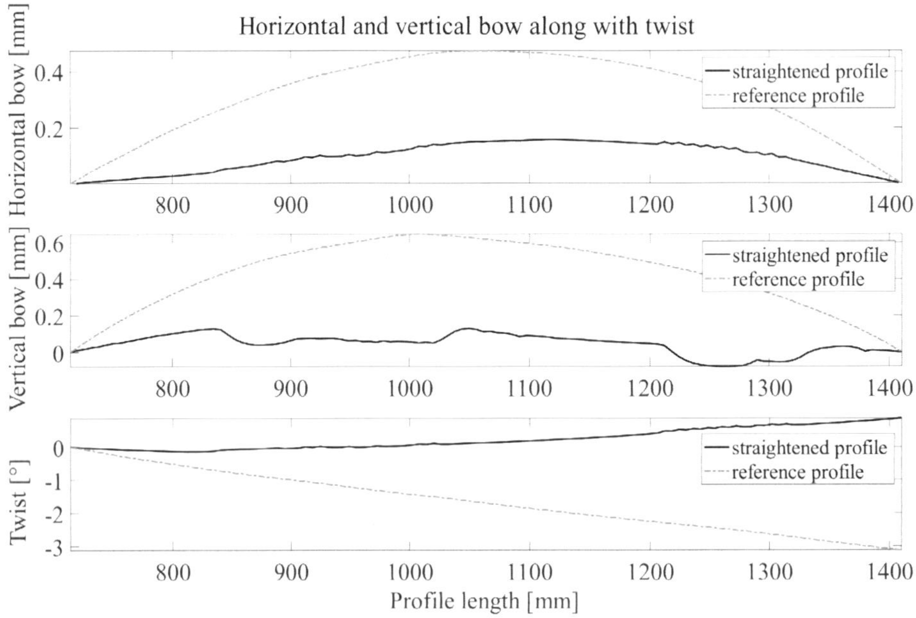

Figure 5. Reduction of profile defects by partial rolling with open-loop control.

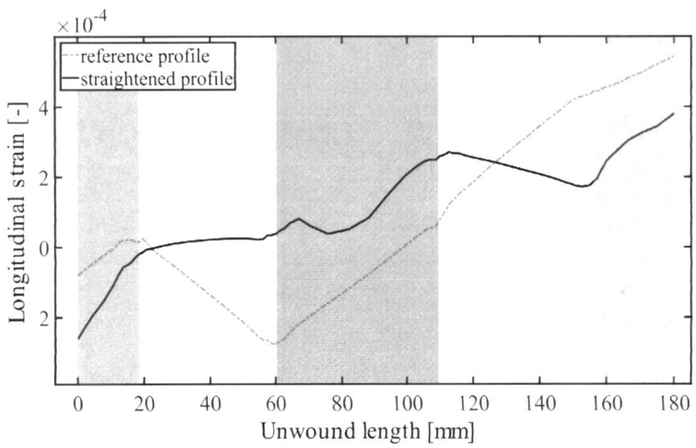

Figure 6. Longitudinal strain distribution in the profile cross-sections of reference and partial rolled profile using an open-loop control

3.2. Closed-loop control

Based on the open-loop control, a closed-loop control is set up. The important difference is a continuous effect of the output value on the input value, and thus a measurable influence on itself. This is achieved by moving the evaluation point of the longitudinal strain 200 mm downstream of the partial rolling station as depicted in Figure 4 (b). This makes it possible to evaluate areas that have already been rolled out and to react to remaining deviations between input and output values.

The input value is the maximum longitudinal strain of the profile cross-section. To pay attention to the dead time of the control system, a Smith-Predictor is used to control the distance of the rolling gaps of the partial rolling system. A Smith-Predictor uses an analytical model to predict the response of the system's output value to a change in actuator input. Therefore, equation (1) is used as it describes the relationship between the roll gap and the introduced longitudinal strain. Output value is the longitudinal strain of the part after being rolled out. In contrast to the open-loop control, the rolls are moved at the same time or increment as the longitudinal strain is determined.

3.2.1. Results. Figure 7 shows the impact of partial rolling on profile defects with closed-loop control. Similar to the open-loop control, a decrease in profile defects can be achieved by using the closed-loop control. The horizontal bow is reduced by roughly 10 %, the vertical bow by 85 % and the twist by 85 % in total. This indicates that plastic longitudinal strains are inserted into the profile at the correct locations. It also shows that the implemented algorithm can decide which rollers need to be moved to homogenize the longitudinal strain in the profile cross-section. It is also able to react to the changing longitudinal strains along the profile.

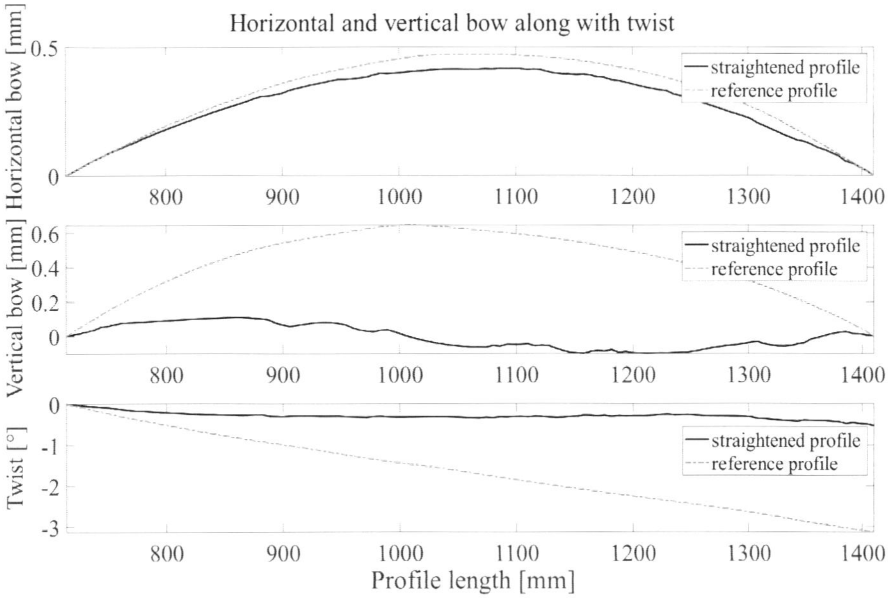

Figure 7. Reduction of profile defects by partial rolling with closed-loop control.

Figure 8 shows the longitudinal strain distribution in the profile cross-section in the stationary area of the reference profile and the profile that is partially rolled. It can be seen that the partial rolling process with the closed-loop control was able to bring the strain distribution to a more linear progression, but not completely homogenize it. The strain distribution is consistent with the expectations associated with the existing profile defects. The vertical bow is almost eliminated due to the nearly linear progression of the longitudinal strain distribution of the partially rolled part, while there is still a lot of horizontal bow due to the compressed shorter flange and the elongated longer flange.

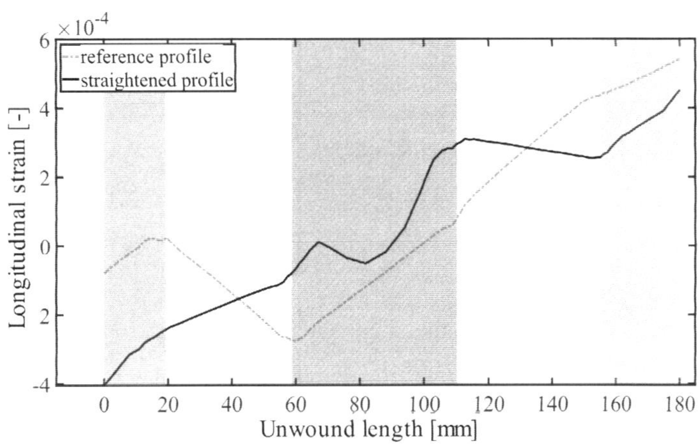

Figure 8. Longitudinal strain distribution in the profile cross-sections of reference and partial rolled profile using a closed-loop control

Compared to the longitudinal strain distribution of the straightened profile with an open-loop control, the closed loop control resulted in a higher deviation of the strain distribution and in a higher horizontal bow. This is probably due to the difference, that the strain distribution of the rolled section is already known during rolling it in the open-loop system, whereas the closed-loop system expects the strain distribution introduced in the roll gap to be the same as that measured 200 mm downstream, which only applies in the stationary area. As the stationary area of the profile is short compared to the distance to the measuring position in the controlled system, it was not possible to completely straighten the profile geometry due to the transient response of the controller.

4. Conclusion and outlook

This paper presented an open- and a closed-loop control for controlling a partial rolling stand for straightening roll formed profiles. It is shown that both control systems can reduce the profile defects by controlling the rolling gap of a partial rolling system. The open-loop control is able to reduce all profile defects by a similar amount, while the closed-loop control in this case can only significantly reduce the vertical bow and the twist. This is probably due to the fact that the stationary area of the investigated profile is not long enough compared to the distance of the strain measurement, so that the closed-loop control does not have enough time to homogenize the longitudinal strain distribution in the profile cross-section.

In future work, the closed-loop controlled system will be simulated with a longer blank to validate the thesis that the closed-loop system is not able to eliminate all profile defects because the blank is not long enough. Furthermore, it will be investigated how long it will take for the closed-loop control to completely straighten the profile. In addition, a partial rolling straightener will be constructed, manufactured and controlled to validate the numerical results with experimental data. Therefore, a measurement system for the detection of the profile bow and twist will be set up and the longitudinal strain distribution will be calculated using the analytical model of Kilz et al [8].

References

[1] Halmos G T 2006 *Roll Froming Handbook* (Boca Raton: Taylor & Francis Group)

[2] Groche P, Güngör B and Kilz J 2023 Straight roll formed profiles through partial rolling *CIRP Annals* **72 (1)** 229-232

[3] Ona K and Jimma T 1983 Experiments into the cold roll-forming of straight asymmetrical channels *Journal of Mechanical Working Technology* **8** 273–291

[4] Abeyrathna B 2014 In-line shape compensation for roll forming through process parameter monitoring *Dissertation* (Melbourne: Deakin University)

[5] Wiebenga J H, Weiss M, Rolfe B and V. D. Boogaard A H 2013 Product defect compensation by robust optimization of cold roll forming process *Journal of Materials Processing Technology* **213** 978–986

[6] Cha W and Kim N 2013 Study on twisting and bowing of roll formed products made of high strength steel *International Journal of Precision Engineering and Manufacturing* **14** 1527–1533

[7] Güngör B and Groche P 2021 Finite element analysis of profile defects in roll forming processes based on longitudinal strain adjustments *WGP Jahreskongress* 56–64

[8] Kilz J, Güngör B, Aign F and Groche P 2023 Profile defects caused by inhomogeneous longitudinal strain distribution in roll forming *International Journal or Material Forming* **16** 36

[9] Güngor B and Kilz J 2023 Längsdehnungsbasiertes Richten durch partielles Auswalzen *12. Fachtagung Walzprofilieren* 46-58

[10] Lange K 1988 Umformtechnik: Handbuch für die Industrie und Wissenschaft *Band 2: Massivumformung* (Berlin: 2nd ed. Springer)

[11] Bui Q V and Ponthot J P 2008 Numerical simulation of cold roll-forming processes *Journal of Materials Processing Technology* **202** 275–282

AUTHOR INDEX